W9-CLO-449

Selected Titles in This Series

(Continued in the back of this publication)

Introduction to the Qualitative Theory of Dynamical Systems on Surfaces

Translations of
MATHEMATICAL MONOGRAPHS

Volume 153

Introduction to the Qualitative Theory of Dynamical Systems on Surfaces

S. Kh. Aranson
G. R. Belitsky
E. V. Zhuzhoma

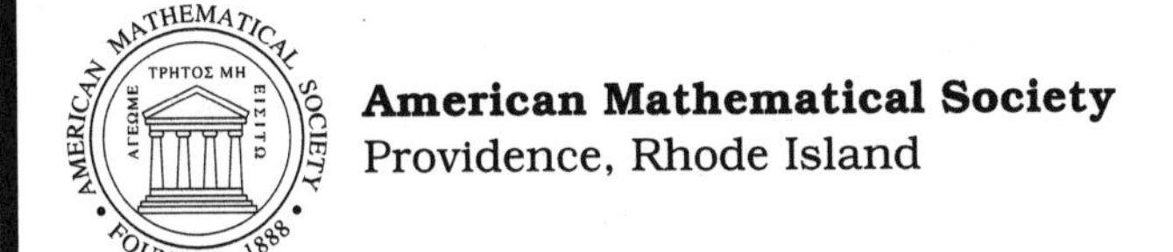

American Mathematical Society
Providence, Rhode Island

С. Х. Арансон, Г. Р. Белицкий, Е. В. Жужома

ВВЕДЕНИЕ
В КАЧЕСТВЕННУЮ ТЕОРИЮ ДИНАМИЧЕСКИХ СИСТЕМ
НА ПОВЕРХНОСТЯХ

Translated by H. H. McFaden from an original Russian manuscript.

1991 *Mathematics Subject Classification.* Primary 58-02, 58F25; Secondary 58F10, 58F21, 34C28, 58F18, 58F36, 34C35, 34D30, 57R30, 54H20.

ABSTRACT. This book is an introduction to the qualitative theory of dynamical systems on manifolds of low dimension (on the circle and on surfaces). Along with classical results, it reflects the most significant achievements in this area obtained in recent times by Russian and Western mathematicians whose work has not yet appeared in the monographic literature. The main emphasis is put on global problems in the qualitative theory of flows on surfaces.

The reader of this book need be familiar only with basic courses in differential equations and smooth manifolds. All the main definitions and notions required for understanding the contents are given in the text.

The book will be useful to mathematicians working in dynamical systems and differential equations, and geometry, and to specialists with a mathematical background who are studying dynamical processes: mechanical engineers, physicists, biologists, and so on.

Library of Congress Cataloging-in-Publication Data
Aranson, S. Kh.
[Vvedenie v kachestvennuîu teoriîu dinamicheskikh sistem na poverkhnostîakh. English]
Introduction to the qualitative theory of dynamical systems on surfaces / S. Kh. Aranson, G. R. Belitsky, E. V. Zhuzhoma; [translator H. H. McFaden].
p. cm.—(Translations of mathematical monographs, ISSN 0065-9282; v. 153)
Includes bibliographical references (p. –).
ISBN 0-8218-0369-7 (alk. paper)
1. Flows (Differentiable dynamical systems) I. Belitskiĭ, Genrikh Ruvimovich. II. Zhuzhoma, E. V. III. Title. IV. Series.
QA614.82.A7313 1996
514′.74—dc20
96-19197
CIP

Printed in the United States of America.

♾ The paper used in this book is acid-free and falls within the guidelines established to ensure permanence and durability.

10 9 8 7 6 5 4 3 2 1   01 00 99 98 97 96

Contents

Foreword

This book is an introduction to the qualitative theory of dynamical systems on manifolds of low dimension (on the circle and on surfaces). Along with classical results, it reflects the most significant achievements in this area obtained in recent times by Russian and foreign mathematicians whose work has not yet appeared in the monographic literature. The main stress here is put on global problems in the qualitative theory of flows on surfaces.

Despite the fact that flows on surfaces have the same local structure as flows on the plane, they have many global properties intrinsic to multidimensional systems. This is connected mainly with the existence of nontrivial recurrent trajectories for such flows. The investigation of dynamical systems on surfaces is therefore a natural stage in the transition to multidimensional dynamical systems.

The reader of this book need be familiar only with basic courses in differential equations and smooth manifolds. All the main definitions and concepts required for understanding the contents are given in the text.

The results expounded can be used for investigating mathematical models of mechanical, physical, and other systems (billiards in polygons, the dynamics of a spinning top with nonholonomic constraints, the structure of liquid crystals, etc.).

In our opinion the book should be useful not only to mathematicians in all areas, but also to specialists with a mathematical background who are studying dynamical processes: mechanical engineers, physicists, biologists, and so on.

CHAPTER 1

Dynamical Systems on Surfaces

§1. Flows and vector fields

1.1. Definitions and examples. A *flow* f^t or a *dynamical system with continuous time* on a manifold $\mathcal{M}$ is defined to be a mapping $f\colon \mathcal{M} \times \mathbb{R} \to \mathcal{M}$ such that

$$1)\ f(m, t_1 + t_2) = f[f(m, t_1), t_2], \qquad m \in \mathcal{M},\ t_1, t_2 \in \mathbb{R},$$
$$2)\ f(m, 0) = m, \qquad m \in \mathcal{M}.$$

A flow f^t is called a C^r-flow ($r \geq 0$) if the mapping f is of smoothness C^r. If in addition the restriction $f|_{\{m\}\times\mathbb{R}} \overset{\text{def}}{=} f^t(m)\colon \mathbb{R} \to \mathcal{M}$ is of smoothness C^{r+1}, then the flow will be called a $C^{r,r+1}$-flow.

It follows from the definition of a C^r-flow that for each fixed $t \in \mathbb{R}$ the mapping $f_t = f(\cdot, t)\colon \mathcal{M} \to \mathcal{M}$ is a C^r-diffeomorphism (a C^0-diffeomorphism is understood to be a homeomorphism). Therefore, a C^r-flow on a manifold $\mathcal{M}$ can be defined to be a one-parameter group of C^r-diffeomorphisms (a C^r-action of the additive group $\mathbb{R}$ on $\mathcal{M}$).

Passing through each point $m \in \mathcal{M}$ is the directed curve $l(m) = \{f(m,t) : -\infty < t < +\infty\}$, called the trajectory (through m). A trajectory $l(m) = \{m\}$ consisting of a single point is called an *equilibrium state* (or rest point, or singular point, or fixed point).

A trajectory homeomorphic to the circle S^1 is said to be *closed*. A trajectory that is not closed and is not an equilibrium state is said to be *nonclosed*.

The positive semitrajectory beginning at a point $m \in \mathcal{M}$ is defined to be the set $l^+(m) = \{f(m,t) : t \geq 0\}$. Similarly, $l^-(m) = \{f(m,t) : t \leq 0\}$ is the negative semitrajectory.

EXAMPLES. 1) f^t is the one-parameter group of rotations of the sphere S^2 for which f_t is the rotation of S^2 about the SN axis (the north pole–south pole) through an angle $t \in \mathbb{R}$ (see Figure 1.1). The points N and S are equilibrium states of the flow f^t. The remaining trajectories are the parallels of the sphere.

2) We represent the two-dimensional torus T^2 as the product $S^1 \times S^1$ of two unit circles. Let the numbers α and β be fixed, and let f_t be the composition of the rotation of the torus along the parallel $\{\cdot\} \times S^1$ through the angle αt and the rotation along the meridian $S^1 \times \{\cdot\}$ through the angle βt (see Figure 1.2). The trajectories of this flow are all closed or all nonclosed in dependence on the numbers α and β (if β/α is rational, then the trajectories are closed, while if β/α is

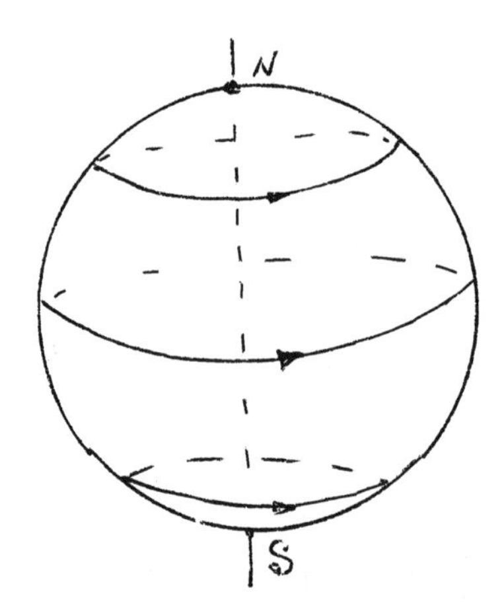

FIGURE 1.1

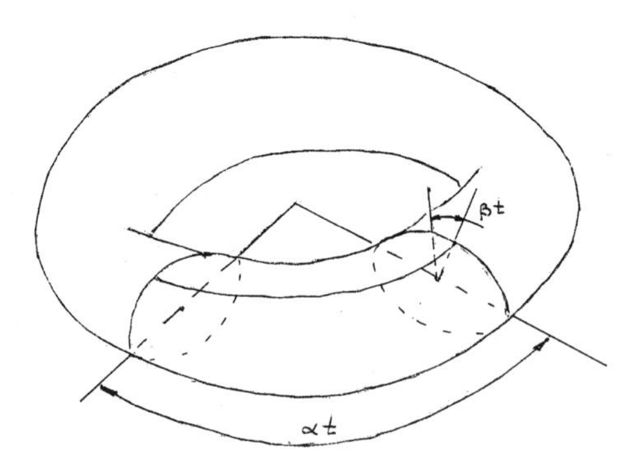

FIGURE 1.2

irrational, then the trajectories are all nonclosed; in the latter case each trajectory is dense on the torus).

1.2. Connection between flows and vector fields. Denote by $T\mathcal{M}$ the tangent space of the manifold $\mathcal{M}$. Recall that a vector field V of class C^r, $r \geq 0$, on $\mathcal{M}$ is defined to be a section $V\colon \mathcal{M} \to T\mathcal{M}$ of the bundle $\pi\colon T\mathcal{M} \to \mathcal{M}$, that is, a mapping V of smoothness class C^r such that $\pi \circ V = \mathrm{id}$.

If f^t is a given C^1-flow or $C^{0,1}$-flow on $\mathcal{M}$, then at each point $m \in \mathcal{M}$ the tangent vector

$$V(m) = \frac{d}{dt} f^t(m) \bigg|_{t=0}$$

to the trajectory $l(m)$ passing through m is defined. The vector $V(m)$ is called the phase velocity of the point m. The correspondence $m \mapsto V(m)$, where m runs through the whole manifold $\mathcal{M}$, is a vector field, which we denote by V_f. If f^t is a $C^{r,r+1}$-flow, then V_f is a C^r-smooth vector field: the field of phase velocities of the flow f^t.

It is known ([**17**], [**67**]) that any vector field V of smoothness class C^r, $r \geq 1$, on a closed manifold $\mathcal{M}$ is the field of phase velocities of some C^r-flow f^t. Therefore, one natural way of specifying a flow on a compact manifold is to specify a vector field.

1.3. Vector fields and systems of differential equations. Each point m_0 of the manifold $\mathcal{M}$ is covered by a local chart $\mathcal{U}$ in which a coordinate system is given by a mapping $(x_1, \dots, x_n)\colon \mathcal{U} \to \mathbb{R}^n$, where $n = \dim \mathcal{M}$. To define a vector field V in the chart $\mathcal{U}$ we must specify n functions $v_1, \dots, v_n\colon \mathcal{U} \to \mathbb{R}$, the components of the field V in the chart $\mathcal{U}$. If f^t is a flow such that $V = V_f$, then

$$v_i(m) = \frac{d}{dt} x_i(f^t m)\bigg|_{t=0}, \qquad m \in \mathcal{U}.$$

This implies that the trajectory $f^t(m) \cap \mathcal{U}$ is a solution of the system of differential equations

$$\dot{x}_i = v_i(x_1, \dots, x_n), \qquad i = 1, \dots, n,$$

with the initial condition $m = (x_1(0), \dots, x_n(0))$. Indeed, it must be verified that the rate of motion of the point $f^t(m)$ at each time t_0 such that $f^{t_0}(m) \in \mathcal{U}$ is equal to $V(f^{t_0}(m))$ (see Figure 1.3). Since f^t is a one-parameter group,

$$\frac{d}{dt} x_i[f^t(m)]\bigg|_{t=t_0} = \frac{d}{d\tau} x_i[f^{t_0+\tau}(m)]\bigg|_{\tau=0} = \frac{d}{d\tau} x_i[f^\tau(f^{t_0}(m))]\bigg|_{\tau=0} = v_i(f^{t_0}(m)).$$

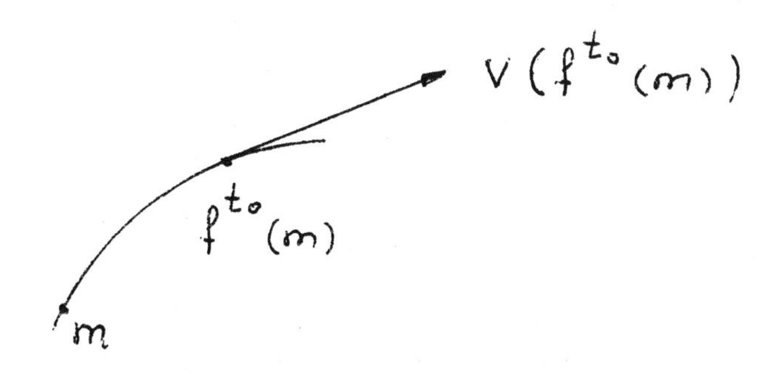

FIGURE 1.3

The field of phase velocities in a local chart is usually given as a system of differential equations.

1.4. Diffeomorphisms of vector fields. Let $\varphi\colon \mathcal{M} \to \mathcal{N}$ be a diffeomorphism, and let V be a vector field on the manifold $\mathcal{M}$. It is known ([**25**], [**47**]) that φ induces a mapping $\varphi_*\colon T\mathcal{M} \to T\mathcal{N}$ of the tangent spaces. The restriction of φ_* to V is a vector field $\varphi_*(V)$ on the manifold $\mathcal{N}$, called the image of the vector field V under the diffeomorphism φ.

§2. Main ways of specifying flows on surfaces

The specification of a flow on a surface depends on and is closely connected with the representation of the surface. For example, the surface can be represented as a set of points in $\mathbb{R}^3$ (the coordinates of which satisfy some equation), or it can be represented as a two-dimensional manifold by means of local charts and compatible coordinates. A surface can also be represented as the quotient space of the plane with respect to some group of transformations. These and other representations enable one to define a flow on a surface in various ways.

2.1. The projection method. Suppose that the equation $f_p(x,y) = 0$ determines $p+1$ disjoint ovals (circles) on the Euclidean plane $\mathbb{R}^2$, where p of the ovals have disjoint interiors and lie inside the last (see Figure 1.4 for $p = 2$). For example, $f_2(x,y) = (x^2+y^2-16)\cdot[(x+2)^2+y^2-1]\cdot[(x-2)^2+y^2-1]$ $(p=2)$. Then the equation $F_p(x,y,z) = f_p(x,y) + z^2 = 0$ determines a closed orientable surface $\mathcal{M}_p$ of genus $p \geq 0$ in $\mathbb{R}^3$ (see Figure 1.5, $p = 2$). If the ovals $f_p(x,y) = 0$ do not have singularities, then the surface $\mathcal{M}_p$ also does not have singularities; therefore, $\Delta = \sqrt{(F_x')^2 + (F_y')^2 + (F_z')^2}\,\Big|_{\mathcal{M}_p} > 0$. Denote by $\vec{n} = \Delta^{-1}\cdot(F_x', F_y', F_z')$ the unit vector normal to $\mathcal{M}_p$.

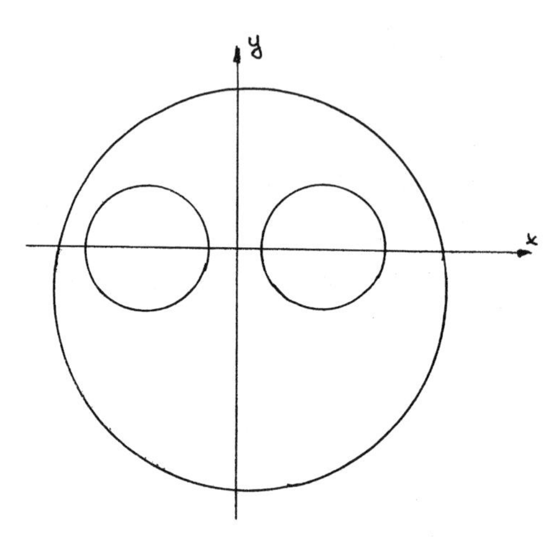

FIGURE 1.4

Associated with an arbitrary vector field $\vec{V}$ on $\mathbb{R}^3$ is the field $\vec{V}_{\tan}$ (the projection of $\vec{V}$ on the tangent plane of the surface $\mathcal{M}_p$) according to the formula

$$\vec{V}_{\tan} = \vec{V} - (\vec{n}\cdot\vec{V})\vec{n}.$$

Since $\vec{n}\cdot\vec{V}_{\tan} = \vec{n}\cdot\vec{V} - (\vec{n}\cdot\vec{V})\vec{n}^2 = 0$, $\vec{V}_{\tan}$ is a vector field on $\mathcal{M}_p$.

It is known that any continuous function on $\mathcal{M}_p$ can be extended to a continuous function on $\mathbb{R}^3$ ([**67**], [**71**]). Therefore, any vector field on $\mathcal{M}_p$ can be extended to $\mathbb{R}^3$. The projection method can thus give us all conceivable vector fields on $\mathcal{M}_p$.

Note that if the flow on $\mathbb{R}^3$ determined by a field $\vec{V}$ is given by the system of differential equations

$$\dot{x} = P(x,y,z),\ \dot{y} = Q(x,y,z),\ \dot{z} = R(x,y,z),$$

then the projected flow on $\mathcal{M}_p$, which is determined by the field $\vec{V}_{\tan}$, is given by the system

$$\begin{cases} \dot{x} = P - \Delta^{-1}(PF_x' + QF_y' + RF_z') \\ \dot{y} = Q - \Delta^{-1}(PF_x' + QF_y' + RF_z') \\ \dot{z} = R - \Delta^{-1}(PF_x' + QF_y' + RF_z'). \end{cases}$$

EXAMPLES. 1) If we project the vector field $\vec{V} = (0,0,-1)$ on the sphere $S^2 : x^2+y^2+z^2 = 1$, then on S^2 we get the field $\vec{V}_{\tan}$ of a flow with two equilibrium

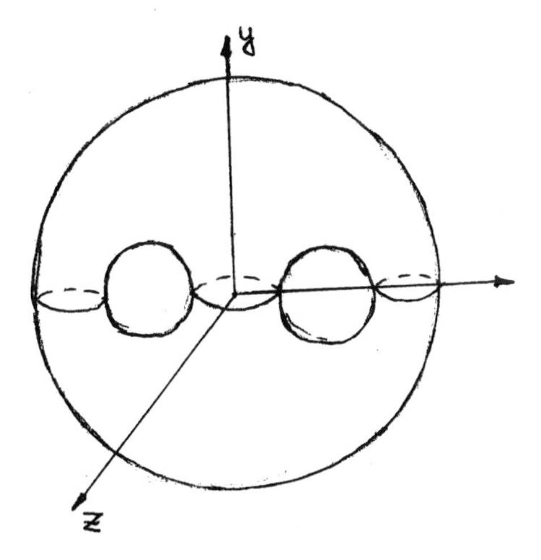

FIGURE 1.5

states, at the north and south poles, and all the remaining trajectories of the flow obtained "flow down" from the north pole to the south pole along meridians (see Figure 1.6).

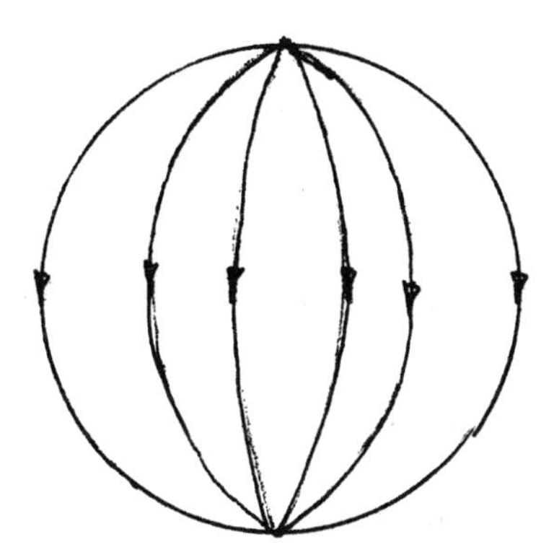

FIGURE 1.6

2) Let us consider the system

$$\dot{x} = -y, \quad \dot{y} = x, \quad \dot{z} = 0$$

in $\mathbb{R}^3$. Each point of the axis Oz is an equilibrium state of this system, and all the remaining trajectories are closed. The sphere S^2 and the torus T^2 determined by the equation $(x^2 + y^2 - 1)(x^2 + y^2 - 1/4) + z^2 = 0$ are integral surfaces of the field $\vec{V} = (-y, x, 0)$; therefore, $\vec{V}_{\tan} = \vec{V}$. The flow on S^2 is represented in Figure 1.1, and the flow on T^2 in Figure 1.7.

REMARK **[68]**. We consider a differential equation

$$F(x, y, \frac{dy}{dx}) = 0 \tag{2.1}$$

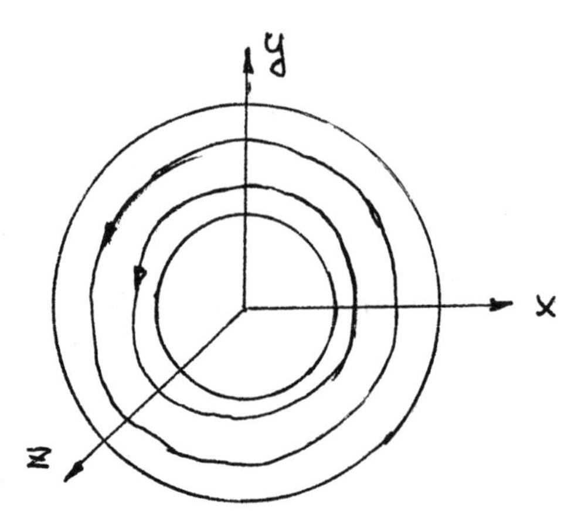

FIGURE 1.7

not solved for the derivative, and we assume that $(F'_x)^2 + (F'_y)^2 + (F'_z)^2|_{F(x,y,z)=0} \neq 0$. Let $\mathcal{M}^2$ be the surface in $\mathbb{R}^3$ given by the equation $F(x, y, z) = 0$. Then $\mathcal{M}^2$ is an integral surface of the system

$$\begin{cases} \frac{dx}{dt} = -F'_z \\ \frac{dy}{dt} = -zF'_z \\ \frac{dz}{dt} = F'_x + zF'_y. \end{cases} \tag{2.2}$$

Indeed, $F'_x(-F'_z) + F'_y(-zF'_z) + F'_z(F'_x + zF'_y) \equiv 0$; therefore, there is a flow f^t on $\mathcal{M}^2$ whose trajectories are solutions of the system (2.2). It is not hard to show that if $F'_z|_{\mathcal{M}^2} \neq 0$, then the projections of the trajectories of f^t on the (x, y)-plane $\mathbb{R}^2$ are solutions of the equation (2.1).

2.2. Systems of differential equations in local charts. Let $(x_1, x_2)\colon \mathcal{U}_1 \to \mathbb{R}^2$ and $(y_1, y_2)\colon \mathcal{U}_2 \to \mathbb{R}^2$ be the coordinates in overlapping local charts $\mathcal{U}_1$ and $\mathcal{U}_2$ of the two-dimensional manifold $\mathcal{M}^2$, and let $m \in \mathcal{U}_1 \cap \mathcal{U}_2$. The coordinates (v_1, v_2) and (w_1, w_2) of the same vector at m in the respective charts $\mathcal{U}_1$ and $\mathcal{U}_2$ are connected by the relations

$$\begin{cases} w_1 = \frac{\partial y_1}{\partial x_1} v_1 + \frac{\partial y_1}{\partial x_2} v_2 \\ w_2 = \frac{\partial y_2}{\partial x_1} v_1 + \frac{\partial y_2}{\partial x_2} v_2, \end{cases} \tag{2.3}$$

or $\vec{W} = \mathcal{J}\vec{V}$, where

$$\mathcal{J} = \begin{vmatrix} \frac{\partial y_1}{\partial x_1} & \frac{\partial y_1}{\partial x_2} \\ \frac{\partial y_2}{\partial x_1} & \frac{\partial y_2}{\partial x_2} \end{vmatrix}$$

is the Jacobian of the transition from the coordinates (x_1, x_2) to the coordinates (y_1, y_2). Thus, if the flow f^t on $\mathcal{M}^2$ is determined in the charts $\mathcal{U}_1$ and $\mathcal{U}_2$ by the corresponding systems of differential equations

$$\begin{cases} \dot{x}_1 = v_1(x_1, x_2) \\ \dot{x}_2 = v_2(x_1, x_2), \end{cases} \qquad \begin{cases} \dot{y}_1 = w_1(y_1, y_2) \\ \dot{y}_2 = w_2(y_1, y_2), \end{cases}$$

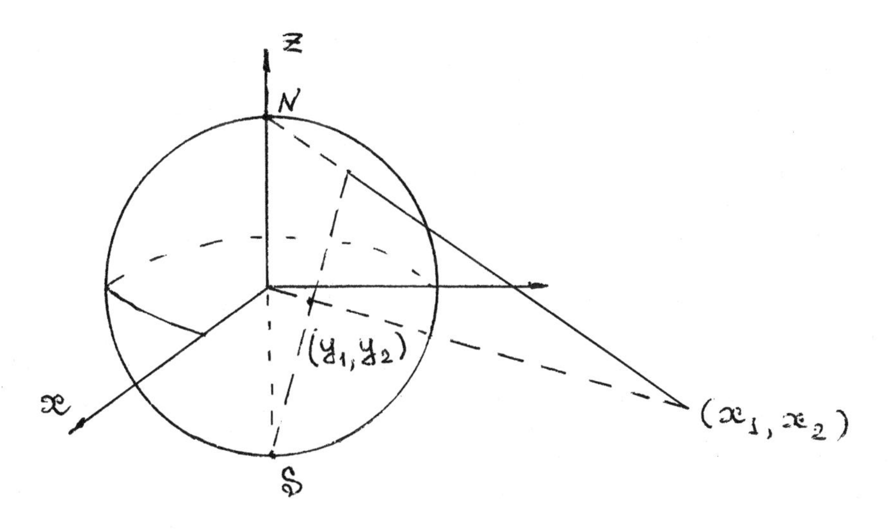

FIGURE 1.8

then (2.3) holds at all points of the intersection $\mathcal{U}_1 \cap \mathcal{U}_2$, and, conversely, if a system of differential equations is given in each local chart, and if (2.3) holds in an intersection of charts, then a vector field is defined on $\mathcal{M}^2$.

EXAMPLE. By means of the stereographic projections (see Figure 1.8) $x\colon \mathcal{U}_N = S^2 \setminus \{N\} \to \mathbb{R}^2$ and $y\colon \mathcal{U}_S = S^2 \setminus \{S\} \to \mathbb{R}^2$ we introduce the respective coordinates (x_1, x_2) and (y_1, y_2) on the local charts $\mathcal{U}_N$ and $\mathcal{U}_S$ of the unit sphere S^2, where $N(0,0,1)$ is the north pole and $S(0,0,-1)$ the south pole of the sphere. The transition from one set of coordinates to the other in the annulus $\mathbb{R}^2 \setminus (0,0)$ is realized by the formulas

$$\begin{cases} x_1 = \frac{y_1}{y_1^2+y_2^2} \\ x_2 = \frac{y_2}{y_1^2+y_2^2}, \end{cases} \qquad \begin{cases} y_1 = \frac{x_1}{x_1^2+x_2^2} \\ y_2 = \frac{x_2}{x_1^2+x_2^2}. \end{cases}$$

The Jacobian of the transition from (x_1, x_2) to (y_1, y_2) has in the coordinates (y_1, y_2) the form

$$\mathcal{J} = \begin{vmatrix} y_2^2 - y_1^2 & -2y_1y_2 \\ -2y_1y_2 & y_1^2 - y_2^2 \end{vmatrix}.$$

In the charts $\mathcal{U}_N$ and $\mathcal{U}_S$ we now write the systems of differential equations

$$\begin{cases} \dot{x}_1 = \frac{x_1}{1+x_1^2+x_2^2} \\ \dot{x}_2 = \frac{x_2}{1+x_1^2+x_2^2}, \end{cases} \qquad \begin{cases} \dot{y}_1 = \frac{y_1(y_1^2+y_2^2)}{1+y_1^2+y_2^2} \\ \dot{y}_2 = \frac{y_2(y_1^2+y_2^2)}{1+y_1^2+y_2^2}. \end{cases}$$

It is not hard to verify that (2.3) holds, and hence a flow f^t is given on S^2. We leave it for the reader to convince himself that f^t is represented in Figure 1.6.

2.3. Specification of a flow with the help of a universal covering. An advantage of this approach is that for surfaces of genus $p \geq 1$ the universal covering is homeomorphic to $\mathbb{R}^2$ with a single coordinate system. After specification of a flow on the universal covering its invariance with respect to the covering transformation group must be verified.

2.3.1. *Transformation groups.* A group Γ of transformations of a topological space $\mathcal{M}$ acts *freely* on $\mathcal{M}$ if any element $\gamma \in \Gamma$ different from the identity does not have fixed points. A freely acting group Γ is said to be discontinuous if each $m \in \mathcal{M}$ has a neighborhood $\mathcal{U} \ni m$ such that $\gamma(\mathcal{U}) \cap \mathcal{U} = \emptyset$ for all $\gamma \in \Gamma$, $\gamma \neq \mathrm{id}$.

It is known [**78**] that if a discontinuous freely acting group Γ of transformations consists of homeomorphisms, then the quotient space $\mathcal{M}/\Gamma$ (each Γ-orbit is identified with a point) is equipped with a topology in which the natural projection $\pi\colon \mathcal{M} \to \mathcal{M}/\Gamma$ is a covering.

2.3.2. *Flows on the torus.* On the Euclidean plane $\mathbb{R}^2$ (with Cartesian coordinates x, y) we consider the group Γ of translations by an integer vector: $(x, y) \mapsto (x+k, y+r)$, $(k, r) \in \mathbb{Z}^2$. It is not hard to verify that Γ acts freely and discontinuously on $\mathbb{R}^2$. To represent the space $\mathbb{R}^2/\Gamma$ we take a suitably chosen fundamental domain F (the closure of a set containing exactly one point from each Γ-orbit) and identify its boundary by means of the action of Γ. We can take F to be the square $0 \le x \le 1$, $0 \le y \le 1$. Opposite sides of F are pasted together under the action of Γ, and we obtain the torus $T^2 = \mathbb{R}^2/\Gamma$, a closed orientable surface of genus 1 (Figure 1.9).

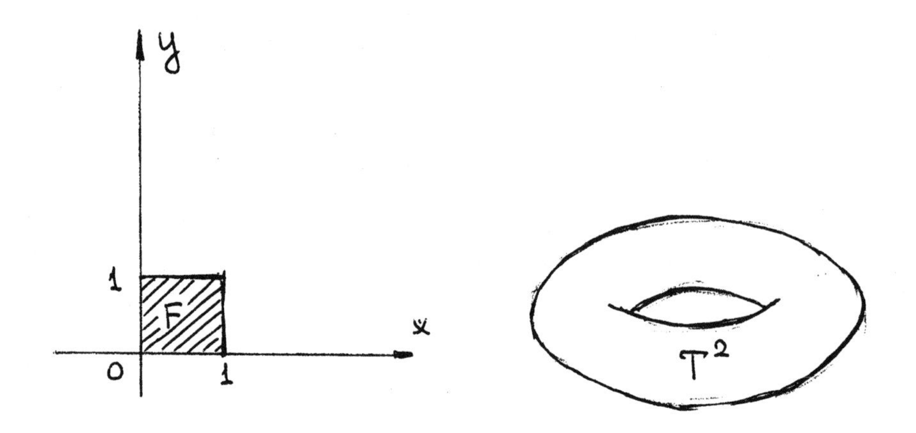

FIGURE 1.9

A dynamical system

$$\begin{cases} \dot{x} = f_1(x, y) \\ \dot{y} = f_2(x, y) \end{cases}$$

on $\mathbb{R}^2$ projects into a dynamical system on T^2 (that is, is a covering system) if and only if $f_1(x, y)$ and $f_2(x, y)$ are periodic functions of period 1 in both arguments. Under the action of π (recall that a covering is a local homeomorphism) the trajectories of a covering flow on $\mathbb{R}^2$ are mapped into the trajectories of the flow on T^2.

EXAMPLE 1 (rational and irrational windings). The flow $\overline{f}^t$ given by the system

$$\begin{cases} \dot{x} = 1 \\ \dot{y} = \mu \end{cases}$$

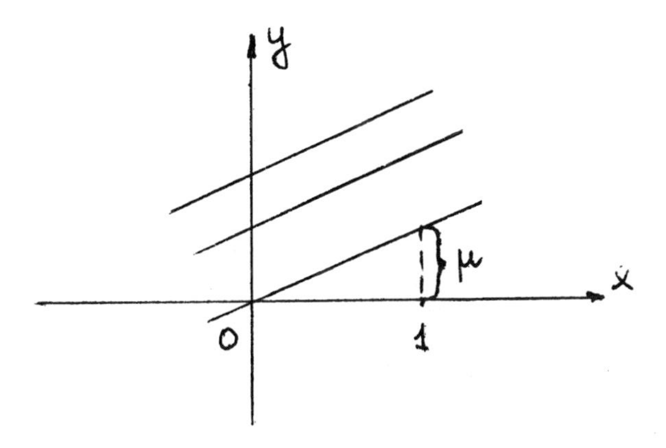

FIGURE 1.10

is a covering flow for a flow f^t on T^2. The integral curves of $\overline{f}^t$ are lines with slope μ (Figure 1.10). The parametric equations of the trajectory $\bar{l}_0$ passing through a point (x_0, y_0) at $t = 0$ have the form $x = x_0 + t$, $y = y_0 + \mu t$.

The behavior of the trajectories on T^2 depends on the number μ. We consider two cases.

a) $\mu = p/q$ is rational.

The trajectory $\bar{l}_0$ passes through the two points (x_0, y_0) and $(x_0 + q, y_0 + p)$ corresponding to $t = 0$ and $t = q$, and these points are mapped by π into a single point on T^2. Consequently, $\pi(\bar{l}_0)$ is a closed trajectory. Conversely, if $\pi(\bar{l}_0)$ is a closed trajectory, then $x_0 + \bar{t} = x_0 + k$ and $y_0 + \mu\bar{t} = y_0 + r$ for some $t = \bar{t} \neq 0$, where $k, r \in \mathbb{Z}$, and thus $\mu = r/k$. Hence, all the trajectories of the flow f^t on T^2 are closed if and only if μ is rational.

b) μ is irrational.

According to a), all the trajectories on T^2 are nonclosed. Moreover, each trajectory is dense in T^2 because each trajectory $\pi(\bar{l}_0)$ intersects the circle $\pi(x = x_0) \overset{\text{def}}{=} S_0$ at the points $\{y_0 + n\mu \pmod 1, n \in \mathbb{Z}\}$, but the latter set is dense in S_0. Indeed, we partition the circle S_0 into k equal half-open intervals of length $1/k$. Since μ is irrational, the points $y_0 + n\mu \pmod 1$, $n \in \mathbb{Z}$, are distinct. Therefore, among the $k+1$ points $y_0 + n\mu \pmod 1$, $n = 0, \dots, k$, there are two, say $y_0 + p\mu \pmod 1$ and $y_0 + q\mu \pmod 1$, that lie in a single half-open interval of length $1/k$. For definiteness we assume that $p > q$, and we set $r = p - q$. Then $0 < r\mu \pmod 1 < 1/k$, and in the sequence of points $y_0 + rn\mu \pmod 1$, $n \in \mathbb{Z}$, the distance between each two adjacent points is $r\mu \pmod 1 < 1/k$. For a given $\varepsilon > 0$ we take a k such that $1/k < \varepsilon$. Then the ε-neighborhood of any point of the circle contains points in the sequence $\{y_0 + rn\mu \pmod 1\}$ (see also [**26**], Chapter 3).

The flow f^t on T^2 is called a *rational* (*irrational*) winding of the torus if μ is rational (irrational); sometimes one refers to a rational (irrational) flow on the torus. In the collection of rational windings we also include the flow on T^2 whose covering flow on $\mathbb{R}^2$ is given by the system $\dot{x} = 0$, $\dot{y} = 1$.

EXAMPLE 2 (impassable grain). The flow

$$\begin{cases} \dot{x} = x^2 + y^2 \\ \dot{y} = \mu(x^2 + y^2) \end{cases}$$

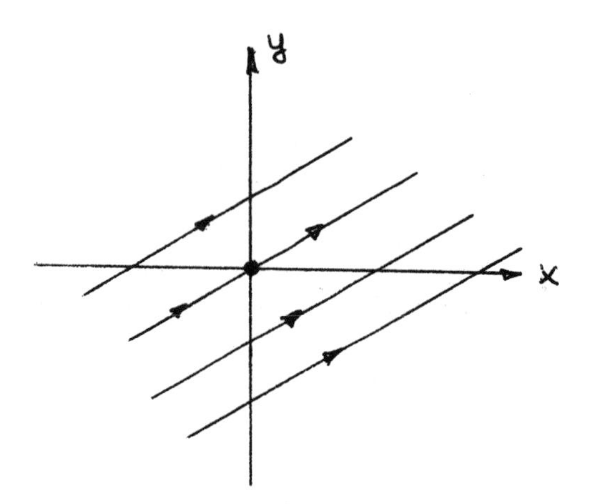

FIGURE 1.11

on $\mathbb{R}^2$ differs qualitatively from the flow in Example 1 in that an equilibrium state with two saddle sectors is "planted" at the origin of coordinates (Figure 1.11). Such an equilibrium state is called an *impassable grain* or *fake saddle.*

A flow on T^2 with an impassable grain is covered by the flow

$$\begin{cases} \dot{x} = \sin^2 \pi x + \sin^2 \pi y \\ \dot{y} = \mu(\sin^2 \pi x + \sin^2 \pi y). \end{cases}$$

A flow with any finite number of impassable grains at specified points can be obtained similarly.

2.3.3. *Flows on closed orientable surfaces of genus* > 1. Denote by Δ the open unit disk of the complex z-plane ($z = x + iy$). The linear fractional transformations of the form

$$z \mapsto \frac{az + b}{\bar{b}z + \bar{a}}, \qquad |a|^2 - |b|^2 = 1,$$

carry Δ into itself and form the group $I(\Delta)$.

The famous uniformization theorem [**77**] asserts that for any closed orientable surface $\mathcal{M}_p$ of genus $p > 1$ there exists a finitely generated subgroup Γ_p of $I(\Delta)$ such that:

1) Γ_p acts freely and discontinuously on Δ;
2) $\mathcal{M}_p = \Delta/\Gamma_p$;
3) the natural projection $\pi\colon \Delta \to \Delta/\Gamma_p = \mathcal{M}_p$ is a universal covering.

A flow

$$\begin{cases} \dot{x} = f_1(x, y) \\ \dot{y} = f_2(x, y) \end{cases}$$

on Δ is covering for some flow on $\mathcal{M}_p$ if and only if $f_i(\gamma(m)) = f_i(m)$, $i = 1, 2$, for all $\gamma \in \Gamma_p$ and all points $m(x, y) \in \Delta$.

An example of a flow on $\mathcal{M}_p$ with some fixed Riemannian metric that is invariant with respect to Γ_p can be obtained as follows. Take a function $\Phi(z)$ that is automorphic with respect to Γ_p, that is, $\Phi(\gamma(z)) = \Phi(z)$ for all $\gamma \in \Gamma_p$ [**78**]. Then the vector field $\vec{V}(z) = \operatorname{grad} \Phi(z)$ is invariant with respect to Γ_p, and hence

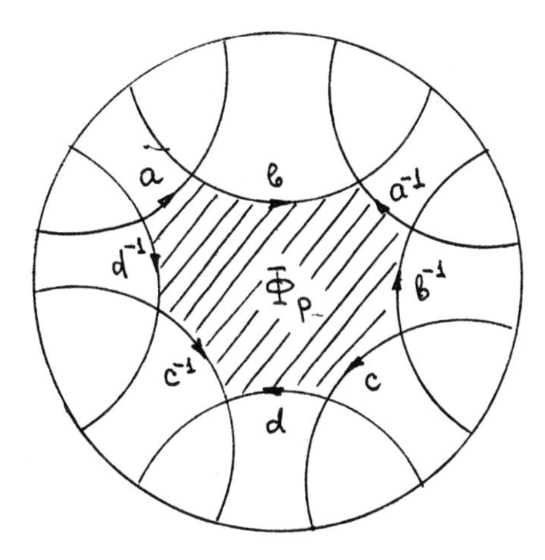

FIGURE 1.12

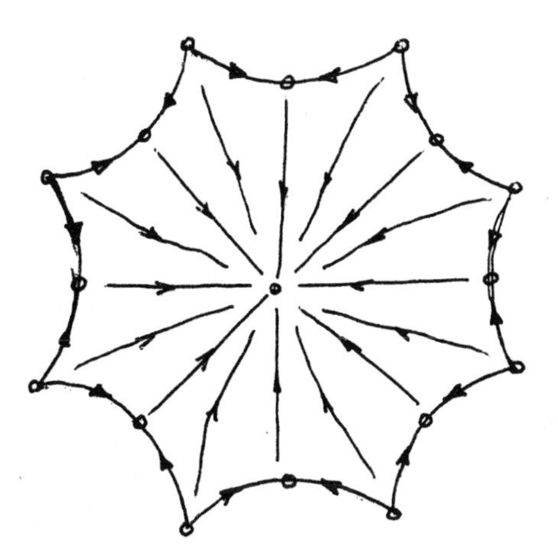

FIGURE 1.13

is covering for some vector field $\vec{V}_p$ on $\mathcal{M}_p$. The field $\vec{V}_p$ generates on $\mathcal{M}_p$ a flow belonging to the class of so-called gradient flows.

EXAMPLE (a polar Morse–Smale flow). As a fundamental domain of the group Γ_p we can take a curvilinear $4p$-gon Φ_p bounded by arcs of Euclidean circles perpendicular to the circle $S_\infty = \partial\Delta$ bounding the disk Δ (see Figure 1.12 for the case $p = 2$), where each side of the polygon Φ_p is identified with one other side of Φ_p by means of Γ_p, and all the vertices of Φ_p are identified with a single point (Figure 1.12 shows an example of identification of the sides of Φ_2). We represent the phase portrait of a flow on Φ_p by locating a stable node at the center of the polygon, a saddle at the midpoint of each arc of the boundary $\partial\Phi_p$, and an unstable node at the vertices (Figure 1.13 for $p = 2$). On the sides of Φ_p this phase portrait "is compatible with" the action of the group Γ_p, and hence it determines the phase portrait of some flow f^t on the surface $\mathcal{M}_p$. The flow f^t has two nodes (one stable and one unstable) and $2p$ saddles, and there are no other equilibrium states nor closed trajectories. The flow constructed (with certain restrictions on the eigenvalues of the equilibrium states) belongs to the class of so-called polar Morse–Smale flows (see Chapter 3 for the exact definition of Morse–Smale flows). This flow is a gradient flow; that is, the corresponding vector field can be represented as the gradient of a function on $\mathcal{M}_p$.

2.4. Specification of a flow with the help of a branched covering.

2.4.1. *Definition of a branched covering.* Denote by δ_k the mapping $z \mapsto z^k$ of the z-plane $\mathbb{C}$, $k \in \mathbb{N}$.

DEFINITION. A transformation $\pi\colon \widetilde{\mathcal{M}} \to \mathcal{M}$ of a two-dimensional manifold $\widetilde{\mathcal{M}}$ into a two-dimensional manifold $\mathcal{M}$ is called a branched covering if any point $m \in \mathcal{M}$ has a neighborhood $\mathcal{U} \ni m$ such that the complete inverse image $\pi^{-1}(\mathcal{U})$ is a union $V_1 \cup V_2 \cup \dots$ of disjoint neighborhoods, and the restriction $\pi|_{V_i}$ is topologically conjugate to some mapping δ_k (that is, there exist homeomorphisms $h\colon \mathcal{U} \to \mathbb{C}$ and $h_i\colon V_i \to \mathbb{C}$ such that $\delta_k \circ h_i|_{V_i} = h \circ \pi|_{V_i}$). The number k is called the *branching index* of the point $z_i = \pi^{-1}(m) \cap V_i$ and is denoted by $k(z_i)$.

We shall consider only (regular) coverings such that all the points in the inverse image $\pi^{-1}(m)$ have the same branching index for any $m \in \mathcal{M}$. For such coverings the *branching order* $k(m)$ of a point $m \in \mathcal{M}$ is defined to be the branching index of any point in $\pi^{-1}(m)$.

A point $m_0 \in \mathcal{M}$ is called a *branch point* if $k(m_0) > 1$. The collection $\mathcal{M}_0 \subset \mathcal{M}$ of branch points is called the *branch set*; $\mathcal{M}_0$ is discrete, and it is finite for compact $\mathcal{M}$.

The number of points in the complete inverse image $\pi^{-1}(x)$, $x \in \mathcal{M} \setminus \mathcal{M}_0$, is called the *multiplicity of the covering.* For an arcwise connected $\mathcal{M}$ this number is independent of the point $x \in \mathcal{M} \setminus \mathcal{M}_0$.

EXAMPLE. In $\mathbb{R}^3$ we consider a surface $\mathcal{M}_{pq}$ of genus p that is symmetric with respect to the axis Oz and such that the axis intersects q "handles" (see Figure 1.14, where $p = 6$ and $q = 3$). Denote by G the group of transformations $\mathbb{R}^3 \to \mathbb{R}^3$ generated by the symmetry with respect to the axis Oz. Then $\mathcal{M}_{pq}$ is invariant under G, and $\mathcal{M}_{pq}/G = \mathcal{M}_0$ is a closed surface of genus $\left[\frac{p-q}{2}\right] + 1$. The surface $\mathcal{M}_0$ can be obtained as follows. Take the part of $\mathcal{M}_{pq}$ lying in the half-space $y \le 0$. This part is a surface F_{pq} with $\left[\frac{p-q}{2}\right] + 1$ handles and a boundary, and it is a fundamental domain of the action of G on $\mathcal{M}_{pq}$. The identification of the boundary points of F_{pq} under the action of G amounts to pasting up the holes in the surface F_{pq}. As a result we get a closed surface $\mathcal{M}_0$ with $\left[\frac{p-q}{2}\right] + 1$ handles.

The natural projection $\mathcal{M}_{pq} \to \mathcal{M}_0$ is a two-sheeted branched covering with $2q$ branch points (the points A, B, C, D, E, and F in Figure 1.15).

2.4.2. *Covering flows.* Let f^t be a smooth flow on a closed orientable surface $\mathcal{M}$, and let $\vec{V}$ be its phase velocity field. Assume that there exists a two-sheeted branched covering $\pi\colon \widetilde{\mathcal{M}} \to \mathcal{M}$, and each branch point coincides with an equilibrium state of f^t. Then the restriction of the flow f^t to the set $\mathcal{M} \setminus \mathcal{M}_0$, where $\mathcal{M}_0$ is the branch set, is a flow f_0^t. Let $\vec{V}_0$ be the phase velocity field of f_0^t. Since the restriction $\pi|_{\widetilde{\mathcal{M}} \setminus \pi^{-1}(\mathcal{M}_0)}\colon \widetilde{\mathcal{M}} \setminus \pi^{-1}(\mathcal{M}_0) \to \mathcal{M} \setminus \mathcal{M}_0$ is an unbranched covering, and since an unbranched covering is a local diffeomorphism in compatible differentiable structures, there is a vector field $\vec{V}_0^*$ on $\widetilde{\mathcal{M}} \setminus \pi^{-1}(\mathcal{M}_0)$ that covers $\vec{V}_0$ (that is, $\pi_*(\vec{V}_0^*) = \vec{V}_0$, where the mapping π_* of the tangent spaces is induced by the local diffeomorphism

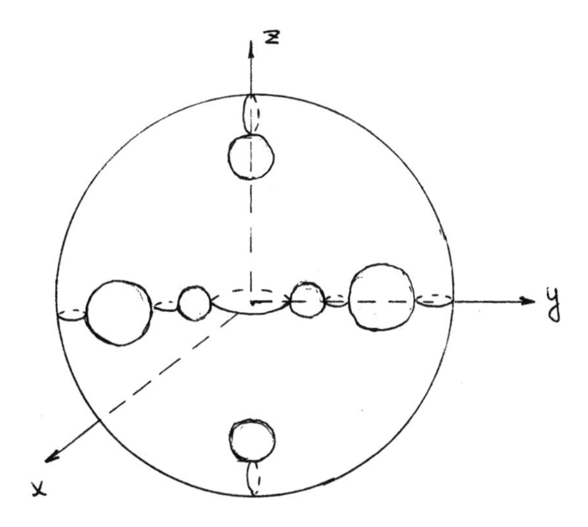

FIGURE 1.14

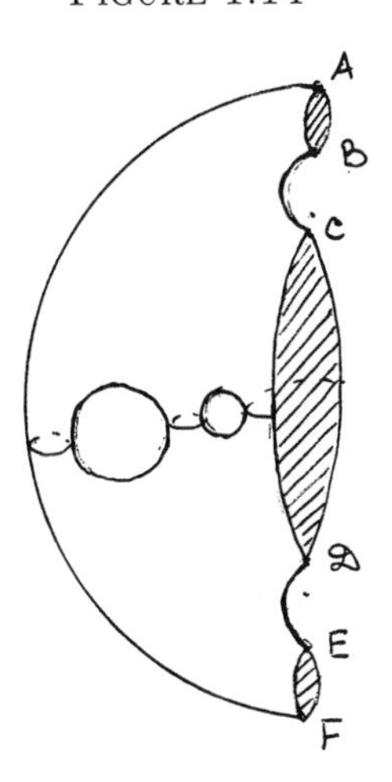

FIGURE 1.15

$\pi|_{\widetilde{\mathcal{M}}\setminus\pi^{-1}(\mathcal{M}_0)}$). We define the field $\vec{V}_0^*$ to be zero at points of $\pi^{-1}(\mathcal{M}_0)$. Since f^t has equilibrium states at points of $\mathcal{M}_0$, it follows from the definition of a branched covering that the vector field $\vec{V}^*$ obtained on $\widetilde{\mathcal{M}}$ is continuous and covers the field $\vec{V}$. Then the flow $\widetilde{f}^t$ on $\widetilde{\mathcal{M}}$ determined by $\vec{V}^*$ is a covering flow for f^t.

2.4.3. *Construction of transitive flows.*

DEFINITION. A flow f^t on $\mathcal{M}$ is said to be *transitive* if f^t has a trajectory that is dense in $\mathcal{M}$.

An irrational winding on T^2 is an example of a transitive flow.

We consider a two-sheeted covering π of the torus T^2 by a pretzel: a closed surface $\mathcal{M}_2$ of genus 2 (see Figure 1.16) with branch points $A, B \in T^2$. This covering is completely analogous to the example described in 2.4.1.

Let f_0^t be an irrational winding on T^2. By moving the points A and B slightly we can ensure that they lie on different tranjectories of the flow f_0^t. Suppose that

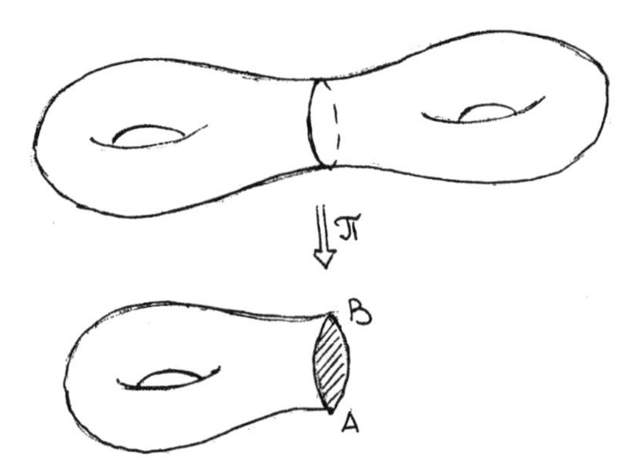

FIGURE 1.16

the flow f^t differs from f_0^t only in the existence of impassable grains at A and B, and let $\widetilde{f}^t$ be a covering flow for f^t (by 2.4.2) on $\mathcal{M}_2$.

It is not hard to see that the dynamical system

$$\begin{cases} \dot{x} = x \\ \dot{y} = -y \end{cases}$$

on $\mathbb{R}^2$ with a unique equilibrium state (a saddle) is a covering flow for the flow

$$\begin{cases} \dot{x} = x^2 + y^2 \\ \dot{y} = 0 \end{cases}$$

(with an impassable grain at the origin) with respect to the two-sheeted branched covering $z \mapsto z^2$ (Figure 1.17). Therefore, the only equilibrium states of the flow $\widetilde{f}^t$ on $\mathcal{M}_2$ are the two saddles at the points $\pi^{-1}(A)$ and $\pi^{-1}(B)$.

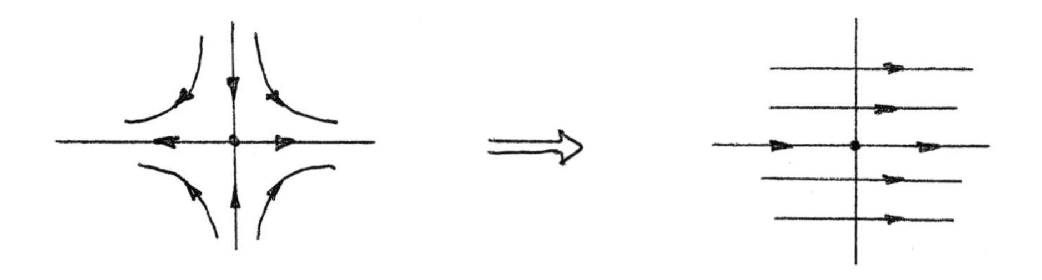

FIGURE 1.17

Any trajectory l of f^t different from an equilibrium state is dense in T^2. Consequently, the complete inverse image $\pi^{-1}(l)$, which consists of two trajectories l_1 and l_2 of the flow $\widetilde{f}^t$, is dense in $\mathcal{M}_2$. It can be shown that in this case one of the trajectories l_1, l_2 is dense in $\mathcal{M}_2$; that is, the flow $\widetilde{f}^t$ is transitive.

Since any closed orientable surface of genus $p \geq 2$ is a two-sheeted covering of the torus T^2, the method described can be used to construct a transitive flow on any $\mathcal{M}_p$, $p \geq 2$.

2.5. The pasting method. Let f_1^t and f_2^t be two flows on the two-dimensional manifolds $\mathcal{M}_1$ and $\mathcal{M}_2$, respectively, and let $B_i \subset \mathcal{M}_i$ be a disk whose boundary intersects trajectories of f_i^t transversally except at the two points $m_1^{(i)}$ and $m_2^{(i)}$, $i = 1, 2$ (see Figure 1.18). Denote by d_i^+ (d_i^-) the arc of the boundary ∂B_i across which trajectories of f_i^t enter (leave) B_i, $i = 1, 2$. We put impassable grains at the points $m_1^{(i)}$ and $m_2^{(i)}$ and denote the resulting flow by $\widehat{f}_i^t$, $i = 1, 2$.

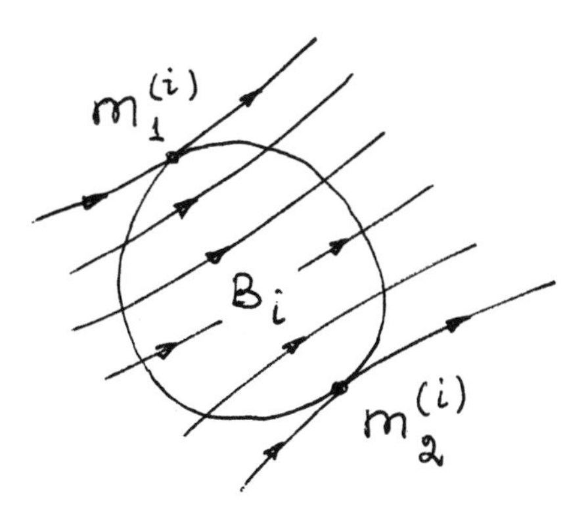

FIGURE 1.18

We now identify the boundaries of the manifolds $\mathcal{M}_1 \setminus B_1$ and $\mathcal{M}_2 \setminus B_2$ with the help of an orientation-reversing homeomorphism $h\colon \partial B_1 \to \partial B_2$ such that $h(d_1^+) = d_2^-$, $h(d_1^-) = d_2^+$, and $h(m_i^{(1)}) = m_i^{(2)}$, $k = 1, 2$. The manifold $\mathcal{M} = \mathcal{M}_1 \# \mathcal{M}_2$ obtained is called a connected sum of the manifolds $\mathcal{M}_1$ and $\mathcal{M}_2$.

The flows $\widehat{f}_1^t$ and $\widehat{f}_2^t$ determine a flow f^t on $\mathcal{M}$ as follows. The points $m_i^{(1)} \overset{h}{\cong} m_i^{(2)}$, $i = 1, 2$, are equilibrium states of the flow f^t. Let $m \in \mathcal{M}_1 \setminus B_1$ and $t \in \mathbb{R}$, $t > 0$. If $(\widehat{f}_1)_\tau(m) \in \mathcal{M}_1 \setminus B_1$ for $0 \le \tau \le t$, then we set $f_t(m) = (\widehat{f}_1)_t(m)$. If $(\widehat{f}_1)_\tau(m) \in B_1$ for some $0 < \tau \le t$, then there is a smallest τ_1 such that $(\widehat{f}_1)_{\tau_1}(m) \in d_1^+$. The trajectory $f^t(m)$ then "passes" to $\mathcal{M}_2 \setminus B_2$; that is, for $\tau_1 < \tau \le t$ sufficiently close to τ_1, we set

$$f_\tau(m) = (\widehat{f}_2)_{\tau-\tau_1}(h \circ \widehat{f}_{1\tau_1}(m)).$$

If $(\widehat{f}_2)_\tau(h \circ \widehat{f}_{1\tau_1}(m)) \in \mathcal{M}_2 \setminus B_2$ for all $0 \le \tau \le t - \tau_1$, then we set $f_t(m) = (\widehat{f}_2)_{t-\tau_1}(h \circ \widehat{f}_{1\tau_1}(m))$. Otherwise, we again pass to $\mathcal{M}_1 \setminus B_1$ via the pasting h. The definition of $f_t(m)$ for $t \in \mathbb{R}$, $t < 0$, and for points $m \in \mathcal{M}_2 \setminus B_2$ is analogous.

Note that the points $m_i^{(2)} = h(m_i^{(1)})$, $i = 1, 2$, are saddles.

EXAMPLE. We take $\mathcal{M}_1 = \mathcal{M}_2$ to be the two-dimensional torus T^2, with irrational windings $f_1^t = f_2^t$. Let $B_1 = B_2 = B$ be such that the points m_1 and m_2 lie on different trajectories. A pasting homeomorphism is constructed in a special way. Let $h_1\colon d^+ \to d^-$ be a trajectorywise homeomorphism (see Figure 1.19); that is, if $m \in d^+$, and with increasing time the trajectory $l(m)$ first intersects the arc d^- at time τ, then $h_1(m) = (f_1)_\tau(m)$. Let $h|_{d^+} = h_1$ and $h|_{d^-} = h_1^{-1}$. The pasting method gives a flow f^t on the connected sum $T^2 \# T^2$ (a surface of genus 2, a pretzel).

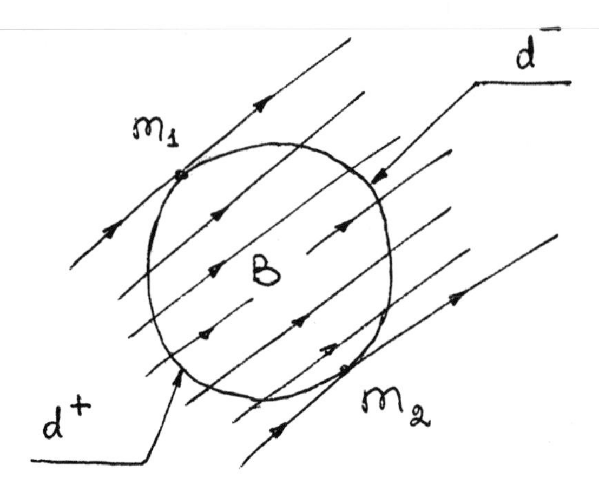

FIGURE 1.19

We leave it as an exercise for the reader to prove that f^t is a transitive flow.

2.6. Suspensions.

2.6.1. *The suspension over a homeomorphism of the circle.* Let f be a homeomorphism of the circle S^1. We use f to paste together the boundary of the cylinder $S^1 \times [0,1]$; that is, we identify the points $(x,1)$ and $(f(x),0)$ in $S^1 \times [0,1]$ for $x \in S^1$. The manifold $\mathcal{M}_f$ obtained is the torus if f preserves the orientation of the circle (since an orientation-preserving homeomorphism is isotopic to the identity), and the Klein bottle if f changes the orientation.

We define a flow on $\mathcal{M}_f$. To do this we represent an arbitrary point of $\mathcal{M}_f$ in the form (x,τ), where $x \in S^1$ and $0 \le \tau < 1$. For $z = (x,\tau) \in \mathcal{M}_f$ and $t \in \mathbb{R}$ we set

$$f_t(z) = (f^{[t+\tau]}(x), \{t+\tau\}),$$

where $[\alpha]$ ($\{\alpha\}$) denotes the integer (fractional) part of a number α (Figure 1.20). The proof of the group property $f_{t_1+t_2} = f_{t_1} \circ f_{t_2}$ is left to the reader.

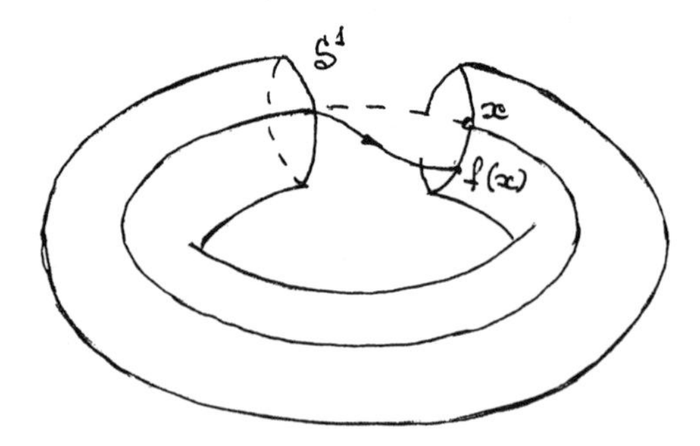

FIGURE 1.20

The flow defined above from the homeomorphism f is called the *suspension* over f and is denoted by $\text{sus}(f)$.

The properties of f and its suspension are closely related, of course. For example, we have

LEMMA 2.1. 1) *If f has a dense orbit, then the flow* $\operatorname{sus}(f)$ *has a dense trajectory (transitivity)*;

2) *if f has a nowhere dense orbit, then* $\operatorname{sus}(f)$ *has a nowhere dense trajectory*;

3) *if f has a periodic orbit, then* $\operatorname{sus}(f)$ *has a periodic trajectory.*

The converse assertions are also valid.

The proof of the lemma follows from the definition of the flow $\operatorname{sus}(f)$ and the fact that any trajectory passing through a point $(x, 0) \in \mathcal{M}_f$ passes through the point $(f^n(x), 0)$ at time $t = n \in \mathbb{Z}$.

EXAMPLE. Let $f = R_\mu$ be the rotation of the circle $x^2 + y^2 = (1/2\pi)^2$ through the angle $2\pi\mu$, $\mu \in \mathbb{R}$. Then $\operatorname{sus}(R_\mu)$ is a rational or irrational winding of the torus, depending on whether μ is rational or irrational.

2.6.2. *The suspension over an exchange of open intervals.* Let $a_1, \dots, a_r \in S^1$ be points labeled successively as the circle S^1 is traversed in the positive direction, $r \geq 2$. Denote by $\Delta_i = (a_i, a_{i+1})$, $i = 1, \dots, r$, the open interval between the points a_i and a_{i+1}, traversed in the positive direction from a_i to a_{i+1}, where $a_{r+1} = a_1$.

Let $b_1, \dots, b_r$ be cyclically labeled points on S^1, distinct from $a_1, \dots, a_r$ in general.

DEFINITION. A transformation $T\colon S^1 \setminus \{a_1, \dots, a_r\} \to S^1 \setminus \{b_1, \dots, b_r\}$ that is one-to-one and is an orientation-preserving homeomorphism on each interval Δ_i, $i = 1, \dots, r$, is called an *exchange of open intervals.*

Associated with each exchange of open intervals is a permutation τ of the numbers $\{1, \dots, r\}$ such that the interval Δ_i is mapped into the interval $(b_{\tau(i)}, b_{\tau(i)+1})$, where $b_{r+1} = b_1$.

REMARK. In ergodic theory [**47**] an exchange is understood to be a one-to-one mapping of the whole circle (or closed bounded interval) into itself that is a shift on each half-open interval.

To construct the suspension over T we first construct the manifold $\mathcal{M}_T$ on which this suspension will be defined.

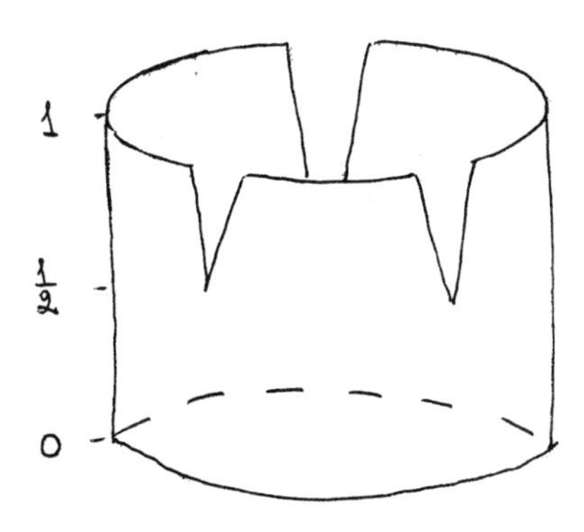

FIGURE 1.21

Denote by $\widehat{\mathcal{M}}$ the manifold obtained by cutting the annulus $S^1 \times [0, 1]$ along the segments $\{a_i\} \times (1/2, 1]$, $i = 1, \dots, r$ (see Figure 1.21). In other words, $\widehat{\mathcal{M}}$ is obtained by attaching disjoint rectangles $\operatorname{cl}(\Delta_i) \times [1/2, 1]$, $i = 1, \dots, r$, to the annulus $S^1 \times [0, 1/2]$ (where $\mathrm{cl}(\Delta_i)$ denotes the closed interval $[a_i, a_{i+1}]$). The side $\{a_i\} \times [1/2, 1]$ of the rectangle $\mathrm{cl}(\Delta_i) \times [1/2, 1]$ ($\mathrm{cl}(\Delta_{i-1}) \times [1/2, 1]$) is called the right (left) side and denoted by r_i (l_i) (see Figure 1.22). We let $l_1 = \{a_1\} \times [1/2, 1] \subset \mathrm{cl}(\Delta_r) \times [1/2, 1]$.

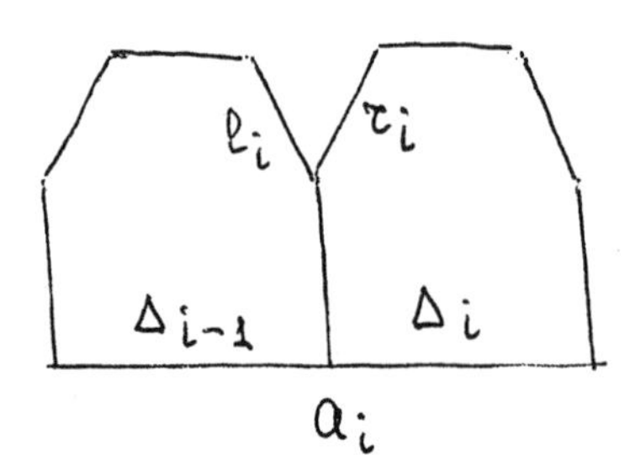

FIGURE 1.22

The side l_j of the manifold $\widehat{\mathcal{M}}$ is pasted together with the side r_k by identifying the points $\{a_j\} \times t \cong \{a_k\} \times t$, $1/2 \leq t \leq 1$, if and only if the segments $T(\Delta_j)$ and $T(\Delta_k)$ are adjacent to each other. The manifold obtained as a result of this pasting is denoted by $\hat{\hat{\mathcal{M}}}$.

By the definition of an exchange of open intervals, each left side in $\widehat{\mathcal{M}}$ is pasted together with one right side, and conversely. Therefore, $\hat{\hat{\mathcal{M}}}$ is a two-dimensional manifold with two boundary components that are homeomorphic to S^1 (Figure 1.23). Speaking loosely, we can say that on one component of the boundary $\partial\hat{\hat{\mathcal{M}}}$ the intervals $\Delta_1, \ldots, \Delta_r$ are arranged in the original order, while the "exchanged" intervals $T(\Delta_1), \ldots, T(\Delta_r)$ lie on the other component of the boundary.

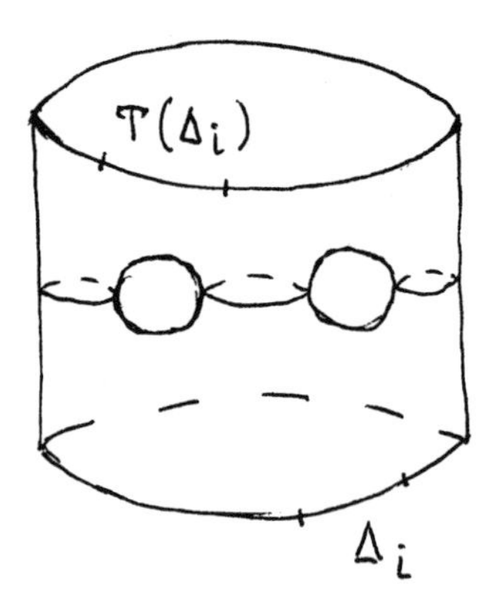

FIGURE 1.23

If in addition we identify the points $(x, 1)$ and $(T(x), 0)$ in $\widehat{\mathcal{M}}$ for $x \in S^2 \setminus \{a_1, \ldots, a_r\}$, then we get a manifold $\overset{\circ}{\mathcal{M}}_T$ of genus $p \geq 2$, with punctures $p_1, \ldots, p_s$ (the punctures correspond to the points $a_1, \ldots, a_r, b_1, \ldots, b_r$; if these points are distinct, then there are $s = 2r$ punctures, but in the general case $r \leq s \leq 2r$).

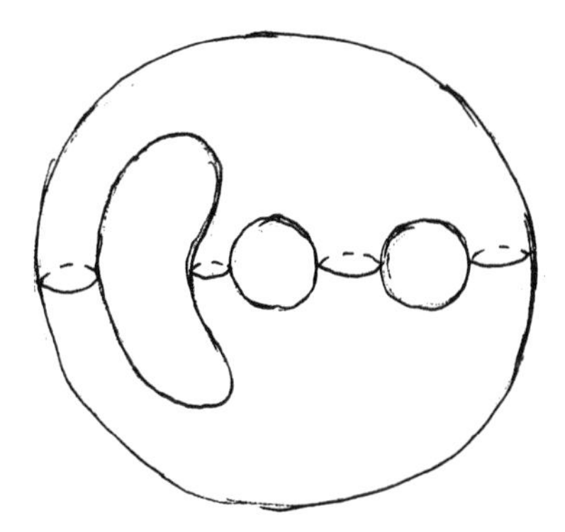

FIGURE 1.24

Let $\mathcal{M}_T$ be the compactification of $\overset{\circ}{\mathcal{M}}_T$. It is clear that $\mathcal{M}_T$ is an orientable closed surface of genus $p \geq 2$ (Figure 1.24).

We now construct the flow $\operatorname{sus}(T)$ on $\mathcal{M}_T$.

On $\mathbb{R}^2$ consider the dynamical system χ^t given by

$$\dot{x} = 0, \qquad \dot{y} = (x^2 + y^2) \cdot [(x-1)^2 + y^2],$$

with two impassable grains at the points $(0,0)$ and $(1,0)$. Denote by V the restriction of the vector field of the flow χ^t to the square $\Pi\colon 0 \leq x \leq 1$, $|y| \leq 1/2$ (see Figure 1.25).

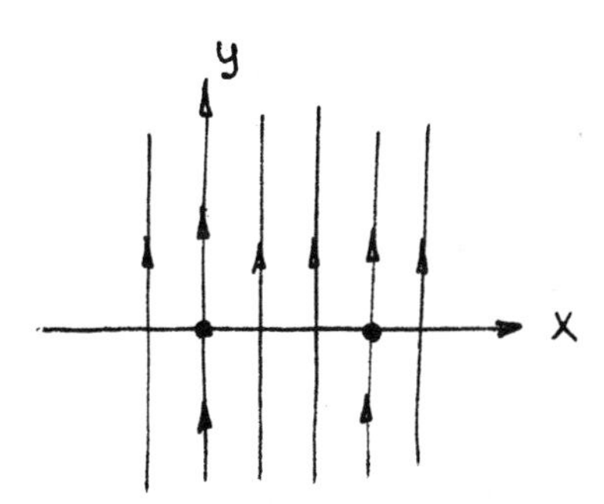

FIGURE 1.25

We represent the manifold $\widehat{\mathcal{M}}$ as the union of the rectangles $\Pi_i = \operatorname{cl}(\Delta_i) \times [0,1]$, $i = 1, \dots, r$. There is a natural distance-preserving diffeomorphism $\psi_i\colon \Pi \to \Pi_i$ carrying the segments $\{x\} \times [-1/2, 1/2]$ into the segments $\{a\} \times [0,1]$ (see Figure 1.26). Then the vector fields $(\psi_i)_*(V)$, $i = 1, \dots, r$, form a vector field $\widehat{V}$ on $\widehat{\mathcal{M}}$. By the definition of the pasting together of the sides l_j and r_k in $\widehat{\mathcal{M}}$, the field $\widehat{V}$ induces a vector field V_T on the manifold $\mathcal{M}_T$.

The flow determined by the vector field V_T on $\mathcal{M}_T$ is called the *suspension over the exchange T of open intervals* and denoted by $\operatorname{sus}(T)$.

We remark that the equilibrium states at the points $p_1, \dots, p_s$ with saddle sectors are the only equilibrium states of the flow $\operatorname{sus}(T)$.

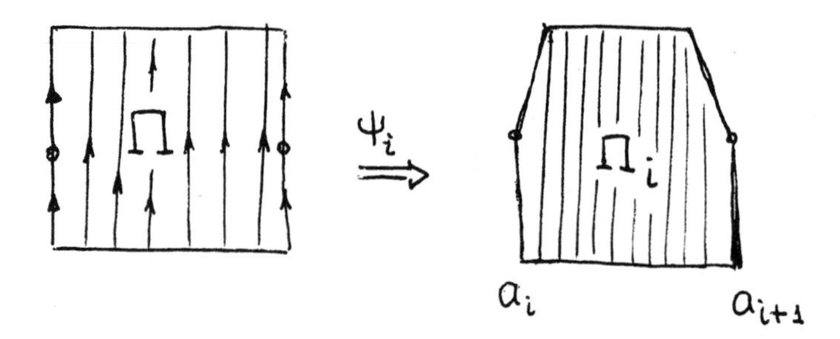

FIGURE 1.26

As in the case of suspensions over homeomorphisms of the circle, the topological properties of an exchange of open intervals are closely related to those of the suspension over it. We have

LEMMA 2.2. *Let T be an exchange of open intervals. Then:*

1) T has a dense semi-orbit if and only if $\operatorname{sus}(T)$ *has a dense semitrajectory;*

2) T has a nowhere dense semi-orbit if and only if $\operatorname{sus}(T)$ *has a nowhere dense semitrajectory different from an equilibrium state;*

3) T has a periodic orbit if and only if $\operatorname{sus}(T)$ *has a closed trajectory.*

The proof follows immediately from the construction of $\operatorname{sus}(T)$, though it is somewhat more complicated than the proof of Lemma 2.1 due to the existence of points at which T is undefined. We leave it to the reader.

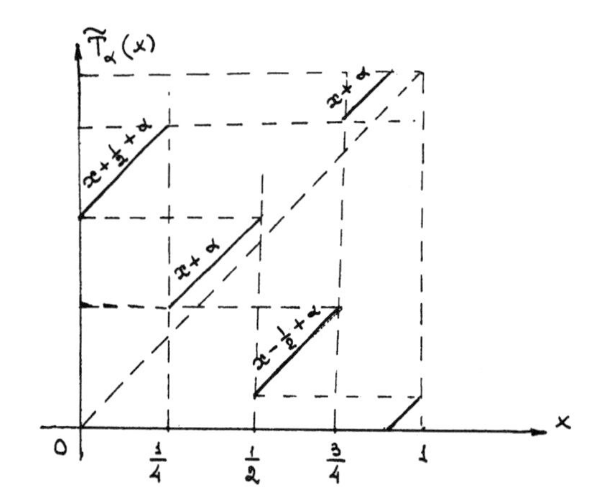

FIGURE 1.27

EXAMPLE (from the paper [**83**]). We represent the circle S^1 as the quotient space of the line $\mathbb{R}$ with respect to the action of the group of translations by integers, or as the interval $[0,1]$ with identification of the endpoints. We consider the exchange $T_\alpha = \widetilde{T}_\alpha \pmod 1$ of the open intervals $(0,1/4)$, $(1/4,1/2)$, $(1/2,3/4)$, and $(3/4,1)$, where $\widetilde{T}_\alpha$ is given graphically in Figure 1.27. It is not hard to see that if α is irrational, then T_α does not have periodic orbits, and, moreover, each orbit is dense in S^1. According to Lemma 2.2, the suspension $\operatorname{sus}(T_\alpha)$ is a transitive flow. Figure 1.28 represents $\operatorname{sus}(T_\alpha)$, where the circles C_1 and C_2 are identified by means of the transformation $x \mapsto x+\alpha \pmod 1$.

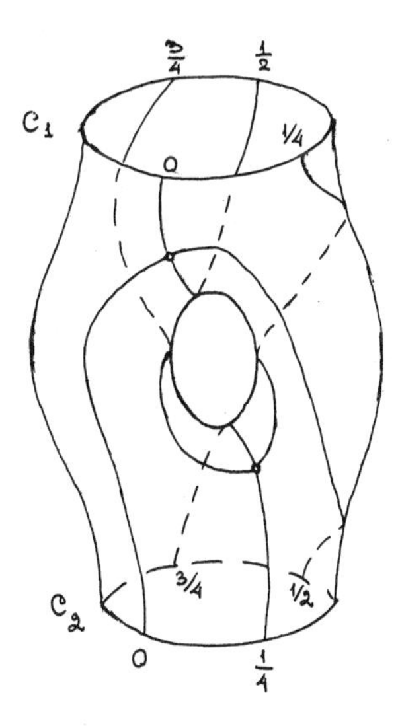

FIGURE 1.28

2.7. Whitney's theorem. Let f^t be a flow on the manifold $\mathcal{M}$. Denote by $\mathrm{Fix}(f^t)$ the set of equilibrium states of f^t. The flow f^t determines on $\mathcal{M} \setminus \mathrm{Fix}(f^t)$ a family of curves: the trajectories of the restriction of f^t to $\mathcal{M} \setminus \mathrm{Fix}(f^t)$. It is clear that this family satisfies definite conditions. In [**110**] and [**111**] Whitney obtained necessary and sufficient conditions for a family of curves to be imbedded in a flow on a locally compact space. For simplicity we present Whitney's result for two-dimensional manifolds (possibly open) without proof.

2.7.1. *The theorem on continuous dependence on the initial conditions.* Let d be a fixed metric on the manifold $\mathcal{M}$. The following result is an immediate consequence of the definition of a flow f^t.

THEOREM 2.1. *For any point $m \in \mathcal{M}$ and any numbers $T > 0$ and $\varepsilon > 0$ there is a number $\delta > 0$ such that if $d(m, \widetilde{m}) < \delta$ and $|t| \leq T$, then*

$$d(f^t(m), f^t(\widetilde{m})) < \varepsilon.$$

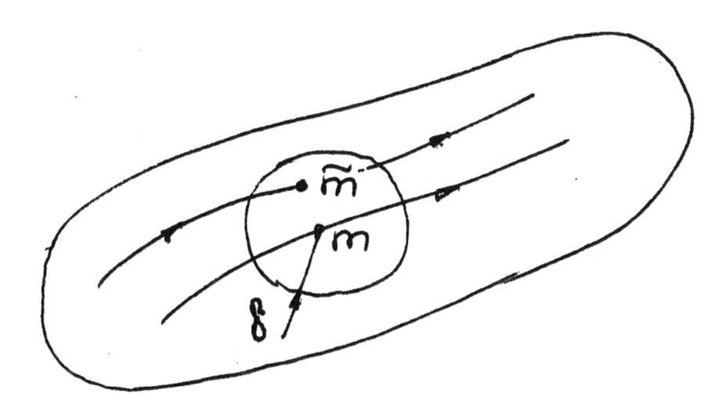

FIGURE 1.29

The condition of geometric continuity is an analogue of the theorem on continuous dependence on the initial conditions.

Let F be a family of disjoint curves on $\mathcal{M}$. For a curve l in F and for points $m_1, m_2 \in l$ denote by $\overset{\frown}{m_1 m_2}$ the arc of l between m_1 and m_2, oriented from m_1 to m_2.

We say that the family F satisfies the condition of *geometric continuity* if for any $\varepsilon > 0$, any curve $l \in F$, and any arc $\overset{\frown}{m_1 m_2} \subset l$ there is a number $\delta > 0$ such that if $d(\widetilde{m}_1, m_1) < \delta$, then there exists an arc $\overset{\frown}{\widetilde{m}_1 \widetilde{m}_2}$ of a curve $\widetilde{l} \in F$ passing through $\widetilde{m}_1$ for which $\overset{\frown}{\widetilde{m}_1 \widetilde{m}_2}$ is in the ε-neighborhood of $\overset{\frown}{m_1 m_2}$ and $d(\widetilde{m}_2, m_2) < \varepsilon$.

The family of trajectories of the restriction of the flow f^t to $\mathcal{M} \setminus \mathrm{Fix}(f^t)$ obviously satisfies the condition of geometric continuity.

2.7.2. *The rectification theorem.* A point of a flow different from an equilibrium state is called a *regular* point.

THEOREM 2.2. *Let $m \in \mathcal{M}$ be a regular point of a C^r-flow f^t $(r \geq 0)$ on a two-dimensional manifold $\mathcal{M}$. Then there exist a neighborhood $\mathcal{U}$ of m and a C^r-diffeomorphism $\mathcal{U} \to \mathbb{R}^2$ carrying the arcs in $\mathcal{U}$ of trajectories into trajectories of the dynamical system*

$$\dot{x} = 1, \qquad \dot{y} = 0$$

on $\mathbb{R}^2$ with preservation of the direction in time.

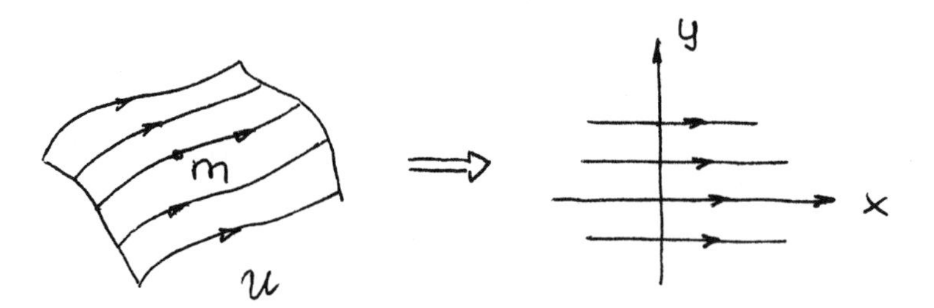

FIGURE 1.30

This theorem is proved in many books on dynamical systems (for example, [**3**], [**26**], [**64**], [**74**]).

A neighborhood $\mathcal{U}$ satisfying Theorem 2.2 will be called a *neighborhood with the structure of a constant field*, and the diffeomorphism $\mathcal{U} \to \mathbb{R}^2$ will be called a *rectifying diffeomorphism.*

DEFINITION. A family $\{l_\alpha\}$ of disjoint curves on a two-dimensional manifold $\mathcal{M}$ is called a *foliation* (or a foliation without singularities) if $\mathcal{M} = \bigcup_\alpha l_\alpha$, and for each point $m \in \mathcal{M}$ there exist a neighborhood $\mathcal{U}$ of m and a homeomorphism $\psi \colon \mathcal{U} \to \mathbb{R}^2$ such that the components of the intesection of the curves l_α with $\mathcal{U}$ are carried by ψ into the lines $y = \mathrm{const}$. The curves l_α are called *leaves*.

Theorem 2.2 shows that the collection of regular trajectories (that is, trajectories different from equilibrium states) of a flow f^t forms a foliation on $\mathcal{M} \setminus \mathrm{Fix}(f^t)$.

It can be shown that the leaves of a foliation satisfy the condition of geometric continuity [**73**].

2.7.3. *Orientability.* We consider two orientable arcs $\overset{\frown}{ab}$ and $\overset{\frown}{cd}$ lying on the manifold $\mathcal{M}$. Denote by $\mathcal{D}(\overset{\frown}{ab}, \overset{\frown}{cd})$ the set of homeomorphisms $H \colon \mathcal{M} \to \mathcal{M}$ such that $H(\overset{\frown}{ab}) = \overset{\frown}{cd}$, $H(a) = c$, and $H(b) = d$.

DEFINITION. The *Fréchet distance* between the arcs $\overset{\frown}{ab}$ and $\overset{\frown}{cd}$ is defined to be the number

$$\sigma(\overset{\frown}{ab}, \overset{\frown}{cd}) = \inf_{H \in \mathcal{D}(\overset{\frown}{ab}, \overset{\frown}{cd})} \sup_{m \in \overset{\frown}{ab}} d(m, H(m)).$$

A sequence of arcs $\overset{\frown}{p_i q_i}$, $i = 1, 2, \ldots$, converges to an arc $\overset{\frown}{pq}$ in the Fréchet sense if $\sigma(\overset{\frown}{pq}, \overset{\frown}{p_i q_i}) \to 0$ as $i \to \infty$.

DEFINITION. A family F of curves on a manifold $\mathcal{M}$ is said to be *orientable* if it is possible to specify an orientation $\mathcal{J}$ on all the curves of F (that is, to give a positive direction on each curve) such that for any arc $\overset{\frown}{pq} \subset L \in F$ oriented in the direction $\mathcal{J}$ and for any sequence of arcs $\overset{\frown}{p_i q_i} \subset L_i \in F$ converging to $\overset{\frown}{pq}$ in the Fréchet sense there is an index i_0 such that for $i \geq i_0$ the orientation on each arc $\overset{\frown}{p_i q_i}$ from p_i to q_i coincides with the orientation $\mathcal{J}$ (where L and L_i are curves of the family F).

According to Theorem 2.1, the family of regular trajectories of a flow is orientable, and the direction of motion along a trajectory as time increases can be taken as the positive direction on a regular trajectory. Thus, the condition that a family of curves forms a foliation (regularity) and the condition of orientability are necessary conditions for the family to be imbedded in a flow. Whitney proved that these conditions are also sufficient. Moreover, he showed that the space on the manifold not occupied by the curves can be filled by equilibrium states.

Let $F = \{L_\alpha\}$ be a family of disjoint curves not necessarily filling the whole manifold $\mathcal{M}$, with the union of the curves in F forming an open subset. The family F is said to be *imbeddable in a flow* if there exists a flow f^t on $\mathcal{M}$ such that each curve in F is a regular trajectory of f^t, and each point in the set $\mathcal{M} \setminus \bigcup_\alpha L_\alpha$ is an equilibrium state of f^t.

THEOREM 2.3 (Whitney's theorem). *Let $F = \{L_\alpha\}$ be a family of disjoint curves on a manifold $\mathcal{M}$ (each curve L_α is the range of an imbedding of the line $\mathbb{R}$ in $\mathcal{M}$). Then F can be imbedded in a flow if and only if F is orientable and regular (that is, it forms a foliation on the set $\bigcup_\alpha L_\alpha$).*

§3. Examples of flows with limit set of Cantor type

Let f^t be a flow on a manifold $\mathcal{M}$, and let $l(m) = \{f(m,t) : -\infty < t < \infty\}$ be the trajectory of f^t passing through a point $m \in \mathcal{M}$.

DEFINITION. The *ω-limit set* of the trajectory $l(m)$ is defined to be the set $\omega[l(m)]$ of points $\widetilde{m}$ such that $f^{t_n}(m) = f(m, t_n) \to \widetilde{m}$ as $n \to +\infty$ for some sequence $\{t_n\}_{n=1}^{\infty}$ of numbers increasing to infinity.

In view of the group property $f^{t_0+t_1} = f^{t_0} \circ f^{t_1}$ the definition of the ω-limit set of $l(m)$ is independent of the choice of the initial point m and depends only on the trajectory itself; that is, $\omega[l(m_1)] = \omega[l(m)] \overset{\text{def}}{=} \omega(l)$ for any point $m_1 \in l(m)$.

If in the above definition we write $\lim_{n\to\infty} t_n = -\infty$ instead of $\lim_{n\to\infty} t_n = +\infty$, then we get the definition of the α-limit set $\alpha[l(m)]$ of the trajectory $l(m)$, which also is independent of the choice of the initial point m: $\alpha(l) = \alpha[l(m)]$, $m \in l$.

The ω- (α-) limit set of a positive (negative) semitrajectory is defined in the obvious way.

The *limit set of a flow* is defined to be the union of the ω-limit sets and the α-limit sets of all its trajectories.

DEFINITION. Let f^t be a flow on a two-dimensional manifold $\mathcal{M}$. The ω- or α-limit set of a trajectory of f^t is called a *limit set of Cantor type* if, with the exception of a finite number (possibly zero) of points, it is locally homeomorphic to the direct product of a Cantor set[1] and a closed bounded interval.

In this section we present the classical examples of Denjoy and Cherry flows, along with a little known example of a flow on a sphere (or on a disk) with limit sets of Cantor type.

3.1. The example of Denjoy.

DEFINITION. A *Denjoy flow* is defined to be a flow f^t without equilibrium states on the two-dimensional torus that has a limit set of Cantor type. We shall obtain a Denjoy flow as the suspension over a Denjoy diffeomorphism of the circle S^1. Let us proceed to the construction of the diffeomorphism.

Denote by $S(\lambda)$ the circle obtained by identifying the endpoints of the interval $[0,\lambda] \subset \mathbb{R}$. In particular, $S^1 = S(1)$. Let $R_\omega \colon x \mapsto x+\omega \pmod 1$ be the rotation of S^1 by some irrational number ω. Take an arbitrary point $x_0 \in S^1$ and let $x_n = R_\omega^n(x_0)$, $n \in \mathbb{Z}$.

Let $a_n = 1/(|n|+2)(|n|+3)$, $n \in \mathbb{Z}$. Then the series $\sum_{-\infty}^{\infty} a_n$ converges; denote its sum by a. We associate with each term a_n of the series an open interval G_n of length a_n. We locate the intervals G_n, $n \in \mathbb{Z}$, on the circle $S(1+a)$ in such a way that they are disjoint and their mutual arrangement corresponds to the mutual arrangement of the points x_n, $n \in \mathbb{Z}$, on S^1.

Since α is irrational, for any distinct points x_i and x_j there are points of the orbit $O(x_0) = \{R_\omega^n(x_0),\ n \in \mathbb{Z}\}$ on each of the arcs of $S^1 \setminus \{x_i, x_j\}$ into which x_i and x_j partition S^1. Therefore, there are intervals G_n between any two intervals G_i and G_j, on both the arcs into which G_i and G_j break up the circle $S(1+a)$. Consequently, the complement of $\cup G_n$ $(n \in \mathbb{Z})$ is a nowhere dense set Ω homeomorphic to the Cantor set.

Denote by $\overset{\circ}{\Omega}$ the subset of points in Ω that are not endpoints of the intervals G_n, $n \in \mathbb{Z}$.

Since the mutual arrangement of the intervals G_n corresponds to the mutual arrangement of the points x_n, $n \in \mathbb{Z}$, there exists an orientation-preserving continuous mapping $h \colon S(1+a) \to S(1) = S^1$ that carries $\mathrm{cl}(G_n)$ into x_n, $n \in \mathbb{Z}$, and is one-to-one on the set $\overset{\circ}{\Omega}$ (Figure 1.31).

We begin the construction of the desired diffeomorphism $f \colon S(1+a) \to S(1+a)$ by constructing its derivative $F \colon S(1+a) \to \mathbb{R}$.

Let $F|_\Omega = 1$. If $G_n = (\alpha_n, \beta_n)$, then we set

$$F(x) = 1 + k_n \cdot \frac{(x-\alpha_n)(\beta_n - x)}{a_n^2}, \qquad x \in G_n,$$

where

$$k_n = 6\left(\frac{a_{n+1}}{a_n} - 1\right).$$

[1]Recall that a Cantor set on a closed bounded interval (or circle) is defined to be a nonempty closed perfect nowhere dense subset of the interval (or circle).

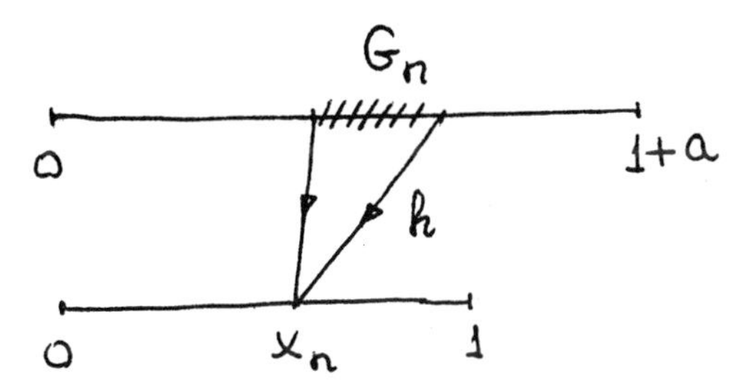

FIGURE 1.31

It can be verified directly that

$$\int_{G_n} F\,dx = a_{n+1}. \tag{3.1}$$

Let us compute the ratio a_{n+1}/a_n. We have that

$$\frac{a_{n+1}}{a_n} = \frac{n+2}{n+1} \qquad \text{for } n \geq 0,$$
$$\frac{a_{n+1}}{a_n} = \frac{|n|+3}{|n|+1} \qquad \text{for } n \leq -1.$$

Then

$$k_n = -\frac{12}{n+4} \qquad \text{for } n \geq 0,$$
$$k_n = \frac{12}{|n|+1} \qquad \text{for } n \leq -1.$$

The quantity $(x-\alpha_n)(\beta_n - x)/a_n^2$ takes a maximal value on the interval $G_n = (\alpha_n, \beta_n)$ at $x = (\alpha_n + \beta_n)/2$. From this and the form of the function $F(x)$ we get the inequalities

$$\begin{aligned} 1/4 \leq 1 - \frac{3}{n+4} \leq F(x) \leq 1, \qquad & n \geq 0, \\ 1 \leq F(x) \leq 1 + \frac{3}{|n|+1} \leq 5/2, \qquad & n \leq -1, \end{aligned} \tag{3.2}$$

where $x \in G_n$. Obviously, $F(x)$ is continuous on the intervals G_n, $n \in \mathbb{Z}$. It follows from (3.2) that $F(x)$ is continuous at the points of Ω. Therefore, $F(x)$ is continuous on $S(1+a)$.

Denote by $\pi\colon \mathbb{R} \to S(1+a)$ a covering of the circle $S(1+a)$ by the line $\mathbb{R}$. It can be assumed without loss of generality that $\pi(0) = \alpha_0$ is the left-hand endpoint of the interval $G_0 = (\alpha_0, \beta_0)$. Let $\overline{F}\colon \mathbb{R} \to \mathbb{R}$ be a covering mapping for F, that is, $\overline{F}(x) = F \circ \pi(x)$, $x \in \mathbb{R}$.

Denote by $\operatorname{meas}(\mathfrak{U})$ the Lebesgue measure of a subset $\mathfrak{U} \subset S(\lambda)$.

Let $\overline{\alpha}_1 \in [0, 1+a]$ be a point such that $\pi(\overline{\alpha}_1) = \alpha_1$, the left-hand endpoint of the interval $G_1 = (\alpha_1, \beta_1)$. We define the mapping $\overline{f}\colon \mathbb{R} \to \mathbb{R}$ by

$$\overline{f}(x) = \overline{\alpha}_1 + \int_0^x \overline{F}(x)\,dx.$$

Then

$$\begin{aligned}\overline{f}[x+(1+a)] &= \overline{\alpha}_1 + \int_0^x \overline{F}(x)\,dx + \int_x^{x+1+a} \overline{F}\,dx \\ &= \overline{f}(x) + \int_{S(1+a)} F\,dx = \overline{f}(x) + \int_\Omega f\,dx + \int_{\cup G_n} F\,dx \\ &= \overline{f}(x) + \operatorname{meas}(\Omega) + \sum_{-\infty}^{\infty} a_n = \overline{f}(x) + 1 + a.\end{aligned}$$

Consequently, the mapping $\overline{f}$ is a covering for some mapping $f\colon S(1+a) \to S(1+a)$ of degree 1. Since $\mathcal{D}\overline{f} = \overline{F}$, it follows that $\mathcal{D}f = F$. By (3.2), f is a C^1-diffeomorphism of the circle $S(1+a)$.

The diffeomorphism f is the desired Denjoy diffeomorphism. Let us consider its properties.

It follows from the construction of the mapping $h\colon S(1+a) \to S^1$ that for any interval $[\alpha, \beta] \subset S(1+a)$ with endpoints $\alpha, \beta \in \mathring{\Omega}$

$$\operatorname{meas}([\alpha, \beta] \cap \Omega) = \operatorname{meas}[h(\alpha), h(\beta)]. \tag{3.3}$$

We show that $f(\Omega) = \Omega$ and that $R_\omega \circ h = h \circ f$. Let $x \in \mathring{\Omega}$ and let I be an index set such that $G_k \subset [\alpha_0, x]$ if and only if $k \in I$. Then $x_k \in [h(\alpha_0), h(x)]$ for $k \in I$. By (3.1), (3.3), and the fact that R_ω is a rotation,

$$\begin{aligned}\operatorname{meas}[\alpha_1, f(x)] &= \int_{[\alpha_0, x]} F\,dx = \int_{\cup G_k} F\,dx + \int_{\Omega\cap[\alpha_0,x]} F\,dx \\ &= \sum_{k\in I} a_{k+1} + \operatorname{meas}[R_\omega h(\alpha_0), R_\omega h(x)].\end{aligned}$$

Since $x \in \mathring{\Omega}$, it follows that $h(x) \notin O(x_0)$ and $R_\omega h(x) \notin O(x_0)$, and hence $h^{-1}R_\omega h(x) \in \mathring{\Omega}$.

It is clear that $x_k \in [R_\omega h(\alpha_0), R_\omega h(x)] = [x_1, R_\omega h(x)]$ if and only if $k-1 \in I$. By the construction of the mapping h, this means that $G_k \in [\alpha_1, h^{-1}R_\omega h(x)]$ if and only if $k-1 \in I$. Consequently (Figure 1.32), (3.3) gives us that

$$\begin{aligned}\operatorname{meas}[\alpha_1, h^{-1}R_\omega h(x)] &= \sum_{k\in I} a_{k+1} + \operatorname{meas}(\Omega \cap [\alpha_1, h^{-1}R_\omega h(x)]) \\ &= \sum_{k\in I} a_{k+1} + \operatorname{meas}[R_\omega h(\alpha_0), R_\omega h(x)].\end{aligned}$$

This implies that $\operatorname{meas}[\alpha_1, h^{-1}R_\omega h(x)] = \operatorname{meas}[\alpha_1, f(x)]$, so $f(x) = h^{-1}R_\omega h(x) \in \mathring{\Omega}$.

The inclusion $f^{-1}(\mathring{\Omega}) \subset \mathring{\Omega}$ is proved similarly. Thus, $f(\mathring{\Omega}) = \mathring{\Omega}$, and the relation $h \circ f = R_\omega \circ h$ holds on the set $\mathring{\Omega}$. It follows from the denseness of $\mathring{\Omega}$ in Ω and the continuity of the mappings f and f^{-1} that $f(\Omega) = f^{-1}(\Omega) = \Omega$, and the relation $h \circ f = R_\omega \circ h$ holds on Ω. This and the monotonicity of f imply that $f(G_n) = G_{n+1}$ for all $n \in \mathbb{Z}$. Since $h(G_n) = x_n$ for $n \in \mathbb{Z}$, the relation $h \circ f = R_\omega \circ h$ holds on the whole circle $S(1+a)$.

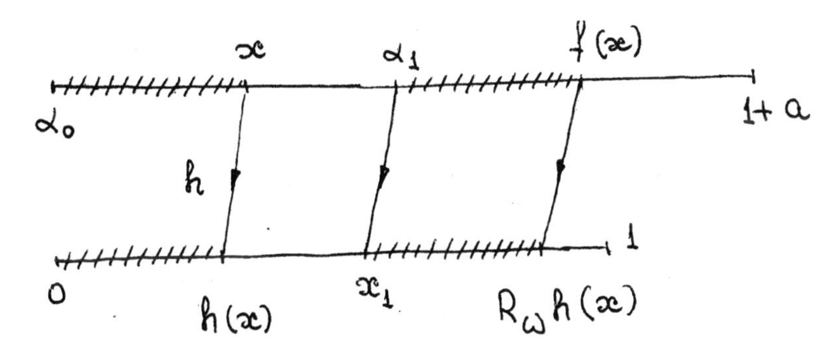

FIGURE 1.32

If $x \in \Omega$, then it follows from the monotonicity of h, the equality $h(\Omega) = S^1$, and the denseness of each orbit of the rotation $R_\omega \colon S^1 \to S^1$ (since ω is irrational) that the limit set of x coincides with Ω. If $x \notin \Omega$, then the orbit $O(x)$ intersects each interval at precisely one point. Since $\operatorname{diam} G_n \to 0$ as $|n| \to \infty$, the limit set of the point x also coincides with Ω. Consequently, the limit set of the Denjoy diffeomorphism f is the Cantor set Ω.

The suspension $\operatorname{sus}(f)$ gives the desired Denjoy flow. It follows from the foregoing that the limit set $\Omega(\operatorname{sus}(f))$ of $\operatorname{sus}(f)$, which consists of the trajectories passing through points of Ω, is locally homeomorphic to the direct product of a closed bounded interval and the Cantor set.

It follows from the denseness of each orbit of the rotation R_ω, the relation $h \circ f = R_\omega \circ h$, and the monotonicity of h that any orbit in Ω with respect to the diffeomorphism f is dense in Ω. Therefore, each trajectory of the flow $\operatorname{sus}(f)$ in the limit set $\Omega(\operatorname{sus}(f))$ is dense in $\Omega(\operatorname{sus}(f))$.

3.2. Cherry flows.

DEFINITION. A *Cherry flow on the torus* T^2 is defined to be a flow f^t satisfying the following conditions:

1) f^t has a limit set Ω of Cantor type containing a nonzero number of equilibrium states $O_1, \dots, O_k$;

2) All the equilibrium states $O_1, \dots, O_k$ are structurally stable saddles;

3) for three (of the four) separatrices of each saddle O_i, $i = 1, \dots, k$, the ω- or α-limit set coincides with Ω.[2]

We describe two Cherry flows with different topological structures. One of them has smoothness C^∞, and the other smoothness C^1. The latter will be needed in Chapter 7.

Both examples are based on a C^∞-flow $\widetilde{f}^t$ on $\mathbb{R}^2$ with the following properties (Figure 1.33):

1) $\widetilde{f}^t$ has two equilibrium states: a stable node at the point $(-1, 0)$ and a saddle at the point $(0, 0)$;

2) in some neighborhood of the boundary of the square $[-2, 2] \times [-2, 2]$ the trajectories of $\widetilde{f}^t$ are straight lines parallel to the axis Ox;

[2]It can be shown that the fourth separatrix of a saddle O_i does not belong to Ω and does not have Ω as its α- or ω-limit set.

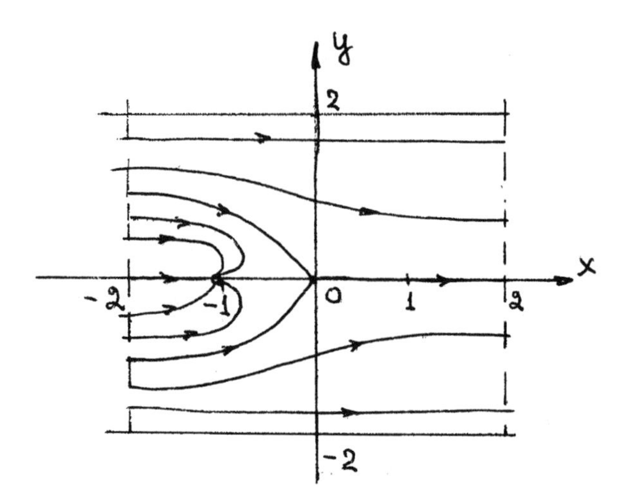

FIGURE 1.33

3) the ω-separatrices of the saddle $O(0,0)$ (that is, the separatrices tending to the saddle as $t \to +\infty$) intersect the line $x = -2$ at the points $(-2, 1)$ and $(-2, -1)$;

4) one α-separatrix of the saddle $O(0,0)$ (that is, a separatrix tending to the saddle as $t \to -\infty$) tends to the node as $t \to +\infty$, while the other α-separatrix coincides with the ray $y = 0$, $x > 0$;

5) the vector field of phase velocities of $\widetilde{f}^t$ is the image of the vector field of the dynamical system $(A) : \dot{x} = x + x^2$, $\dot{y} = -y$ (Figure 1.34) under the action of a C^∞-diffeomorphism $\varphi\colon \mathbb{R}^2 \to \mathbb{R}^2$ for which the arc $C(x \leq 1)$ of the circle $x^2 + y^2 = 4$ passes into the segment $x = -2$, $-2 \leq y \leq 2$, the arcs d_1 and d_2 of trajectories of the flow (A) (see Figure 1.34) pass into the arcs $y = 2$, $-2 \leq x \leq 2$ and $y = -2$, $-2 \leq x \leq 2$ of trajectories of $\widetilde{f}^t$, respectively, and the segment of the line $x = 2$ between d_1 and d_2 passes into the segment $x = 2$, $-2 \leq y \leq 2$.

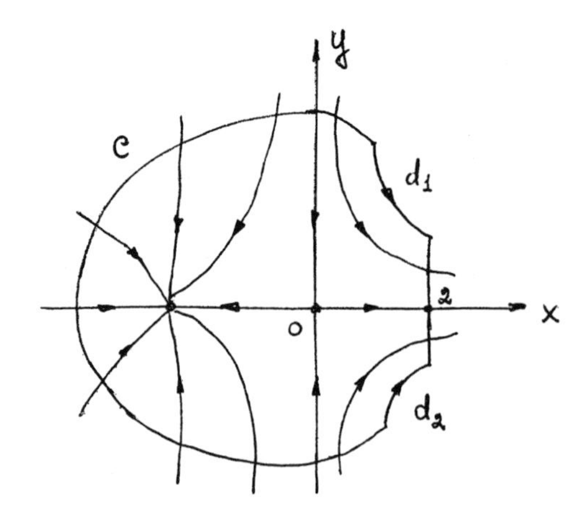

FIGURE 1.34

It is not hard to see that the trajectories of the dynamical system (A) intersect the arc C transversally.

Denote by $\vec{V}$ the restriction of the vector field of phase velocities of $\widetilde{f}^t$ to the square $[-2,2] \times [-2,2]$. If we identify opposite sides of this square, then we get a torus T^2 with a vector field $\vec{V}_0$. Suppose that the field $\vec{V}_0$ determines a flow f_0^t on T^2 that is given by the dynamical system (A_0):

$$\dot{x} = P(x,y), \qquad \dot{y} = Q(x,y),$$

where $P(x,y)$ and $Q(x,y)$ are C^∞-functions, and are 1-periodic in each variable. We consider the dynamical system (A_μ) given by

$$\dot{x} = P(x,y) - \mu Q(x,y), \qquad \dot{y} = Q(x,y) + \mu P(x,y).$$

It follows from results in Chapters 5 and 6 that there is a $\mu = \mu_*$ such that the dynamical system (A_{μ_*}) does not have closed trajectories, and its limit set is locally homeomorphic to the product of a closed bounded interval and the Cantor set, with the exception of two points that are equilibrium states. The dynamical system (A_{μ_*}) gives the desired C^∞-Cherry flow on the torus (Figure 1.35). The construction presented is an insignificant modification of the construction of the C^∞-Cherry flow in [**39**].

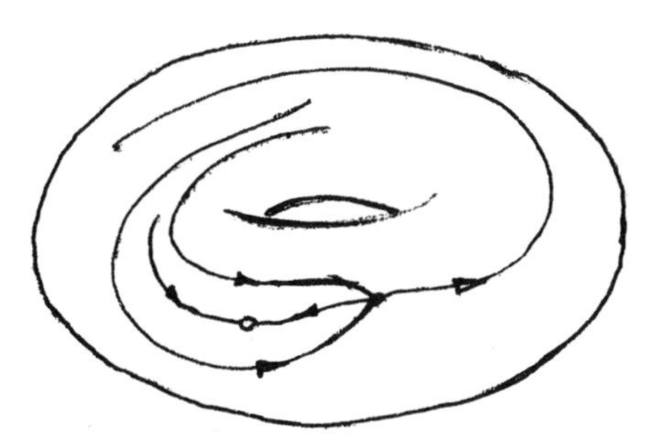

FIGURE 1.35

REMARK. In [**81**] Cherry constructed such a flow with analytic smoothness by starting out from an analytic dynamical system with properties analogous to those of the system (A_0).

We can show the existence of an analytic Cherry flow by using the concept of the rotation number, introduced in Chapters 5 and 6. To do this we take two parameters μ_1 and μ_2 such that the rotation numbers of the dynamical systems (A_{μ_1}) and (A_{μ_2}) are distinct, and we approximate these systems by analytic dynamical systems $(\widetilde{A}_1)$ and $(\widetilde{A}_2)$ in such a way that their rotation numbers are also distinct. Furthermore, we make sure that $(\widetilde{A}_1)$ and $(\widetilde{A}_2)$ each have two equilibrium states: a structurally stable node and a structurally stable saddle. In the space of analytic systems we join the systems $(\widetilde{A}_1)$ and $(\widetilde{A}_2)$ by a family $(\widetilde{A}_\alpha)$, where $1 \leq \alpha \leq 2$. Since the rotation number depends continuously on the parameter, there is an $\alpha = \overline{\alpha}$ such that the dynamical system $(\widetilde{A}_{\overline{\alpha}})$ has an irrational rotation number and two equilibrium states of the above type. Then $(\widetilde{A}_{\overline{\alpha}})$ gives us the desired analytic Cherry flow.

We now construct a C^1-Cherry flow by using the dynamical system (A_0) and a Denjoy flow. In addition to the properties 1)–5) satisfied by the system (A_0) we require that (A_0) also have the following properties (Figure 1.36):

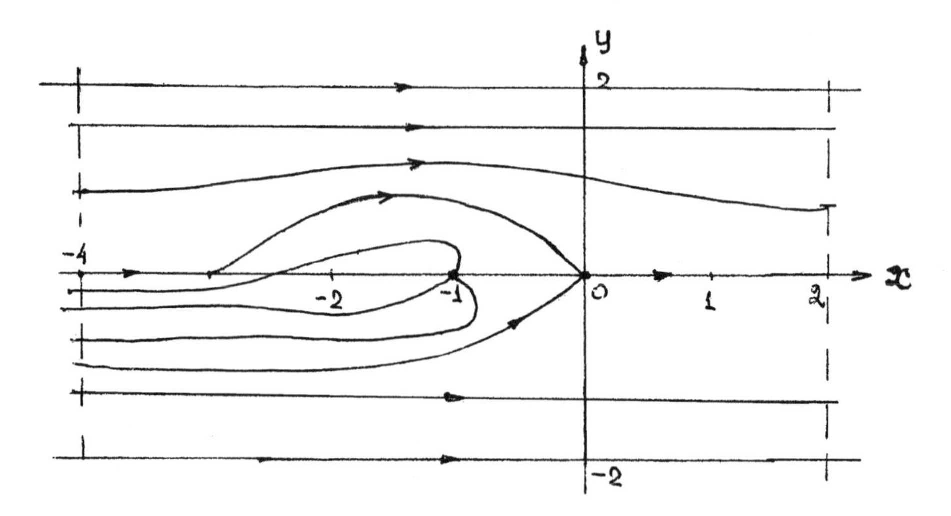

FIGURE 1.36

6) in some neighborhood of the boundary of the rectangle $[-4, 2] \times [-2, 2]$ the trajectories of the flow $\widetilde{f}^t$ are lines parallel to the axis Ox;

7) any trajectory intersecting the line $x = -4$ at a point $(-4, \theta)$, $0 < \theta \leq 2$, intersects the line $x = 2$ at the point $(2, \theta)$ as time increases;

8) the ray $y = 0$, $x \leq -4$ is an ω-separatrix of the saddle O that passes through the point $(-2, 1)$.

Let g^t be a C^1-Denjoy flow on the torus T^2. Denote by $\Omega(g^t)$ the limit set of g^t. By the construction of a Denjoy flow, the set $T^1 \setminus \Omega(g^t)$ consists of a single simply connected component w. We take a point $m \in \Omega(g^t)$ on the boundary of w. Then there is a neighborhood $\mathfrak{U}$ of m that is partitioned by the component of $l(m) \cap \mathfrak{U}$ containing m into two parts $\mathfrak{U}^+$ and $\mathfrak{U}^-$, one of which (say $\mathfrak{U}^-$) does not meet $\Omega(g^t)$ (Figure 1.37).

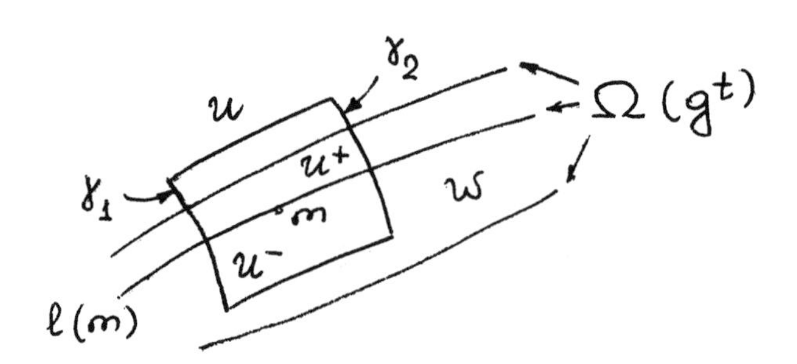

FIGURE 1.37

It can be assumed without loss of generality that $\mathfrak{U}$ is a neighborhood with the structure of a constant field bounded by two arcs of trajectories and by two arcs γ_1 and γ_2 (Figure 1.37) transversal to the trajectories of g^t, and there exists a rectifying C^1-diffeomorphism $\chi\colon \mathfrak{U} \to \mathbb{R}^2$ carrying $\mathfrak{U}$ into the rectangle $[-4, 2] \times [-2, 2]$ such that $\chi(\gamma_1) = \{x = -4,\ -2 \leq y \leq 2\}$, $\chi(\gamma_2) = \{x = 2,\ -2 \leq y \leq 2\}$, and $\chi(\mathfrak{U}^-) = [-4, 2] \times [-2, 0]$.

We replace the vector field of phase velocities of the flow g^t in the neighborhood $\mathcal{U}$ by the vector field $(\chi^{-1})_*(\vec{V})$, where $\vec{V}$ is the restriction of the vector field of the flow $\widetilde{f}^t$ to the rectangle $[-4, 2] \times [-2, 2]$. The vector field obtained on T^2 is denoted by $\vec{W}$. Let f^t be the flow induced by the field $\vec{W}$. Then f^t is the desired C^1-Cherry flow (Figure 1.37). It follows from the property 7) that the limit set of f^t includes a saddle, a node, and a set coinciding with the limit set of the Denjoy flow outside the neighborhood $\mathcal{U}$.

3.3. An example of a flow on the sphere. We consider two disks $\mathcal{D}_1$ and $\mathcal{D}_2$ bounded by circles S_1 and S_2 of unit length. Each circle will be represented as the interval $[0, 1]$ with endpoints identified ($S_i \overset{\text{def}}{=} [0, 1]/\sim$, $i = 1, 2$). Suppose that on each disk $\mathcal{D}_i$, $i = 1, 2$, we are given the same foliation F_i, $i = 1, 2$, with two singularities of "thorn" type and with leaves transversal to the boundary $\partial\mathcal{D}_i = S_i$, $i = 1, 2$ (Figure 1.38). Furthermore, the points $x, 1 - x \in S_i$ lie on a single leaf for any $x \in S_i$, $i = 1, 2$.

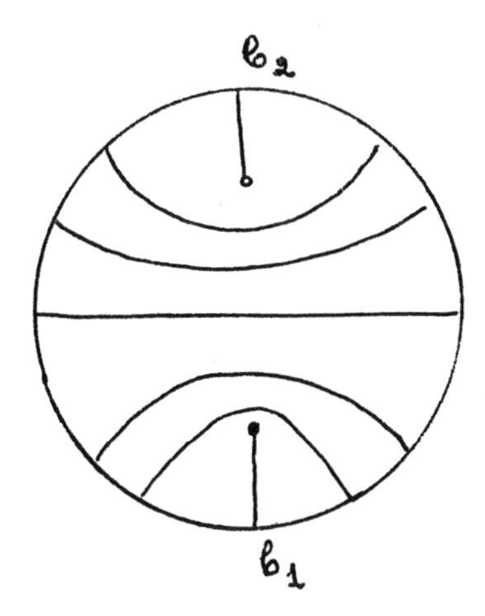

FIGURE 1.38

We identify the boundaries of $\mathcal{D}_1$ and $\mathcal{D}_2$ with the help of the shift $R_\mu \colon S_1 \to S_2$ ($x \mapsto x + \mu \pmod 1$), where μ is an irrational number. As a result of pasting together the two disks we get the sphere S^2. The foliations F_1 and F_2 form a foliation F on S^2 such that (since μ is irrational) each of its leaves is dense on the sphere and none are compact. (Indeed, if from a point $x \in \partial\mathcal{D}_1$ we move along a leaf first on $\mathcal{D}_1$ and then on $\mathcal{D}_2$, then we hit the point $x - \mu \pmod 1 \in \partial\mathcal{D}_1$; but each orbit of the homeomorphism of the circle of the form $x \mapsto x - \mu \pmod 1$ is dense on the circle).

The foliation F can also be obtained as follows. It is not hard to verify that the quotient space of the Euclidean plane $\mathbb{R}^2$ by the action of the group G of homeomorphisms of the form

$$\begin{cases} x \mapsto (-1)^k x + r \\ y \mapsto (-1)^k y + s \end{cases}$$

is homeomorphic to the sphere S^2, that is, $S^2 \cong \mathbb{R}^2/G$. We can take as a fundamental polygon of G the rectangle $0 \le x \le 1/2$, $0 \le y \le 1$ with sides identified under the action of G as shown in Figure 1.39. The natural projection $\pi\colon \mathbb{R}^2 \to \mathbb{R}^2/G = S^2$ is a universal branched covering with four branch points $\pi(m/2, n/2) \in S^2$, $m, n \in \mathbb{Z}$, each with branching order two.

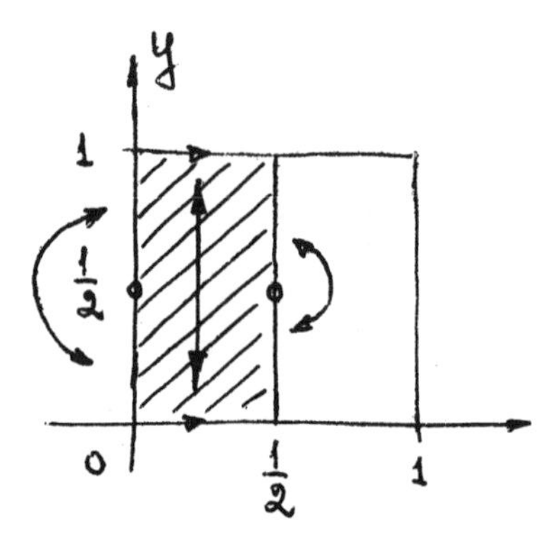

FIGURE 1.39

The family of lines $y = -\mu x + \lambda$, $\lambda \in \mathbb{R}$, forms a foliation on $\mathbb{R}^2$ which projects into the foliation F on S^2. The lines passing through the points with integer and half-integer coordinates project into leaves containing singularities of "thorn" type.

We remark that the foliation F is obtained as the image of an irrational winding (without taking into account the direction in time) of the torus T^2 under a two-sheeted branched covering $T^2 \to S^2$ with four branch points of index two.

We take a leaf L of F and denote by x_n, $n \in \mathbb{N}$, the points where L intersects S_1. With the respective points x_n we associate numbers $a_n > 0$ such that the series $\sum_{n=1}^{\infty} a_n$ converges. The subsequent construction is an operation of "blowing up" the leaf L and is analogous to the similar operation in the construction of a Denjoy flow.

With each number a_n, $n \in \mathbb{N}$, we associate an open interval G_n of length a_n, and we arrange the intervals G_n, $n \in \mathbb{N}$, on the circle $S(1+a)$, where $a = \sum_{n=1}^{\infty} a_n$, corresponding to the way the points x_n, $n \in \mathbb{N}$, are arranged on S_1, with the G_n disjoint.

Since the set of x_n, $n \in \mathbb{N}$, is dense in S_1, the set $\Omega_1 = S(1+a) \setminus \cup G_n$ $(n \in \mathbb{N})$ is a Cantor set. Denote by $\overset{\circ}{\Omega}_1 \subset \Omega_1$ the subset of points that are not endpoints of the intervals G_n, $n \in \mathbb{N}$.

In view of the distribution of the G_n, $n \in \mathbb{N}$, there exists a continuous orientation-preserving mapping $h\colon S(1+a) \to S_1$ carrying $\mathrm{cl}(G_n)$ into x_n, $n \in \mathbb{N}$, and one-to-one on the set $\overset{\circ}{\Omega}_1$.

Let $\widetilde{\mathfrak{D}}_1 \subset \mathbb{R}^2$ be the disk bounded by the circle $S(1+a)$. We construct on $\widetilde{\mathfrak{D}}_1$ a foliation $\widetilde{F}_1$ transversal to the boundary $\partial\widetilde{\mathfrak{D}}_1 = S(1+a)$ and having two singularities of "thorn" type. For definiteness and for simplicity we assume that the leaf L does not contain a singularity, although it will be clear from what follows how to modify the construction in the opposite case.

Denote by $b_1, b_2 \in S_1$ points lying on leaves of F_1 that contain singularities of "thorn" type (Figure 1.38). In view of our assumption, $b_1, b_2 \notin L$. We require that for the foliation $\widetilde{F}_1$ the points $h^{-1}(b_1), h^{-1}(b_2) \in \overset{\circ}{\Omega}_1$ lie on leaves with singularities of "thorn" type. Suppose that the points $c_1, c_2 \in S_1$ do not lie on the leaf L and are joined by a leaf of the foliation F_1; then the points $h^{-1}(c_1), h^{-1}(c_2) \in \overset{\circ}{\Omega}_1$ are also joined by a leaf of the foliation $\widetilde{F}_1$ (recall that h is one-to-one on $\overset{\circ}{\Omega}_1$). If the points $x_i, x_j \in L \cap S_1$ are joined by a leaf of F_1 (that is, these points belong to a component of the intersection $L \cap \mathcal{D}_1$), then we require that each point of the interval $\mathrm{cl}(G_i)$ be joined by a leaf of $\widetilde{F}_1$ to some point of the interval $\mathrm{cl}(G_j)$ (Figure 1.40). Since h_* is an orientation-preserving continuous mapping, a foliation $\widetilde{F}_1$ with the required properties exists.

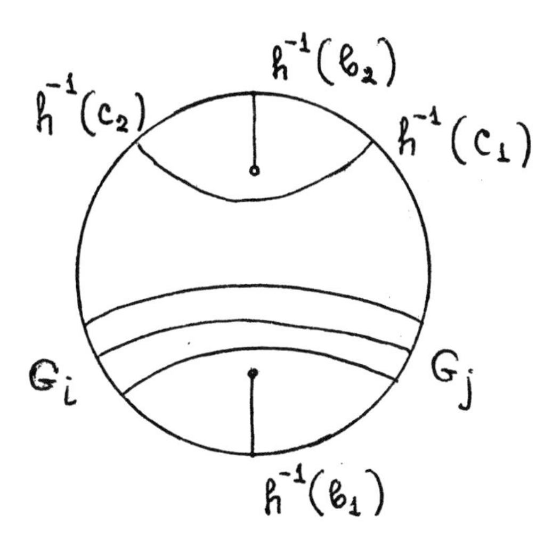

FIGURE 1.40

We now realize similarly an operation of "blowing up" the leaf L for the foliation F_2 on the disk $\mathcal{D}_2$. We get a foliation on the disk $\widetilde{\mathcal{D}}_2$ bounded by the circle $S(1+a)$. Because of the equality $R_\mu(L \cap S_1) = L \cap S_2$ and the fact that h is an orientation-preserving continuous mapping, there is a homeomorphism $f\colon S(1+a) \to S(1+a)$ such that

$$R_\mu \circ h = h \circ f \tag{3.4}$$

(see §2.5 in Chapter 5 for a rigorous construction of the homeomorphism f).

We identify the boundaries of the disks $\widetilde{\mathcal{D}}_1$ and $\widetilde{\mathcal{D}}_2$ with the help of f. The foliations $\widetilde{F}_1$ and $\widetilde{F}_2$ form a foliation $\widetilde{F}$ on the sphere S^2 obtained. Denote by $\widetilde{\Omega}$ the set of leaves of $\widetilde{F}$ passing through points in $\Omega_1 \subset S(1+a) \subset S^2$. It is clear that $\widetilde{\Omega}$ is locally homeomorphic to the product of a closed bounded interval and the Cantor set.

It follows from (3.4) that $\widetilde{\Omega}$ is the union of all the self-limit leaves of $\widetilde{F}$ (a leaf L_0 is called a self-limit leaf if for any point $m_0 \in L_0$ and any segment Σ transversal to the leaves and passing through m_0 the intersection $L_0 \cap \Sigma$ has m_0 as an accumulation point). All the remaining leaves of $\widetilde{F}$ are proper; that is, the intrinsic topology of each of them coincides with the topology induced by the topology of the sphere S^2 (consequently, these leaves are not self-limits).

Denote by $\overset{\circ}{F}$ the restriction of $\widetilde{F}$ to the set $S^2 \setminus \Omega$. The family $\overset{\circ}{F}$ of curves is orientable. Therefore, by Theorem 2.3 (Whitney's theorem), $\overset{\circ}{F}$ can be imbedded in a flow f^t on S^2, and each point of $\widetilde{\Omega}$ is an equilibrium state of f^t. The limit set of any trajectory lying in $S^2 \setminus \widetilde{\Omega}$ coincides with $\widetilde{\Omega}$. Consequently, f^t is the desired C^0-flow on the sphere with a limit set of Cantor type. The flow f^t can be constructed to be also of smoothness class C^∞ (see Chapter 7).

In conclusion we note that a foliation analogous to $\widetilde{F}$ can be obtained from the diffeomorphism $f_0 \colon S^2 \to S^2$ constructed by R. V. Plykin in the paper *Sources and sinks of A-diffeomorphisms of surfaces* (Mat. Sb. **94 (136)** (1974), 243–264; English transl. in Math. USSR Sb. **23** (1974)). The unstable manifolds of the points of the hyperbolic attractor of f_0 form a family of curves that can be imbedded in a foliation analogous to the one constructed. See also Plykin's survey, *On the geometry of hyperbolic attractors of smooth cascades* (Uspekhi Mat. Nauk **39** (1984), no. 6, 75–113; English transl. in Russian Math. Surveys **39** (1984)).

§4. The Poincaré index theory

One of the basic concepts in the qualitative theory of dynamical systems is the concept of the Poincaré index. In this section we study the index theory, omitting proofs and following the exposition [**68**] of Poincaré. We regard this approach as more acceptable in the treatment of continuous flows and foliations than the currently prevalent approach based on the concepts of the rotation of a vector field and the degree of a mapping.

4.1. Contact-free segments and cycles. Suppose that f^t is a given flow on a two-dimensional manifold $\mathcal{M}$.

DEFINITION. A segment $\Sigma \subset \mathcal{M}$ (that is, the range of an imbedding of the interval $[0,1]$ in $\mathcal{M}$) is called a *contact-free segment* or a *transversal* of the flow f^t if for any point $m \in \Sigma \setminus \partial\Sigma$ there exist a neighborhood $\mathcal{U} \ni m$ with the structure of a constant field and a rectifying diffeomorphism $\varphi \colon \mathcal{U} \to \mathbb{R}^2$ (with smoothness the same as that of the flow) such that $\varphi(m) = (0,0) \in \mathbb{R}^2$ and $\varphi(\Sigma \cap \mathcal{U}) = \{0\} \times [-1,1] \subset \mathbb{R}^2$ (we remark that φ carries the trajectories of f^t in $\mathcal{U}$ into the lines $y = \text{const}$).

A simple closed curve C (the image of an imbedding of a circle in $\mathcal{M}$) is called a *contact-free cycle* or a *closed transversal* of the flow if its arcs are contact-free segments.

In the definition of a contact-free segment Σ it is possible that $\partial\Sigma = \emptyset$. In this case Σ will be called an *open contact-free segment.*

The next result follows directly from the rectification theorem.

LEMMA 4.1. *Through each regular point of a flow there passes a contact-free segment.*

It is not hard to give an example of a flow for which no regular point has a contact-free cycle passing through it (for example, a flow on a sphere with two saddles and four centers).

4.2. The index of a nondegenerate cycle in a simply connected domain. Let f^t be a flow on a two-dimensional manifold $\mathcal{M}$, suppose that the simple closed curve $C \subset \mathcal{M}$ bounds a simply connected domain $\mathcal{D}^-$ in $\mathcal{M}$, and let $\mathcal{D}^+ = \mathcal{M} \setminus (C \cup \mathcal{D}^-)$.

DEFINITION. We say that a trajectory l of f^t *is tangent to* C at a point $m \in C$ if there exists a neighborhood $\mathcal{U} \ni m$ such that:

1) m is the only point common to C and the component d of $l \cap \mathcal{U}$ containing m;

2) the arc d either lies in $\mathcal{D}^- \cup \{m\}$ (in this case m is a *point of interior tangency*; see Figure 1.41, a) or $d \subset \mathcal{D}^+ \cup \{m\}$ (in this case m is a *point of exterior tangency*; see Figure 1.41, b).

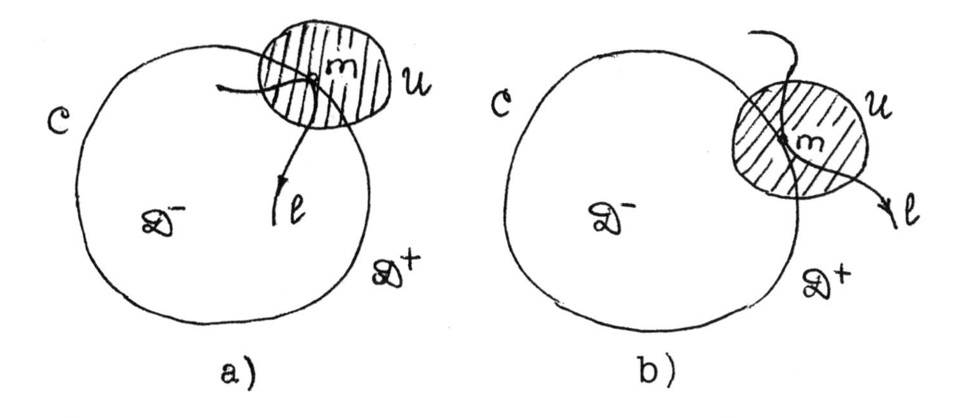

FIGURE 1.41

According to Poincaré, the simple closed curve C is called a *nondegenerate cycle* (of the flow f^t) if:

1) C does not contain equilibrium states of f^t;

2) C contains at most finitely many points $m_1, \dots, m_k$ of tangency to trajectories of f^t (there may be none);

3) each arc of the complement $C \setminus \{m_1, \dots, m_k\}$ is an open contact-free segment (Figure 1.42).

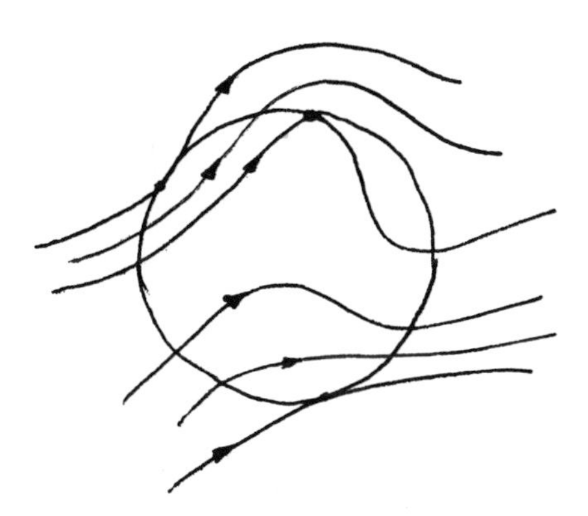

FIGURE 1.42

Recall that an isotopy of a manifold $\mathcal{N}$ into $\mathcal{M}$ is defined to be a mapping $\Phi\colon \mathcal{N}\times[0,1]\to\mathcal{M}$ such that for any $t\in[0,1]$ the mapping $\varphi_t\colon \mathcal{N}\to\mathcal{M}$, $x\mapsto\Phi(x,t)$, is an imbedding. If $\mathcal{N}=S^1$, then we say that the curve $\varphi_0(S^1)$ is isotopic to the curve $\varphi_1(S^1)$ by means of the connecting family of curves $\varphi_t(S^1)$, $0\le t\le 1$.

LEMMA 4.2. *Suppose that there are no equilibrium states of the flow f^t on the simple closed curve $C_0\subset\mathcal{M}$. Then there exists an isotopic nondegenerate cycle C_1 of f^t, and f^t does not have equilibrium states on any of the curves in the connecting family C_t, $0\le t\le 1$.*

This can be proved from the rectification theorem and theorems on reduction to general position [**40**].

DEFINITION. The *index* $j(C,f^t)$ of a nondegenerate cycle C with respect to the flow f^t is defined to be the number

$$j(C,f^t)=\tfrac{1}{2}(2-k_{\text{ex}}+k_{\text{in}}),$$

where k_{ex} is the number of points of exterior tangency to C of trajectories of f^t, and k_{in} the number of points of interior tangency.

It follows from topological considerations and from the theorem on continuous dependence on the initial conditions that the number $-k_{\text{ex}}+k_{\text{in}}$ is even, and hence the index is an integer.

The definition of the index gives us the next result directly.

COROLLARY 4.1. *If C is a contact-free cycle for the flow f^t, then* $j(C,f^t)=1$

The proof follows from $k_{\text{ex}}=k_{\text{in}}=0$. □

4.3. The index of an isolated equilibrium state. Let m_0 be an isolated equilibrium state of the flow f^t. This means that there exists a simply connected neighborhood $\mathcal{U}(m_0)$ of m_0 in which there are no equilibrium states of f^t besides m_0. By Lemma 4.2, there is a nondegenerate cycle $C\subset\mathcal{U}(m_0)$ that contains m_0 in its interior.

DEFINITION. The *index $j(m_0,f^t)$ of an isolated equilibrium state m_0* of a flow f^t is defined to be the index of a nondegenerate cycle lying in a simply connected neighborhood of the equilibrium state and containing m_0 in its interior.

The next lemma shows that the definition of the index of an equilibrium state does not depend on the choice of nondegenerate cycle satisfying the conditions of the definition.

LEMMA 4.3. *Let C_0 and C_1 be isotopic nondegenerate cycles of the flow f^t, and suppose that that each curve of a connecting family C_t, $0\le t\le 1$, does not pass through equilibrium states of f^t. Then $j(C_0,f^t)=j(C_1,f^t)$.*

EXAMPLES. 1) the index of a topological node or focus is equal to $+1$ (Figure 1.43, a, b);

2) the index of a topological saddle with four separatrices is equal to -1 (Figure 1.43, c).

THEOREM 4.1 (Bendixson). *Suppose that an isolated equilibrium state m_0 has h hyperbolic sectors, and e elliptic sectors. Then its index is equal to $j(m_0,f^t)=\frac{1}{2}(2-h+e)$.*

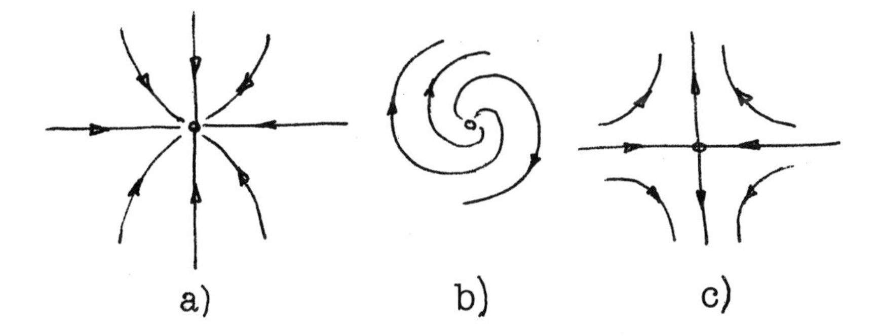

FIGURE 1.43

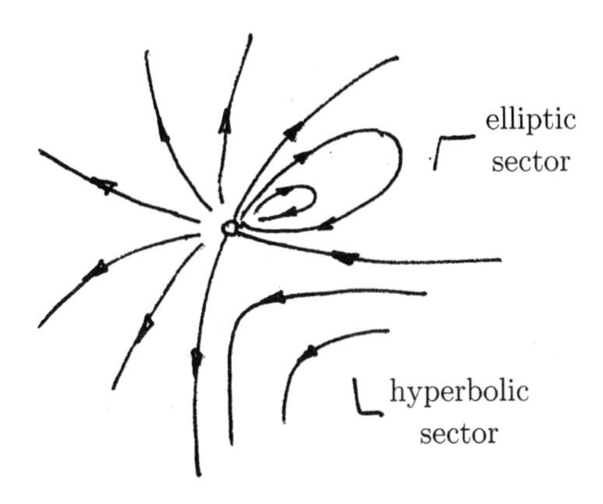

FIGURE 1.44

There is a proof of Theorem 4.1 in the book [**3**] (English pp. 511–515).

The following result can be proved by induction.

THEOREM 4.2 (the sum of the indices). *Suppose that a nondegenerate cycle C of a flow f^t bounds a simply connected domain $\mathcal{D}$ on a two-dimensional manifold $\mathcal{M}$, and suppose that $\mathcal{D}$ contains finitely many equilibrium states of f^t. Then the index of C with respect to f^t is equal to the sum of the indices of the equilibrium states in $\mathcal{D}$.*

COROLLARY 4.2. *If the index of a nondegenerate cycle bounding a simply connected domain $\mathcal{D}$ is nonzero, then there is at least one equilibrium state in $\mathcal{D}$. In particular, a simply connected domain bounded by a contact-free cycle* (*according to Corollary* 4.1) *contains at least one equilibrium state.*

COROLLARY 4.3. *Suppose that a closed trajectory l of the flow f^t bounds a simply connected domain $\mathcal{D}$ on $\mathcal{M}$, and that f^t has only isolated equilibrium states. Then $\mathcal{D}$ contains at least one equilibrium state.*

PROOF. By Lemma 4.2, there exists a nondegenerate cycle C isotopic to l such that each curve in the family connecting l and C does not pass through equilibrium states of f^t (Figure 1.45). Since l is closed, the number of points of exterior tangency to C of trajectories is equal to the number of points of interior

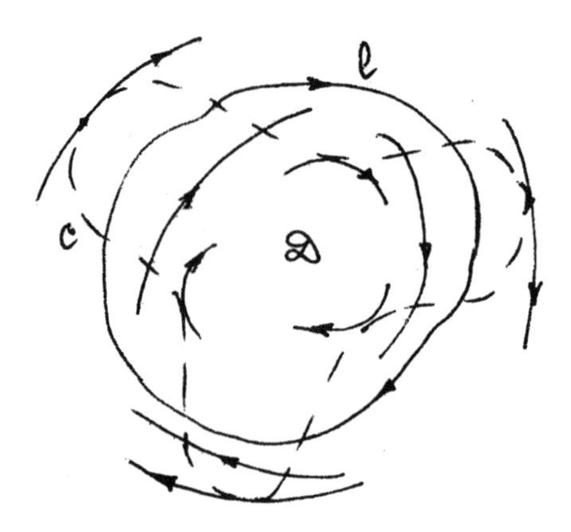

FIGURE 1.45

tangency, and therefore $j(C, f^t) = 1$. The required assertion follows from this and Corollary 4.2. □

4.4. The Euler characteristic and the Poincaré index. Recall that the Euler characteristic $\chi(\mathcal{M})$ of a closed orientable two-dimensional manifold $\mathcal{M}$ of genus $p \geq 0$ is equal to $\chi(\mathcal{M}) = 2 - 2p$. If $\mathcal{M}$ is nonorientable, then $\chi(\mathcal{M}) = 2 - p$.

THEOREM 4.3 (Poincaré). *Let f^t be a flow with finitely many equilibrium states $m_1, \dots, m_k$ on a closed two-dimensional manifold $\mathcal{M}$. Then $\sum_{i=1}^k j(m_i, f^t) = \chi(\mathcal{M})$.*

COROLLARY 4.4. *Assume the conditions of Theorem 4.3, and let $\mathcal{M}_q \subset \mathcal{M}$ be a connected orientable submanifold of genus $q \geq 0$ bounded by finitely many contact-free cycles $C_1, \dots, C_r$ for the flow f^t. Then*

$$\sum j(m_{i_n}, f^t) = 2 - 2q - r,$$

where the summation is over all equilibrium states in $\mathcal{M}_q$.

PROOF. $\mathcal{M}_q$ can be imbedded in a closed orientable two-dimensional manifold $\widetilde{\mathcal{M}}_q$ of genus q ($\widetilde{\mathcal{M}}_q$ is obtained by "attaching" disks to $\mathcal{M}_q$ along the contact-free cycles $C_1, \dots, C_r$), and a flow $\widetilde{f}^t$ can be defined on $\widetilde{\mathcal{M}}_q$ that has r nodes and the equilibrium states of f^t that lie in $\mathcal{M}_q$. The required assertion follows from Theorem 4.3 and the fact that the index of a node is equal to 1. □

COROLLARY 4.5. *Assume the conditions of Corollary 4.4, and suppose that f^t does not have equilibrium states on $\mathcal{M}_q$. Then $\mathcal{M}_q$ is homeomorphic to an annulus.*

PROOF. It follows from Corollary 4.4 that $2 - 2q = r \geq 1$. Since $q \geq 0$ is an integer, we have that $q = 0$ and $r = 2$.

4.5. Connection between the index and the orientability of foliations. In the preceding subsections the presence of a motion along the trajectories of the flow as time changes did not play an essential role. Therefore, the definitions of a nondegenerate cycle and of the index carry over in their entirety to the same concepts for foliations. All the preceding results also are preserved, except that "singularities of the foliation" should replace equilibrium states. An essential difference for foliations is the possibility of nondegenerate cycles and certain singularities being nonintegral. For example, singularities with indices $1/2$ and $-1/2$ are represented in Figure 1.46 a and b, respectively. It is clear that the index being nonintegral is connected with the foliations being nonorientable. In this subsection we investigate this connection in greater detail. First we give a definition of an orientable nondegenerate cycle, which is not homotopic to zero in general.

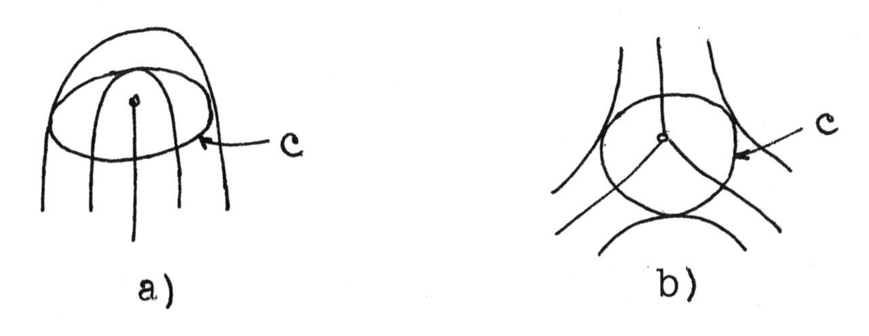

FIGURE 1.46

Let F be a foliation on a surface $\mathcal{M}$ with the set $\mathfrak{s}(F)$ of singularities, and let $C \subset \mathcal{M} \setminus \mathfrak{s}(F)$ be a simple closed curve with only finitely many points of tangency to leaves of F. Such a curve, with a direction (taken to be positive) introduced on it is called an *orientable nondegenerate cycle.*

If the surface $\mathcal{M}$ is orientable, then the nondegenerate cycle C has a neighborhood $\mathcal{D}_C$ homeomorphic to an annulus, and C divides it into two components $\mathcal{D}_C^-$ and $\mathcal{D}_C^+$, which are also homeomorphic to an annulus (Figure 1.47). Denote by $\mathcal{D}_C^-$ (respectively, $\mathcal{D}_C^+$) the component of $\mathcal{D}_C \setminus C$ to the left (right) upon traversing C in the positive direction.

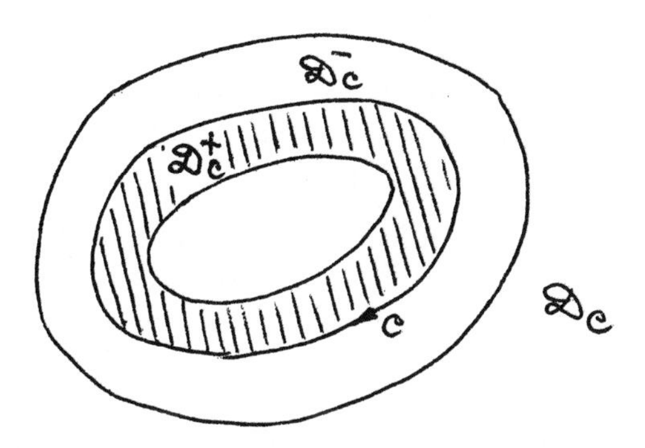

FIGURE 1.47

In order for the following definitions to agree with the preceeding ones in the case when C bounds a simply connected domain in $\mathcal{M}$ we denote by $\mathcal{D}_C^-$ the component belonging to the simply connected domain, and we orient the curve C correspondingly.

For the oriented nondegenerate cycle C we introduce the concepts of points of interior and exterior tangency in a way completely analogous to the corresponding concepts in §4.2, except that the domains $\mathcal{D}^-$ and $\mathcal{D}^+$ are replaced by $\mathcal{D}_C^-$ and $\mathcal{D}_C^+$.

DEFINITION. The index of an orientable nondegenerate cycle C with respect to a foliation F is defined to be the number

$$j(C, F) = \tfrac{1}{2}(2 - k_{\text{ex}} + k_{\text{in}}),$$

where k_{ex} (k_{in}) is the number of points of exterior (interior) tangency to C of the leaves of F.

If an orientable nondegenerate cycle $\widetilde{C}$ coincides with C as a set, but its orientation is opposite to that of C, then

$$j(\widetilde{C}, F) = 2 - j(C, F).$$

This implies that whether the index is an integer does not depend on the orientation of the nondegenerate cycle.

The index of any nondegenerate cycle with respect to a flow is an integer. Therefore, if there exists an orientable nondegenerate cycle with nonintegral index with respect to some foliation F, then the family of leaves lying in $\mathcal{M} \setminus \mathfrak{s}(F)$ cannot be imbedded in a flow. But if the index of any orientable nondegenerate cycle with respect to F is an integer, then the family of leaves lying in $\mathcal{M} \setminus \mathfrak{s}(F)$ can be imbedded in a flow.

Lemma 4.3, which is valid for foliations, enables us to give a definition of the index of any simple orientable closed curve, not necessarily having only finitely many tangencies to the leaves of the foliation (for example, a curve consisting of contact-free segments and arcs of leaves). Namely, the index of a simple orientable closed curve $C_0 \subset \mathcal{M}$ (with respect to a foliation F) not passing through singularities of F is defined to be the index $j(C_1, F)$ of any orientable nondegenerate cycle $C_1 \subset \mathcal{M}$ isotopic to C_0 and such that every curve in a connecting family C_t, $0 \le t \le 1$, does not contain singularities of F. The isotopy from C_0 to C_1 carries the orientation of C_0 in the natural way into the orientation of C_1. According to Lemma 4.2, such a nondegenerate cycle exists, and by Lemma 4.3 the definition of the index $j(C_0, F)$ does not depend on the concrete nondegenerate cycle C_1. These arguments are valid for the index of a curve with respect to a flow, of course.

4.6. An example of a foliation that is locally but not globally orientable. In this section we present a foliation $\widehat{F}$ on a closed orientable surface $\mathcal{M}_3$ of genus $p = 3$ that has only two singularities of saddle type with six separatrices each. Thus, each singularity has index $j = -2$ and $\widehat{F}$ is locally orientable in a neighborhood of each singularity. However, the family of regular curves of $\widehat{F}$ is not orientable and cannot be imbedded in a flow. It will be clear from the construction how the foliation $\widehat{F}$ can be constructed to be transitive.

Let us begin with a foliation F_0 with five singularities on a sphere: four "thorns", and one "tripod". We consider two disks $\mathcal{D}_1$ and $\mathcal{D}_2$ bounded by circles S_1 and S_2, and the foliations F_1 and F_2 on these disks are pictured in Figure 1.48. We paste $\mathcal{D}_1$ and $\mathcal{D}_2$ together along S_1 and S_2 with the help of a homeomorphism $S_1 \to S_2$ in such a way that the foliation F_0 obtained on the sphere S^2 does not have leaves joining two singularities nor closed (compact) leaves (we can take

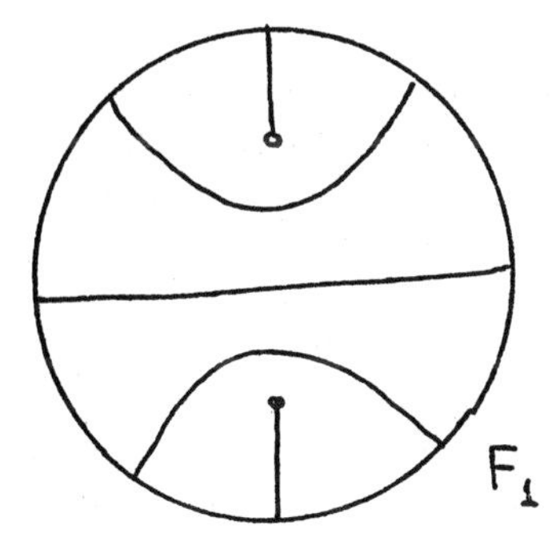

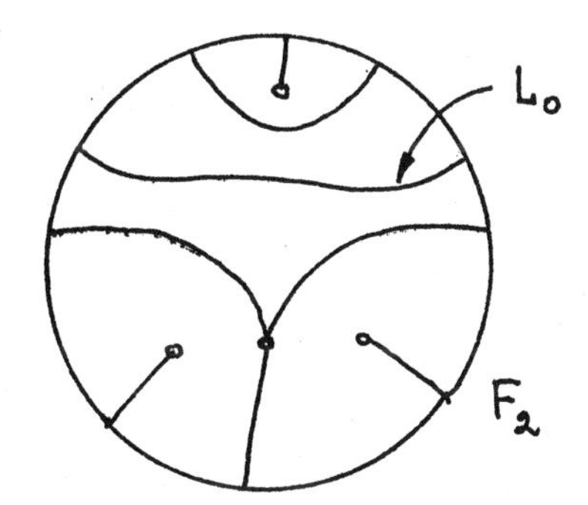

FIGURE 1.48

the pasting homeomorphism to be such that we get a transitive foliation F_0, and therefore all the subsequent foliations will also be transitive).

On the disk $\mathcal{D}_2$ we take a leaf L_0 of F_2 dividing $\mathcal{D}_2$ into two domains $\mathcal{D}_{21}$ and $\mathcal{D}_{22}$, one of which, say $\mathcal{D}_{21}$, contains one singularity of thorn type, while the other ($\mathcal{D}_{22}$) contains a tripod and two thorns (Figure 1.48). The boundary $\partial\mathcal{D}_{22}$ of $\mathcal{D}_{22}$ consists of the leaf L_0 and a segment transversal to the leaves of F_2. Denote by λ_0 the simple closed curve into which $\partial\mathcal{D}_{22}$ passes when $\mathcal{D}_1$ and $\mathcal{D}_2$ are pasted together. Then λ_0 divides S^2 into two disks, which we also denote by $\mathcal{D}_1$ and $\mathcal{D}_2$. We introduce an orientation on λ_0 such that the disk $\mathcal{D}_2$ (which contains the tripod and two thorns) is to the left upon moving along λ_0 in the positive direction. It is not hard to see that the index of the oriented curve λ_0 with respect to the foliation F_0 is equal to the index of a singularity of thorn type; that is, $j(\lambda_0, F_0) = 1/2$ (Figure 1.49).

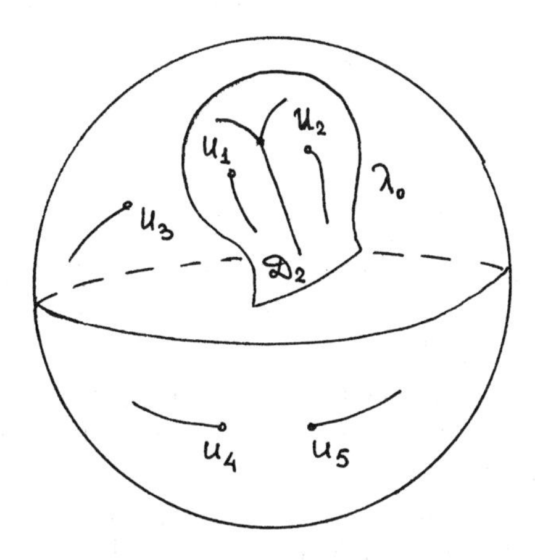

FIGURE 1.49

Denote by $u_1, u_2 \in \mathcal{D}_2$ and $u_3, u_4, u_5 \in \mathcal{D}_1$ the thorns of the foliation F_0 (see Figure 1.49). There exists a two-sheeted branched covering $\pi_0 \colon T = \mathcal{M}_2 \to S^2$ with the four branch points u_1, u_2, u_4, and u_5 (each of index two). We describe the covering π_0 and its construction in greater detail. Let $\mathcal{U}_{12} \subset \mathcal{D}_2$ and $\mathcal{U}_{45} \subset \mathcal{D}_1$ be arcs without self-intersections joining the respective pairs u_1, u_2 and u_4, u_5 and not passing through any other singularities of F_0. Moreover, $\mathcal{U}_{12}$ and $\mathcal{U}_{45}$ can be assumed to be transversal to F_0. Cutting S^2 along $\mathcal{U}_{12}$ and $\mathcal{U}_{45}$, we get a surface $\mathcal{M}_{0,2}$ of genus 0 with two holes. We imbed $\mathcal{M}_{0,2}$ in the three-dimensional Euclidean space $\mathbb{R}^3$ with coordinates (x, y, z) in such a way that $\mathcal{M}_{0,2}$ intersects the axis Oz only at the points u_1, u_2, u_4, and u_5, and each component of the boundary $\partial\mathcal{M}_{0,2}$ is invariant with respect to the rotation $\gamma_z\colon \mathbb{R}^3 \to \mathbb{R}^3$ about the axis Oz through a $180°$ angle. The transformation γ_z has the form $\overline{x} = -x$, $\overline{y} = -y$, $\overline{z} = z$. Then the surfaces $\mathcal{M}_{0,2}$ and $\gamma_z(\mathcal{M}_{0,2})$ have a common boundary, and the union $\mathcal{M}_{0,2} \cup \gamma_z(\mathcal{M}_{0,2})$ is a two-dimensional torus T. By construction, the quotient space $T/G(\gamma_z)$ of T by the group $G(\gamma_z)$ generated by the element γ_z is homeomorphic to S^2, and the natural projection $\pi_0\colon T \to T/G(\gamma_z) \cong S^2$ is a two-sheeted branched covering with the four branch points u_1, u_2, u_4, and u_5.

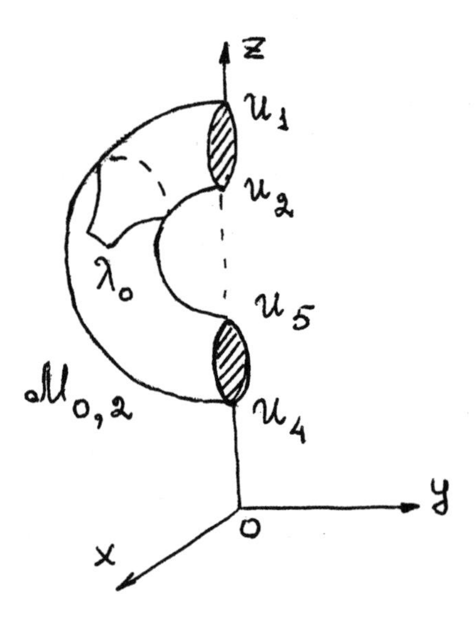

FIGURE 1.50

On T there exists a foliation F covering the foliation F_0. Two tripods at the points $\pi^{-1}(Y) = \{Y_1, Y_2\}$ and two thorns at the points $\pi_0^{-1}(u_3) = \{v_1, v_2\} \in T$ make up the singularities of F.

Since $j(\lambda_0, F_0) = 1/2$, both the curves λ_1 and λ_2 in the complete inverse image $\pi_0^{-1}(\lambda_0)$ also have index $1/2$ with respect to F when equipped with the orientation induced by the covering π_0.

It follows from the construction of the covering π_0 that the curves λ_1 and λ_2 are not homotopic and hence not homologous to zero; that is, each of them separates the torus T (Figure 1.51). Therefore, there exist arcs $d_1, d_2 \subset T \setminus \{\lambda_1\}$ not passing through singularities of F and joining the respective pairs Y_1, v_1 and Y_2, v_2 of points. The arcs d_1 and d_2 can be constructed to be transversal to the foliation F. Next, we proceed as above, cutting T along d_1 and d_2 and constructing a two-sheeted branched covering $\widehat{\pi}\colon \mathcal{M}_3 \to T$ with the four branch points Y_1, Y_2, v_1, and v_2 (Figure 1.52). The foliation $\widehat{F}$ on $\mathcal{M}_3$ covering F has only two singularities: each being a saddle with six separatrices. However, each of the closed curves $\widehat{\lambda}_1$ and $\widehat{\lambda}_2$ in the complete inverse image $\pi^{-1}(\lambda_1)$ has a nonintegral index with respect

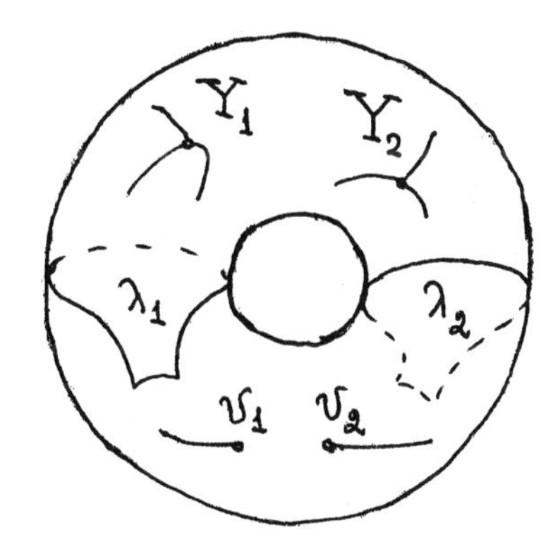

FIGURE 1.51

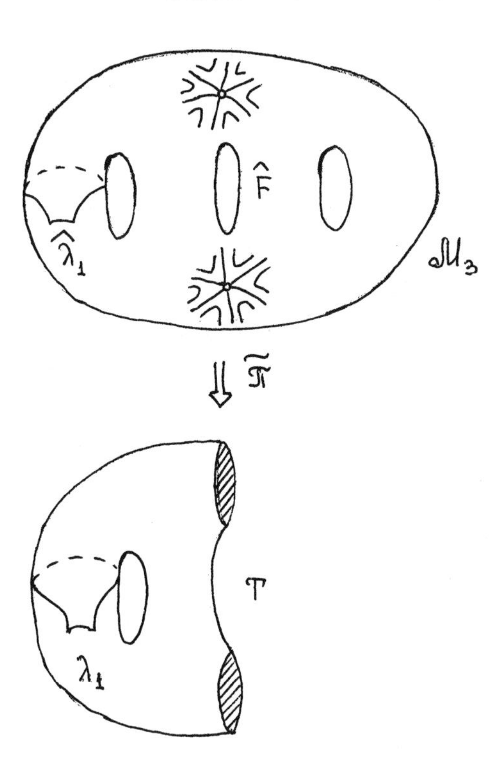

FIGURE 1.52

to the foliation $\widehat{F}$. Therefore, the leaves of $\widehat{F}$ are not globally orientable; that is, the family of regular leaves of $\widehat{F}$ cannot be imbedded in a flow.

Remark. About a result of Èl′sgol′ts. Suppose that a flow f^t on a closed two-dimensional manifold $\mathcal{M}$ has only isolated equilibrium states. According to Theorem 4.3, the sum of the indices of all the equilibrium states of f^t is equal to the Euler characteristic $\chi(\mathcal{M})$, and the number $|\chi(\mathcal{M})|$ gives a lower bound for the sum of the moduli of the indices of all the equilibrium states. If the equilibrium states are of the simplest type, that is, are stable and unstable nodes or saddles with four separatrices, then the lower bound of the number of equilibrium states can be made concrete.

Let us consider a flow f^t having equilibrium states only of the simplest type. Define k_0 (k_2) to be the number of unstable (stable) nodes of f^t, and k_1 to be the number of saddles with four separatrices. Since the index of a node is equal to 1, while the index of a saddle is equal to -1, we have that

$$(k_0 + k_2) - k_1 = \chi(\mathcal{M}).$$

We consider the case of a closed orientable two-dimensional manifold of genus $p \geq 1$. Then

$$(k_0 + k_2) - k_1 = 2 - 2p \leq 0.$$

This implies that the smallest number k_1 of saddles is equal to $2 - 2p$, and then $k_0 = k_2 = 0$. Thus,

$$k_0 \geq 0, \qquad k_1 \geq 2 - 2p, \qquad k_2 \geq 0.$$

If $\mathcal{M}$ is the sphere S^2, then $(k_0 + k_2) - k_1 = 2$. Therefore, the smallest value of the sum $k_0 + k_2$ is equal to 2, and then $k_1 = 0$. It can be shown that

$$k_0 \geq 1, \qquad k_1 \geq 0, \qquad k_2 \geq 1.$$

Analogous estimates can be obtained for nonorientable $\mathcal{M}$. All the estimates are sharp.

The result described was obtained with the use of Betti numbers by L. È. Èl′sgol′ts (*An estimate of the number of singular points of a dynamical system given on a manifold*, Mat. Sb. **26 (68)** (1950), 215–223), and later reproved in [**100**].

CHAPTER 2

Structure of Limit Sets

In this chapter we prove the classical theorems of Maĭer (a criterion for recurrence) and Cherry (on the closure of a recurrent trajectory), and we present Maĭer's estimate of the number of independent nontrivial recurrent semitrajectories. The chapter concludes with an exposition of the Poincaré–Bendixson theory, a catalogue of limit sets and minimal sets, and an investigation of the structure of quasiminimal sets.

§1. Initial concepts and results

1.1. The long flow tube theorem, and construction of a contact-free cycle.

The rectification theorem (sometimes called the flow tube theorem) generalizes to the case when a finite arc of a regular trajectory is considered instead of a single regular point.

THEOREM 1.1 (the long flow tube theorem). *Let d be a compact arc of a regular trajectory (that is, a trajectory different from an equilibrium state) of a C^r-flow f^t $(r \geq 0)$, and suppose that d does not form a closed curve. Then there exists a neighborhood $\mathcal{U}$ of d and a C^r-diffeomorphism $\psi\colon \mathcal{U} \to \mathbb{R}^2$ carrying the arcs in $\mathcal{U}$ of trajectories of f^t into trajectories of the dynamical system $\dot{x} = 1$, $\dot{y} = 0$.*

A proof follows from the rectification theorem and the theorem on continuous dependence on the initial conditions. The books [**3**] and [**26**] contain the detailed proof, and we omit it.

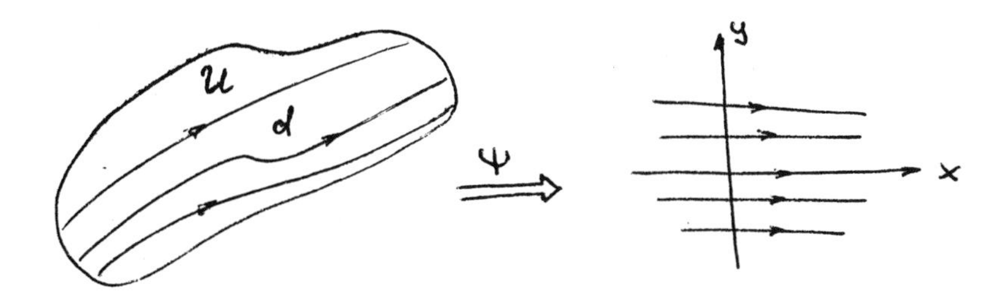

FIGURE 2.1

Just as in the rectification theorem, the neighborhood $\mathcal{U}$ in Theorem 1.1 will be called a *neighborhood with the structure of a constant field*, and the diffeomorphism ψ a *rectifying diffeomorphism*.

Let f^t be a flow on a two-dimensional manifold $\mathcal{M}$.

DEFINITION. An open domain $\mathcal{U} \subset \mathcal{M}$ bounded by two contact-free segments Σ_1 and Σ_2 and two arcs d_1 and d_2 of trajectories of f^t such that the union $d_1 \cup d_2 \cup \Sigma_1 \cup \Sigma_2$ is a simple closed curve will be called an *open flow box* if $\mathcal{U}$ does not contain equilibrium states of f^t (Figure 2.2). The union of the domain $\mathcal{U}$ with the curve $d_1 \cup d_2 \cup \Sigma_1 \cup \Sigma_2$ will be called a *flow box* (or *closed flow box*).

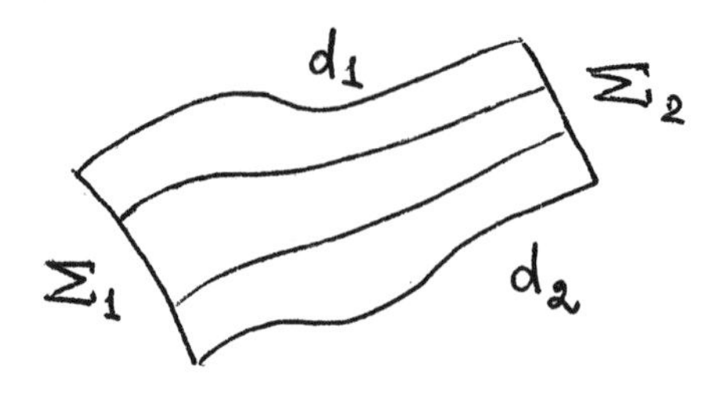

FIGURE 2.2

LEMMA 1.1. *Let γ be either a regular point or a compact arc of a regular trajectory of the flow f^t, and suppose that γ does not form a closed curve in the latter case. Then there exists a neighborhood $\mathcal{U} \supset \gamma$ that is an open flow box.*

PROOF. This follows immediately from the rectification theorem and the long flow tube theorem. □

LEMMA 1.2. *Suppose that the trajectory l of the flow f^t intersects a contact-free segment Σ at more than one point. Then there exists a contact-free cycle that intersects l.*

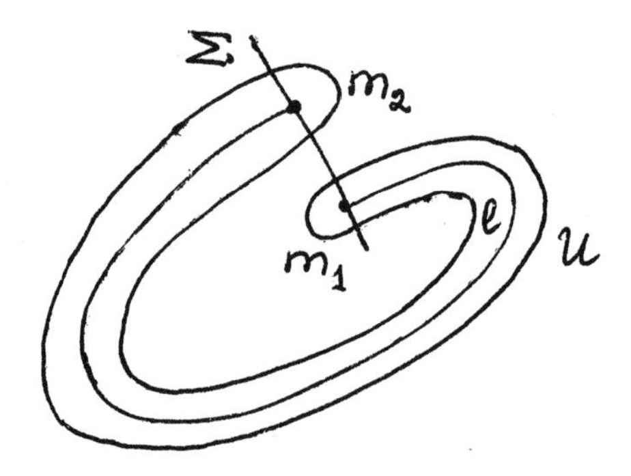

FIGURE 2.3

PROOF. Let m_1 and m_2 be successive points of intersection of l and Σ with respect to time (Figure 2.3). Then the arc γ of l between m_1 and m_2 does not intersect Σ. According to Theorem 1.1, there exists a neighborhood $\mathcal{U}$ of γ in which the trajectories of f^t are arranged as a family of parallel lines (Figure 2.3). Therefore, $\mathcal{U}$ contains a contact-free segment $\widehat{\Sigma}$ with endpoints $m_1, m_2' \in \Sigma$, and it can be assumed that $\widehat{\Sigma}$ is tangent to Σ at the endpoints (Figure 2.4).

The union of $\widehat{\Sigma}$ with the subsegment $\overline{m_1 m_2'} \subset \Sigma$ (the subsegment of Σ between m_1 and m_2') gives the desired contact-free cycle, which intersects l at least at the point m_1. □

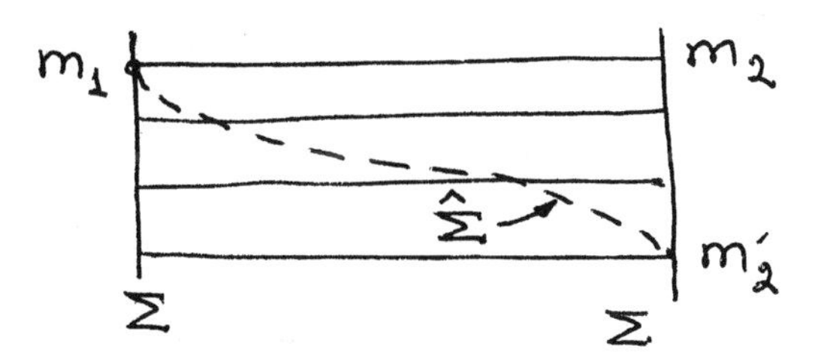

FIGURE 2.4

COROLLARY 1.1. *If the trajectory l intersects some neighborhood with the structure of a constant field in at least two disjoint arcs, then there exists a contact-free cycle intersecting l.*

COROLLARY 1.2. *If the ω- (α-) limit set of a nonclosed trajectory l contains a regular point, then there exists a contact-free cycle that intersects l.*

PROOF. We draw a contact-free segment Σ through a regular point $m \in \omega(l)$ $(\alpha(l))$. The last condition implies that l intersects Σ more than once, and the required assertion follows from Lemma 1.2. □

1.2. The Poincaré mapping. Let $\overset{\frown}{m_1 m_2}$ be the arc with endpoints m_1 and m_2 on a trajectory l of a flow f^t, and let Σ_1 and Σ_2 be disjoint contact-free segments passing through m_1 and m_2, respectively, such that $\overset{\frown}{m_1 m_2} \cap (\Sigma_1 \cup \Sigma_2) = \{m_1, m_2\}$ (Figure 2.5). For definiteness it will be assumed that $m_2 \in l^+(m_1)$; that is, the point m_2 is hit upon moving along l from m_1 with increasing time.

By Theorem 1.1, there exists a neighborhood $\Sigma \subset \Sigma_1$ of m_1 on Σ_1 such that for any $m \in \Sigma$ the positive semitrajectory $l^+(m)$ intersects Σ_2 without first intersecting Σ_1. Denote by $\widetilde{m}$ the first point where $l^+(m)$ intersects Σ_2 (Figure 2.5).

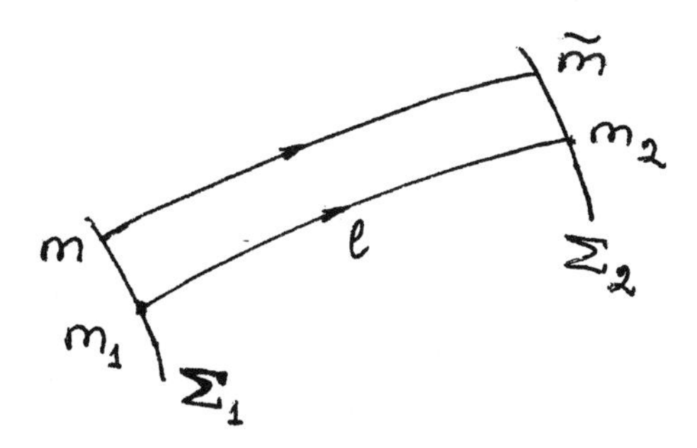

FIGURE 2.5

DEFINITION. The mapping $P \overset{\text{def}}{=} P(m, \Sigma) \colon \Sigma_1 \to \Sigma_2$ assigning the point $\widetilde{m} \in \Sigma_2$ to a point $m \in \Sigma$ according to the rule above is called the *Poincaré mapping* (induced by the flow f^t).

Theorem 1.1 gives us

LEMMA 1.3. *Let $P = P(m, \Sigma) \colon \Sigma_1 \to \Sigma_2$ be the Poincaré mapping of the contact-free segment Σ_1 into the contact-free segment Σ_2 induced by a C^r-flow f^t. Then P is a C^r-diffeomorphism of Σ onto its range.*

Suppose now that $m_1 = m_2$ and $\Sigma_1 = \Sigma_2$; that is, l is a closed trajectory of f^t. It follows from the theorem on continuous dependence on the initial conditions and the rectification theorem that for points $m \in \Sigma = \Sigma_1 = \Sigma_2$ sufficiently close to $m_0 = m_1 = m_2$ the semitrajectories $l^+(m) \setminus \{m\}$ going out from them intersect Σ (Figure 2.6). Let $\widetilde{m}$ be the first point where $l^+(m) \setminus \{m\}$ intersects Σ upon moving away from m along $l^+(m)$ with increasing time.

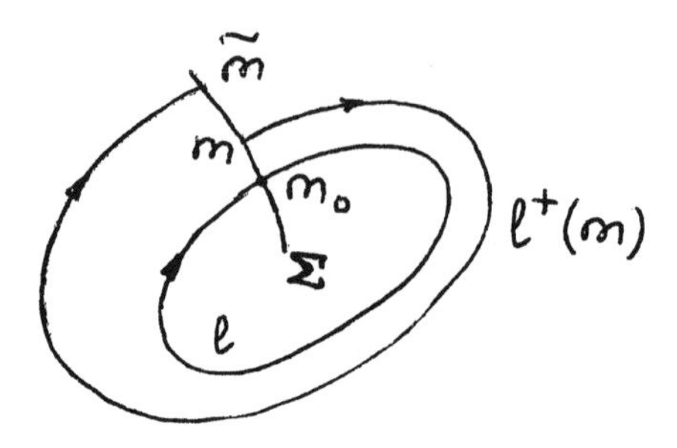

FIGURE 2.6

The mapping $P = P(l, \Sigma)$ assigning the point $\widetilde{m}$ to a point m is called the *Poincaré mapping for the closed trajectory* l.

LEMMA 1.4. *For a closed trajectory l the Poincaré mapping $P(l, \Sigma)$ induced by a C^r-flow is a C^r-diffeomorphism in the domain where it is defined.*

PROOF. The mapping $P(l, \Sigma)$ can be represented as a composition of mappings $P_1 : \Sigma \to \widetilde{\Sigma}$ and $P_2 : \widetilde{\Sigma} \to \Sigma$ (see Figure 2.7), each of them a C^r-diffeomorphism by Lemma 1.3. □

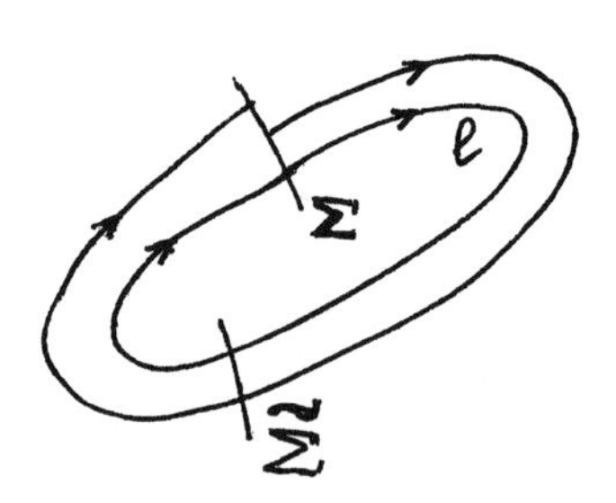

FIGURE 2.7

If $m_1 \neq m_2$, but $\Sigma_1 = \Sigma_2 = \Sigma$, then the definition of the Poincaré mapping $P : \Sigma_1 \to \Sigma_2$ is analogous. Namely, if the semitrajectory $l^+(m)$, $m \in \Sigma$, intersects Σ again after the point m, then we set $P(m) = \widetilde{m}$, where $\widetilde{m}$ is the first point where $l^+(m) \setminus \{m\}$ intersects Σ with increasing time.

1.3. The limit sets. Let f^t be a flow on a closed surface $\mathcal{M}$, and let $l(m)$ be the trajectory of f^t passing through a point $m \in \mathcal{M}$.

LEMMA 1.5. *The ω- (α-) limit set of the trajectory $l(m)$ is nonempty, connected, closed, and invariant (that is, consists of whole trajectories).*

PROOF. The closedness (compactness) of $\mathcal{M}$ implies that the limit set $\omega(l(m))$ $(\alpha(l(m)))$ is nonempty. The fact that it is closed follows immediately from the definition of a limit set (see §3 in Chapter 1). If $m_1, m_2 \in \omega(l(m))$ $(\alpha(l(m)))$ are two distinct points, then any neighborhoods of them are joined by an arc of the trajectory $l(m)$, which is an arcwise connected set. This implies that $\omega(m)$ is connected. The invariance of $\omega(l(m))$ follows from the continuous dependence of trajectories on the initial conditions. □

If l is an equilibrium state or a closed trajectory, then $\omega(l) = \alpha(l) = l$.

LEMMMA 1.6. *Suppose that the ω- (α-) limit set of the trajectory l contains a closed trajectory l_0. Then $\omega(l)$ $(\alpha(l)) = l_0$.*

PROOF. Assume for definiteness that $l_0 \subseteq \omega(l)$ and $\mathcal{M}$ is an orientable surface (the proof for a nonorientable surface is left to the reader as an exercise). If $l = l_0$, then the assertion is obvious, so we consider the case when $l \neq l_0$.

Since $\mathcal{M}$ is orientable, l_0 has a neighborhood $\mathcal{U}$ homeomorphic to an annulus such that l_0 separates $\mathcal{U}$ into two domains. It follows from the definition of f^t and the continuity of f^t that the set $\mathrm{Fix}(f^t)$ of equilibrium states is closed, and hence separated from l_0. We take the neighborhood $\mathcal{U}$ small enough that it does not contain equilibrium states.

Through some point $m_0 \in l_0$ we draw a contact-free segment Σ such that $\Sigma \subset \mathcal{U}$ and l_0 intersects Σ only at m_0. Since $l_0 \subseteq \omega(l)$, l intersects Σ at a point $m_1 \neq m_0$ lying in the domain $\Sigma_0 \subset \Sigma$ of the Poincaré mapping $P(l_0, \Sigma)\colon \Sigma_0 \to \Sigma$ (Figure 2.8).

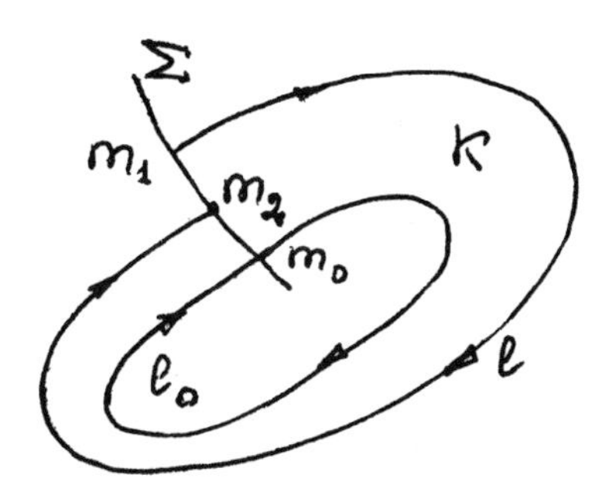

FIGURE 2.8

Let $m_2 = P(m_1)$. By the theorem on continuous dependence on the initial conditions, if m_1 is sufficiently close to m_0, then the arc $\overset{\frown}{m_1m_2}$ of l between m_1 and m_2 lies in $\mathcal{U}$. Since l_0 separates $\mathcal{U}$, $\overset{\frown}{m_1m_2}$ belongs to a single component of $\mathcal{U} \setminus l_0$, and hence m_1 and m_2 are in one of the segments in $\Sigma \setminus m_0$. Therefore, the simple closed curve $\gamma_0 = \overset{\frown}{m_1m_2} \cup \overline{m_1m_2}$ does not intersect l_0 and lies in $\mathcal{U}$, where $\overline{m_1m_2}$ is the subsegment of Σ with endpoints m_1 and m_2.

The naturally oriented curve l_0 has index $+1$ with respect to the flow f^t (it follows from the proof of Lemma 1.2 that γ_0 is isotopic to a contact-free cycle, which has index $+1$ in view of Corollary 4.1 in Chapter 1). If γ_0 bounded a simply connected domain $\mathcal{D}_0$ in $\mathcal{U}$, then $\mathcal{D}_0 \subset \mathcal{U}$ would contain an equilibrium state by Corollary 4.2 in Chapter 1, and this contradicts the choice of the neighborhood $\mathcal{U}$.

Therefore, the curve γ_0 in $\mathfrak{U}$ is not homotopic to zero and together with l_0 bounds an annular domain $K \subset \mathfrak{U}$ (Figure 2.8).

Denote by Σ^+ and Σ^- the components of the set $\Sigma \setminus m_0$, and let $m_1, m_2 \in \Sigma^+$. There are two possibilities: 1) the positive semitrajectory $l^+(m_2)$ does not intersect Σ^- in some neighborhood of m_0; 2) $l^+(m_2)$ intersects Σ^- at points arbitrarily close to m_0.

In case 1) we show that m_2 lies on Σ between m_1 and m_0. Assume not; that is, assume that m_1 lies between m_2 and m_0. Then $l^+(m_2)$ leaves the annulus K and cannot hit it after that since on $\overline{m_1 m_2} \subset \Sigma$ all the positive semitrajectories go out of K (Figure 2.9). But in case 1) this means that $l^+(m_2)$ does not intersect Σ in some neighborhood of m_0, which contradicts the inclusion $m_0 \in \omega(l^+(m_2))$.

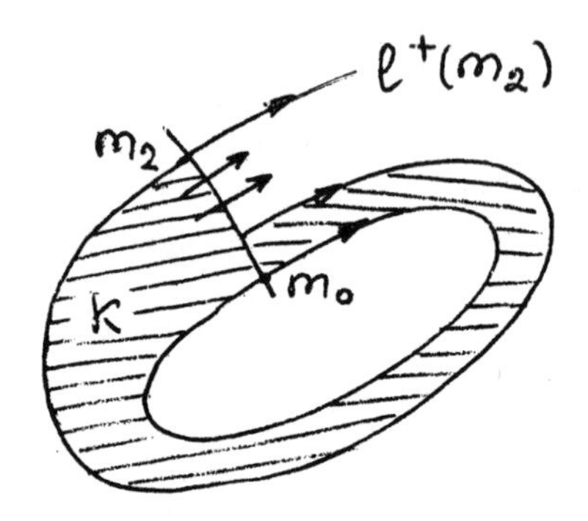

FIGURE 2.9

Let us prove the lemma in case 1). Thus, m_2 lies on Σ between m_0 and m_1, and hence $m_2 \in \Sigma_0$ (the domain of the Poincaré mapping P). We show that the point $m_3 = P(m_2)$ lies on Σ between m_2 and m_0. Indeed, the semitrajectory $l^+(m_2)$, upon entering K, cannot leave it because on the segment $\overline{m_1 m_2}$ all the positive semitrajectories go into K (see Figure 2.8). For the same reason, $m_3 \notin \overline{m_1 m_2}$, and hence $m_3 \in \overline{m_1 m_0} \subset \Sigma$.

Continuing this iteration process, we get a sequence of points $m_k \in \Sigma$, $k \in \mathbb{N}$, with the following properties: a) $m_{k+1} \in l^+(m_k)$; b) the point m_{k+1} lies on Σ between m_k and m_0 (monotonicity); c) $l^+(m_1)$ does not intersect Σ at any points other than m_k, $k \in \mathbb{N}$ (Figure 2.10).

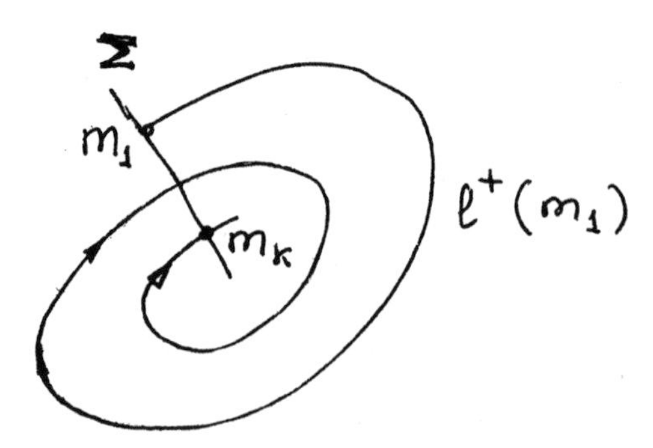

FIGURE 2.10

In view of the property b), the sequence m_k, $k \in \mathbb{N}$, has a limit $m_* \in \Sigma$. Since $m_0 \in \omega(l^+(m_1))$, we have that $m_* = m_0$. Moreover, according to the properties b) and c), m_0 is the unique accumulation point of the sequence m_k, $k \in \mathbb{N}$.

Assume that a trajectory $l_* \neq l_0$ is in the ω-limit set $\omega(l)$. Since $l^+(m_1) \subset K$, $l_* \subset K$. It follows from the theorem on continuous dependence of trajectories on the initial conditions and the long flow tube theorem (Theorem 1.1) that l_* intersects the segment $\Sigma \cap K$ at some point $\widetilde{m} \neq m_0$, but it follows from $l_* \subset \omega(l)$ that $\widetilde{m}$ must be an accumulation point of the set $l^+(m_1) \cap \Sigma = \cup\{m_k\}$, $k \in \mathbb{N}$. This contradiction (to the fact that m_0 is the unique accumulation point of the sequence m_k, $k \in \mathbb{N}$) proves the lemma in case 1).

Let us consider case 2). It follows from the preceding arguments that in this case the semitrajectory $l^+(m_2)$ leaves K and does not hit K after that. In a way completely analogous to the construction of K we construct an annulus $\widetilde{K}$ bounded by an arc $\overset{\frown}{\widetilde{m}_1\widetilde{m}_2}$ of the semitrajectory $l^+(m_2)$ and by the segment $\widetilde{m}_1\widetilde{m}_2 \subset K$ (Figure 2.11). Since $l_0 \subseteq \omega(l)$, the semitrajectory $l^+(\widetilde{m}_2)$ must enter the annulus $\widetilde{K}$, and hence the point $\widetilde{m}_2$ on Σ^- lies between m_1 and m_0. The rest of the proof is entirely analogous to that in case 1), with K, m_1, and m_2 replaced by $\widetilde{K}$, $\widetilde{m}_1$, and $\widetilde{m}_2$, respectively. □

The next result is an immediate consequence of the proof of Lemma 1.6 and the theorem on continuous dependence of trajectories on the initial conditions.

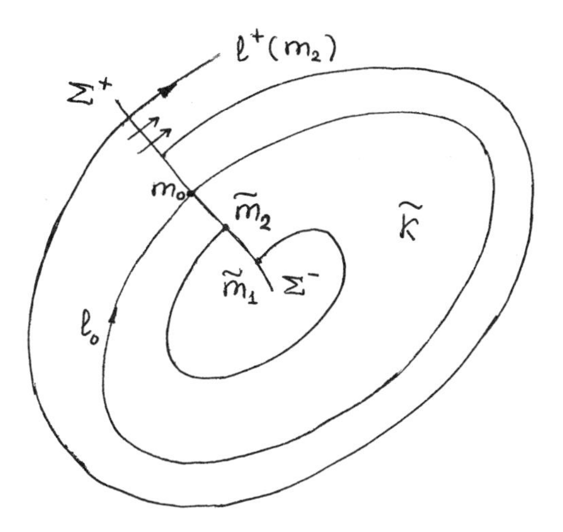

FIGURE 2.11

COROLLARY 1.3. *Suppose that the conditions of Lemma* 1.6 *hold and fix a point* $m \in l$. *There exists a neighborhood* $\mathfrak{U}(m)$ *of* m *such that the* ω- (α-) *limit set of any trajectory passing through* $\mathfrak{U}(m)$ *coincides with* l_0.

1.4. Minimal sets.

DEFINITION. A *minimal set* of a flow is defined to be a nonempty closed invariant set not containing proper closed invariant subsets.

The simplest examples of minimal sets are equilibrium states and closed trajectories. An irrational winding on the torus provides an example of a minimal set that coincides with the torus.

In the limit set $\Omega(\text{sus}(f))$ of a Denjoy flow (see §3.1 in Chapter 1) each trajectory is dense in $\Omega(\text{sus}(f))$. Therefore, this limit set is a minimal set. Further, it is nowhere dense on the torus and has the local structure of the product of a closed bounded interval and the Cantor set.

LEMMA 1.7. *Let N be a minimal set for a flow f^t on a compact manifold $\mathcal{M}$. Then:*

1) *either $N = \mathcal{M}$ or N is nowhere dense in $\mathcal{M}$;*

2) *each trajectory in N is dense in N, and, moreover, N coincides with the ω- (α-) limit set of any of its trajectories.*

PROOF. It follows from the invariance of N and the theorem on continuous dependence of trajectories on the initial conditions that the boundary ∂N (a point x_0 is in ∂N if any neighborhood of it contains a point in N and a point not in N) of N is also invariant. Since N is a minimal set, either $\partial N = N$ or $\partial N = \emptyset$. Consequently, either N does not contain interior points (and then is nowhere dense in $\mathcal{M}$), or N coincides with the set $\operatorname{int} N$ of its interior points. In the last case $N = \operatorname{int} N$ is open and closed, so that $N = \mathcal{M}$.

According to Lemma 1.5, the ω- (α-) limit set of any trajectory is invariant and closed. This and the definition of a minimal set give us the assertion 2).

DEFINITION. A trajectory l of a flow f^t is said to be recurrent in the Birkhoff sense (or B-recurrent) if for any $\varepsilon > 0$ there is a number $T = T(\varepsilon) > 0$ such that the whole of l is contained in the ε-neighborhood of any arc of l of time length T.

An equilibrium state and a closed trajectory of any flow are B-recurrent trajectories. Each trajectory of the minimal set of a Denjoy flow is B-recurrent, as is each trajectory of an irrational winding of the torus. B-recurrent trajectories belong to a minimal set of the flow in all the examples given. It can be shown that any trajectory of a compact minimal set is B-recurrent. If a trajectory l is B-recurrent, then its closure is a compact minimal set (these results are proved in the book [**61**]).

We remark that the trajectories of a noncompact minimal set can fail to be B-recurrent. For example, any trajectory of the flow $\dot{x} = 1$, $\dot{y} = 0$ on $\mathbb{R}^2$ is a minimal set, but is not B-recurrent.

It is known [**61**] that every compact invariant set of a flow contains some minimal set. From this and Lemma 1.5, the ω- (α-) limit set of any trajectory of a flow on a closed surface contains at least one minimal set.

1.5. Nonwandering points. Let f^t be a flow on a manifold $\mathcal{M}$. We recall that f_t denotes the mapping $\mathcal{M} \to \mathcal{M}$ that carries each point $m \in \mathcal{M}$ along the trajectory $l(m)$ of f^t to the point corresponding to the time $t \in \mathbb{R}$ (the shift by the time t along the trajectory).

DEFINITION. A point $m \in \mathcal{M}$ is called a *nonwandering point* of a flow f^t if for any neighborhood $\mathcal{U}(m)$ of m and any number $T > 0$ there is a $|t| \geq T$ such that $f_t[\mathcal{U}(m)] \cap \mathcal{U}(m) \neq \emptyset$.

A point that is not nonwandering is called a *wandering point* of f^t.

The set of nonwandering points of a flow f^t is denoted by $NW(f^t)$.

An equilibrium state of any flow is a nonwandering point, as is each point of a closed trajectory.

LEMMA 1.8. *Suppose that f^t is a flow on a compact manifold $\mathcal{M}$. The set of nonwandering points of f^t is nonempty, closed, and invariant (that is, consists of whole trajectories). Moreover, $NW(f^t)$ contains the ω- and α-limit sets of any trajectory of f^t.*

PROOF. By definition, a point $m \in \mathcal{M}$ is wandering if there exist a neighborhood $\mathcal{U}(m)$ of m and a number $T_0 > 0$ such that $f_t[\mathcal{U}(m)] \cap \mathcal{U}(m) = \emptyset$ for all $|t| \geq T_0$. This implies that the set of wandering points is open. Consequently, the set of nonwandering points is closed.

The invariance of the set $NW(f^t)$ follows from the theorem on continuous dependence of trajectories on the initial conditions.

Let us consider the limit set $\omega(l)$ of a trajectory l of f^t. We take an $x \in \omega(l)$ and a point $m \in l$. By the definition of the ω-limit set, there is a sequence of numbers t_k, $k \in \mathbb{N}$, with $t_k \to \infty$ such that $f_{t_k}(m) \to x$ as $k \to \infty$. Therefore, the points $f_{t_k}(m)$ lie in any given neighborhood $\mathcal{U}(x)$ of x for sufficiently large indices $k \geq k_0$. Since t_k increases to infinity, for any $T > 0$ there is a number t_{k_1} with index $k_1 \geq k_0$ such that $t \overset{\text{def}}{=} t_{k_1} - t_{k_0} \geq T$. Then $f_t[\mathcal{U}(x)] \cap \mathcal{U}(x) \ni f_{t_{k_1}}(m) = f_t[f_{t_0}(m)]$; that is, $f_t[\mathcal{U}(x)] \cap \mathcal{U}(x) \neq \emptyset$. This proves that $\omega(l) \subset NW(f^t)$. It can be shown similarly that $\alpha(l) \subset NW(f^t)$ for any trajectory l. For a compact manifold $\omega(l)$ $(\alpha(l)) \neq \emptyset$ (Lemma 1.5), and hence $NW(f^t) \neq \emptyset$. □

Trajectories belonging to the nonwandering set are called *nonwandering* trajectories. The remaining trajectories are said to be *wandering*.

§2. The theorems of Maĭer and Cherry

2.1. Definitions of recurrence. Let f^t be a flow on a surface $\mathcal{M}$, and let l be a trajectory of f^t.

DEFINITION. A trajectory l (a positive semitrajectory l^+) is said to be P^+ *recurrent* or *recurrent in the positive direction* if it belongs to its own ω-limit set, that is, $l \subseteq \omega(l)$ $(l^+ \subseteq \omega(l^+))$.

A trajectory l (a negative semitrajectory l^-) is said to be P^- *recurrent* or *recurrent in the negative direction* if it belongs to its own α-limit set: $l \subseteq \alpha(l)$ $(l^- \subseteq \alpha(l^-))$.

A trajectory is said to be *recurrent* if it is P^+ and P^- recurrent (another common term for this is Poisson-stable).

If at least one positive (negative) semitrajectory of l is P^+ (P^-) recurrent, then l itself and any one of its positive (negative) semitrajectories are also P^+ (P^-) recurrent.

An equilibrium state and a closed trajectory are recurrent trajectories. Recurrent trajectories and semitrajectories different from these are called *nontrivial* recurrent trajectories and semitrajectories. Each trajectory of an irrational winding of the torus and each trajectory in the minimal set of a Denjoy flow are nontrivial recurrent trajectories (even B-recurrent).

It follows from Lemma 1.7 that any B-recurrent trajectory is recurrent.

There are non-B-recurrent recurrent trajectories in a Cherry flow (see §3.2 in Chapter 1). Namely, any regular trajectory in the limit set of a Cherry flow is recurrent in one of the directions, but is not B-recurrent, because the limit set of a Cherry flow contains an equilibrium state (a saddle), and hence is not a minimal set.

If an impassable grain is put on one of the trajectories in an irrational winding of the torus (see §2.3 in Chapter 1), then in the resulting flow each regular trajectory becomes non-B-recurrent and recurrent in one direction (or both).

Recall that $\mathrm{cl}(l)$ denotes the topological closure of a trajectory l ($\mathrm{cl}(l) \stackrel{\mathrm{def}}{=} l \cup \partial l$). An immediate consequence of the definition of the ω- (σ-) limit set is that

$$\omega(l) \cup \alpha(l) \subseteq \mathrm{cl}(l).$$

If l is P^+ recurrent, then $l \subseteq \omega(l)$. Since the set $\omega(l)$ is closed, $\mathrm{cl}(l) \subseteq \omega(l)$. Therefore, for a P^+ recurrent trajectory

$$\omega(l) = \mathrm{cl}(l). \tag{2.1}$$

Similarly, if l is P^- recurrent, then

$$\alpha(l) = \mathrm{cl}(l). \tag{2.2}$$

If l is a recurrent trajectory, then

$$\omega(l) = \alpha(l) = \mathrm{cl}(l). \tag{2.3}$$

For any trajectory l we have $l \subseteq \mathrm{cl}(l)$, so the equalities (2.1) and (2.2) are equivalent to P^+ and P^- recurrence, respectively.

DEFINITION. A P^+ or P^- nontrivial recurrent trajectory is said to be *locally dense* (*exceptional*) if its closure contains interior points (does not contain interior points).

Local denseness means that the trajectory is dense in some domain of the manifold. Being exceptional means that the trajectory and its closure are nowhere dense on the manifold.

Each trajectory of an irrational winding f_0^t of the torus is locally dense. We obtain a more nontrivial example from f_0^t if on one of the trajectories we put an impassable grain and then "blow up" this grain to form a disk $\mathcal{D}$, deforming the trajectories of f_0^t in an obvious way (Figure 2.12). Then each trajectory in $T^2 \setminus \mathcal{D}$ is locally dense and dense in $T^2 \setminus \mathcal{D}$.

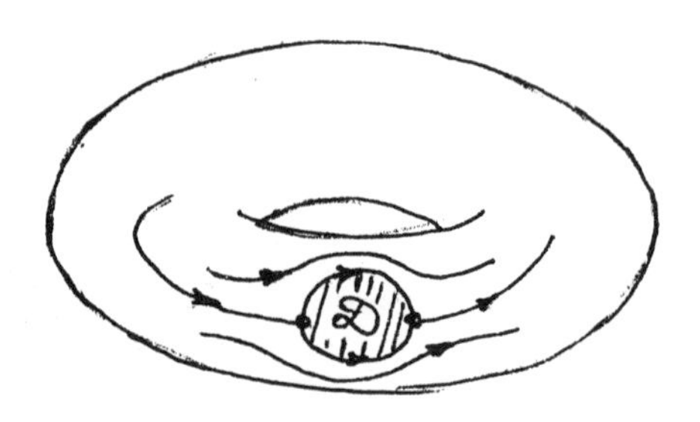

FIGURE 2.12

There are exceptional recurrent trajectories in Denjoy and Cherry flows.

LEMMA 2.1. *Suppose that a P^+ (P^-) nontrivial recurrent trajectory l intersects a contact-free segment or cycle Σ at an interior point. Then l intersects Σ in a countable perfect set of points.*[1]

PROOF. The set $l \cap \Sigma$ is perfect because l is nonclosed and is its own limit set.

Let $\dots, t_\nu, t_{\nu'}, \dots$ be the successive times when l intersects Σ, where the index ν belongs to some set of numbers. Since each point in Σ is regular, and hence lies in a neighborhood with the structure of a constant field, it follows that the numbers t_ν are isolated in $\mathbb{R}$. What is more, the quantity $|t_\nu - t_{\nu'}|$ is bounded below by a positive constant. Therefore, the set of values $\dots, t_\nu, t_{\nu'}, \dots$ is countable. □

COROLLARY 2.1. *Suppose that the P^+ (P^-) nontrivial recurrent trajectory l intersects a neighborhood $\mathcal{U}$ with the structure of a constant field. Then $l \cap \mathcal{U}$ consists of a countable set of disjoint arcs of l.*

2.2. The absence of nontrivial recurrent semitrajectories on certain surfaces. All the results of this subsection are based on the following simple result.

LEMMA 2.2. *Let f^t be a flow on a surface $\mathcal{M}$, and let C be a contact-free cycle for f^t. If C intersects some trajectory at more than one point, then C does not separate $\mathcal{M}$ (that is, the set $\mathcal{M} \setminus C$ is connected).*

PROOF. Since C is a contact-free cycle, it follows from the rectification theorem that C has a neighborhood $\mathcal{U}$ homeomorphic to an annulus and such that $\mathcal{U} \setminus C$ consists of two components $\mathcal{U}_1$ and $\mathcal{U}_2$. If it is assumed that C separates $\mathcal{M}$ into two submanifolds $\mathcal{M}_1$ and $\mathcal{M}_2$, then one component, say $\mathcal{U}_1$, lies in $\mathcal{M}_1$, and the other ($\mathcal{U}_2$) in $\mathcal{M}_2$. On the other hand, all the trajectories of f^t intersect the contact-free cycle C, passing from one component, say $\mathcal{U}_2$, into the other ($\mathcal{U}_1$). According to the assumption of the lemma, there is a path (along an arc of a trajectory) from $\mathcal{U}_1$ to $\mathcal{U}_2$ that does not intersect C (Figure 2.13), and this contradicts the assumption that C separates $\mathcal{M}$ into the two components $\mathcal{M}_1$ and $\mathcal{M}_2$. □

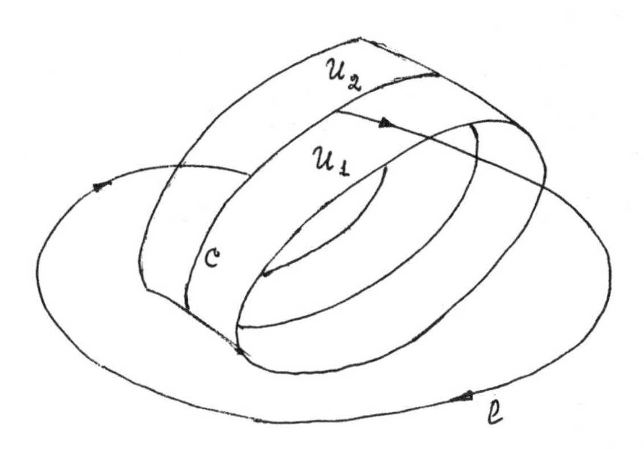

FIGURE 2.13

[1] Recall that a set N is said to be perfect if any point $x \in N$ is an accumulation point of the rest of the set, $N \setminus \{x\}$.

LEMMA 2.3. *Let l be a P^+ (P^-) nontrivial recurrent trajectory. Then:*
1) *there exists a contact-free cycle intersecting l;*
2) *any contact-free cycle intersecting l does not separate the surface.*

PROOF. This follows from Lemmas 1.2, 2.1, and 2.2. □

REMARK. It follows from the proof of Lemma 1.2 that a contact-free cycle can be drawn through any given point of the trajectory l.

LEMMA 2.4. *Let f^t be a flow on the surface $\mathcal{M}$, and let $G \subset \mathcal{M}$ be a submanifold homeomorphic to the sphere S^2 with finitely many holes (perhaps none). (In particular, if $G = S^2$, then $\mathcal{M} = S^2$). In this case G cannot entirely contain a nontrivial recurrent semitrajectory.*

PROOF. Assume the contrary. Then, by Lemma 2.3, there exists a contact-free cycle $C \subset G$ that does not separate G. But this contradicts the Jordan curve theorem, which says that any simple closed curve separates a sphere with finitely many holes (perhaps none) into two domains. □

It follows from Lemma 2.4 that there are no flows on the sphere with nontrivial recurrent semitrajectories (this was proved in [**36**]). According to §§2.3 and 2.4 in Chapter 1, all other closed orientable surfaces have transitive flows, and hence flows with nontrivial recurrent semitrajectories. Among the closed nonorientable surfaces there are two on which such flows are absent, namely, the projective plane (a closed nonorientable surface of genus $p = 1$) and the Klein bottle (a closed nonorientable surface of genus $p = 2$). For the projective plane this follows from the fact that the sphere is a two-sheeted covering for it. As for the Klein bottle, the torus is a two-sheeted covering for it on which there can be nontrivial recurrent semitrajectories, and hence the absence of such semitrajectories on the Klein bottle necessitates a special treatment. In the absence of equilibrium states on the Klein bottle this proposition was proved by Kneser [**90**], while in the presence of equilibrium states it was proved independently in 1969 by Markley [**95**] and Aranson [**8**]. In 1978 the result was reproved by Gutierrez [**85**]. At the same time, on any nonorientable closed surface of genus $p \geq 3$ there exists a transitive flow (the construction of such flows is left to the reader as an exercise). We sum up these facts as a lemma.

LEMMA 2.5. *The following three surfaces do not have flows with nontrivial recurrent semitrajectories:*

1) *the sphere (an orientable surface of genus 0);*
2) *the projective plane (a nonorientable surface of genus 1);*
3) *the Klein bottle (a nonorientable surface of genus 2).*

All other closed surfaces (orientable or not) have flows with nontrivial recurrent semitrajectories (even transitive flows).

PROOF. A flow having nontrivial recurrent trajectories can be constructed on a closed nonorientable surface of genus $p \geq 3$ by starting from a Denjoy flow on the torus and attaching the necessary number of Möbius bands in the domain $T^2 \backslash \Omega(f^t)$, where $\Omega(f^t)$ is the minimal set of the Denjoy flow f^t.

We prove that on the projective plane $\mathbb{R}P^2$ there are no flows with nontrivial recurrent semitrajectories. Assume the contrary: suppose that the flow f^t on $\mathbb{R}P^2$ has such semitrajectories. Denote by $\widetilde{\pi}\colon S^2 \to \mathbb{R}P^2$ a two-sheeted (unbranched) covering, and by $\widetilde{f}^t$ the flow on S^2 covering f^t. Our assumption implies that there is a nonclosed semitrajectory $\widetilde{l}^{()}$ of $\widetilde{f}^t$ on S^2 whose limit set contains a regular point. By Lemmas 1.2 and 2.2, there exists on S^2 a simple closed curve not separating S^2, and this contradicts the Jordan curve theorem.

We show that the Klein bottle K^2 does not have flows with nontrivial recurrent semitrajectories. Of the three proofs in [**8**], [**95**], and [**85**] we present the proof of Gutierrez [**85**].

Assume the contrary: let f^t be a flow on K^2 with a nontrivial recurrent semitrajectory l^+. In view of Lemma 1.2, there exists a contact-free cycle C for f^t that intersects l^+. Since C is a contact-free cycle, it has a cylindrical neighborhood in K^2 (that is, C is a two-sided closed curve). Repeating the proof of Lemma 2.2, we show that C does not separate K^2. Therefore, $K^2 \setminus C$ is an annulus with two boundary components C_1 and C_2, and to obtain a Klein bottle from $K^2 \setminus C$ we must paste C_1 and C_2 together by means of an orientation-preserving homeomorphism $\varphi\colon C_1 \to C_2$ (Figure 2.14).

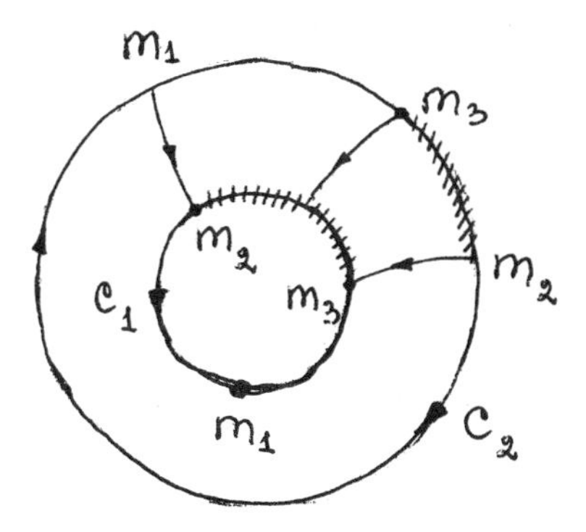

FIGURE 2.14

Let $m_1, m_2, m_3 \in l^+$ be successive points of intersection of l^+ with C, and let $\overline{m_1m_2}$ be the arc of C containing m_3. By virtue of the identification φ the points m_1, m_2, and m_3 are arranged on C_1 and C_2 as shown in Figure 2.14. This implies that the semitrajectory $l^+(m_3)$ intersects C only on the arc $\overline{m_3m_2} \subset \overline{m_1m_2}$, which does not contain m_1. But this contradicts the recurrence of the semitrajectory l^+, whose ω-limit set contains m_1. □

2.3. The Cherry theorem on the closure of a recurrent semitrajectory. The purpose of this subsection is to prove the following theorem of Cherry [**80**].

THEOREM 2.1. *The topological closure* cl(l) *of a P^+ (P^-) nontrivial recurrent semitrajectory l contains a continuum of nontrivial recurrent trajectories, each of which is dense in* cl(l) *and contains l in its limit set.*

We first prove some auxiliary statements.

DEFINITION. Let $\mathfrak{U}$ be a neighborhood with the structure of a constant field. Two sets $\mathfrak{U}_1, \mathfrak{U}_2 \subset \mathfrak{U}$ are said to be *not coupled with respect to* $\mathfrak{U}$ if $\mathfrak{U}$ does not contain arcs of trajectories intersecting both $\mathfrak{U}_1$ and $\mathfrak{U}_2$ (Figure 2.15).

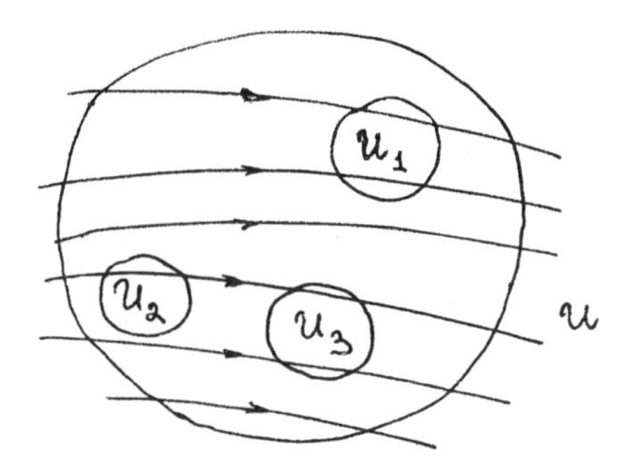

FIGURE 2.15

In Figure 2.15 the sets $\mathcal{U}_1$, $\mathcal{U}_2$ and the sets $\mathcal{U}_1$, $\mathcal{U}_3$ are not coupled with respect to $\mathcal{U}$, but $\mathcal{U}_2$ and $\mathcal{U}_3$ are coupled with respect to $\mathcal{U}$.

By a closed disk we mean a set homeomorphic to the closed unit disk. Let $\operatorname{int} \mathcal{D}$ be the interior of a set $\mathcal{D}$. In what follows we take l to be a positive semitrajectory and denote it by l^+.

LEMMA 2.6. *Let $\mathcal{U}$ be a neighborhood with the structure of a constant field intersecting the P^+ nontrivial recurrent semitrajectory l^+. Then for any closed disk $\mathcal{D} \subset \mathcal{U}$ with $\operatorname{int} \mathcal{D} \cap l^+ \neq \emptyset$ there exist closed disks $\mathcal{D}_0, \mathcal{D}_1 \subset \mathcal{D}$ that are not coupled with respect to $\mathcal{U}$ and such that $\operatorname{int} \mathcal{D}_i \cap l^+ \neq \emptyset$ for $i = 0, 1$.*

PROOF. It follows from Lemma 2.1 and Corollary 2.1 that $\operatorname{int} \mathcal{D}$ intersects l^+ in a countable set of disjoint arcs of l^+, and this yields the required assertion. □

Fix a metric on $\mathcal{M}$, and denote by $\mathcal{U}_\varepsilon(m)$ the ε-neighborhood ($\varepsilon > 0$) of a point $m \in \mathcal{M}$.

If a flow f^t is given on $\mathcal{M}$, then for a fixed number $t \in \mathbb{R}$ and a subset N we denote by $f_t(N)$ the shift of the set N along trajectories of f^t by the time t.

LEMMA 2.7. *Fix a point m_0 on a P^+ nontrivial recurrent semitrajectory l^+, and let $\mathcal{D}$ be a closed disk whose interior intersects l^+. Then for any numbers $\eta > 0$ and $T > 0$ there exists a closed disk $\Phi(\mathcal{D}, \eta, T) \subset \mathcal{D}$ such that:* a) $\operatorname{int} \Phi(\mathcal{D}, \eta, T) \cap l^+ \neq \emptyset$; b) $f_t[\Phi(\mathcal{D}, \eta, T)] \subset \mathcal{U}_\eta(m_0)$ *for some $t = -t_1 < -T$ and $t = t_2 > T$, and $\Phi(\mathcal{D}, \eta, T)$ can have an arbitrarily small diameter.*

PROOF. Since l^+ is a P^+ recurrent semitrajectory, there exists a $t_1 > T$ such that $f_{t_1}(m_0) \in \operatorname{int} \mathcal{D}$. By the theorem on continuous dependence of trajectories on the initial conditions, there is a neighborhood $\mathcal{U}_{\eta_1}(m_1)$ of the point $m_1 = f_{t_1}(m_0)$ such that $f_{-t_1}(\mathcal{U}_{\eta_1}(m_1)) \subset \mathcal{U}_\eta(m_0)$ (Figure 2.16). Similarly, there exists a $t_2 > T$ such that $f_{t_2}(m_1) \in \mathcal{U}_\eta(m_0)$ and $f_{t_2}[\mathcal{U}_{\eta_2}(m_1)] \subset \mathcal{U}_\eta(m_0)$ for some neighborhood $\mathcal{U}_{\eta_2}(m_1)$.

We take a closed disk $\Phi(\mathcal{D}, \eta, T)$ encircling the point m_1 and lying in $\mathcal{U}_{\eta_1}(m_1) \cap \mathcal{U}_{\eta_2}(m_1)$. Then $f_t[\Phi(\mathcal{D}, \eta, T)] \subset \mathcal{U}_\eta(m_0)$ for $t = -t_1 < -T$ and $t = t_2 > T$. Since $m_1 \in \operatorname{int} \Phi(\mathcal{D}, \eta, T) \cap l^+$, it follows that $\operatorname{int} \Phi(\mathcal{D}, \eta, T) \cap l^+ \neq \emptyset$. Obviously, $\Phi(\mathcal{D}, \eta, T)$ can have arbitrarily small diameter. □

PROOF OF THEOREM 2.1. We take numerical sequences $\eta_1 > \eta_2 > \cdots > \eta_n > \ldots$ with $\eta_n \to 0$ and $0 < T_1 < T_2 < \cdots < T_n < \ldots$ with $T_n \to \infty$, a

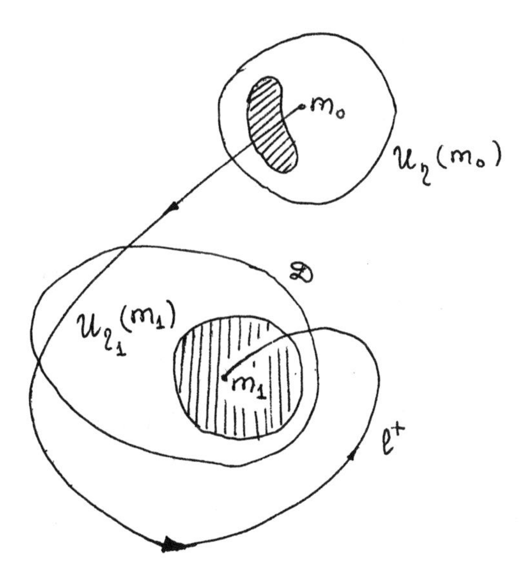

FIGURE 2.16

point $m_0 \in l^+$, and a neighborhood $\mathcal{U}$ with the structure of a constant field such that $\mathcal{U} \cap l^+ \neq \emptyset$ and $m_0 \notin \mathrm{cl}(\mathcal{U}) \stackrel{\mathrm{def}}{=} \mathcal{D}$. By Lemma 2.7, there exists a closed disk $\Phi(\mathcal{D}, \eta_1, T_1) \subset \mathcal{D}$ of diameter less than η_1. According to Lemma 2.6, there are closed disks $\mathcal{D}_0, \mathcal{D}_1 \subset \Phi(\mathcal{D}, \eta_1, T_1)$ that are not coupled with respect to $\mathcal{U}$ and such that $\operatorname{int} \mathcal{D}_i \cap l^+ \neq \emptyset$ for $i = 0, 1$ (Figure 2.17). Again using Lemma 2.7 for the numbers η_2 and T_2 and the closed disks $\mathcal{D}_0$ and $\mathcal{D}_1$, we get closed disks $\Phi(\mathcal{D}_i, \eta_2, T_2) \subset \mathcal{D}_i$ $(i = 0, 1)$ with diameter less than η_2, and so on. This process of obtaining closed disks is pictured in Figure 2.18. As a result we have a family of closed disks $\mathcal{D}_{i_1 \ldots i_n}$, $i_k \in \{0, 1\}$, with the following properties:

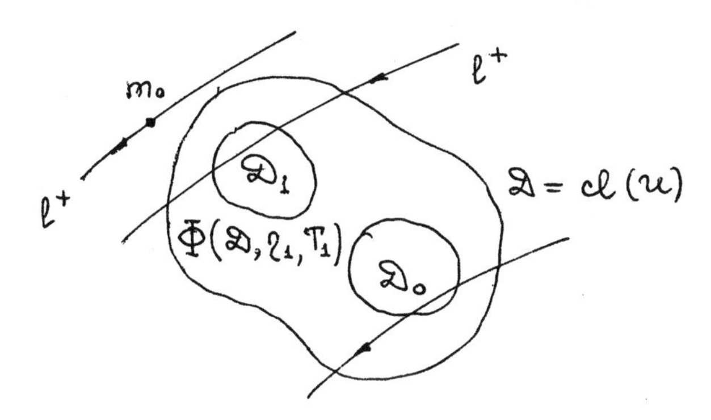

FIGURE 2.17

1) $\mathcal{D}_{i_1 \ldots i_{n-1} i_n} \subset \mathcal{D}_{i_1 \ldots i_{n-1}}$ for both the values $i_n = 0$ and $i_n = 1$;

2) the diameter of $\mathcal{D}_{i_1 \ldots i_n}$ is less than η_n;

3) for $\mathcal{D}_{i_1 \ldots i_n}$ there exist $t'_n < -T_n$ and $t''_n > T_n$ such that $f_{t'_n}(\mathcal{D}_{i_1 \ldots i_n}) \subset \mathcal{U}_{\eta_0}(m_0)$ and $f_{t''_n}(\mathcal{D}_{i_1 \ldots i_n}) \subset \mathcal{U}_{\eta_n}(m_0)$;

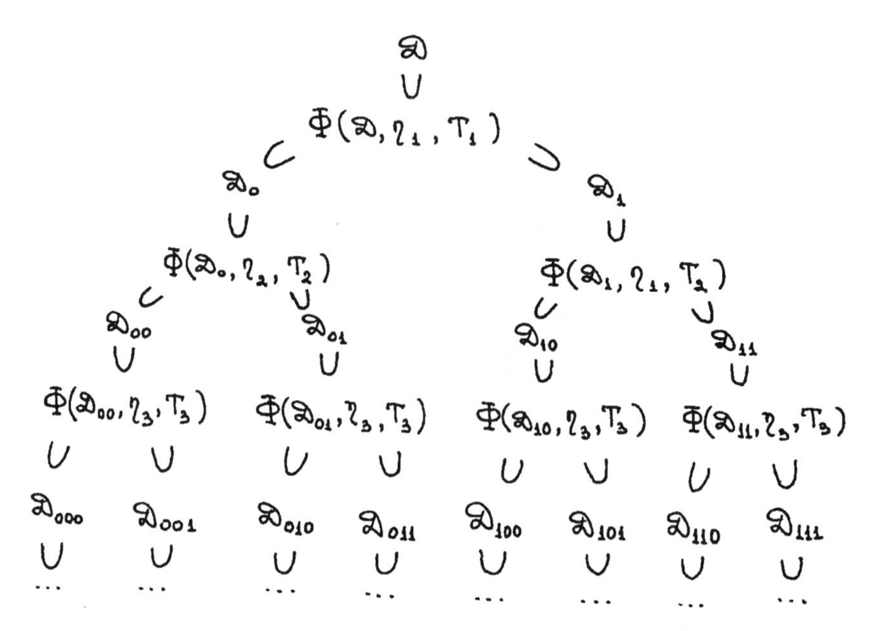

FIGURE 2.18

4) the closed disks $\mathcal{D}_{i_1\dots i_n}$ and $\mathcal{D}_{j_1\dots j_n}$ are not coupled with respect to $\mathcal{U}$ if at least one of the corresponding indices is not the same, $i_k \neq j_k$.

With each sequence $i_1, \dots, i_n, \dots$ of 0's and 1's we associate the sequence of closed disks

$$\mathcal{D}_{i_1}, \mathcal{D}_{i_1 i_2}, \dots, \mathcal{D}_{i_1\dots i_n}, \dots, \tag{2.4}$$

which by 1) and 2) is a nested sequence of closed sets with $\operatorname{diam}(\mathcal{D}_{i_1\dots}) \to 0$. By Cantor's theorem, the intersection $\cap\mathcal{D}_{i_1\dots i_n}$ of all the disks in the sequence (2.4) consists of a single point $Q_{i_1\dots i_n\dots}$.

It follows from the property 4) that different sequences $i_1, \dots, i_n, \dots$ ($i_k \in \{0, 1\}$) determine different points $Q_{i_1\dots i_n\dots}$. Since the set of all possible sequences of 0's and 1's has the cardinality of a continuum, so does the set of points $Q_{i_1\dots}$.

We consider some point $Q = Q_{i_1\dots i_n\dots}$. By construction, $\operatorname{int}\mathcal{D}_{i_1\dots i_n} \cap l^+ \neq \emptyset$, and hence $Q \in \operatorname{cl}(l^+)$. This and the continuous dependence of trajectories on the initial conditions give us that the trajectory $l(Q)$ through Q lies in $\operatorname{cl}(l^+)$. Since the ω- (α-) limit set is closed, we then get that

$$\omega(l(Q)) \cup \alpha(l(Q)) \subset \operatorname{cl}(l(Q)) \subset \operatorname{cl}(l^+). \tag{2.5}$$

By the property 3), $f_{t'_n}(Q) \in \mathcal{U}_{\eta_n}(m_0)$ and $f_{t''_n}(Q) \in \mathcal{U}_{\eta_n}(m_0)$. Since $\eta_n \to 0$, it follows that $m_0 \in \omega(l(Q)) \cap \alpha(l(Q))$. The ω- (α-) limit set is invariant, so $l^+ \subset \omega(l(Q)) \cap \alpha(l(Q))$, and, consequently,

$$\operatorname{cl}(l^+) \subset \omega(l(Q)) \cap \alpha(l(Q)). \tag{2.6}$$

Comparing (2.5) and (2.6), we get that

$$\operatorname{cl}(l^+) = \operatorname{cl}(l(Q)) = \omega(l(Q)) = \alpha(l(Q));$$

that is, $l(Q)$ is a recurrent trajectory (see (2.3) in §2.1) which is dense in $\mathrm{cl}(l^+)$. The inclusion $l^+ \subset \mathrm{cl}(l(Q))$ implies that $l(Q)$ is a nonclosed trajectory.

It remains to show that the set of trajectories $l(Q)$ has the cardinality of a continuum.

We take two different points $Q_{i_1 i_2 \dots}$ and $Q_{j_1 j_2 \dots}$ and let $i_n \neq j_n$ be the first distinct indices. Then in view of the property 4) the respective points $Q_{i_1 i_2 \dots}$ and $Q_{j_1 j_2 \dots}$ lie in closed disks $\mathcal{D}_{i_1 \dots i_n}$ and $\mathcal{D}_{j_1 \dots j_n}$ that are not coupled with respect to $\mathcal{U}$, and hence they belong to different arcs of trajectories in the neighborhood $\mathcal{U}$. Thus, $\mathcal{U}$ contains a continuum of arcs of trajectories on which the points $Q = Q_{i_1 i_2 \dots}$ lie. According to Corollary 2.1, each nontrivial recurrent trajectory intersects $\mathcal{U}$ in a countable family of arcs; therefore, the set of distinct trajectories $l(Q)$, $Q = Q_{i_1 \dots}$, has the cardinality of a continuum. □

2.4. The Maĭer criterion for recurrence. Let f^t be a flow on an orientable compact surface $\mathcal{M}_p$ of genus $p \geq 1$. If l is a P^+ (P^-) nontrivial recurrent trajectory, then it contains a regular point in its ω- (α-) limit set and lies in the limit set of some trajectory (indeed, $l \subset \omega(l)$ ($\alpha(l)$), and any point of l is regular). Maĭer [**55**] proved that these necessary conditions for recurrence are also sufficient.

To prove Maĭer's theorem we need the following result.

LEMMA 2.8. *Let $\mathcal{M}_p$ be a compact orientable surface of genus $p \geq 1$.*

a) *If the boundary $\partial\mathcal{M}_p$ consists of a single component K, then any $2p$ simple (that is, without self-intersections) disjoint arcs with endoints in K combine to distinguish a flat simply connected domain on $\mathcal{M}_p$.*

b) *If $\partial\mathcal{M}_p$ consists of two components K_1 and K_2, then any $2p+1$ simple disjoint arcs joining K_1 and K_2 combine to distinguish a flat simply connected domain on $\mathcal{M}_p$.*

c) *If $\partial\mathcal{M}_p$ consists of two components K_1 and K_2, and there is a countable family of disjoint simple arcs joining K_1 and K_2 on $\mathcal{M}_p$, then there exist two arcs in this family that bound a simply connected domain on $\mathcal{M}_p$.*

PROOF. We prove a) for a surface $\mathcal{M}_1$ of genus $p = 1$ (a torus with a hole). If a simple arc d_1 with endpoints on $\partial\mathcal{M}_1$ does not bound a disk on $\mathcal{M}_1$, then cutting $\mathcal{M}_1$ along d_1 gives an annulus K^2. Let d_2 be a simple arc on $\mathcal{M}_1$ that is disjoint from d_1 and has endpoints on $\partial\mathcal{M}_1$. When $\mathcal{M}_1$ is cut, the arc d_2 passes into an arc $d_2^1 \subset K^2$ with endpoints on ∂K^2. It follows from geometric considerations that either the endpoints of d_2^1 lie on the same component of ∂K^2, and d_2^1 (together with ∂K^2) distinguishes a simply connected domain on K^2, or the endpoints of d_2^1 lie on different components of ∂K^2, and then cutting the annulus along d_2^1 gives a disk. Thus, a) is proved for a torus with a hole ($p = 1$). We suppose that the assertion has been proved for surfaces of genus $1, \dots, p-1 \geq 1$, and we prove it for a surface $\mathcal{M}_p$ of genus $p \geq 2$.

Let $d_1, \dots, d_{2p}$ be disjoint simple arcs on $\mathcal{M}_p$ with endpoints on $\partial\mathcal{M}_p$. If one of these arcs separates $\mathcal{M}_p$ into two surfaces $\mathcal{M}_{p_1}$ and $\mathcal{M}_{p_2}$ of respective genera p_1 and p_2, then in view of the equality $p_1 + p_2 = p$ one of the surfaces $\mathcal{M}_{p_i}$ has at least $2p_i$ disjoint simple arcs that combine to distinguish a simply connected domain on $\mathcal{M}_{p_i}$ (and hence on $\mathcal{M}_p$), because of the induction hypothesis for $p_i < p$.

Suppose that none of the arcs $d_1, \dots, d_{2p}$ separates $\mathcal{M}_p$. Cutting $\mathcal{M}_p$ along d_1, we obtain a compact surface $\mathcal{M}'_{p-1}$ of genus $p-1$ with two boundary components K_1 and K_2. Two cases are possible: 1) the endpoints of all the arcs $d_2, \dots, d_{2p}$

belong to a single component, say K_1; 2) at least one of $d_2, \dots, d_{2p}$ (say d_2) joins K_1 to K_2. In case 1) we use a disk to paste closed one hole of the surface $\mathcal{M}'_{p-1}$ along the component K_2. On the resulting surface $\mathcal{M}_{p-1}$ of genus $p-1$ there are $2p-1$ disjoint simple arcs with endpoints on $K_1 = \partial\mathcal{M}_{p-1}$. By the induction hypothesis, these $2p-1 > 2p-2$ arcs distinguish a simply connected domain on $\mathcal{M}_{p-1}$. In case 2) we cut $\mathcal{M}'_{p-1}$ along d_2. We get a surface $\mathcal{M}_{p-1}$ of genus $p-1$ with a single boundary component and $2p-2$ holes that, by the induction hypothesis, distinguish a simply connected domain. This proves a).

The assertion b) follows from a), since cutting $\mathcal{M}_p$ along one arc yields a surface of genus p with a single boundary component and $2p$ disjoint simple arcs.

According to b), cutting $\mathcal{M}_p$ along $2p+1$ arcs leads to a flat simply connected domain on which a countable family of disjoint simple arcs is left. This gives us c). □

Following Gutierrez, we present the

DEFINITION. Let Σ be a contact-free segment or a contact-free cycle for a flow f^t, and let l be a trajectory of f^t that intersects Σ at points $m_1, m_2 \in l \cap \Sigma$. The arc $\overset{\frown}{m_1 m_2}$ of l with endpoints m_1 and m_2 is called a Σ-*arc* if $\overset{\frown}{m_1 m_2} \cap \Sigma = \{m_1, m_2\}$, that is, $\overset{\frown}{m_1 m_2}$ intersects Σ only at the endpoints m_1 and m_2 (Figure 2.19).

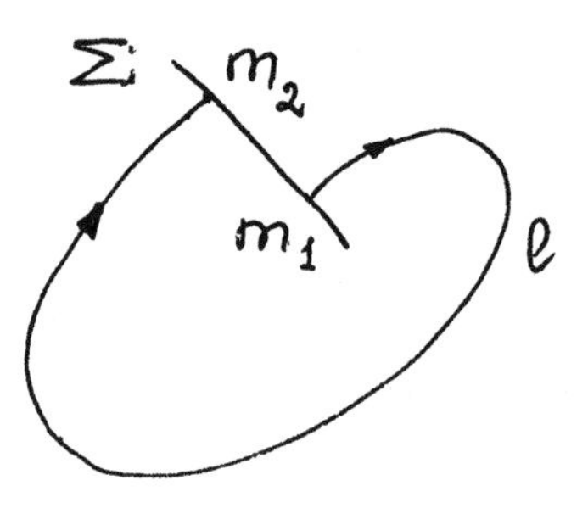

FIGURE 2.19

Suppose that a flow is given on a compact surface $\mathcal{M}$.

THEOREM 2.2 (the Maĭer criterion for recurrence). *If a nonclosed trajectory l has a regular point m_0 in its ω- (α-) limit set and lies in the limit set of some trajectory l', then l is a P^+ (P^-) recurrent trajectory.*

PROOF. We draw a contact-free segment Σ through m_0. For definiteness assume that $m_0 \in \omega(l)$. Then Σ contains a sequence of points $m_i \to m_0$ (as $i \to \infty$) corresponding to an unbounded monotonically increasing sequence of times when l intersects Σ.

The part of Σ between $p, q \in \Sigma$ is denoted by $\overline{pq}$.

We show that for any Σ-arc $\overset{\frown}{pq}$ of the trajectory l the simple closed curve $\overset{\frown}{pq} \cup \overline{pq}$ cannot bound a simply connected domain on $\mathcal{M}$ (it is not homotopic to zero), and it cannot even separate $\mathcal{M}$ into two submanifolds (it is not homologous to zero). Indeed, otherwise no trajectory (including l') can intersect $\overline{pq}$ in more than two points (Figure 2.20). But this contradicts the condition $p, q \in \omega(l')$ ($\alpha(l')$).

From what has been proved it follows that the genus of $\mathcal{M}$ is at least 1.

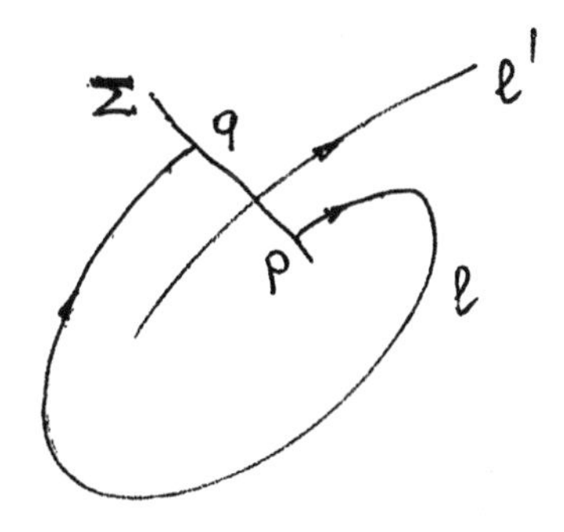

FIGURE 2.20

We now assume that the theorem is false, that is, l is not a P^+ recurrent trajectory. Then there is an interval $\mathfrak{I}$ on Σ that contains the point $m_1 \in l \cap \Sigma$ and does not contain any other points of the intersection $l^+(m_1) \cap \Sigma$ (Figure 2.21). Passing to a subsequence and changing the notation if necessary, we can assume that all the points in $\{m_i\}_2^\infty$ belong to a single component of $\Sigma \setminus \{m_0\}$ and form a monotone sequence; that is, for each $i = 2, 3, \ldots$ the point m_{i+1} lies on Σ between m_i and m_0.

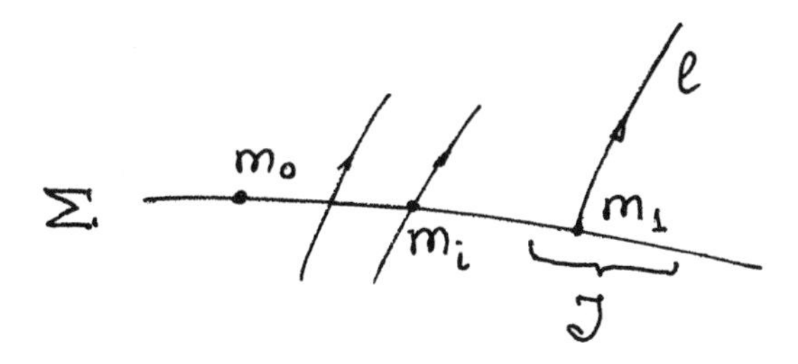

FIGURE 2.21

Assume for definiteness that $l \subset \omega(l')$ (the proof is analogous in the case $l \subset \alpha(l')$).

We show that there exist a subsequence of points $m_{j_k} \in \{m_i\}_2^\infty$, $k = 1, 2, \ldots$, and a sequence $\{l'_{j_k}\}_{k=1}^\infty$ of disjoint arcs of l' such that:

1) for $k \in \mathbb{N}$ the endpoints of l'_{j_k} lie on $\mathfrak{I}$ and are the only points where l'_{j_k} intersects $\mathfrak{I}$;

2) l'_{j_k} intersects Σ between m_{j_k} and m_{j_k+1}, but does not intersect Σ between m_0 and m_{j_k+1} (Figure 2.22).

Indeed, since $m_1 \in \omega(l')$, there exists an arc l'_{j_1} of l' satisfying 1). By the theorem on continuous dependence of trajectories on the initial conditions, the arc l'_{j_1} can be chosen to intersect Σ between m_0 and m_2. This implies the existence of a point $m_{j_1} \in \{m_i\}_2^\infty$ satisfying 2). If we have already constructed arcs $l'_{j_1}, \ldots, l'_{j_k}$, $k \geq 1$, and points $m_{j_1}, \ldots, m_{j_k}$ satisfying the required conditions, then we again get from the inclusion $m_1 \in \omega(l')$ and the theorem on continuous dependence of trajectories on the initial conditions that there is an arc $l'_{j_{k+1}} \subset l'$ that does not intersect the arcs $l'_{j_1}, \ldots, l'_{j_k}$ but intersects Σ between m_0 and m_{j_k+1}. Since $l'_{j_{k+1}}$ is compact, there is a point $m_{j_{k+1}} \in \{m_i\}_2^\infty$ such that $l'_{j_{k+1}}$ intersects Σ between

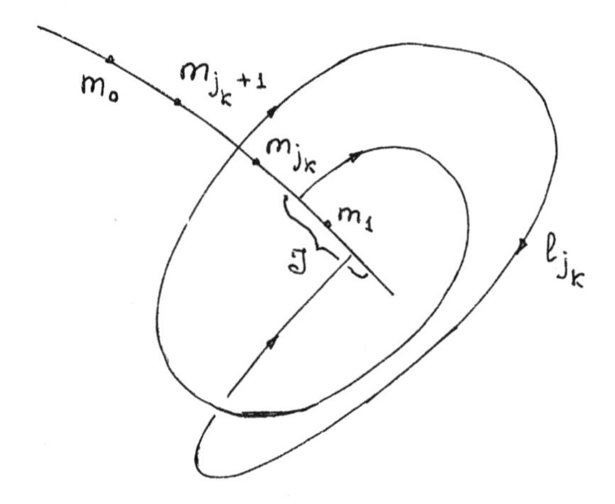

FIGURE 2.22

$m_{j_{k+1}}$ and $m_{j_{k+1}+1}$ but does not intersect Σ between m_0 and $m_{j_{k+1}+1}$. Continuing in this way, we obtain the required family $\{l'_{j_k}\}_{k=1}^{\infty}$ of arcs.

Let us cut the surface $\mathcal{M}$ along $\mathcal{J}$, and denote by $\mathcal{M}'$ the resulting surface with boundary K. The arcs l'_{j_k}, $k \in \mathbb{N}$, pass into arcs (which we also denote by l'_{j_k}) on $\mathcal{M}'$ with endpoints on K. According to Lemma 2.8, for some k_0 the family of arcs $l'_{j_1}, \dots, l'_{j_{k_0}}$ distinguishes a simply connected flat domain G on $\mathcal{M}'$, and, moreover, we can assume without loss of generality that $l'_{j_{k_0}}$ is a part of the boundary of G. Since $l'_{j_{k_0}}$ intersects Σ between $m_{j_{k_0}}$ and $m_{j_{k_0}+1}$, while the remaining arcs of the family distinguishing a simply connected domain do not intersect Σ between m_0 and $m_{j_{k_0}}$, it follows that one of $m_{j_{k_0}}$ and $m_{j_{k_0}+1}$ lies interior to G (Figure 2.23). By assumption, the semitrajectory $l^+(m_1)$ has at most one point of intersection with the boundary of G (possibly at m_1). Therefore, the whole of $l^+(m_1)$ must lie in the flat simply connected domain G (as must m_0). Then any Σ-arc $\overset{\frown}{pq}$ of $l^+(m_1)$ intersecting Σ only at the endpoints p and q combines with the segment $\overline{pq} \subset \Sigma$ to bound a disk on $\mathcal{M}$, and this is impossible. □

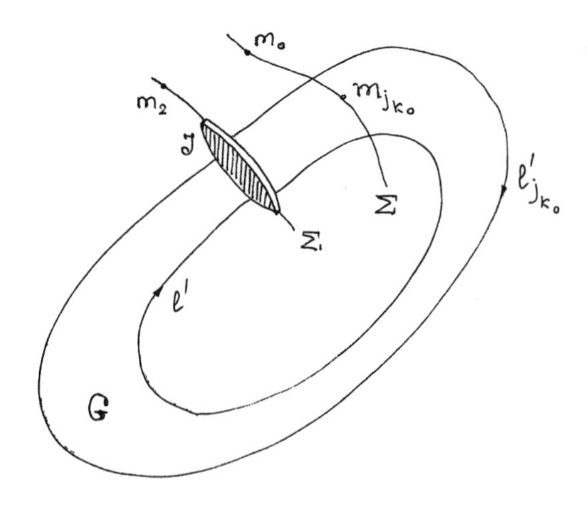

FIGURE 2.23

2.5. The Maĭer estimate for the number of independent nontrivial recurrent semitrajectories.

THEOREM 2.3. *Let a flow be given on a compact surface $\mathcal{M}$. If a P^+ (P^-) nontrivial recurrent trajectory l' contains a P^+ or P^- nontrivial recurrent trajectory l in its ω- (α-) limit set, then l' is contained in the ω- (α-) limit set of l.*

PROOF. For definiteness we assume that l and l' are P^+ recurrent trajectories (the proofs are analogous in the other three cases P^+–P^-, P^-–P^+, and P^-–P^-).

According to Lemma 2.3, there exists a contact-free cycle C that is not homologous to zero and intersects l. Since $l \subset \omega(l')$, it follows that $C \cap l' \neq \emptyset$.

We take a point $m_1 \in l' \cap C$ and assume that the theorem is false. Then there is an interval $\mathcal{J} \subset C$ such that $m_1 \in \mathcal{J}$ and $\mathcal{J} \cap l = \emptyset$. Since $C \cap l \neq \emptyset$ and l is a P^+ nontrivial recurrent trajectory, there is a point $m_0 \in \omega(l)$ on C. The fact that $m_1 \in \mathcal{J}$ implies that l' intersects $\mathcal{J}$ in a countable set of points. An argument analogous to that in the proof of Theorem 2.2 then shows that either a positive semitrajectory $l^+ \subset l$ intersects $\mathcal{J}$, or l lies in a flat simply connected domain, which is impossible. □

Theorem 2.3 means that if one nontrivial recurrent semitrajectory is a limit for another nontrivial recurrent semitrajectory, then the second is a limit for the first.

DEFINITION. Nontrivial recurrent semitrajectories $l_1^{()}$ and $l_2^{()}$ are said to be *independent* if they are not limits for each other; that is, neither lies in the limit set of the other.

According to Theorem 2.3, the relation of dependence of nontrivial recurrent semitrajectories is an equivalence relation, and hence the set of nontrivial recurrent semitrajectories of a flow is broken up into equivalence classes. The following theorem of Maĭer gives an upper estimate for the number of these equivalence classes.

THEOREM 2.4. *On a closed orientable surface $\mathcal{M}_p$ of genus $p \geq 0$ there cannot be more than p independent nontrivial recurrent semitrajectories.*

PROOF. According to Lemma 2.4, there are no nontrivial recurrent semitrajectories on a surface $\mathcal{M}_p = S^2$ of genus $p = 0$. Therefore, let $p \geq 1$.

Assume the contrary, that $\mathcal{M}_p$ has $p+1$ independent nontrivial recurrent semitrajectories $l_1^{()}, \dots, l_{p+1}^{()}$. Since these semitrajectories are independent, we can draw through any point $m_i \in l_i^{()}$ a contact-free segment Σ_i, $i \in \{1, \dots, p+1\}$, that is disjoint from $l_j^{()}$ for $i \neq j$. We take a Σ_i-arc $\overset{\frown}{A_iB_i}$ of the semitrajectory $l_i^{()}$, $k = 1, \dots, p+1$, and we form the simple closed curve $K_i = \overset{\frown}{A_iB_i} \cup \overline{A_iB_i}$, where $\overline{A_iB_i} \subset \Sigma_i$ is the part of Σ_i between A_i and B_i.

We show that $l_i^{()}$ intersects $\overline{A_iB_i} \subset \Sigma_i$, $i = 1, \dots, p$, at infinitely many points. For since A_i lies in the limit set of the semitrajectory $l_i^{()}$, the assumption $\operatorname{int} \overline{A_iB_i} \cap l_i^{()} = \emptyset$ implies that $l_i^{()}$ intersects a segment of the complement $\Sigma_i \setminus \overline{A_iB_i}$ at points arbitrarily close to A_i (Figure 2.24). It follows from the orientability of $\mathcal{M}_p$ and the theorem on continuous dependence of trajectories on the initial conditions that $l_i^{()}$ intersects $\overline{A_iB_i}$ at points arbitrarily close to B_i (infinitely many times).

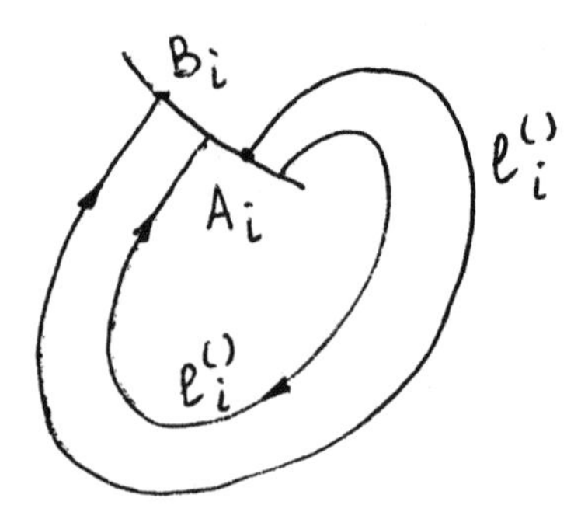

FIGURE 2.24

Since $l_i^{()}$ intersects $\overline{A_iB_i} \subset K_i$ infinitely many times, each curve $K_1, \dots, K_p$ does not separate $\mathcal{M}_p$, and, moreover, cutting $\mathcal{M}_p$ along the curves $K_1, \dots, K_p$ leads to a surface $\mathcal{M}_{0,2p}$ of genus 0 with $2p$ holes.

The surface $\mathcal{M}_{0,2p}$ is homeomorphic to a flat $(2p-1)$-connected domain and contains a nontrivial recurrent semitrajectory $l_{p+1}^{()}$, which contradicts Lemma 2.4. □

The estimate of the maximal number of independent nontrivial recurrent semitrajectories in Theorem 2.4 cannot be improved because on any orientable closed surface $\mathcal{M}_p$ of genus $p \geq 1$ there exists a flow with p independent nontrivial recurrent trajectories. Such a flow can be constructed by starting from a Denjoy flow f^t on the torus and attaching $p-1$ tori with a hole to the domain $T^2 \setminus \Omega(f^t)$, each of which also has a Denjoy flow given on it; here $\Omega(f^t)$ is the minimal set of f^t (Figure 2.25).

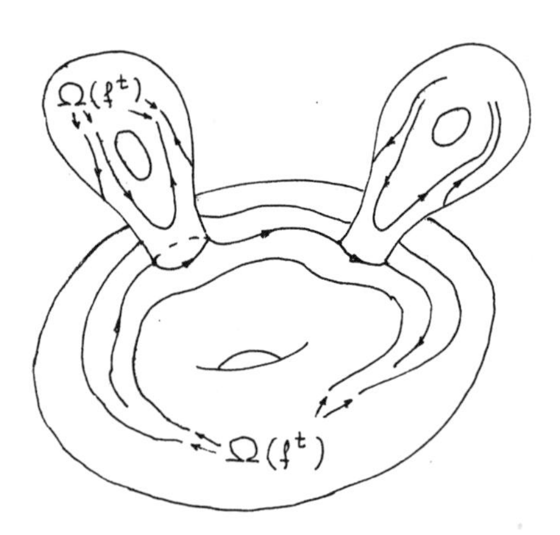

FIGURE 2.25

§3. The Poincaré–Bendixson theory

By the Poincaré–Bendixson theory we understand the investigation of the possible behavior of individual semitrajectories (and trajectories) and a description of their limit sets. At the end of the section we give a catalogue of limit sets of trajectories and a catalogue of minimal sets.

3.1. The Poincaré–Bendixson theorem.

THEOREM 3.1. *If the ω- (α-) limit set of a nonclosed positive (negative) semitrajectory l^+ (l^-) of a flow on the sphere S^2 does not contain equilibrium states, then $\omega(l^+)$ (respectively, $\alpha(l^-)$) is a closed trajectory.*

PROOF. Assume the contrary. Then there is a nonclosed trajectory l_1 in $\omega(l^+)$. Since $\omega(l_1) \neq \emptyset$, it follows from the inclusion $\omega(l_1) \subset \omega(l)$ and the condition of the theorem that $\omega(l_1)$ contains a regular point. Then Theorem 2.2 gives us that l_1 is a P^+ nontrivial recurrent trajectory. This contradicts Lemma 2.4, by which there are no such trajectories on the sphere. □

3.2. Bendixson extensions. Denote by $\Sigma(m)$ a contact-free segment drawn through a regular point $m \in \mathcal{M}$ of a flow f^t. The point m divides $\Sigma(m)$ into two segments $\Sigma_L(m)$ and $\Sigma_R(m)$ (left and right), as shown in Figure 2.26.

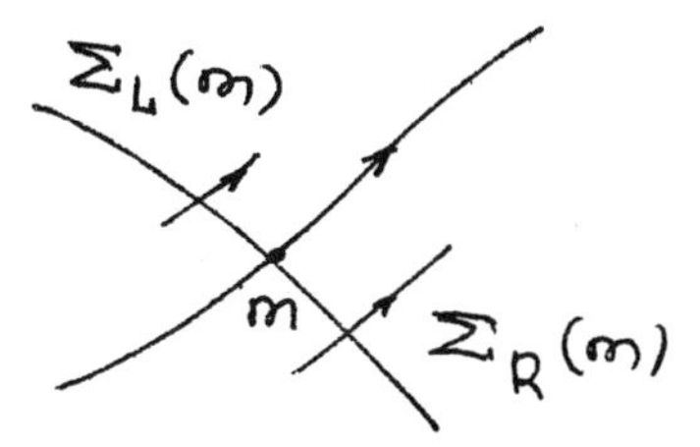

FIGURE 2.26

Let m_0 be an isolated equilibrium state of f^t. Denote by $\mathcal{U}(m_0)$ a neighborhood of m_0 diffeomorphic to an open disk with smooth boundary.

We consider a semitrajectory l^+ with $\omega(l^+) = m_0$.

DEFINITION. The semitrajectory l^+ is said to be *extendible to the right* (*left*) *with respect to the neighborhood* $\mathcal{U}(m_0)$ if for any $m \in l^+ \cap \mathcal{U}(m_0)$ there exists a contact-free segment $\Sigma(m) \subset \mathcal{U}(m_0)$ such that any positive semitrajectory beginning on $\Sigma_R(m)$ (respectively, on $\Sigma_L(m)$) leaves $\mathcal{U}(m_0)$ as the time increases (Figure 2.27).

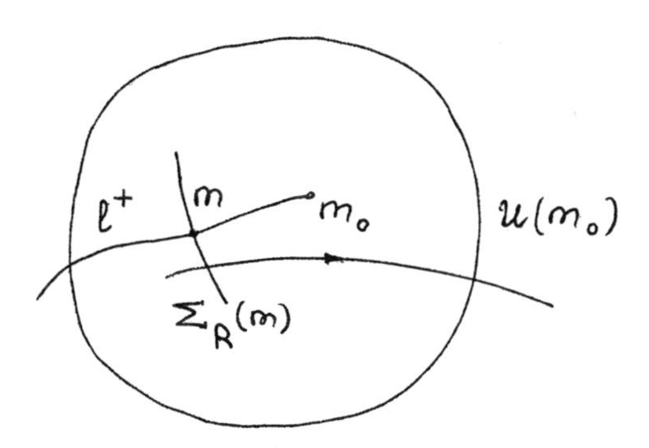

FIGURE 2.27

There is an analogous definition for a negative semitrajectory l^- with $\alpha(l^-) = m_0$.

Obviously, if a semitrajectory $l^{()}$ is extendible with respect to a neighborhood $\mathfrak{U}(m_0)$, then it is extendible with respect to any neighborhood $\mathfrak{U}'(m_0) \subset \mathfrak{U}(m_0)$.

THEOREM 3.2. *Let m_0 be an isolated equilibrium state, and let $\mathfrak{U}(m_0)$ be a neighborhood of m_0 not containing other equilibrium states. Then there are only finitely many semitrajectories tending to m_0 and extendible with respect to $\mathfrak{U}(m_0)$.*

PROOF. Assume the contrary. Suppose that there are infinitely many semitrajectories $l_n^{()}$, $n \in \mathbb{N}$, that are extendible with respect to $\mathfrak{U}(m_0)$. We can assume without loss of generality that all the $l_n^{()}$ are positive semitrajectories, and $\omega(l_n^+) = m_0$. For each semitrajectory l_n^+ there is a last (with increasing time) point A_n where l_n^+ intersects the boundary $\partial\mathfrak{U}(m_0)$ (Figure 2.28). Let A be an accumulation point of the sequence A_n, $n \in \mathbb{N}$. It follows from the continuous dependence of the trajectories on the initial conditions that $l^+(A)$ cannot leave $\mathfrak{U}(m_0)$. Therefore, there is a point $m \in l^+(A) \cap \mathfrak{U}(m_0)$, and through it we draw a contact-free segment $\Sigma(m) \subset \mathfrak{U}(m_0)$. Since A is an accumulation point of the sequence $\{A_n\}_1^\infty$, $\Sigma(m)$ intersects infinitely many of the semitrajectories l_n^+.

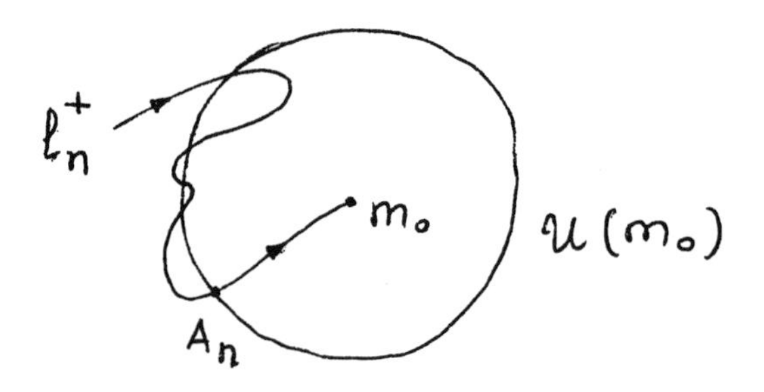

FIGURE 2.28

Let m_1, m_2, and m_3 be the last (with increasing time) points where the respective semitrajectories $l_{n_1}^+ = l^+(m_1)$, $l_{n_2}^+ = l^+(m_2)$, and $l_{n_3}^+ = l^+(m_3)$ intersect $\Sigma(m)$, and suppose that m_2 lies on $\Sigma(m)$ between m_1 and m_3. The semitrajectories $l^+(m_1)$ and $l^+(m_3)$, the point m_0, and the segment $\overline{m_1m_3} \subset \Sigma(m)$ bound a domain $\mathcal{D}$ in $\mathfrak{U}(m_0)$ (a "Bendixson bag"), and $l^+(m_2)$ enters $\mathcal{D} \subset \mathfrak{U}(m_0)$ and cannot leave it again (Figure 2.29). This contradicts the extendibility of l_{n_2}. □

Let m_0 be an equilibrium state, and let l^+ and l_1^- be semitrajectories such that $m_0 = \omega(l^+) = \alpha(l_1^-)$.

DEFINITION. The semitrajectory l_1^- is called a *Bendixson extension of l^+ to the right (to the left) with respect to the neighborhood $\mathfrak{U}(m_0)$* if for any points $m \in l^+ \cap \mathfrak{U}(m_0)$ and $m_1 \in l_1^- \cap \mathfrak{U}(m_0)$ there exist contact-free segments $\Sigma(m)$, $\Sigma(m_1) \subset \mathfrak{U}(m_0)$ such that any semitrajectory $l^+(\widetilde{m})$ with $\widetilde{m} \in \Sigma_R(m)$ (respectively, $\widetilde{m} \in \Sigma_L(m)$) intersects $\Sigma_R(m_1)$ (respectively, $\Sigma_L(m_1)$) without first leaving $\mathfrak{U}(m_0)$, and the first point where $l^+(\widetilde{m})$ intersects $\Sigma_R(m_1)$ (respectively, $\Sigma_L(m_1)$) tends to m_1 as $\widetilde{m} \to m$ (Figure 2.30).

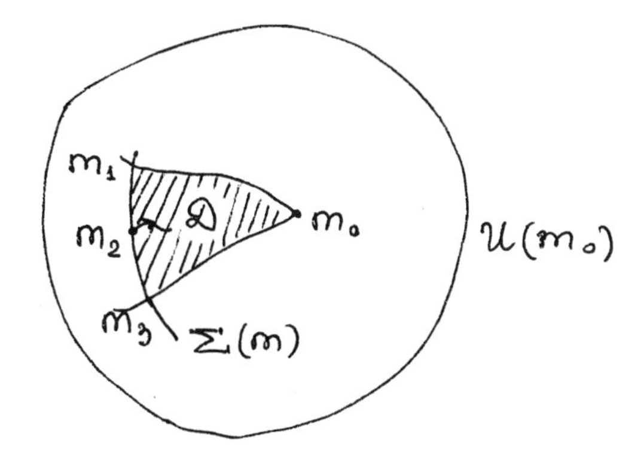

FIGURE 2.29

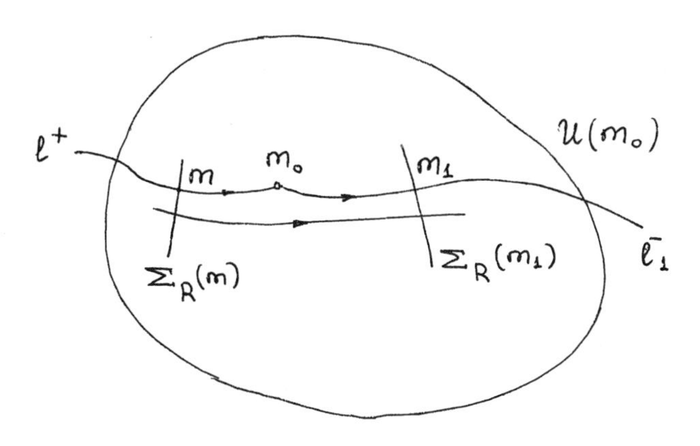

FIGURE 2.30

The semitrajectory l_1^- is called a *Bendixson extension of* l^+ *to the right* (*to the left*) if there exists a neighborhood $\mathcal{U}(m_0)$ such that l_1^- is a Bendixson extension of l^+ to the right (to the left) with respect to any neighborhood $\mathcal{U}'(m_0) \subset \mathcal{U}(m_0)$.

A Bendixson extension is defined similarly for the negative semitrajectory l_1^- with $\alpha(l_1^-) = m_0$. In particular, the semitrajectory l^+ is a Bendixson extension of l_1^- (to the right) with respect to $\mathcal{U}(m_0)$ in Figure 2.30.

LEMMA 3.1. *Suppose that a semitrajectory* l^+ *with* $\omega(l^+) = m_0$ *is extendible to the right* (*to the left*) *with respect to a neighborhood* $\mathcal{U}(m_0)$ *not containing equilibrium states other than* m_0. *Then there exists a negative semitrajectory* l_1^- *with* $\alpha(l_1^-) = m_0$ *that is a Bendixson extension of* l^+ *to the right* (*to the left*) *with respect to* $\mathcal{U}(m_0)$.

PROOF. Suppose that l^+ is extendible to the right with respect to $\mathcal{U}(m_0)$. We take a point $m \in l^+ \cap \mathcal{U}(m_0)$ and a contact-free segment $\Sigma(m) \subset \mathcal{U}(m_0)$ satisfying the definition of extendibility of the semitrajectory l^+. Then all the semitrajectories $l^+(\widetilde{m})$ with $\widetilde{m} \in \Sigma_R(m)$ leave $\mathcal{U}(m_0)$.

Assume that $m_i \to m$, $m_i \in \Sigma_R(m)$, and let A_i be the first point where $l^+(m_i)$ intersects $\partial\mathcal{U}(m_0)$ (Figure 2.31). Denote by A an accumulation point of the sequence of points A_i, $i \in \mathbb{N}$. We show that the negative semitrajectory $l^-(A)$ does not intersect $\Sigma(m)$. Since the neighborhood $\mathcal{U}(m_0)$ is simply connected, all the semitrajectories $l^+(m_i)$ beginning with some index i_0 intersect $\Sigma(m)$ only at m_i, $i \geq i_0$. If we assume that $l^-(A)$ intersects $\Sigma(m)$ at a point m_*, then $m_i \to m_*$, and hence $m_* = m$, which is impossible.

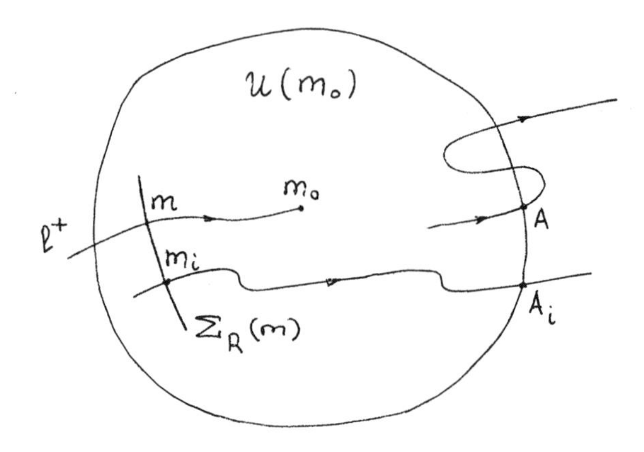

FIGURE 2.31

It follows from the theorem on continuous dependence of trajectories on the initial conditions and from the inclusions $\overset{\frown}{m_iA_i} \subset \mathfrak{U}(m_0)$, $i \in \mathbb{N}$, that $l^-(A) \subset \mathfrak{U}(m_0)$, and hence $\alpha(l^-(A)) \subset \mathfrak{U}(m_0)$. By what was proved earlier, $\Sigma(m) \cap \alpha(l^-(A)) = \emptyset$. This and the simple connectedness of $\mathfrak{U}(m_0)$ give us that $\alpha(l^-(A))$ does not contain regular points; therefore, $\alpha(l^-(A)) = m_0$. Since trajectories depend continuously on the initial conditions, $l^-(A) = l_1^-$ is a Bendixson extension of l^+ to the right. □

3.3. Separatrices of an equilibrium state.

DEFINITION. A positive semitrajectory l^+ is called an *ω-separatrix of an equilibrium state* m_0 if $\omega(l^+) = m_0$ and if for any point $m \in l^+$ there is an $\varepsilon > 0$ such that for any δ-neighborhood $\mathfrak{U}_\delta$ ($\delta > 0$) of m there exists a semitrajectory going out of $\mathfrak{U}_\delta$ that leaves the ε-neighborhood of l^+ with increasing time (Figure 2.32).

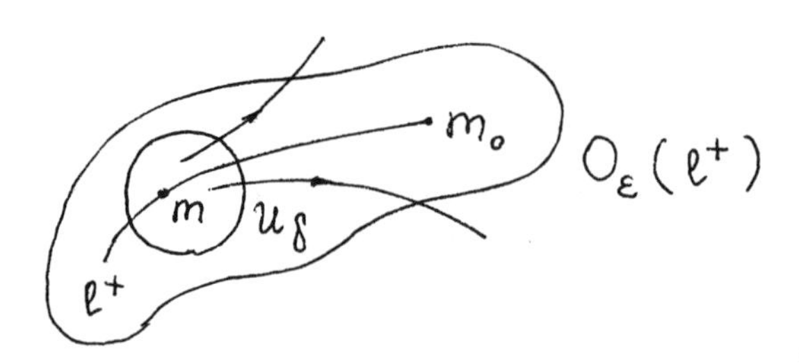

FIGURE 2.32

The definition of an α-separatrix l^- of an equilibrium state m_0 with $\alpha(l^-) = m_0$ is analogous.

For sufficiently small $\delta > 0$ the component $\widehat{l}$ of $l^+ \cap \mathfrak{U}_\delta$ containing m divides $\mathfrak{U}_\delta$ into two domains $\mathfrak{U}_{\delta,L}$ and $\mathfrak{U}_{\delta,R}$ as shown in Figure 2.33.

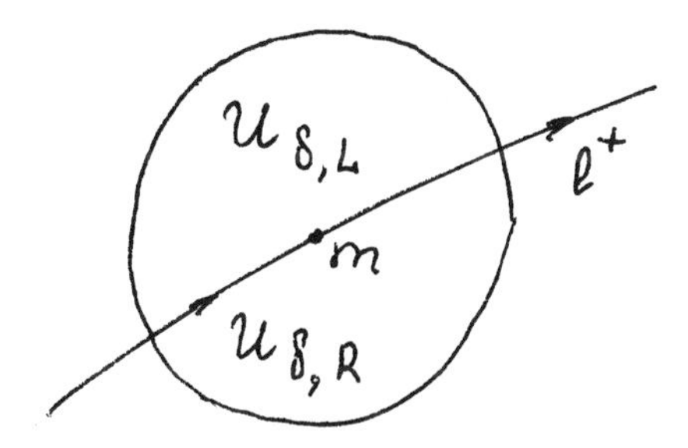

FIGURE 2.33

If in the preceding definition there exists for any sufficiently small $\delta > 0$ a semitrajectory going out of $\mathfrak{U}_{\delta,R}$ ($\mathfrak{U}_{\delta,L}$) and leaving the ε-neighborhood of l^+ with increasing time, then l^+ is called a *right-sided* (*left-sided*) *separatrix.* A separatrix that is right-sided and left-sided is said to be *two-sided.*

LEMMA 3.2. *A semitrajectory l^+ (l^-) tending to an isolated equilibrium state m_0 is an ω- (α-) separatrix (right-sided or left-sided) if and only if there exists a neighborhood $\mathfrak{U}(m_0)$ of m_0 with respect to which l^+ (l^-) is extendible (to the right or to the left).*

PROOF. This follows from the above definition and the theorem on continuous dependence of trajectories on the initial conditions. □

The next result is a consequence of Theorem 3.2 and Lemmas 3.1 and 3.2.

LEMMA 3.3. *An arbitrary ω- (α-) separatrix of an isolated equilibrium state has a Bendixson extension to the right or to the left that is an α- (ω-) separatrix of the same equilibrium state.*

In the next lemma we give a sufficient condition for a semitrajectory tending to an isolated equilibrium state to be a separatrix, along with a sufficient condition for an equilibrium state to have separatrices.

LEMMA 3.4. *Let m_0 be an isolated equilibrium state.*

a) *If $m_0 = \omega(\widetilde{l})$ ($\alpha(\widetilde{l})$), where $\widetilde{l}$ is a regular trajectory, and $\widetilde{l}$ lies in the limit set of some semitrajectory $l^{()}$, then $\widetilde{l}$ is an ω- (α-) separatrix of m_0, and a Bendixson extension of it (to the right or to the left) together with m_0 also lies in the limit set of $l^{()}$.*

b) *If m_0 lies in the ω-limit set of some nonclosed trajectory l, but $m_0 \neq \omega(l)$, then there exist an ω-separatrix l_1 and an α-separatrix l_2 of m_0 such that $l_1 \cup \{m_0\} \cup l_2 \subset \omega(l)$, and l_2 is a Bendixson extension (to the right or to the left) of l_1.*

PROOF. a) We take an ε-neighborhood $O_\varepsilon(m_0)$ of m_0 containing no equilibrium states other than m_0 and such that there are points of $\widetilde{l}$ outside $O_\varepsilon(m_0)$. For definiteness we assume that $m_0 = \omega(\widetilde{l})$ and $\widetilde{l} \subset \omega(l)$.

We draw a contact-free segment $\Sigma(m) \subset O_\varepsilon(m_0)$ through a point $m \in \widetilde{l} \cap O_e(m_0)$. Since $m \in \omega(l)$, at least one of the segments $\Sigma_L(m)$ and $\Sigma_R(m)$ is intersected by a semitrajectory $l^+ \subset l$ at points arbitrarily close to m. Suppose that l^+ intersects $\Sigma_R(m)$ at points arbitrarily close to m.

We show that any semitrajectory $l^+(\widetilde{m})$, $\widetilde{m} \in \Sigma_R(m)$, either leaves $O_\varepsilon(m_0)$ or tends to m_0, that is, $\omega(l^+(\widetilde{m})) = m_0$. Assume the opposite. Then $\omega(l^+(\widetilde{m})) \subset$

$O_\varepsilon(m_0)$, and $\omega(l^+(\widetilde{m}))$ contains a regular point, through which we draw a contact-free segment $\mathcal{J} \subset O_\varepsilon(m_0)$. Two cases are possible: 1) $l^+(\widetilde{m})$ is a closed trajectory; 2) $l^+(\widetilde{m})$ is a nonclosed trajectory.

In case 1) $l^+(\widetilde{m})$ bounds a simply connected domain $\mathcal{D}$ in $O_\varepsilon(m_0)$. By the choice of the neighborhood $O_\varepsilon(m_0)$, $\widetilde{l}$ has points outside $O_\varepsilon(m_0)$, and hence $\mathcal{D}$ does not contain m_0 (m_0 "is joined by" an arc of $\widetilde{l}$ to the boundary $\partial O_\varepsilon(m_0)$, and $\widetilde{l} \cap l^+(\widetilde{m}) = \emptyset$). On the other hand, $\mathcal{D}$ contains an equilibrium state by virtue of Corollary 4.3 in Chapter 1, and this contradicts the fact that $O_\varepsilon(m_0)$ does not contain equilibrium states other than m_0.

In case 2) the semitrajectory $l^+(\widetilde{m})$ intersects $\mathcal{J}$ infinitely many times. An arbitrary $\mathcal{J}$-arc $\overset{\frown}{pq}$ of $l^+(\widetilde{m})$ together with the segment $\overline{pq} \subset \mathcal{J}$ forms a simple closed curve γ bounding a simply connected domain $\mathcal{D}$ in $O_\varepsilon(m_0)$ (Figure 2.34). Since an ω-limit set is invariant, $\omega(l^+(\widetilde{m}))$ has a regular point not lying on $\widetilde{l}^+(m) \subset \widetilde{l}$ (in particular, this point can lie on $\widetilde{l}^-(m) \cap O_\varepsilon(m_0)$). Therefore, we take the contact-free segment $\mathcal{J}$ to be disjoint from $\widetilde{l}^+(m)$. Then $\widetilde{l}^+(m)$ does not intesect γ, and hence $\mathcal{D}$ does not contain m_0. On the other hand, the index of γ with respect to the flow is equal to 1 (see §4.5 in Chapter 1), and in view of Corollary 4.2 in Chapter 1 there is at least one equilibrium state in $\mathcal{D} \subset O_\varepsilon(m_0)$, which contradicts the absence of equilibrium states in $O_\varepsilon(m_0) \setminus \{m_0\}$.

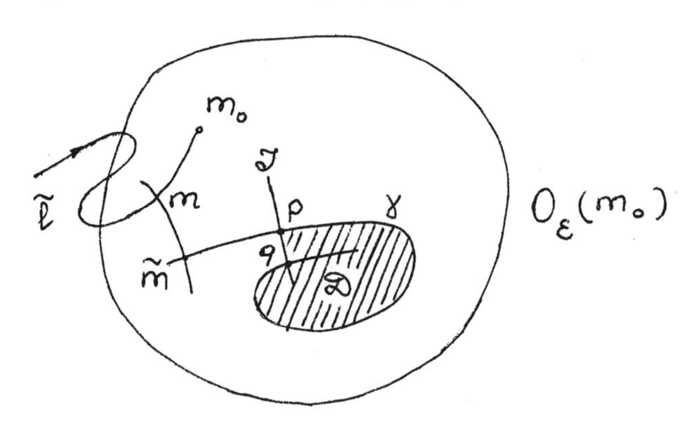

FIGURE 2.34

We now show that any semitrajectory $l^+(\widetilde{m})$ with $\widetilde{m} \in \Sigma_R(m)$ leaves the neighborhood $O_\varepsilon(m_0)$. Assume the contrary. Then it follows from what has been proved that $\omega(l^+(\widetilde{m})) = m_0$, and the closed curve $\widetilde{l}^+(m) \cup l^+(\widetilde{m}) \cup \{m_0\} \cup m\widetilde{m}$, where $m\widetilde{m} \subset \Sigma(m)$ is the subsegment of $\Sigma(m)$ with endpoints m and $\widetilde{m}$, bounds a "Bendixson bag" $\widehat{\mathcal{D}}$. Since l^+ intersects $\Sigma_R(m)$ at points arbitrarily close to m, l^+ enters $\widehat{\mathcal{D}}$ across the segment $m\widetilde{m} \subset \Sigma(m)$ and cannot go out of $\widehat{\mathcal{D}} \subset O_\varepsilon(m_0)$. This contradicts the inclusion $\widetilde{l} \subset \omega(l)$ and the fact that there are points of $\widetilde{l}$ outside $O_\varepsilon(m_0)$. The contradiction shows that $\widetilde{l}$ is an ω-separatrix. It follows from the definition of Bendixson extendibility that the Bendixson extension of $\widetilde{l}^+$ (to the right) belongs to $\omega(l)$.

We prove the assertion b). Since $m_0 \neq \omega(l)$ by assumption, there exists an ε-neighborhood $O_\varepsilon(m_0)$ with points of $\omega(l)$ outside it and no equilibrium states other than m_0 inside it.

Since $m_0 \in \omega(l)$, there is a sequence of points $m_k \in l \cap O_\varepsilon(m_0)$ that tends to m_0 as $k \to \infty$. The semitrajectories $l^+(m_k)$, $k \in \mathbb{N}$, leave $O_\varepsilon(m_0)$ because there are points of $\omega(l)$ outside $O_\varepsilon(m_0)$. Denote by m_k^+ the first point where $l^+(m_k)$ intersects $\partial O_\varepsilon(m_0)$, and let m^+ be an accumulation point of the sequence $\{m_k^+\}$ (Figure 2.35).

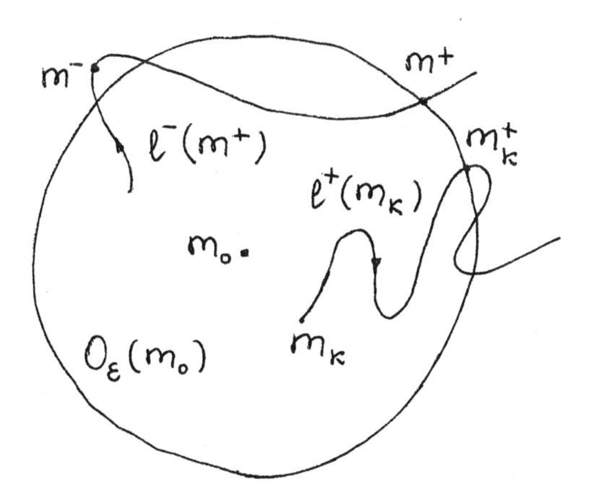

FIGURE 2.35

We show that the negative semitrajectory $l^-(m^+)$ does not leave $O_\varepsilon(m_0)$. Assume the contrary. Then there is an arc $\overset{\frown}{m^-m^+}$ of $l^-(m^+)$ such that $m^- \notin O_\varepsilon(m_0)$ (Figure 2.35). By the long flow tube theorem, there exists a neighborhood $\mathcal{U}$ of $\overset{\frown}{m^-m^+}$ with the structure of a constant flow. Therefore, all the trajectories intersecting $\mathcal{U}$ go out of $O_\varepsilon(m_0)$ as time decreases unboundedly (Figure 2.36). We take the neighborhood $\mathcal{U}$ small enough that $m_0 \notin \mathcal{U}$. For sufficiently large k the point m_k^+ is in $\mathcal{U}$. Since the arc $\overset{\frown}{m_k m_k^+}$ lies in $O_\varepsilon(m_0)$, it follows that $m_k \in \mathcal{U}$. On the other hand, $m_k \notin \mathcal{U}$ for sufficiently large k because $m_0 \notin \mathcal{U}$ and $m_k \to m_0$. This contradiction proves that $l^-(m^+) \subset O_\varepsilon(m_0)$, and hence $\alpha(l^-(m^+)) \subset \operatorname{cl} O_\varepsilon(m_0)$.

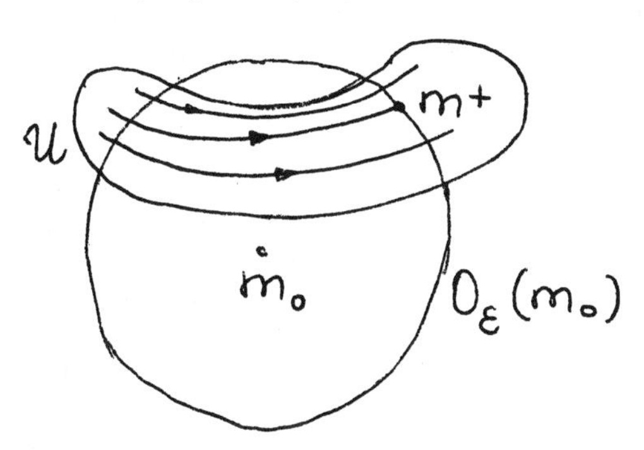

FIGURE 2.36

Let us show that $m_0 = \alpha(l^-(m^+))$. Assume not. Then $\alpha(l^-(m^+))$ contains a regular point. Since m^+ is an accumulation point for the points $m_k^+ \in l^+$, $l^-(m^+)$ lies in the limit set of l^+. According to Theorem 2.2 (the Maĭer criterion for recurrence), if $l^-(m^+)$ is a nonclosed trajectory, then $l^-(m^+)$ is a nontrivial recurrent semitrajectory lying in $O_\varepsilon(m_0)$, which contradicts Lemma 2.4. If $l^-(m^+)$ is a closed trajectory, then it bounds a simply connected domain $\mathcal{D}$ in $O_\varepsilon(m_0)$. Since the semitrajectories $l^+(m_k)$ leave $O_\varepsilon(m_0)$, m_0 cannot be in $\mathcal{D}$. But by virtue of Corollary 4.3 in Chapter 1 there is at least one equilibrium state interior to $\mathcal{D}$, and this contradicts the fact that $O_\varepsilon(m_0)$ does not contain equilibrium states other than m_0. The contradiction proves that $\alpha(l^-(m^+)) = m_0$, and b) now follows from a). □

The conditions and assertions of Lemma 3.4 are illustrated in Figure 2.37.

FIGURE 2.37

3.4. The Bendixson theorem on equilibrium states.

THEOREM 3.3. *If m_0 is an isolated equilibrium state, then either any neighborhood of m_0 contains a closed trajectory with m_0 inside it, or there exists a semitrajectory tending to m_0.*

PROOF. We take a neighborhood $\mathcal{U}(m_0)$ of m_0 that is diffeomorphic to an open disk and does not contain equilibrium states other than m_0, and we assume that $\mathcal{U}(m_0)$ does not contain a closed trajectory with m_0 inside it.

Let us first show that there exists a positive or negative semitrajectory lying entirely in $\mathcal{U}(m_0)$. Assume not. We take a sequence of points $m_k \to m_0$, $k \in \mathbb{N}$, and denote by m_k^- and m_k^+ the first points where the trajectory $l(m_k)$ intersects $\partial\mathcal{U}(m_0)$ when we move along $l(m_k)$ in the negative and positive directions, respectively (Figure 2.38). Such points exist in view of our assumption. Denote by m^- an accumulation point of the sequence $\{m_k^-\}_1^\infty$. By our assumption, there is an arc $\overset{\frown}{pq}$ of $l(m^-)$ whose endpoints p and q lie outside $\mathcal{U}(m_0)$ (Figure 2.38). According to the long flow tube theorem, there exists a neighborhood W of $\overset{\frown}{pq}$ with the structure of a constant field. Therefore, all the semitrajectories $l^-(m)$ and $l^+(m)$ with $m \in W \cap \mathcal{U}(m_0)$ leave $\mathcal{U}(m_0)$. It is clear that W can be taken small enough that $m_0 \notin W$. Since m^- is an accumulation point of the sequence $\{m_k^-\}_1^\infty$, for sufficiently large k the arcs $\overset{\frown}{m_k^- m_k^+} \subset l(m_k)$ intersect W, and hence $m_k \in W$. This contradicts the facts that $m_k \to m_k$ and $m_0 \notin W$.

Thus, there exists a point $m \in \mathcal{U}(m_0)$ such that the semitrajectory $l^{()}(m)$ lies entirely in $\mathcal{U}(m_0)$. For definiteness we take $l^{()}(m)$ to be the positive semitrajectory, that is, $l^{()}(m) = l^+(m) \overset{\text{def}}{=} l^+$. Then $\omega(l^+) \subset \operatorname{cl}\mathcal{U}(m_0)$.

If $\omega(l^+) = m_0$, then the theorem is proved. Assume that $\omega(l^+)$ contains a regular point $\widetilde{m}$. We prove that $\omega(l^+(\widetilde{m})) = m_0$, that is, $l^+(\widetilde{m})$ is the desired semitrajectory. Assume the contrary. Then $\omega(l^+(\widetilde{m}))$ contains a regular point. If

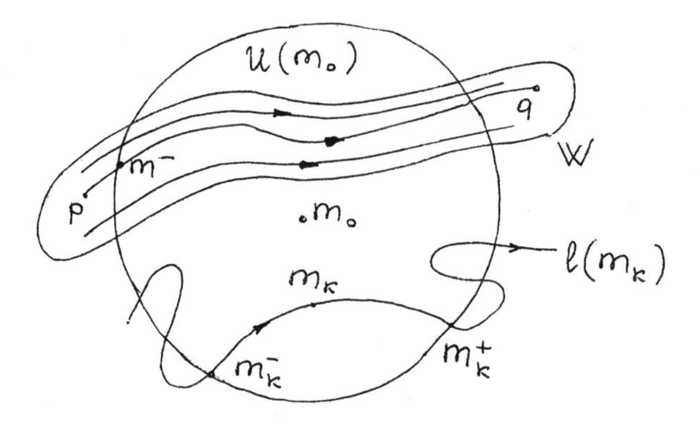

FIGURE 2.38

$l^+(\widetilde{m})$ is a nonclosed semitrajectory, then we have a contradiction to Theorem 2.2 and Lemma 2.4. If $l^+(\widetilde{m})$ is a closed trajectory, then by the invariance of the ω-limit set and the inclusions $\omega(l^+(\widetilde{m})) \subset \omega(l^+) \subset \operatorname{cl}\mathcal{U}(m_0)$, the trajectory $l^+(\widetilde{m}) = \omega(l^+(\widetilde{m}))$ lies in $\operatorname{cl}\mathcal{U}(m_0)$ and bounds a simply connected domain $\mathcal{D} \subset \mathcal{U}(m_0)$ in the disk $\operatorname{cl}\mathcal{U}(m_0)$. Since there are no closed trajectories in the neighborhood $\mathcal{U}(m_0)$ with m_0 inside them, it follows that $m_0 \notin \mathcal{D}$. But in view of Corollary 4.3 in Chapter 1 the domain $\mathcal{D}$, and hence the neighborhood $\mathcal{U}(m_0)$, contains an equilibrium state other than m_0, which contradicts the choice of $\mathcal{U}(m_0)$. □

We consider an isolated equilibrium state m_0 and a neighborhood $\mathcal{U}(m_0)$ diffeomorphic to an open disk and not containing equilibrium states other than m_0. Suppose that the semitrajectories $l_1^{()}$ and $l_2^{()}$ tend to m_0 (as $t \to +\infty$ or $t \to -\infty$), and both have points outside $\mathcal{U}(m_0)$. Denote by m_1 and m_2 the last points $l_1^{()}$ and $l_2^{()}$ have in common with the boundary $\partial\mathcal{U}(m_0)$ (Figure 2.39).

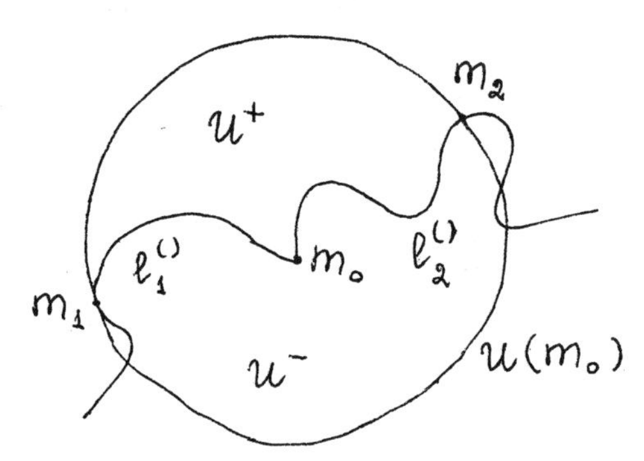

FIGURE 2.39

The curve consisting of $l_1^{()}(m_1), l_2^{()}(m_2) \subset \mathcal{U}(m_0)$ and the point m_0 separates $\mathcal{U}(m_0)$ into two domains $\mathcal{U}^-$ and $\mathcal{U}^+$, called the *curvilinear sectors bounded by the semitrajectories* $l_1^{()}$ *and* $l_2^{()}$ *in the neighborhood* $\mathcal{U}(m_0)$. In this notation we give the following definition.

DEFINITION. The curvilinear sector $\mathcal{U}^*$ in $\mathcal{U}(m_0)$ bounded by the semitrajectories $l_1^{()}$ and $l_2^{()}$ is called a *hyperbolic or saddle sector in the neighborhood* $\mathcal{U}(m_0)$

if $l_1^{()}$ and $l_2^{()}$ are separatrices of the equilibrium state m_0 that are Bendixson extensions of each other (to the right or to the left), and any trajectory $l(m)$, $m \in \mathfrak{U}^*$, leaves $\mathfrak{U}^*$ with increasing or decreasing time.

There are definitions of parabolic and elliptic sectors; however, we omit them and confine ourselves to an illustration (Figure 2.40). See [**3**], §17 of Chapter 8, for the precise definitions.

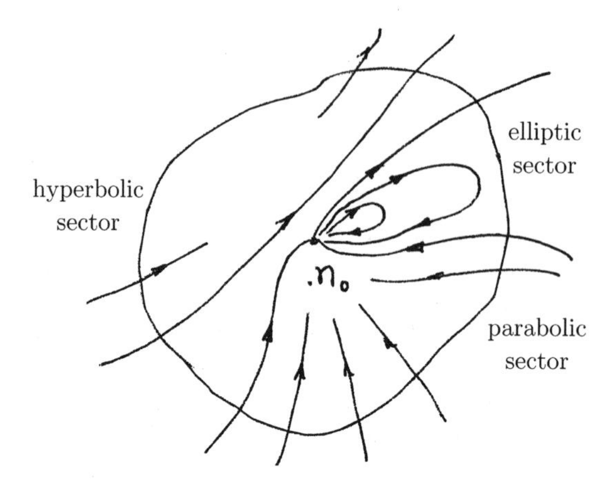

FIGURE 2.40

If the equilibrium state m_0 has only finitely many separatrices tending to it, then there exists a neighborhood $\mathfrak{U}(m_0)$ such that the number of hyperbolic, elliptic, and parabolic sectors is the same in all neighborhoods $\mathfrak{U}'(m_0) \subset \mathfrak{U}(m_0)$. Therefore, in this case we can refer to the number of hyperbolic, elliptic, and parabolic sectors of the equilibrium state.

CONVENTION. *Unless a statement to the contrary is made, each equilibrium state below will be assumed to be isolated, and to have only finitely many separatrices (possibly none) tending to it.*

DEFINITION. An equilibrium state to which a finite nonzero number of separatrices tend is called a *topological saddle* if it has only hyperbolic sectors (Figure 2.41).

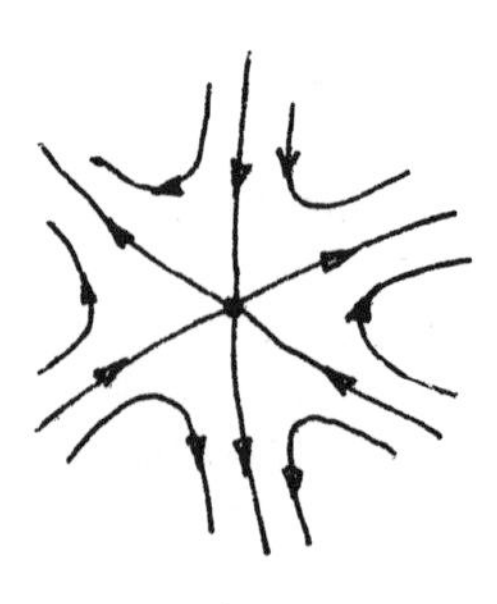

FIGURE 2.41

A topological saddle with four hyperbolic sectors will sometimes simply be called a *saddle* (or structurally stable saddle).

Lemma 3.4 gives us

COROLLARY 3.1. *Equilibrium states of a transitive flow on a closed surface are topological saddles.*

We note that such a flow can have impassable grains: topological saddles with two hyperbolic sectors (see §2.3.2 in Chapter 1).

3.5. One-sided contours. Consider a flow on a two-dimensional manifold. A trajectory of it that is both an α-separatrix of an equilibrium state and an ω-separatrix of an equilibrium state (possibly the same equilibrium state) will be called a *separatrix joining equilibrium states*, or a *trajectory connecting equilibrium states*, or a *separatrix going from an equilibrium state to an equilibrium state* (Figure 2.42).

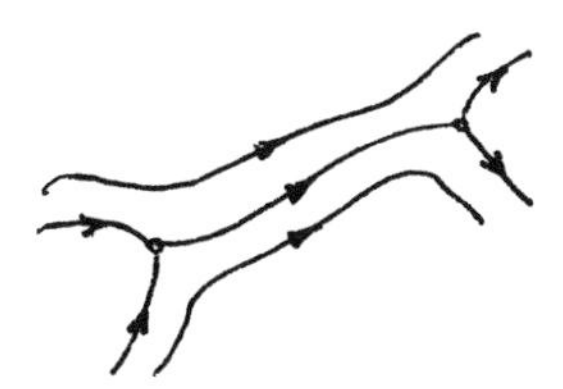

FIGURE 2.42

DEFINITION. A family of separatrices $l_1, \dots, l_s$ joining equilibrium states, together with equilibrium states $m_1, \dots, m_s$ such that $\alpha(l_i) = m_i$ $(i = 1, \dots, s)$ and $\omega(l_j) = m_{j+1}$ $(j = 1, \dots, s)$ (where $m_{s+1} = m_1$) is called a *right-sided* (*left-sided*) *contour* if for all $i = 1, \dots, s$ the separatrix l_{i+1} (where $l_{s+1} = l_1$) is a Bendixson extension of l_i to the right (to the left).

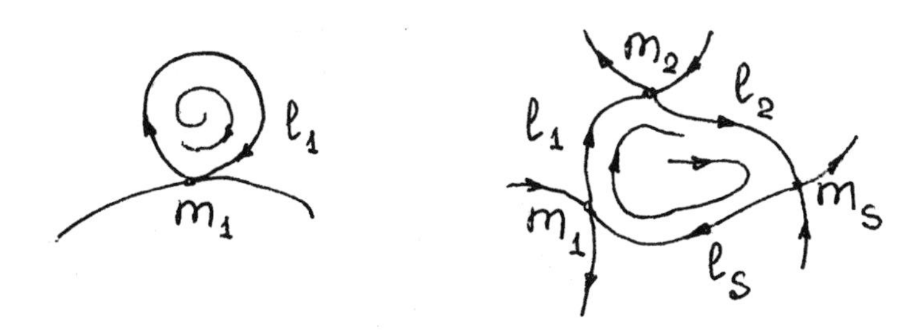

FIGURE 2.43

A right-sided or left-sided contour is called a *one-sided contour.*

In particular, a one-sided contour can consist of a single equilibrium state and a single separatrix (a separatrix loop).

LEMMA 3.5. *Let l be a nonclosed trajectory of a flow on a compact orientable surface $\mathcal{M}$. In this case:*

a) *if the ω- (α-) limit set of l contains a one-sided contour K, then $\omega(l)$ ($\alpha(l)$) $=$ K;*

b) *if the* ω- (α-) *limit set of* l *consists of finitely many equilibrium states and finitely many separatrices joining equilibrium states, then* $\omega(l)$ ($\alpha(l)$) *is a one-sided contour.*

The proof of a) is completely analogous to that of Lemma 1.6, in which it is shown that if $\omega(l)$ ($\alpha(l)$) contains a closed trajectory, then $\omega(l)$ ($\alpha(l)$) is a closed trajectory.

The assertion b) follows from Lemma 3.4 and a). □

3.6. Lemmas on the Poincaré mapping. Let f^t be a flow on a compact two-dimensional manifold $\mathcal{M}$, and let Σ_1 and Σ_2 be disjoint contact-free segments for f^t such that the Poincaré mapping $P\colon \Sigma_1 \to \Sigma_2$ is defined at some point of Σ_1 (see §1.2).

Assume that P is undefined at the endpoints of Σ_1, and let $\mathrm{Dom}(P)$ be the domain of P.

According to the theorem on continuous dependence of trajectories on the initial conditions, if $P[\mathrm{Dom}(P)] \subset \mathrm{int}\,\Sigma_2$ (that is, the image of any point $m \in \mathrm{Dom}(P)$ is not an endpoint of Σ_2), then the set $\mathrm{Dom}(P)$ is open in Σ_1 (Figure 2.44). Consequently, $\mathrm{Dom}(P)$ is a union of at most countably many disjoint open intervals, which we call the *components* of $\mathrm{Dom}(P)$.

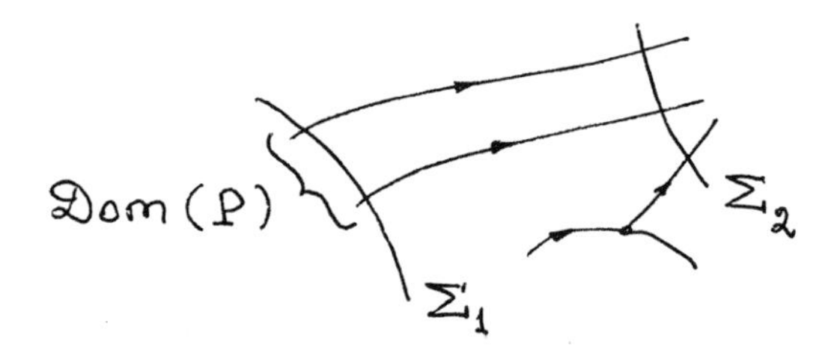

FIGURE 2.44

The set $\mathrm{Dom}(P)$ and the concept of the components of it are defined similarly for a contact-free cycle C of f^t and for the Poincaré mapping $P\colon C \to C$ (Figure 2.45).

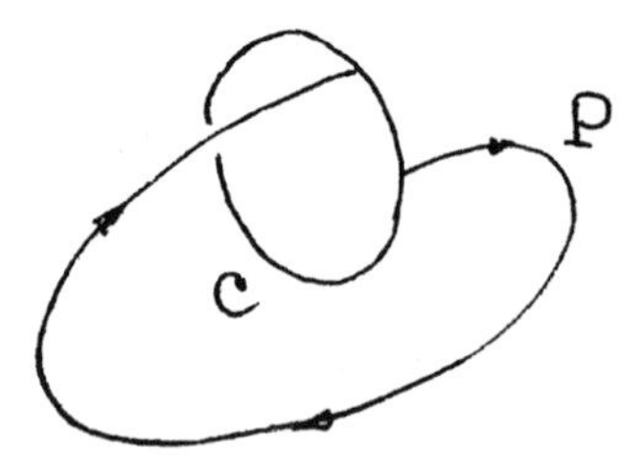

FIGURE 2.45

We orient each contact-free segment Σ and each contact-free cycle C (a positive direction is introduced); for Σ the orientation introduced is such that we hit the component $\Sigma_R(m)$ when we move in the positive direction from any point $m \in \Sigma$ (see §3.2 for the definition of $\Sigma_R(m)$). For points $a, b \in \Sigma$ (C) we denote by (a, b)

the interval of the contact-free segment (cycle) Σ (C) traversed from a to b in the positive direction.

For $a, b \in \Sigma$ let $\overline{ab}$ be the interval of Σ with endpoints a and b.

LEMMA 3.6. *Let Σ_1 and Σ_2 be disjoint contact-free segments for a flow f^t on a compact two-dimensional manifold, and suppose that the Poincaré mapping $P\colon \Sigma_1 \to \Sigma_2$ has nonempty domain* $\mathrm{Dom}(P)$. *Assume that* $P[\mathrm{Dom}(P)] \subset \mathrm{int}\,\Sigma_2$, *and let (m_1, m_2) be a component of the set* $\mathrm{Dom}(P)$. *Then:*

a) *the semitrajectories $l^+(m_1)$ and $l^+(m_2)$ are ω-separatrices;*

b) *there exists a sequence $l^+(m_1) = l_{10}, l_{11}, \dots, l_{1k_1}$ of separatrices such that l_{1s} is a Bendixson extension of $l_{1\,s-1}$ to the right $(s = 1, \dots, k_1)$, and the separatrices l_{1i} $(i = 0, \dots, k_1 - 1)$ do not intersect Σ_2, but l_{1k_1} does intersect Σ_2;*

c) *there exists a sequence $l^+(m_2) = l_{20}, l_{21}, \dots, l_{2k_2}$ of separatrices such that l_{2s} is a Bendixson extension of $l_{2\,s-1}$ to the left $(s = 1, \dots, k_2)$, and the separatrices l_{2j} $(j = 0, \dots, k_2 - 1)$ do not intersect Σ_2, but l_{2k_2} does intersect Σ_2;*

d) *the image of the interval (m_1, m_2) under the action of P is an interval $\overline{\overline{m}_1\overline{m}_2} \subset \Sigma_2$, where $\overline{m}_i$ is the first point where l_{ik_i} intersects Σ_2 with increasing time, $i = 1, 2$;*

e) *for points $a, b \in (m_1, m_2)$ the interval $(a, b) \subset \Sigma_1$, the arcs $\overset{\frown}{aP}(a)$ and $\overset{\frown}{bP}(b)$ of the respective semitrajectories $l^+(a)$ and $l^+(b)$, and the interval $\overline{P(a)P(b)} \subset \Sigma_2$ bound an open simply connected domain in $\mathcal{M}$ (Figure* 2.46).

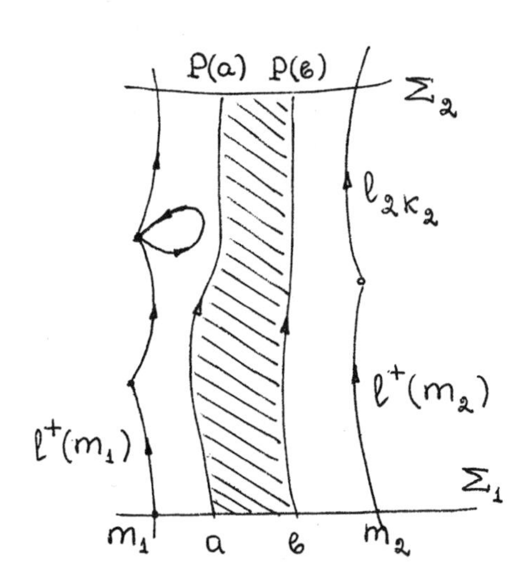

FIGURE 2.46

PROOF. a) Since (m_1, m_2) is a component of the set $\mathrm{Dom}(P)$, $l^+(m_1)$ and $l^+(m_2)$ do not intersect the contact-free segment Σ_2.

We show that $\omega(l^+(m_1))$ is an equilibrium state. Assume not. Suppose that $\omega(l^+(m_1))$ contains a regular point A. We consider first the case when $l^+(m_1)$ is a nonclosed semitrajectory. Obviously, $A \notin \Sigma_2$, and there is an arc $\overset{\frown}{mP}(m)$ of a semitrajectory $l^+(m)$ with $m \in (m_1, m_2)$ that does not contain A. Through A we draw a contact-free segment $\Sigma(A)$ disjoint from $\Sigma_2 \cup \overset{\frown}{mP}(m)$, and we take an arbitrary $\Sigma(A)$-arc $\overset{\frown}{pq}$ of $l^+(m_1)$. Such an arc exists, because $l^+(m_1)$ intersects $\Sigma(A)$ in a countable set of points. In view of Lemma 1.2 the simple closed curve $\overset{\frown}{pq} \cup \overline{pq}$ can be approximated by a contact-free cycle C that intersects $l^+(m_1)$, and it follows from the disjointness of $\Sigma(A)$ and $\Sigma_2 \cup \overset{\frown}{mP}(m)$ that C can be constructed to be disjoint from $\Sigma_2 \cup \overset{\frown}{mP}(m)$. We take such a cycle C (Figure 2.47).

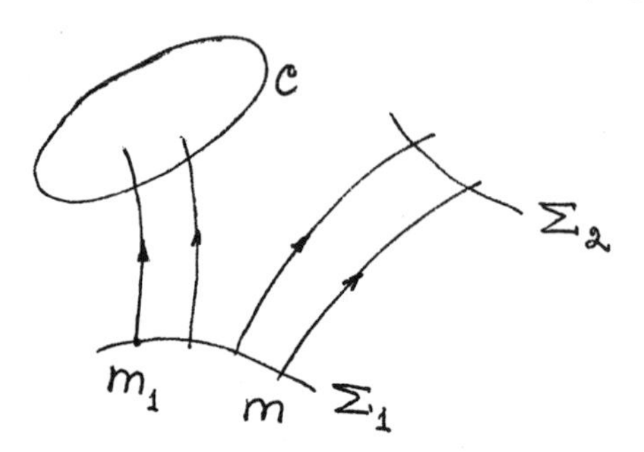

FIGURE 2.47

Let us break up the closed interval $\mathrm{cl}(m_1, m) \overset{\text{def}}{=} [m_1, m]$ into two sets: $R_1 = \{x \in [m_1, m] \mid$ the semitrajectory $l^+(x)$ intersects C without first intersecting the segment $\Sigma_2\}$, and $R_2 = \{x \in [m_1, m] \mid l^+(x)$ intersects Σ_2 without first intersecting the cycle $C\}$. Since $m_1 \in R_1$ and $m \in R_2$, it follows that $R_1 \neq \emptyset$ and $R_2 \neq \emptyset$. It then follows from the theorem on continuous dependence of trajectories on the initial conditions that R_1 and R_2 are relatively open in the topology of the segment $[m_1, m]$. It is clear that $R_1 \cup R_2 = \emptyset$. From this and the connectedness of $[m_1, m]$, there exists a point $m^* \in (m_1, m)$ such that $l^+(m^*)$ is disjoint from C and Σ_2, which contradicts the fact that all the semitrajectories $l^+(\widetilde{m})$ with $\widetilde{m} \in (m_1, m_2)$ intersect Σ_2.

If $l^+(m_1)$ is a closed trajectory disjoint from Σ_2, then for points $\widetilde{m} \in (m_1, m_2)$ sufficiently close to m_1 the semitrajectories $l^+(\widetilde{m})$ first intersect the contact-free segment Σ_1, and then the segment Σ_2. This contradicts the definition of the Poincaré mapping P (see §1.2).

Thus, $\omega(l^+(m_1))$ is an equilibrium state m_0. Let $\mathfrak{U}(m_0)$ be a neighborhood of m_0 disjoint from Σ_2. It follows from the theorem on continuous dependence of trajectories on the initial conditions that for points $m \in (m_1, m_2)$ sufficiently close to m_1 the semitrajectories $l^+(m)$ enter $\mathfrak{U}(m_0)$ without first intersecting Σ_2, and then must leave $\mathfrak{U}(m_0)$ in order to intersect Σ_2. Therefore, $l^+(m_1)$ is an ω-separatrix of the equilibrium state m_0 (Lemma 3.2). The proof that $l^+(m_2)$ is also an ω-separatrix is completely analogous, and the assertion a) is proved.

We prove the assertion b). It follows from the foregoing that $l^+(m_1)$ is extendible to the right. By Lemma 3.3, there exists an α-separatrix l_{11} of m_0 that is a Bendixson extension of the ω-separatrix $l_{10} = l^+(m_1)$ to the right. If l_{11} intersects Σ_2, then b) is proved. Suppose that l_{11} is disjoint from Σ_2. Repeating the proof of a) with no fundamental changes, and using the definition of a Bendixson extension, we show that $\omega(l_{11})$ is an equilibrium state, and l_{11} is an ω-separatrix (that is, l_{11} is a separatrix joining equilibrium states). Continuing in this way, we get a sequence $l_{10}, l_{11}, \dots, l_{1i}, \dots$ of separatrices, where l_{1i} is a Bendixson extension of $l_{1\,i-1}$ to the right for each $i \geq 1$.

We show that l_{1i} intersects Σ_2 for some index $i \geq 1$ (which concludes the proof of b)). Assume the contrary. By our convention about the finiteness of the number of separatrices of any equilibrium state and the finiteness of the number of equilibrium states on the compact manifold $\mathcal{M}$, we then get that l_{1i} contains the semitrajectory $l_{10} = l^+(m_1)$ for some $i \geq 1$. Consequently, the separatrices $l_{10}, l_{11}, \dots, l_{1i}$ and their limit sets (the equilibrium states which they join) form a right-sided contour. This contour is disjoint from Σ_2 by assumption, so for points $m \in (m_1, m_2)$ sufficiently close to m_1 the semitrajectories $l^+(m)$ intersect Σ_1 without first intersecting Σ_2 as time increases, and this contradicts the definition of the Poincaré mapping $P\colon \Sigma_1 \to \Sigma_2$. The contradiction proves b).

The proof of c) is analogous.

We prove d). On its domain P is a homeomorphism (Lemma 1.3), so $P[(m_1, m_2)]$ is an open interval $m_1' m_2' \subset \Sigma_2$. On the other hand, it follows from the definition of the Bendixson extension that $P(m) \to \overline{m}_i$ as $m \to m_i$, $m \in (m_1, m_2)$ ($i = 1$, 2), where $\overline{m}_i$ is the first point where the separatrix l_{ik_i} intersects Σ_2. This implies that $\overline{m}_1 = m_1'$ and $\overline{m}_2 = m_2'$.

The assertion d) follows from the fact that the domain bounded by the curve

$$(a, b) \cup \overset{\frown}{aP}(a) \cup \overset{\frown}{bP}(b) \cup \overline{P(a)P(b)}$$

can be covered by finitely many long flow tubes. $\square$

We now consider a contact-free cycle C for a flow f^t given on a compact orientable two-dimensional manifold $\mathcal{M}$, and we assume that the Poincaré mapping $P\colon C \to C$ induced by f^t has nonempty domain $\mathrm{Dom}(P)$. Recall that for any point $m \in \mathrm{Dom}(P)$ the positive semitrajectory $l^+(m) \setminus \{m\}$ intersects C, and the first such point of intersection is denoted by $P(m)$. The set $\mathrm{Dom}(P)$ is open and is the union of an at most countable family of disjoint open intervals which we call the components of $\mathrm{Dom}(P)$.

LEMMA 3.7. *Let C be a contact-free cycle for a flow f^t on a compact orientable two-dimensional manifold $\mathcal{M}$, and let the Poincaré mapping $P\colon C \to C$ induced by f^t have nonempty domain* $\mathrm{Dom}(P)$. *Let (m_1, m_2) be a component of* $\mathrm{Dom}(P)$. *Then:*

a) *the semitrajectories $l^+(m_1)$ and $l^+(m_2)$ are ω- separatrices;*

b) *there exists a sequence $l^+(m_1) = l_{10}, l_{11}, \dots, l_{1k_1}$ of separatrices such that l_{1s} is a Bendixson extension of $l_{1\,s-1}$ to the right ($s = 1, \dots, k_1$), and the separatrices l_{1i} ($i = 0, \dots, k_1 - 1$) do not intersect C, but l_{1k_1} does intersect C;*

c) *there exists a sequence $l^+(m_2) = l_{20}, l_{21}, \dots, l_{2k_2}$ of separatrices such that l_{2s} is a Bendixson extension of $l_{2\,s-1}$ to the left ($s = 1, \dots, k_2$), and the separatrices l_{2j} ($j = 0, \dots, k_2 - 1$) do not intersect C, but l_{2k_2} does intersect C;*

d) *the image of (m_1, m_2) under the action of P is the interval $(\overline{m}_1, \overline{m}_2)$, where $\overline{m}_i$ is the first point where l_{ik_i} intersects C, $i = 1$, 2;*

e) *for any endpoints $a, b \in (m_1, m_2)$ of an interval $(a, b) \subset (m_1, m_2)$ the closed curve $(a, b) \cup \overset{\frown}{aP}(a) \cup \overset{\frown}{bP}(b) \cup (P(a), P(b))$ bounds a simply connected open domain in $\mathcal{M}$, where $\overset{\frown}{aP}(a)$ and $\overset{\frown}{bP}(b)$ are arcs of the respective semitrajectories $l^+(a)$ and $l^+(b)$.*

PROOF. This can be proved according to the scheme used to prove Lemma 3.6. We omit the proof and leave it to the reader as an exercise. □

LEMMA 3.8. *Suppose that a flow f^t on a closed orientable two-dimensional manifold $\mathcal{M}$ induces a Poincaré mapping $P\colon C \to C$ with nonempty domain* $\mathrm{Dom}(P)$ *on a contact-free cycle C. Then* $\mathrm{Dom}(P)$ *has finitely many components.*

PROOF. Assume the contrary. Let (a_i, b_i) $(i = 1, 2, \dots)$ be the family of components of the set $\mathrm{Dom}(P)$. In view of Lemma 3.7 the semitrajectories $l^+(a_i)$ and $l^+(b_i)$ $(i \in \mathbb{N})$ are ω-separatrices (Figure 2.48). Since each of a_i and b_i $(i \in \mathbb{N})$ is a boundary point of at most two components of $\mathrm{Dom}(P)$, each ω-separatrix in the collection $\{l^+(a_i), l^+(b_i), i \in \mathbb{N}\}$ is repeated at most twice; therefore, the collection has countably many distinct ω-separatrices. This contradicts our convention that f^t has finitely many equilibrium states on the compact manifold, and that each equilibrium state has finitely many (perhaps none) separatrices. □

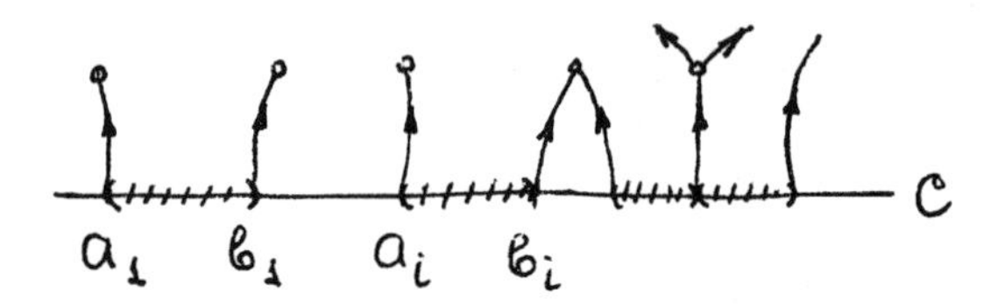

FIGURE 2.48

Lemma 3.8 generalizes to flows f^t that can have infinitely many equilibrium states. It is clear that in this situation the lemma is not true in general, because there are examples of flows with countably many impassable grains for which the semitrajectories tending to the impassable grains separate $\mathrm{Dom}(P)$ into countably many components. We proceed to a generalization of Lemma 3.8 after introducing the necessary definitions.

Suppose that the flow f^t, which can have infinitely many equilibrium states, induces a Poincaré mapping $P\colon C \to C$ with nonempty domain $\mathrm{Dom}(P)$ on the contact-free cycle C.

DEFINITION. Two points $x_1, x_2 \in C$ are said to be *Gutierrez-equivalent* (written as $x_1 \overset{\mathrm{G}}{\sim} x_2$) if there exist points $a, b \in \mathrm{Dom}(P)$ such that (a, b) contains x_1 and x_2, and the closed curve $[a, b] \cup \overset{\frown}{aP}(a) \cup \overset{\frown}{bP}(b) \cup [P(a), P(b)]$ bounds a simply connected domain on the manifold (Figure 2.49).

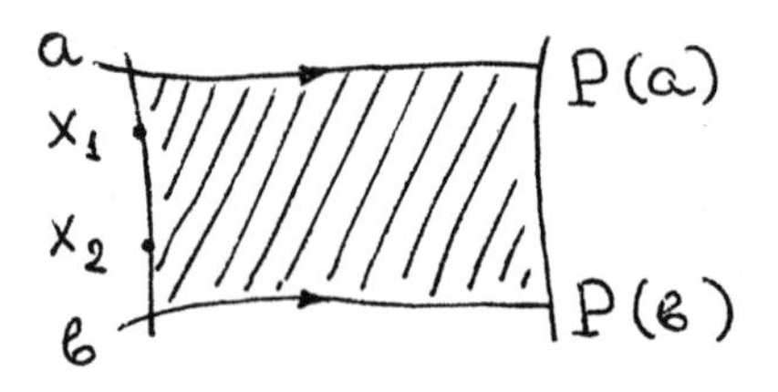

FIGURE 2.49

The relation of the Gutierrez equivalence is an ordinary equivalence relation, and the set of points of C is broken up into disjoint equivalence classes. It follows immediately from the definition and from the long flow tube theorem that an equivalence class with more than one point is an open subset of C. Such a Gutierrez equivalence class will be called an *open class*. The union of the open equivalence classes will be denoted by $\mathcal{D}(C, P, \sim)$. Obviously, $\mathrm{Dom}(P) \subset \mathcal{D}(C, P, \sim)$. In view of Lemma 3.7 e) a component of $\mathrm{Dom}(P)$ belongs to a single open equivalence class. The set $\mathcal{D}(C, P, \sim)$ is open and is the union of an at most countable family of open intervals which we call its components.

LEMMA 3.9. *Let f^t be a flow (having infinitely many equilibrium states in general) on a compact orientable two-dimensional manifold $\mathcal{M}$, and suppose that f^t induces a Poincaré mapping $P\colon C \to C$ with nonempty domain on a contact-free cycle C. Then:*

a) *each open Gutierrez equivalence class is an open interval and is a component of the set $\mathcal{D}(C, P, \sim)$;*

b) *the number of open Gutierrez equivalence classes is finite.*

PROOF. The assertion a) follows immediately from the definition of the Gutierrez equivalence.

We prove b). Assume the contrary: suppose that the family of open equivalence classes is countable. By the definition of the Gutierrez equivalence, each open class contains a point where P is defined. Let a_i ($i \in \mathbb{N}$) be points of $\mathrm{Dom}(P)$ that are in distinct open equivalence classes. The contact-free cycle C does not separate $\mathcal{M}$ (Lemma 2.2), so cutting $\mathcal{M}$ along C leads to a manifold $\mathcal{M}'$ with two components K_1 and K_2 of the boundary $\partial\mathcal{M}'$ that are joined by the arcs $\overset{\frown}{a_i P(a_i)}$. It follows from Lemma 2.8 that there are two arcs $\overset{\frown}{a_j P(a_j)}$ and $\overset{\frown}{a_k P(a_k)}$ that bound a simply connected domain on $\mathcal{M}'$. This contradicts the fact that a_j and a_k belong to different equivalence classes. □

3.7. Description of quasiminimal sets. Let f^t be a flow on a compact two-dimensional manifold $\mathcal{M}$, and assume that it has nontrivial recurrent semitrajectories.

DEFINITION. A set $N \subset \mathcal{M}$ is called a *quasiminimal* set of the flow f^t if it is the (topological) closure of a nontrivial recurrent semitrajectory.

According to Theorem 2.1, the topological closure of a nontrivial recurrent semitrajectory $l^{()}$ contains a continuum of nontrivial recurrent trajectories that are dense in $\mathrm{cl}(l^{()})$. Therefore, a quasiminimal set is the topological closure of a nontrivial recurrent trajectory (and can be so defined).

THEOREM 3.4. *On a compact two-dimensional manifold $\mathcal{M}$ let f^t be a flow having nontrivial recurrent semitrajectories, and let N be a quasiminimal set of f^t. Then N is invariant (consists of whole trajectories) and can contain only the following trajectories:*

1) *nontrivial recurrent trajectories;*

2) *ω- (α-) separatrices that are P^- (P^+) nontrivial recurrent trajectories;*

3) *separatrices joining equilibrium states;*

4) *equilibrium states, to each of which at least one ω-separatrix in N tends, and at least one α-separatrix in N.*

PROOF. Let $N = \mathrm{cl}(l)$, where l is a nontrivial recurrent trajectory. Since $\mathrm{cl}(l) = \omega(l) = \alpha(l)$ (see (2.3), §2.1), the invariance of N follows from that of an ω- (α-) limit set.

If $m_0 \in N$ is an equilibrium state, then $m_0 \in \omega(l)$, and Lemma 3.4 implies 4).

Consider a regular trajectory $\widetilde{l} \subset N$. The trajectory $\widetilde{l}$ cannot be closed, for otherwise $\widetilde{l} = \omega(l)$ in view of Lemma 1.6, and this contradicts the fact that l is a nontrivial recurrent trajectory. If the ω- (α-) limit set of $\widetilde{l}$ contains a regular point, then by the Maĭer theorem (Theorem 2.2), $\widetilde{l}$ is a P^+ (P^-) nontrivial recurrent trajectory. If $\omega(\widetilde{l})$ ($\alpha(\widetilde{l})$) is an equilibrium state, then it follows from the inclusion $\widetilde{l} \subset \omega(l)$ and Lemma 3.4 that $\widetilde{l}$ is an ω- (α-) separatrix. □

To prove Theorem 3.5 we need the following result.

LEMMA 3.10. *Let $l^{()}$ be an exceptional nontrivial recurrent semitrajectory* (*see* §2.1 *for the definition*), *and let C be a contact-free cycle intersecting $l^{()}$. Then $\mathrm{cl}(l^{()} \cap C)$ is a Cantor set on C* (*that is, a perfect, nowhere dense, closed subset of C*).

PROOF. According to Lemma 2.1, $l^{()} \cap C$ is a perfect set, and hence so is $\mathrm{cl}(l^{()} \cap C)$. The fact that $\mathrm{cl}(l^{()} \cap C)$ is nowhere dense follows immediately from the definition of an exceptional semitrajectory $l^{()}$. □

Since the Cantor set is perfect and nowhere dense, its complement consists of countably many open intervals called the *adjacent intervals* of the Cantor set.

THEOREM 3.5. *Let f^t be a flow on a closed orientable two-dimensional manifold $\mathcal{M}$. If the ω- (α-) limit set of a trajectory l contains a nontrivial recurrent trajectory $\widetilde{l}$ of f^t, then $\omega(l)$ ($\alpha(l)) = \mathrm{cl}(\widetilde{l})$.*

PROOF. For definiteness we assume that $\widetilde{l} \subset \omega(l)$, and we prove that $\omega(l) = \mathrm{cl}(\widetilde{l})$.

Since $\mathrm{cl}(\widetilde{l}) = \omega(\widetilde{l})$, it follows from $\widetilde{l} \subset \omega(l)$ that $\mathrm{cl}(\widetilde{l}) = \omega(\widetilde{l}) \subset \omega(l)$. It remains to prove that $\omega(l) \subset \mathrm{cl}(\widetilde{l})$.

If $\widetilde{l}$ is a locally dense trajectory, then $\mathrm{cl}(\widetilde{l})$ contains interior points to which l comes arbitrarily close. Consequently, there are points on l lying in $\mathrm{cl}(\widetilde{l}) = \omega(\widetilde{l})$. Therefore, $l \subset \omega(\widetilde{l})$ and $\omega(l) \subset \omega(\widetilde{l}) = \mathrm{cl}(\widetilde{l})$.

We now consider the case of an exceptional trajectory $\widetilde{l}$. If $l \subset \mathrm{cl}(\widetilde{l}) = \omega(\widetilde{l})$, then the theorem is proved, so we assume that $l \not\subset \mathrm{cl}(\widetilde{l})$. By Lemma 2.3, there exists a contact-free cycle C intersecting $\widetilde{l}$, and hence also l. According to Lemma 3.10, $\Omega \stackrel{\mathrm{def}}{=} \mathrm{cl}(\widetilde{l} \cap C)$ is a Cantor set. Therefore, the complement $C \setminus \Omega$ consists of countably many disjoint open intervals—the adjacent intervals of Ω. By our assumption that $l \not\subset \mathrm{cl}(\widetilde{l})$, the trajectory l intersects C in the adjacent intervals. We write the adjacent intervals $G_1, G_2, \dots$ in the order of their intersection with l, beginning from some fixed time as we move in the positive direction.

Since $\widetilde{l}$ intersects C in a set that is dense in Ω and since $\widetilde{l} \subset \omega(l)$, l intersects a countable family of distinct adjacent intervals (in particular, l is a nonclosed trajectory).

We denote by $P\colon C \to C$ the Poincaré mapping induced by f^t. Obviously, $l \cap C \subset \mathrm{Dom}(P)$. Since the number of components of the domain $\mathrm{Dom}(P)$ is finite

(Lemma 3.8), the number of adjacent intervals containing endpoints of components of $\mathrm{Dom}(P)$ is also finite.

We show that l cannot intersect an adjacent interval infinitely many times. Assume the contrary. Since l intersects a countable family of distinct adjacent intervals, our assumption implies that there is a countable family of adjacent intervals $G_{i_1}, G_{i_2}, \ldots$ (not necessarily distinct) with the following property: each G_{i_k} contains at least two points $x'_{i_k}, x''_{i_k} \in l \cap C$ such that $P(x'_{i_k})$ and $P(x''_{i_k})$ belong to different adjacent intervals G'_{i_k} and G''_{i_k} (Figure 2.50), and the intervals $G_{i_1}, G_{i_2}, \ldots$ are distinct. The existence of the families $\{G_{i_1}, G_{i_2}, \ldots\}$ and $\{G'_{i_1}, G'_{i_2}, \ldots\}$ means either the existence of an adjacent interval containing countably many endpoints of components of $\mathrm{Dom}(P)$, or the existence of a countable family of distinct adjacent intervals, each containing an endpoint of a component of $\mathrm{Dom}(P)$. This contradicts the finiteness of the number of components of $\mathrm{Dom}(P)$, and the contradiction proves that l cannot intersect an adjacent interval infinitely many times.

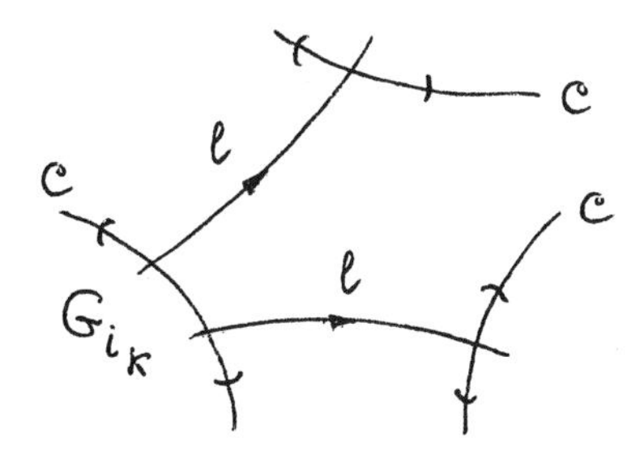

FIGURE 2.50

We can prove similarly that the adjacent intervals $G_{n_0}, G_{n_0+1}, \ldots$ are distinct beginning with some index n_0.

This and the fact that only finitely many adjacent intervals contain endpoints of components of $\mathrm{Dom}(P)$ imply the existence of an index $i_0 \geq n_0$ such that the adjacent intervals $G_{i_0}, G_{i_0+1}, \ldots$ (which are successively intersected by l) are distinct and belong to $\mathrm{Dom}(P)$. It is clear that $P(G_j) = G_{j+1}$ for $j \geq i_0$. It can be assumed without loss of generality that $i_0 = 1$ and $G_1 \neq G_i$ for $i \geq 2$ (that is, l does not intersect G_1 after a certain fixed time moment). The last point where l intersects G_1 as time increases is denoted by x_0.

The equality $P(G_i) = G_{i+1}$ $(i \geq 1)$ implies that the set $\mathfrak{Q} \stackrel{\text{def}}{=} \cup l^+(m)$, $m \in G_1$, is homeomorphic to the strip $G_1 \times [0, \infty)$. Since $l^+(x_0)$ intersects each adjacent interval $G_1, G_2, \ldots$ at only one point, the ω-limit set of the semitrajectory $l^+(x_0)$ does not lie in $\mathfrak{Q}$.

Let $x \in \omega(l^+(x_0)) = \omega(l)$. Since $\mathfrak{Q}$ is homeomorphic to $G_1 \times [0, \infty)$, and since $l^+(x_0) \subset \mathfrak{Q}$ and $x \notin \mathfrak{Q}$, there is a sequence of points m_k, $k \in \mathbb{N}$, belonging to the boundary $\partial\mathfrak{Q}$ and tending to x as $k \to \infty$. All the trajectories in $\partial\mathfrak{Q}$ belong to $\omega(\widetilde{l})$ because $\widetilde{l} \cap C$ is dense in Ω. Therefore, $m_k \in \omega(\widetilde{l})$, and hence $x \in \omega(\widetilde{l})$. This proves the required inclusion $\omega(l) \subset \omega(\widetilde{l})$. □

3.8. Catalogue of limit sets. In this subsection we describe all the ω-limit sets of individual trajectories of a flow. The same description applies in the case of α-limit sets, of course.

THEOREM 3.6. *On a closed orientable surface let f^t be a flow with finitely many equilibrium states and finitely many separatrices, and let $\omega(l)$ be the ω-limit set of a trajectory l of f^t. Then one of the following possibilities is realized:*

1) *$\omega(l)$ is a single equilibrium state (Figure 2.51 a));*

2) *$\omega(l)$ is a single closed trajectory (Figure 2.51 b));*

3) *$\omega(l)$ is a single one-sided contour (Figure 2.51 c));*

4) *$\omega(l)$ is a quasiminimal set containing only trajectories of the type 1)–4) in Theorem 3.4, any nontrivial recurrent semitrajectory in $\omega(l)$ is dense in $\omega(l)$, and, moreover, either a) each nontrivial recurrent semitrajectory in $\omega(l)$ is locally dense, or b) each nontrivial recurrent semitrajectory in $\omega(l)$ is exceptional.*

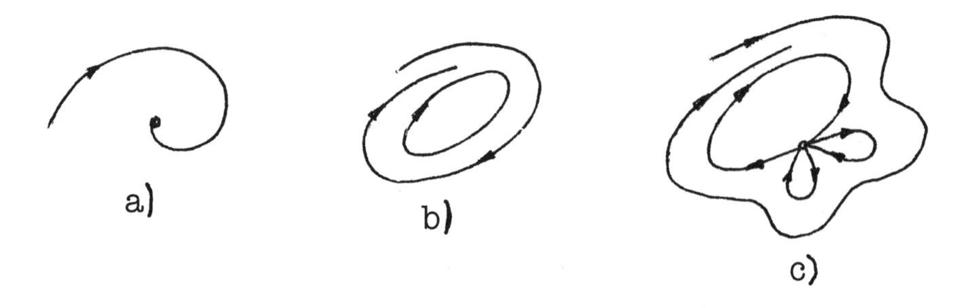

FIGURE 2.51

PROOF. If $\omega(l)$ consists solely of equilibrium states, then $\omega(l)$ is a single equilibrium state because an ω-limit set is connected (Lemma 1.5).

If $\omega(l)$ contains a closed trajectory, then $\omega(l)$ is a single closed trajectory according to Lemma 1.6.

The possibilities 1) and 2) hold if l is an equilibrium state or a closed trajectory, respectively.

It remains to consider the case when l is nonclosed and $\omega(l)$ contains a regular point on a nonclosed trajectory, that is, $\omega(l)$ contains a nonclosed trajectory.

If the ω- and α-limit sets of any nonclosed trajectory in $\omega(l)$ consist solely of equilibrium states, then it follows from Lemmas 3.4 and 3.5 that $\omega(l)$ is a single one-sided contour.

For a nonclosed trajectory $\widetilde{l} \subset \omega(l)$ suppose that the set $\omega(\widetilde{l}) \cup \alpha(\widetilde{l})$ contains a regular point. By the Maĭer theorem (Theorem 2.2), $\widetilde{l}$ is a P^+ or P^- nontrivial recurrent trajectory. The inclusion $\widetilde{l} \subset \omega(l)$ and the closedness of an ω-limit set (Lemma 1.5) imply that $\mathrm{cl}(\widetilde{l}) \subset \omega(l)$. According to Cherry's theorem (Theorem 2.1), $\mathrm{cl}(\widetilde{l})$ contains a nontrivial recurrent trajectory $\widehat{l}$. Therefore, $\widehat{l} \subset \omega(l)$. By Theorem 3.5, $\omega(l) = \mathrm{cl}(\widehat{l})$; that is, $\omega(l)$ is a quasiminimal set.

Let l_1^+ (l_1^-) be a P^+ (P^-) nontrivial recurrent semitrajectory in the quasiminimal set $\omega(l) = \mathrm{cl}(\widehat{l})$. Since the equalities $\mathrm{cl}(\widehat{l}) = \omega(\widehat{l}) = \alpha(\widehat{l})$ hold for the recurrent trajectory $\widehat{l}$, l_1^+ (l_1^-) is contained in the limit set of $\widehat{l}$. According to Theorem 2.3, $\widehat{l}$ is contained in the ω- (α-) limit set of the semitrajectory l_1^+ (l_1^-). Consequently, l_1^+ (l_1^-) is dense in $\omega(l)$.

Again using Theorem 2.3, we get the assertions 4, a) and 4, b). □

The example in §3.3 of Chapter 1 of a flow on the sphere shows that Theorem 3.6 is false without the assumption that there are finitely many equilibrium states.

3.9. Catalogue of minimal sets. In this subsection we present all possible minimal sets of flows on closed surfaces.

THEOREM 3.7. *On a closed orientable surface $\mathcal{M}$ let f^t be a flow with finitely many equilibrium states and finitely many separatrices, and let N be a minimal set of f^t. Then one of the following possibilities is realized:*

1) *N is a single equilibrium state;*

2) *N is a single closed trajectory;*

3) *N consists of nontrivial recurrent trajectories, each dense in N, and either* a) *N coincides with the whole of $\mathcal{M}$, and $\mathcal{M}$ is the torus T^2 in this case, or* b) *N is nowhere dense in $\mathcal{M}$, consists of exceptional nontrivial recurrent trajectories, and is locally homeomorphic to the direct product of a closed bounded interval and the Cantor set.*

PROOF. By Lemma 1.7, a minimal set coincides with the ω- (α-) limit set of any of its trajectories. Therefore, the possible types of N are included in the list 1)–4) in Theorem 3.6.

An equilibrium state and a closed trajectory are minimal sets.

A one-sided contour contains an equilibrium state, which is a nonempty closed invariant subset, and hence it is not a minimal set.

If N is a quasiminimal set, then by minimality it must not contain equilibrium states, and hence separatrices. Therefore, in the last case N consists of nontrivial recurrent (in both directions) trajectories, each dense in N.

According to Lemma 1.7, either N is nowhere dense in $\mathcal{M}$, or $N = \mathcal{M}$. If N is nowhere dense, then by Theorem 3.6, the quasiminimal and minimal set N consists solely of exceptional nontrivial recurrent trajectories. It follows from Lemma 3.10 that N is locally homeomorphic to the product of a closed bounded interval and the Cantor set. If $N = \mathcal{M}$, then by minimality the flow f^t on $\mathcal{M}$ does not have equilibrium states. According to Theorem 4.3 in Chapter 1, the sum of the indices of the equilibrium states is equal to the Euler characteristic $\chi(\mathcal{M}) = 2 - 2p$, where p is the genus of the surface. In the case $N = \mathcal{M}$ we get that $\chi(\mathcal{M}) = 2 - 2p = 0$, so that $p = 1$. Consequently, the closed surface $\mathcal{M}$ is the two-dimensional torus. □

A flow on a manifold M is said to be minimal if M itself is a minimal set of the flow. According to Theorem 3.7, the torus is the unique compact orientable surface admitting minimal flows (for example, irrational flows). The analogue of minimal flows for surfaces of higher genus is provided by flows in which every one-dimensional trajectory (that is, every trajectory different from a fixed point) is dense. Following Gardiner, we call such flows *highly transitive.*

Examples of highly transitive flows on genus-2 surfaces were constructed in §2 of Chapter 1.

§4. Quasiminimal sets

Of all the limit sets of individual trajectories the most complicated are quasiminimal sets. These are the only limit sets containing a continuum of trajectories. Quasiminimal sets in Denjoy and Cherry flows (see §3, Chapter 1) give an idea of their possible structure. In such flows a quasiminimal set in a neighborhood of a regular point has the structure of the product of a closed bounded interval and the Cantor set, and is nowhere dense on the surface. A quasiminimal set of a transitive flow coincides with the whole manifold. Theorem 3.6 lists all possible trajectories that can lie (or necessarily lie) in a quasiminimal set. In this section we study its structure in greater detail.

4.1. An estimate of the number of quasiminimal sets. Let f^t be a flow on a closed orientable two-dimensional manifold $\mathcal{M}$, and suppose that a quasiminimal set N_i of f^t is the topological closure of a nontrivial recurrent trajectory l_i $(i = 1, 2)$. If $N_1 \subset N_2$, then since l_2 is dense in N_2 (Theorem 3.6), the trajectory l_1 lies in the limit set of l_2. According to Theorem 2.3, l_2 lies in the limit set of l_1, and therefore $N_1 = N_2$. Thus, of any two distinct quasiminimal sets of a flow one does not contain the other as a proper subset.

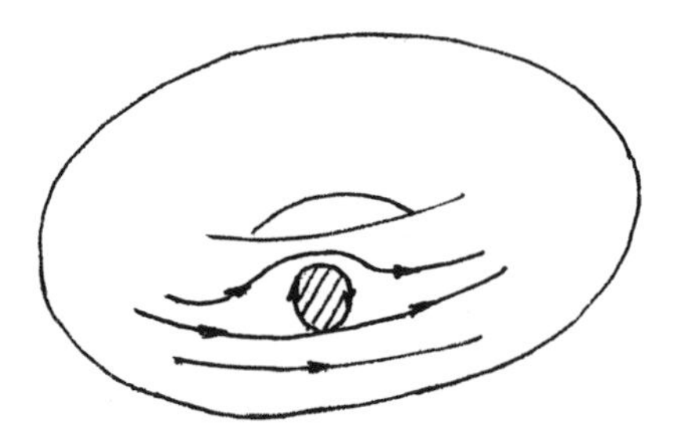

FIGURE 2.52

Quasiminimal sets can intersect. We give a schematic description of such a flow. Starting from an irrational winding on the torus, we construct a transitive flow f_1^t on a torus T_1 with a hole, where ∂T_1 consists of an equilibrium state and a separatrix loop (Figure 2.52). We take the same flow f_2^t on a torus T_2 with a hole, and identify ∂T_1 with ∂T_2 by means of a suitable homeomorphism (in particular, this homeomorphism must paste together the equilibrium states). The result is a pretzel (a closed orientable surface of genus 2) on which the flows f_1^t and f_2^t form a flow f^t with two quasiminimal sets, which intersect in an equilibrium state and a separatrix loop (Figure 2.53).

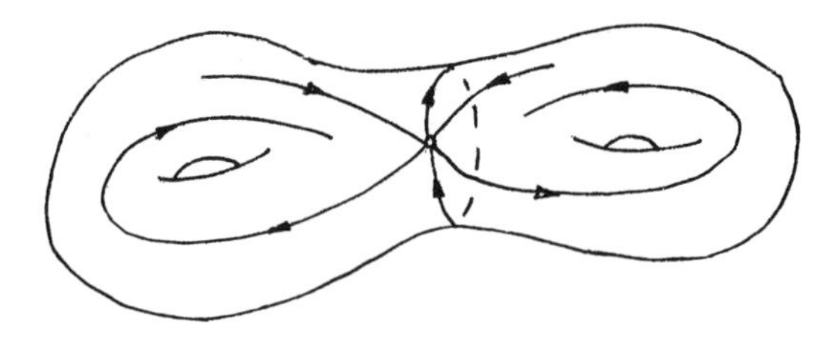

FIGURE 2.53

THEOREM 4.1. *Let f^t be a flow on a closed orientable two-dimensional manifold $\mathcal{M}$. Then:*

1) *any nontrivial recurrent semitrajectory or trajectory belongs to exactly one quasiminimal set of f^t;*

2) *the number $q(f^t)$ of quasiminimal sets of f^t is finite and does not exceed the genus of $\mathcal{M}$;*

3) *two quasiminimal sets can intersect only in equilibrium states and separatrices going from an equilibrium state to an equilibrium state.*

PROOF. According to Theorem 3.6, 4), each nontrivial recurrent semitrajectory is dense in the quasiminimal set containing it. Therefore, such a semitrajectory belongs to exactly one quasiminimal set, and two quasiminimal sets cannot intersect in a nontrivial recurrent semitrajectory. It then follows from Theorem 3.4 that two quasiminimal sets can intersect only in equilibrium states and separatrices joining equilibrium states. The assertions 1) and 3) are proved.

We prove 2). In each quasiminimal set we take a nontrivial recurrent semitrajectory. It follows from 3) that these semitrajectories are independent. According to Theorem 2.4, the number of independent nontrivial recurrent semitrajectories is finite and does not exceed the genus p of the surface $\mathcal{M}$. Therefore, $q(f^t) \leq p$. □

Remarks. The estimates of Aranson, Markley, and Levitt. Aranson's estimate ([**8**], 1969; reproved by Markley in 1970 [**96**]) of the number of independent nontrivial recurrent semitrajectories for flows on closed nonorientable surfaces implies the estimate

$$q(f^t) \leq \left[\frac{p-1}{2}\right]$$

for the number $q(f^t)$ of quasiminimal sets of flows on such surfaces, where p is the genus of the surface, and $[x]$ is the integer part of a number x. This estimate cannot be improved; that is, on a closed nonorientable surface of genus $p \geq 1$ there is a flow with exactly $[\frac{p-1}{2}]$ quasiminimal sets. In particular, there are no flows with quasiminimal sets on the projective plane and the Klein bottle.

In his dissertation [**93**] Levitt considered foliations $\mathcal{F}$ on a compact orientable surface $\mathcal{M}$ (with boundary in general) that have singularities only of saddle type, including thorns, and he obtained the estimate

$$q(\mathcal{F}) \leq \left[\frac{3p+b+e-2}{2}\right]$$

for the number $q(\mathcal{F})$ of quasiminimal sets of such foliations, where p is the genus of $\mathcal{M}$, b is the number of boundary components, and e is the number of thorns. We note that the existence of at least one singularity is assumed for a foliation on the torus T^2.

4.2. A family of special contact-free cycles.

LEMMA 4.1. *On a closed orientable two-dimensional manifold $\mathcal{M}$ let f^t be a flow, and suppose that $N_1, \dots, N_k$ are (distinct) quasiminimal sets of f^t. Then there exists a family of disjoint contact-free cycles $C_1, \dots, C_k$ such that: 1) $C_i \cap N_i \neq \emptyset$, $i = 1, \dots, k$; 2) $C_i \cap N_j = \emptyset$ for $i \neq j$, $i, j \in \{1, \dots, k\}$; 3) for any $i = 1, \dots, k$ either $C_i \cap N_i = C_i$ or $C_i \cap N_i$ is a Cantor set.*

PROOF. In the quasiminimal set N_i we take a nontrivial recurrent trajectory l_i, $i = 1, \dots, k$. Since $l_1, \dots, l_k$ are independent, there exist contact-free segments $\Sigma_1, \dots, \Sigma_k$ with the following properties: a) $\Sigma_i \cap l_i \neq \emptyset$ for $i = 1, \dots, k$; b) $\Sigma_i \cap l_j = \emptyset$ for $i \neq j$, $i, j \in \{1, \dots, k\}$; c) if l_i is locally dense, then $\Sigma_i = \mathrm{cl}(\Sigma_i \cap l_i)$, that is, $\Sigma_i \cap l_i$ is dense in Σ_i.

By a), there is a Σ_i-arc $\overset{\frown}{p_i q_i}$ of the trajectory l_i for each index $i = 1, \dots, k$. Then the family of simple closed curves $\widehat{C}_i = \overset{\frown}{p_i q_i} \cup \overline{p_i q_i}$ $(i = 1, \dots, k)$, where $\overline{p_i q_i} \subset \Sigma_i$, satisfies the following conditions: I) $\widehat{C}_i \cap N_i \neq \emptyset$ for $i = 1, \dots, k$; II) $\widehat{C}_i \cap N_j = \emptyset$ for $i \neq j$, $i, j \in \{1, \dots, k\}$. This and the proof of Lemma 1.2 give us that each curve $\widehat{C}_i$ can be approximated by a contact-free cycle C_i in such a way that the family of contact-free cycles $C_1, \dots, C_k$ satisfies the assertions 1) and 2). Moreover, if a nontrivial recurrent trajectory is locally dense, then by c) the corresponding contact-free cycle can be constructed so that $C_i = \mathrm{cl}(C_i \cap l_i)$, and hence so that $C_i = C_i \cap N_i$. If a nontrivial recurrent trajectory l_i is exceptional, then by the equality $N_i = \mathrm{cl}(l_i)$ and Lemma 3.10, the set $N_i \cap C_i = \mathrm{cl}(l_i \cap C_i)$ is a Cantor set. $\square$

DEFINITION. Let $N_1, \dots, N_k$ be quasiminimal sets of a flow f^t. A family of disjoint contact-free cycles $C_1, \dots, C_k$ of f^t satisfying Lemma 4.1 is called a *special family*. If $k = 1$, then a contact-free cycle C_1 constructed for the quasiminimal set N_1 and satisfying the assertion 3) of Lemma 4.1 is called a *special contact-free cycle.*

According to Lemma 4.1, if f^t has nontrivial recurrent trajectories, then there is a special family of contact-free cycles for f^t.

4.3. Partition of a contact-free cycle. Let f^t be a flow on a compact orientable surface $\mathcal{M}$, and let N be a quasiminimal set of f^t. According to Lemma 2.3, there exists a contact-free cycle C intersecting at least one nontrivial recurrent semitrajectory lying in N, and hence all such semitrajectories. We define a partition ξ_N of C into closed disjoint subsets (elements).

DEFINITION. The elements of the partition ξ_N of the contact-free cycle C are the closures of the intervals in $C \setminus N$ and the points not belonging to these intervals.

We remark that in view of Lemma 2.1 the intersection $N \cap C$ is a perfect set; that is, each point in $N \cap C$ is not isolated. From this it follows that the closures of any two distinct intervals in $C \setminus N$ are disjoint. Therefore, the elements of ξ_N are disjoint closed subsets of C.

EXAMPLES. 1) Let f^t be an irrational winding on the torus T^2, and let C be an arbitrary contact-free cycle for f^t. Then $N = T^2$, and all the elements of ξ_N are one-point sets.

2) By using an irrational winding it is possible to construct a flow f^t on T^2 with a quasiminimal set N homeomorphic to T^2 with finitely many disks removed. There exists a contact-free cycle C for f^t such that $C \setminus N$ contains finitely many intervals (Figure 2.54). Thus, ξ_N consists of finitely many closed intervals and the points of the contact-free cycle C not belonging to these closed intervals.

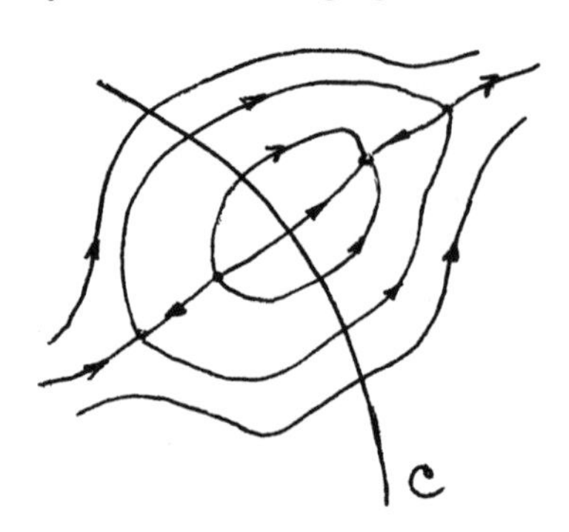

FIGURE 2.54

3) Let f^t be a Denjoy flow on T^2, and let C be a contact-free cycle for f^t. Then $C \cap N$ is a Cantor set, where N is the unique quasiminimal and minimal set of the Denjoy flow. The partition ξ_N contains a countable family of closed intervals (the closures of the adjacent intervals of the Cantor set $C \cap N$) and a continuum of points not belonging to the closures of the adjacent intervals and forming a nowhere dense subset of C.

By Lemma 4.1, for any quasiminimal set N there is a contact-free cycle C (a special contact-free cycle) such that the partition ξ_N either is the partition in example 1) (that is, the partition into one-point subsets), or is the partition in example 3).

4.4. The Gardiner types of partition elements. Suppose that a flow f^t on a compact orientable surface $\mathcal{M}$ has a contact-free cycle C and induces a Poincaré mapping $P\colon C \to C$ with domain $\mathrm{Dom}(P)$.

We first define the mappings P_l and P_r on the endpoints of the components of $\mathrm{Dom}(P)$. Let $(a, b) \subset \mathrm{Dom}(P)$ be a component of $\mathrm{Dom}(P)$. According to Lemma 3.7, the semitrajectory $l^+(a)$ is an ω-separatrix, and there exists a sequence of separatrices $l_{10} = l^+(a), l_{11}, \dots, l_{1k_1}$ such that l_{1s} is a Bendixson extension of $l_{1\,s-1}$ to the right ($s = 1, \dots, k_1$), the separatrices l_{1i} ($i = 0, \dots, k_1 - 1$) do not intersect C, and l_{1k_1} does intersect C (Figure 2.55; the arrow along C indicates the positive direction on the curve). Denote by $\overline{a}$ the first point where l_{1k_1} intersects C as time increases. We define P_r on all the left-hand endpoints of the components of $\mathrm{Dom}(P)$ by setting $P_r(a) = \overline{a} \in C$.

Using Lemma 3.7, we define P_l on all the right-hand endpoints of the components of $\mathrm{Dom}(P)$ similarly (see Figure 2.55).

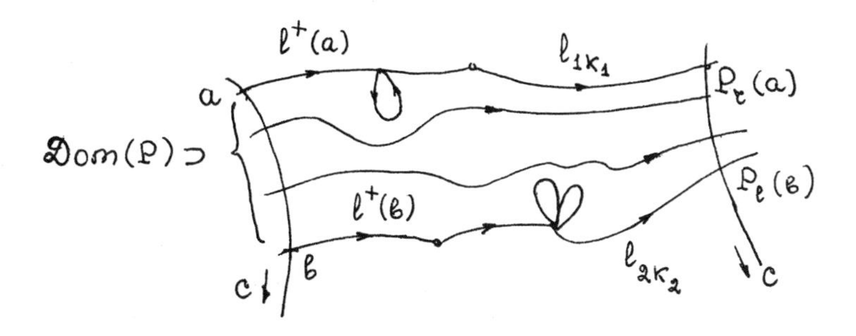

FIGURE 2.55

We remark that some points $x_0 \in C$ can be simultaneously left- and right-hand endpoints of different components of $\mathrm{Dom}(P)$, and both $P_l(x_0) = P_r(x_0)$ and $P_l(x_0) \neq P_r(x_0)$ are possible (Figure 2.56).

If $m \in C$ lies in $\mathrm{Dom}(P)$, then we set $P_l(m) = P_r(m) = P(m)$.

Suppose now that f^t has a quasiminimal set N, and C is a contact-free cycle for f^t that intersects N. We write an arbitrary element η of the partition ξ_N as an interval $[\eta_l, \eta_r]$, with the understanding that $\eta_l = \eta_r$ if η is a one-point set. Since nontrivial recurrent trajectories in N are dense, such trajectories intersect C at points arbitrarily close to $\eta_l \in N$ on the left. Nontrivial recurrent trajectories intersect C at points in $\mathrm{Dom}(P)$. According to Lemma 3.8, the number of components of $\mathrm{Dom}(P)$ is finite, and hence two cases are possible: 1) η_l is the right-hand

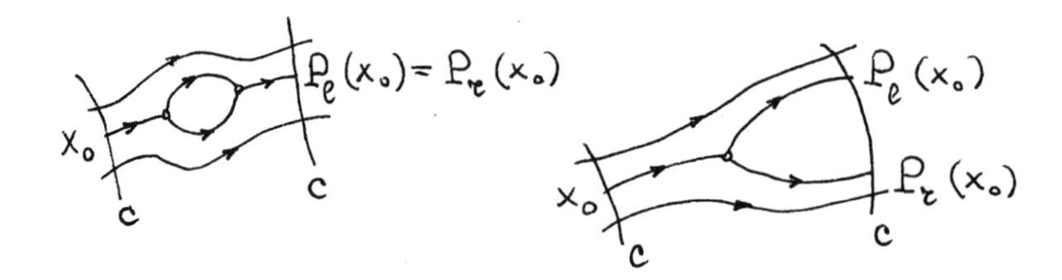

FIGURE 2.56

endpoint of some component of Dom(P) (Figure 2.57); or 2) $\eta_l \in$ Dom(P). In both cases $P_l(\eta_l)$ is defined, and in view of the orientability of $\mathcal{M}$ and the theorem on continuous dependence of trajectories on the initial conditions, the nontrivial recurrent trajectories in N intersect C at points arbitrarily close to $P_l(\eta_l)$ on the left. Therefore, $P_l(\eta_l)$ is the left-hand endpoint of an element of ξ_N.

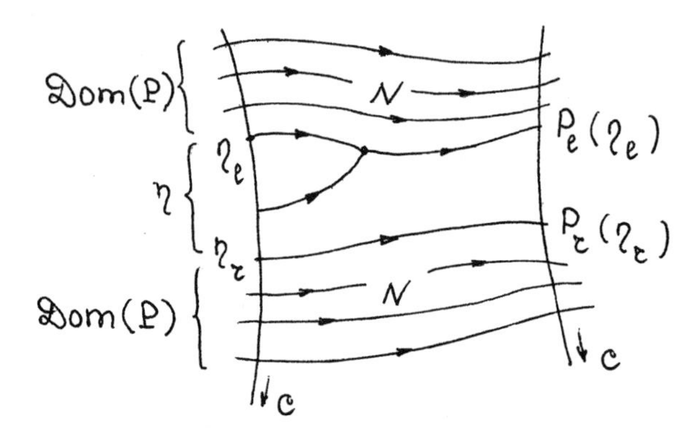

FIGURE 2.57

An analogous definition applies to the point $P_r(\eta_r)$, which is the right-hand endpoint of some element of ξ_N.

Following [**52**], we give the

DEFINITION. An element $\eta = [\eta_l, \eta_r] \in \xi_N$ is called an *element of type* 1 if the interval $[P_l(\eta_l), P_r(\eta_r)]$ (possibly a single point) is an element of ξ_N. All the remaining elements of the partition ξ_N are called *elements of type* 2.

Figures 2.58 and 2.59 show elements of types 1 and 2, respectively.

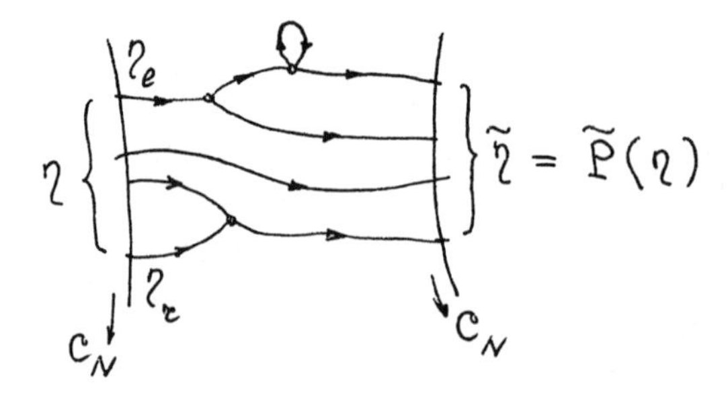

FIGURE 2.58

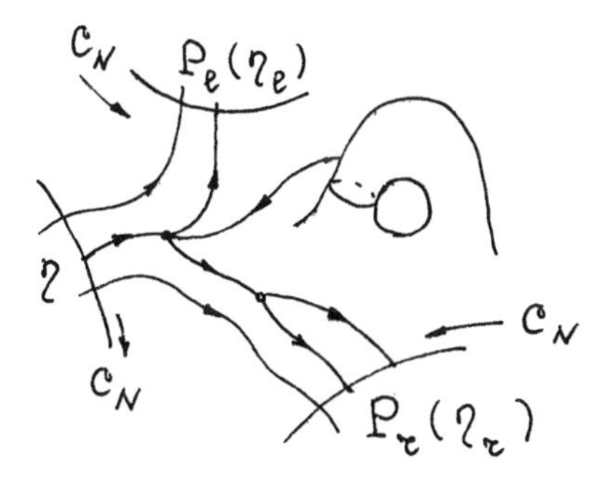

FIGURE 2.59

We examine more closely an element $\eta = [\eta_l, \eta_r] \in \xi_N$ of type 2. By definition, $P_l(\eta_l)$ and $P_r(\eta_r)$ belong to different elements of ξ_N, which we denote by η^{-1} and η^{+1}, respectively. Let $\eta^{-1} = [\eta_l^{-1}, \eta_r^{-1}]$ and $\eta^{+1} = [\eta_l^{+1}, \eta_r^{+1}]$. Since $\mathcal{M}$ is orientable, $\eta_l^{-1} = P_l(\eta_l)$ and $\eta_r^{+1} = P_r(\eta_r)$. It follows from the invariance of a quasi-minimal set N and the denseness in N of nontrivial recurrent trajectories that there exist elements $\eta^{-2} = [\eta_l^{-2}, \eta_r^{-2}]$, $\eta^{+2} = [\eta_l^{+2}, \eta_r^{+2}] \in \xi_N$ such that $P_l(\eta_l^{+2}) = \eta_l^{+1} \in \eta^{+1}$ and $P_r(\eta_r^{-2}) = \eta_r^{-1} \in \eta^{-1}$ (Figure 2.60).

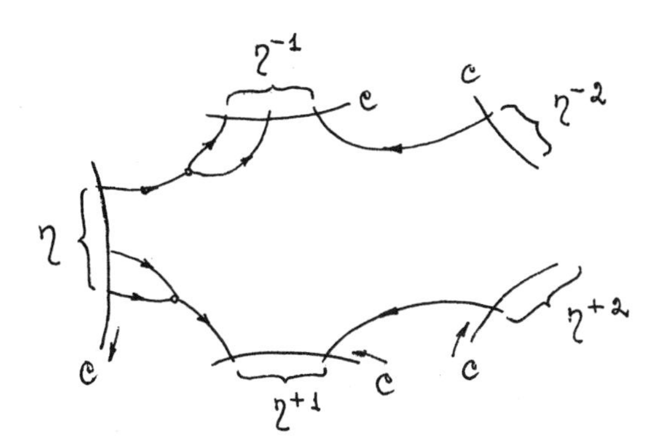

FIGURE 2.60

We show that $\eta \neq \eta^{+2}$. Assume the contrary. Then $P_l(\eta_l^{+2}) = P_l(\eta_l) = \eta_l^{+1} \in \eta^{+1}$. This and the fact that $P_r(\eta_r) \in \eta^{+1}$ give us that η is an element of type 1. The contradiction proves that $\eta \neq \eta^{+2}$.

The inequality $\eta \neq \eta^{-2}$ is proved similarly.

Since $\eta \neq \eta^{+2}$, $\eta \neq \eta^{-2}$, and $\eta^{-1} \neq \eta^{+1}$, both the elements η^{-2} and η^{+2} have type 2.

The equality $\eta^{-2} = \eta^{+2}$ is possible, and in this case we get the *saddle set* $\eta^{-2}, \eta^{-1}, \eta, \eta^{+1}, \eta^{+2}$ of elements of ξ_N (Figure 2.61).

Suppose that $\eta^{-2} \neq \eta^{+2}$. Then there exist elements $\eta^{-3}, \eta^{+3} \in \xi_N$ such that $P_l(\eta_l^{-2}) = \eta_l^{-3} \in \eta^{-3}$ and $P_r(\eta_r^{+2}) = \eta_r^{+3} \in \eta^{+3}$. If $\eta^{-3} = \eta^{+3}$, then we get the saddle set $\eta^{-3}, \eta^{-2}, \eta^{-1}, \eta, \eta^{+1}, \eta^{+2}, \eta^{+3}$ (Figure 2.62).

If $\eta^{-3} \neq \eta^{+3}$, then there exist elements $\eta^{-4}, \eta^{+4} \in \xi_N$ such that $P_l(\eta_l^{+4}) = \eta_l^{+3} \in \eta^{+3}$ and $P_r(\eta_r^{-4}) = \eta_r^{-3} \in \eta^{-3}$.

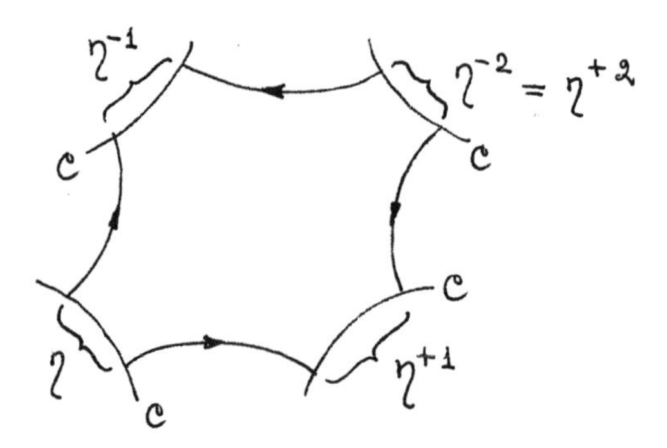

FIGURE 2.61

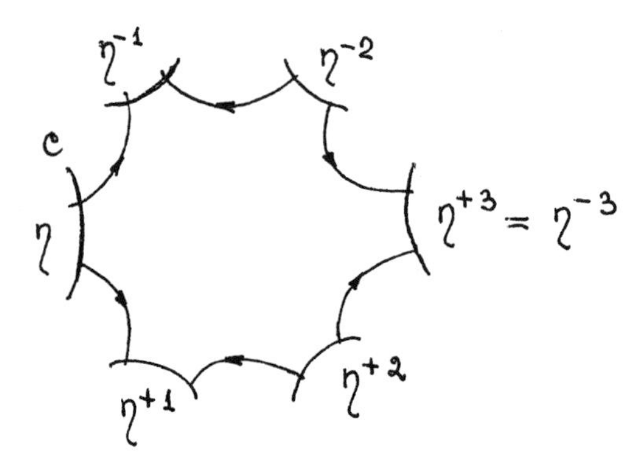

FIGURE 2.62

Similarly, the equality $\eta^{+2} = \eta^{+4}$ implies that $P_l(\eta_l^{+2}) = P_l(\eta_l^{+4}) = \eta_l^{+3} \in \eta^{+3}$ and $P_r(\eta_r^{+2}) = \eta_r^{+3} \in \eta$, which contradicts the fact that η^{+2} has type 2. Therefore, $\eta^{+2} \neq \eta^{+4}$. If we assume that $\eta = \eta^{+4}$, then $\eta^{-2} = \eta^{+2}$ (but we are considering the case $\eta^{-2} \neq \eta^{+2}$), and therefore $\eta \neq \eta^{+4}$. The inequalities $\eta \neq \eta^{-4}$ and $\eta^{-2} \neq \eta^{-4}$ are proved in the same way. It can again be shown as before that η^{-4} and η^{+4} are elements of type 2.

If $\eta^{-4} \neq \eta^{+4}$, then we continue the process described above. Let us show that this process is finite, that is, that there is a $k \in \mathbb{N}$ such that $\eta^{-k} = \eta^{+k}$. Assume not. Then there is an infinite sequence

$$(*) \qquad \ldots, \eta^{-k}, \ldots, \eta^{-2}, \eta^{-1}, \eta, \eta^{+1}, \eta^{+2}, \ldots, \eta^{+k}, \ldots$$

of elements of ξ_N constructed by the process above. It can be proved by induction that the elements with even superscripts are of type 2 (we leave it as an exercise for the reader to supply the details). If $\eta^{+2i} = \eta^{+2j}$ (for definiteness let $i < j$), then by the simple relations $2i - (i + j) = i - j$ and $2j - (i + j) = -(i - j)$ we get that $\eta^{i-j} = \eta^{-(i-j)}$, which contradicts the assumption that the sequence $(*)$ is infinite. An analogous contradiction follows from the equalities $\eta^{-2i} = \eta^{-2j}$ and $\eta^{-2j} = \eta^{+2i}$, $i, j \in \mathbb{N}$. Thus, the elements η^{2s}, $s \in \mathbb{N}$, of type 2 in $(*)$ are distinct.

We show that there are finitely many elements of type 2 in ξ_N. Indeed, suppose that $\eta = [\eta_l, \eta_r] \in \xi_N$ is an element of type 2. Since nontrivial recurrent trajectories are dense in N, each of η_l and η_r either belongs to an open Gutierrez equivalence class, or lies on the boundary of an open Gutierrez equivalence class (see §3.6). Denote by $O(\eta_l)$ (respectively, $O(\eta_r)$) an open Gutierrez equivalence class that either is adjacent to the point η_l (η_r) or contains η_l (η_r).

We take points $a \in O(\eta_l)$ and $b \in O(\eta_r)$ through which there pass nontrivial recurrent trajectories in the set N. Then $[a, b]$ contains the interval $[\eta_l, \eta_r]$. Assume that $O(\eta_l) = O(\eta_r)$; that is, a and b belong to a single open Gutierrez equivalence class. By the definition of Gutierrez equivalence, the closed curve $[a, b] \cup \overset{\frown}{aP}(a) \cup \overset{\frown}{bP}(b) \cup [P(a), P(b)]$ bounds a disk $\mathfrak{D}$. Since η is an element of type 2, there are points in N on $[P(a), P(b)]$, and hence $[P(a), P(b)]$ must intersect nontrivial recurrent trajectories in N (Figure 2.63). Entering $\mathfrak{D}$ with decreasing time, these trajectories cannot leave $\mathfrak{D}$ as $t \to -\infty$, and this contradicts Lemma 2.4, according to which nontrivial recurrent semitrajectories cannot lie entirely in $\mathfrak{D}$. The contradiction proves that $O(\eta_l) \neq O(\eta_r)$. The fact that there are finitely many elements of type 2 now follows from the fact that there are finitely many open Gutierrez equivalence classes (Lemma 3.9).

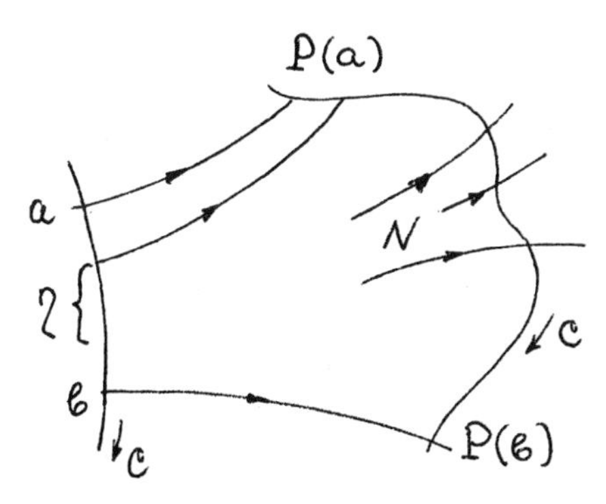

FIGURE 2.63

Thus, the sequence (∗) constructed by the process above is finite. The collection

$$\eta^{-k}, \dots, \eta^{-1}, \eta, \eta^{+1}, \dots, \eta^{+k}$$

of elements obtained is called a *saddle set*. The index of this saddle set is equal to $1 - k$. Figures 2.61 and 2.62 picture saddle sets of indices -1 and -2, respectively.

We remark that the elements of type 2 with even superscripts in a saddle set are all distinct and different from η, with the possible exception of the extreme ones η^{-k} and η^{+k}. This is not so in general for elements with odd superscripts (see Figure 2.64).

We sum up the arguments of this subsection in the form of a lemma.

LEMMA 4.2. *Let ξ_N be the partition of a contact-free cycle C. Then:*
1) *there are finitely many elements of type 2 in ξ_N;*
2) *any saddle set in ξ_N is finite;*
3) *in any saddle set*

$$\eta^{-k}, \dots, \eta^{-1}, \eta = \eta^0, \eta^{+1}, \dots, \eta^{+k}$$

all the elements with even superscripts are elements of type 2, and are distinct with the possible exception of the extreme elements η^{-k} and η^{+k};

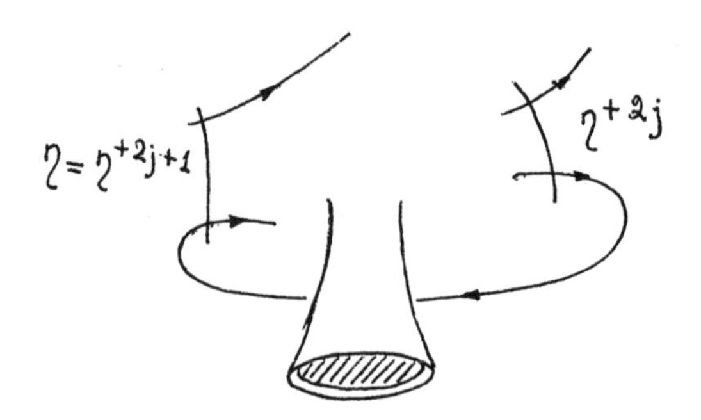

FIGURE 2.64

4) *each endpoint* η_l *or* η_r *of any element* $[\eta_l, \eta_r] \in \xi_N$ *is an accumulation point for the points lying in the intersection with* C *of the nontrivial recurrent trajectories in the quasiminimal set* N;

5) *each endpoint* η_l *or* η_r *of any element* $[\eta_l, \eta_r] \in \xi_N$ *either belongs to an open Gutierrez equivalence class or is a boundary point of an open Gutierrez equivalence class.*

4.5. The structure theorem. In this subsection we study the global structure of a flow on a quasiminimal set. We first give some definitions.

Let $f\colon \mathcal{M} \to \mathcal{M}$ be a transformation of a manifold $\mathcal{M}$ into itself with domain $\mathrm{Dom}(f) \subset \mathcal{M}$. Assume that all the iterates $f^n(z)$, $n \in \mathbb{Z}$, are defined for some point $z \in \mathcal{M}$; that is, $z \in \mathrm{Dom}(f^n)$ for $n \in \mathbb{Z}$ (we set $f^0 = \mathrm{id}$). The set $O(z) = \{f^n(z), n \in \mathbb{Z}\}$ is called the *nonsingular orbit* of z with respect to the transformation f.

DEFINITION. A transformation $f\colon \mathcal{M} \to \mathcal{M}$ is said to be *minimal* if it has at least one nonsingular orbit, and each nonsingular orbit of f is dense in $\mathcal{M}$.

We now assume that $\mathcal{M}$ is orientable.

DEFINITION. A transformation $f_1\colon \mathcal{M} \to \mathcal{M}$ is *topologically semiconjugate* to a transformation $f_2\colon \mathcal{M} \to \mathcal{M}$ if there exists a continuous orientation-preserving mapping $h\colon \mathcal{M} \to \mathcal{M}$ such that $h \circ f_1(z) = f_2 \circ h(z)$ at all points $z \in \mathrm{Dom}(f_1)$.

It follows from the definition that $h[\mathrm{Dom}(f_1)] \subset \mathrm{Dom}(f_2)$.

If in this definition we require that h be a homeomorphism, then the transformations f_1 and f_2 are *topologically conjugate.*

We say that f_1 is topologically semiconjugate (conjugate) to f_2 by means of h.

THEOREM 4.2 (structure theorem). *Let* f^t *be a flow on a closed orientable two-dimensional manifold* $\mathcal{M}$, *and let* $N_1, \dots, N_k$ *be quasiminimal sets of* f^t. *Then there exist disjoint open connected sets* $V_1, \dots, V_k$ *and contact-free cycles* $C_1, \dots, C_k$ *for* f^t *such that the following conditions hold:*

1) *either* $C_i \cap N_i = C_i$ *or* $C_i \cap N_i$ *is a Cantor set in* C_i, $i = 1, \dots, k$, *and* $C_i \cap N_j = \emptyset$ *for* $i \neq j$ *(a special family of contact-free cycles)*;

2) *if* $P_i\colon C_i \to C_i$ *is the Poincaré mapping induced on* C_i *by* f^t $(i = 1, \dots, k)$, *then*

a) P_i *is topologically conjugate to a minimal exchange of open intervals* $T_i\colon S^1 \to S^1$ *in the case* $C_i \cap N_i = C_i$, *and*

b) P_i *is topologically semiconjugate to a minimal exchange of open intervals* $T_i\colon S^1 \to S^1$ *by means of a continuous mapping* $h_i\colon C_i \to S^1$ *in the case when*

$C_i \cap N_i$ is a Cantor set, where the closure of each adjacent interval of $C_i \cap N_i$ is carried by h_i into a point, and the restriction of h_i to the complement of the closures of all the adjacent intervals is a homeomorphism onto its range;

3) *for each $i = 1, \dots, k$ the domain V_i contains C_i and all the nontrivial recurrent semitrajectories in the quasiminimal set N_i;*

4) *the boundary ∂V_i of V_i can be made up only of whole trajectories, equilibrium states, and finitely many contact-free segments, and, moreover, there do not exist arcs of trajectories with endpoints on ∂V_i and interior points in V_i.*

PROOF. According to Lemma 4.1, there exists a special family of contact-free cycles $C_1, \dots, C_k$ satisfying 1).

We fix an index $i \in \{1, \dots, k\}$ and prove 2). Let ξ_i be the partition of the contact-free cycle C_i. We recall that the elements of ξ_i are closures of intervals in the set $C_i \setminus N_i$ and points not belonging to these intervals.

Denote by ξ_i' the family of elements $\eta \in \xi_i$ such that the interval $[\eta_l, \eta_r]$ lies in a single open Gutierrez equivalence class. It follows from Lemma 4.2, 5) and the finiteness of the number of open Gutierrez equivalence classes (Lemma 3.9) that the complement $\xi_i \setminus \xi_i'$ has finitely many elements.

We remark that in general the complement $\xi_i \setminus \xi_i'$ can contain elements of type 1 in addition to elements of type 2 (Figure 2.65).

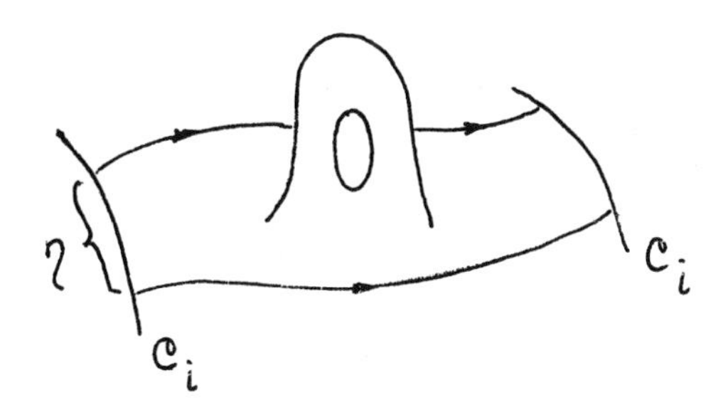

FIGURE 2.65

We define the mapping $\widetilde{P}_i \colon \xi_i' \to \xi_i$ as follows. For an element $\eta = [\eta_l, \eta_r] \in \xi_i'$ we set $\widetilde{P}_i(\eta) = [P_l(\eta_l), P_r(\eta_r)] \in \xi_i$, where $P \colon C_i \to C_i$ is the Poincaré mapping induced on C_i by f^t.

We introduce an equivalence relation $\sim(i)$ by regarding points belonging to a single element of ξ_i as equivalent. Then the quotient manifold $C_i/{\sim}(i)$ is homeomorphic to the circle S^1, and the natural projection $h_i \colon C_i \to C_i/{\sim}(i)$ is either a Cantor function[2] (if $C_i \cap N_i$ is a Cantor set), or a homeomorphism (if $C_i \cap N_i = C_i$ and all the elements of ξ_i are one-point sets).

The mapping $\widetilde{P}_i \colon \xi_i' \to \xi_i$ induces a transformation $T_i \colon S^1 \to S^1$ by means of the natural projection $h_i \colon C_i \to S^1 = C_i/{\sim}(i)$. This and the definition of $\widetilde{P}_i$ imply the commutative diagram

$$\begin{array}{ccc} C_i & \xrightarrow{P_i} & C_i \\ {\scriptstyle h_i}\downarrow & & \downarrow{\scriptstyle h_i} \\ S^1 & \xrightarrow{T_i} & S^1 \end{array}$$

[2]More precisely, a covering $\mathbb{R}^1 \to \mathbb{R}^1$ over h_i is a Cantor function.

at points belonging to elements in ξ_i'. Since the number of elements in $\xi_i \setminus \xi_i'$ is finite, T_i is undefined at only finitely many points. Since trajectories depend continuously on the initial conditions, T_i is a homeomorphism from its domain onto its range. Therefore, T_i is an exchange of open intervals.

The partition ξ_i contains one-point elements through which nontrivial recurrent trajectories pass. Consequently, nonsingular orbits of T_i pass through the points of the circle $C_i/\sim(i) = S^1$ that correspond to these elements. The denseness in N_i of any nontrivial recurrent semitrajectory lying in $\mathrm{N_i}$ implies the denseness in S^1 of any nonsingular orbit of T_i. Therefore, T_i is a minimal exchange of open intervals.

If $C_i \cap N_i = C_i$, then all the elements of ξ_i are one-point sets, and h_i is a homeomorphism. If $C_i \cap N_i$ is a Cantor set, then the complement $C_i \setminus (C_i \cap N_i)$ is a countable union of open intervals G_k, $k \in \mathbb{N}$. By definition, the topological closure $\operatorname{cl} G_k$ of each adjacent interval G_k is an element of ξ_i, and hence is mapped into a point under the action of h_i. Each point not in $\operatorname{cl} G_k$, $k \in \mathbb{N}$, is an element of ξ_i. Thus, h_i is one-to-one on the set $C_i \setminus \bigcup_{k=1}^{\infty} \operatorname{cl} G_k$. It follows from the continuous dependence of trajectories on the initial conditions that the restriction of h_i to $C_i \setminus \bigcup_{k=1}^{\infty} \operatorname{cl} G_k$ is a homeomorphism onto its range. The assertion 2) is proved.

It remains to construct open domains $V_1, \dots, V_k$ satisfying the conditions 3) and 4).

We again fix an index $i \in \{1, \dots, k\}$ and construct the domain V_i. Denote by $O_1, \dots, O_r$ the open Gutierrez equivalence classes on C_i with respect to the mapping P_i. By the definition of an open equivalence class, for any points $m_1, m_2 \in O_j$ $(j \in \{1, \dots, r\})$ there exist points $a, b \in O_j$ in $\operatorname{Dom}(P_i)$ such that $\overline{m_1 m_2} \subset \overline{ab} \subset O_j$, and the simple closed curve $\overline{ab} \cup \overset{\frown}{aP_i(a)} \cup \overset{\frown}{bP_i(b)} \cup \overline{P_i(a)P_i(b)}$ bounds an open disk $\mathcal{D}(a,b)$ on the manifold.

Let $\mathcal{D}(O_j) = \cup \mathcal{D}(a,b)$ and $O_j' = \cup \overline{P_i(a)P_i(b)}$, where in both expressions the union is over all points $a, b \in \operatorname{Dom}(P_i) \cap O_j$. According to Lemma 3.9, each open class O_j is an open interval (a_j, b_j). Therefore, O_j' is also an open interval (a_j', b_j'). We remark that $O_1', \dots, O_r'$ are open Gutierrez equivalence classes with respect to the mapping P_i^{-1}.

Each endpoint of the open equivalence class O_j is an endpoint of a component of $\operatorname{Dom}(P_i)$, so in view of Lemma 3.6 the semitrajectories $l^+(a_j)$ and $l^+(b_j)$ are ω-separatrices that intersect C_i only at a_j and b_j, respectively. Similarly, the semitrajectories $l^-(a_j')$ and $l^-(b_j')$ are α-separatrices, and $l^-(a_j') \cap C_i = \{a_j'\}$ and $l^-(b_j') \cap C_i = \{b_j'\}$.

Let $V(O_j) = O_j \cup O_j' \cup \mathcal{D}(O_j) \cup l^+(a_j') \cup l^+(b_j') \cup l^+(a_j) \cup l^+(b_j)$ (Figure 2.66), and let $\widetilde{V}_i = \bigcup_{j=1}^{r} V(O_j)$.

It follows from Theorem 3.6, 4) that the set $\widetilde{V}_i$ contains all nontrivial recurrent semitrajectories in the quasiminimal set N_i.

Two cases are possible: 1) $C_i = \bigcup_{j=1}^{r} \operatorname{cl}(O_j)$; 2) $C_i \neq \bigcup_{j=1}^{r} \operatorname{cl}(O_j)$.

In case 1) the set $V_i \overset{\text{def}}{=} \widetilde{V}_i$ is an open set, and its boundary ∂V_i satisfies the assertion 4).

In the case 2) the set $C_i \setminus \bigcup_{j=1}^{r} \operatorname{cl}(O_j)$ is made up of finitely many intervals $\Sigma_1, \dots, \Sigma_\nu \subset C_i$. For each interval Σ_j we construct an open disk $\mathcal{D}(\Sigma_j)$ as shown in Figure 2.67 so that the parts of the boundary $\partial\mathcal{D}(\Sigma_j)$ not in $\widetilde{V}_i$ are transversal

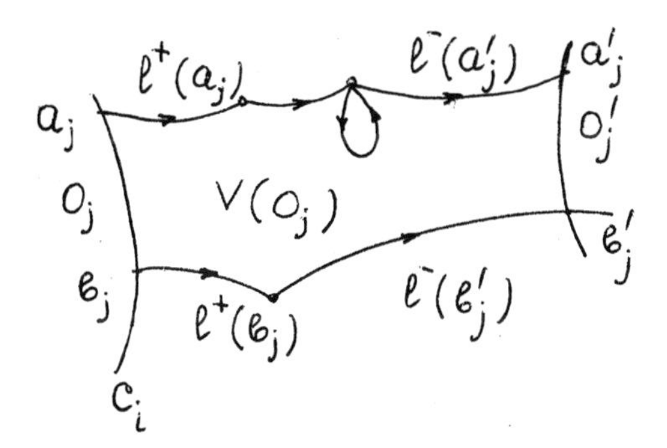

FIGURE 2.66

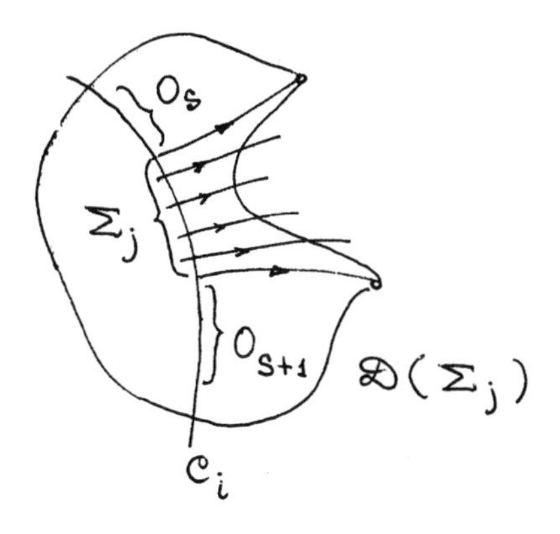

FIGURE 2.67

to the flow. Let $V_i = \widetilde{V}_i \bigcup_{j=1}^{\nu} \mathcal{D}(\Sigma_j)$. The set V_i is open, and its boundary ∂V_i satisfies the assertion 4).

We go through the construction described for all $i = 1, \ldots, k$. By choosing $\mathcal{D}(\Sigma_j)$ small enough we can make the domains $V_1, \ldots, V_k$ disjoint. □

CHAPTER 3

Topological Structure of a Flow

In the first section of this chapter we give the basic concepts of the qualitative theory of the theory of flows—topological and smooth equivalence, classification, and so on. In the second section we prove that in a certain sense a flow on a surface breaks up into flows that do not have nontrivial recurrent trajectories and irreducible flows. In this section we give a special decomposition of a flow which is due to Levitt and is based on cutting a surface by closed transversals. In the next section we show that an irreducible flow either is highly transitive or is obtained from a highly transitive flow by a blowing-up operation. The fourth section is devoted to flows without nontrivial recurrent trajectories. For such flows we single out a family of (singular) trajectories that determine the qualitative structure of the flow, and we study the components of the complement of the singular trajectories on the surface. In the conclusion we introduce a metric on the space of flows, turning that space into a metric and topological space and enabling us to investigate important classes of flows from the point of view of such topological concepts as openness, denseness, and so on.

§1. Basic concepts of the qualitative theory

Classification problems occupy a significant place in the theory of dynamical systems. Corresponding areas of the theory of dynamical systems are determined by the equivalence relation used in the classification. The basic equivalence relation in the qualitative theory of flows is topological equivalence. In the framework of this theory two flows are taken to be the same (equivalent) if the spaces of their trajectories have the same topological structure. In this section we give the main definitions in the qualitative theory of flows.

1.1. Topological and smooth equivalence.

DEFINITION. Two flows f^t and g^t on a manifold $\mathcal{M}$ are said to be *topologically equivalent* if there exists a homeomorphism $h\colon \mathcal{M} \to \mathcal{M}$ carrying each trajectory of one flow into a trajectory of the other.

If h preserves the direction in time (the positive direction) on trajectories, then f^t and g^t are said to be *topologically orbitally equivalent.*

Any two rational windings on the torus can serve as examples of topologically orbitally equivalent flows (and, in particular, topologically equivalent flows) (Figure 3.1). On the other hand, the flows on an annulus pictured in Figure 3.2 are not topologically equivalent (and, in particular, not topologically orbitally equivalent).

These two concepts are the basic concepts in the qualitative theory of flows (dynamical systems with continuous time).

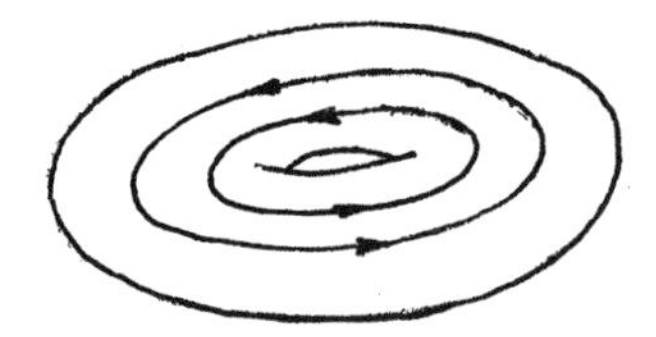
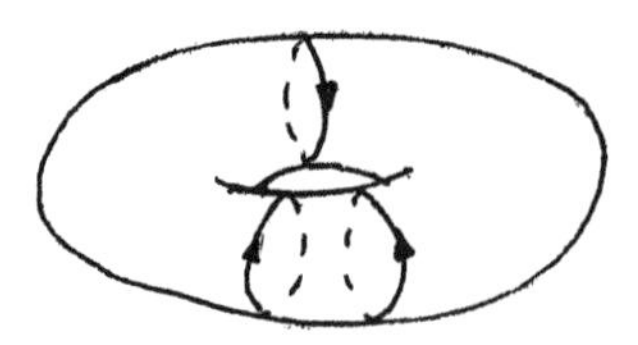

FIGURE 3.1

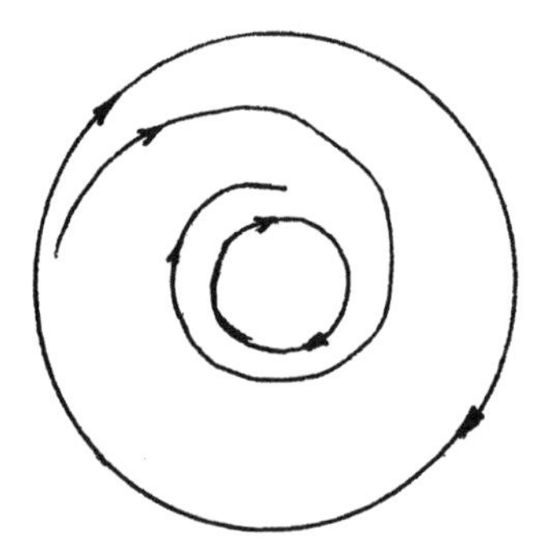
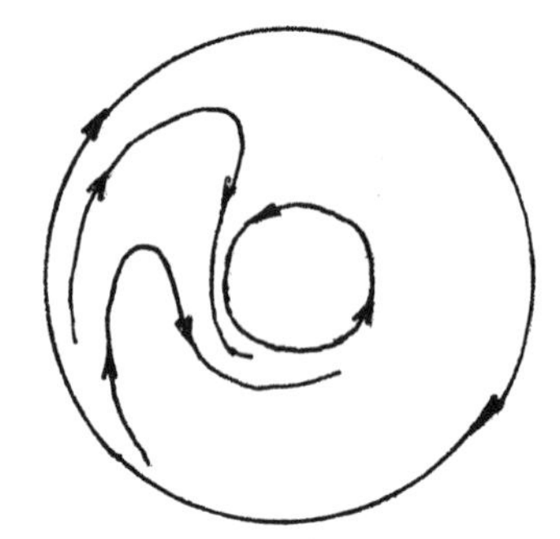

FIGURE 3.2

Two flows are said to have the same topological (qualitative) structure if they are topologically or topologically orbitally equivalent.

Topological equivalence has the properties of reflexivity, symmetry, and transitivity. Therefore, the collection of all dynamical systems breaks up into disjoint classes of systems with the same topological structure.

In establishing the topological equivalence of flows the question naturally arises of local topological equivalence.

DEFINITION. Two flows f^t and g^t on a manifold $\mathcal{M}$ are said to be *locally topologically equivalent at the respective points* $m_1, m_2 \in \mathcal{M}$ if there exist neighborhoods $\mathcal{U}(m_1)$ and $\mathcal{U}(m_2)$ of m_1 and m_2 and a homeomorphism $h\colon \mathcal{U}(m_1) \to \mathcal{U}(m_2)$ such that $h(m_1) = m_2$ and h carries arcs in $\mathcal{U}(m_1)$ of trajectories of f^t into arcs in $\mathcal{U}(m_2)$ of trajectories of g^t.

If h preserves the direction in time on arcs, then the flows f^t and g^t are said to be *locally topologically orbitally equivalent* at the respective points m_1 and m_2.

According to the rectification theorem (Theorem 2.2 in Chapter 1), any flows are locally topologically orbitally equivalent at regular points.

If in the above definition m_1 and m_2 are understood to be invariant compact sets for the respective flows f^t and g^t and it is required in addition that the homeomorphism h carry each trajectory in m_1 into a trajectory in m_2, then we get the definition of *local topological equivalence of flows* f^t *and* g^t *on the sets* m_1 *and* m_2.

In particular, the sets m_1 and m_2 can be equilibrium states, closed trajectories, and contours made up of equilibrium states and separatrices joining them. Accordingly, we refer to local topological equivalence of equilibrium states, closed trajectories, and contours.

The famous Grobman–Hartman theorem ([**27**], [**74**]) asserts that a hyperbolic equilibrium state of a C^1-flow on a finite-dimensional manifold is locally topologically orbitally equivalent to an equilibrium state of a flow given by a linear vector field.

The contours pictured in Figure 3.3 are not locally topologically equivalent.

FIGURE 3.3

The flows in Figure 3.4 are not topologically equivalent, although their equilibrium states are locally topologically orbitally equivalent.

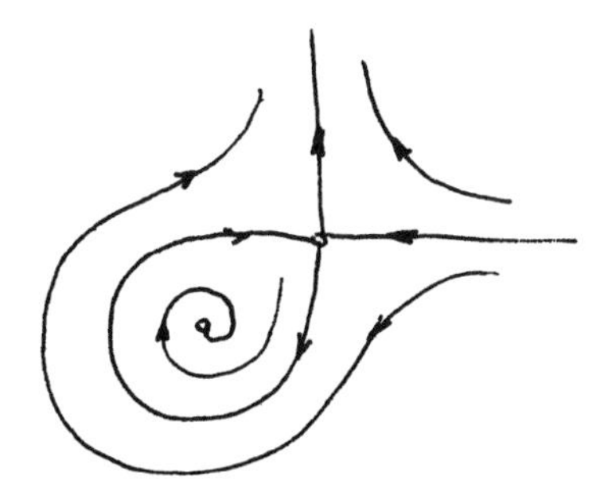
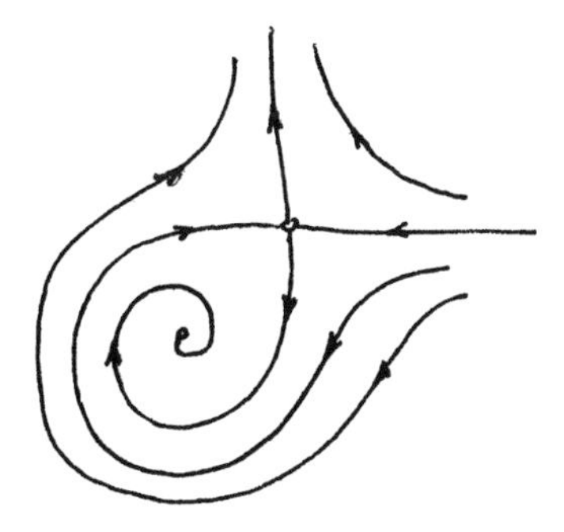

FIGURE 3.4

The concept of smooth equivalence is a generalization of topological equivalence (a C^0-diffeomorphism is understood to be a homeomorphism).

DEFINITION. Two C^r-flows f^t and g^t $(r \geq 0)$ on a manifold $\mathcal{M}$ are said to be *C^k-smoothly equivalent* $(0 \leq k \leq r)$ if there exists a C^k-diffeomorphism $h\colon \mathcal{M} \to \mathcal{M}$ carrying each trajectory of one flow into a trajectory of the other.

If h preserves the direction in time on trajectories, then the flows f^t and g^t are said to be *C^k-smoothly orbitally equivalent.*

The definitions of local C^k-smooth equivalence and local C^k-smooth orbital equivalence are analogous to the definition of local equivalence.

According to the rectification theorem, any regular points of a C^r-flow are locally C^r-smoothly equivalent.

1.2. Invariants. One of the main problems in the qualitative theory is to single out quantities, characteristics, or properties of a dynamical system that coincide for topologically equivalent dynamical systems and indicate by their difference that dynamical systems belong to different topological equivalence classes. Such quantities (characteristics, properties) are called *topological invariants of a dynamical system.*

For example, the number of quasiminimal sets of a flow is a topological invariant.

Invariants of smooth equivalence are called *smooth invariants*. For example, the characteristic numbers of the linear parts for equilibrium states of a C^1-flow are smooth invariants that are not topological invariants (if the flow has equilibrium states).

There is no substantive universal topological invariant for all flows. A class of flows is usually singled out in a special way, and then a topological invariant is introduced for dynamical systems of this class.

For example, the number of equilibrium states and the number of closed trajectories are topological invariants in the class of flows with finitely many equilibrium states and closed trajectories.

For a given class $\mathfrak{N}$ of dynamical systems we say that a topological invariant is *complete* if two arbitrary dynamical systems in $\mathfrak{N}$ are topologically equivalent precisely when this topological invariant is the same for the systems.

For instance, in the class of minimal flows on the torus the Poincaré rotation number is a complete topological invariant up to recomputation with the help of a unimodular integer matrix (§1 in Chapter 6).

1.3. Classification. Let us consider some class $\mathfrak{N}$ of dynamical systems. A *topological classification* of the dynamical systems in $\mathfrak{N}$ is defined to be a solution of the following two problems: a) find a complete topological invariant for the dynamical systems in $\mathfrak{N}$; 2) the realization problem.

A *realization* is defined to be a determination of the admissible values of a topological invariant, and the construction, from a given admissible topological invariant, of a dynamical system in $\mathfrak{N}$ with that topological invariant.

A C^r-smooth ($r \geq 1$) classification of dynamical systems in the given class is defined similarly.

Let $\mathfrak{N}$ be the set of minimal flows on the torus. As mentioned earlier (and as will be proved in §1 of Chapter 6), up to recomputation with the help of a unimodular integer matrix, the Poincaré rotation number is a complete topological invariant of a flow in $\mathfrak{N}$. The rotation number of any flow in $\mathfrak{N}$ is an irrational number. Conversely, for any irrational number $\mu \in \mathbb{R}$ there exists a minimal flow on the torus with Poincaré rotation number equal to μ. Thus, both topological classification problems are solved for the set of minimal flows on the torus.

§2. Decomposition of a flow

In this section we prove the existence of a decomposition of a flow into flows with simpler topological structure. Namely, each flow in the decomposition either does not contain nontrivial recurrent semitrajectories or contains only one quasiminimal set, and it does not admit further nontrivial decomposition. Our presentation follows [**83**]. Such a decomposition can also be obtained from the structure theorem.

In the conclusion we give a decomposition of Levitt [**91**] that differs essentially from the one mentioned above and is based on cutting the manifold along closed transversals of the flow.

2.1. Characteristic curves of a quasiminimal set. Let f^t be a flow with a quasiminimal set N on an orientable closed surface $\mathcal{M}$. According to Lemma 4.1 in Chapter 2, there exists a contact-free cycle C for f^t that intersects all nontrivial recurrent semitrajectories in N, with either $N \cap C = C$ or $N \cap C$ a Cantor set on C.

Let ξ_N be the corresponding partition of the contact-free cycle C. By virtue of Lemma 4.2, each of the endpoints η_l and η_r of any element $\eta = [\eta_l, \eta_r] \in \xi_N$ either belongs to an open Gutierrez equivalence class or is a boundary point of such a class. Denote by $\Gamma(\eta_l)$ (respectively, $\Gamma(\eta_r)$) an open class which either contains the point η_l (η_r) or is such that $\eta_l \in \partial\Gamma(\eta_l)$ (respectively, $\eta_r \in \partial\Gamma(\eta_r)$).

Let $\eta = [\eta_l, \eta_r] \in \xi_N$ be an element of type 1. Since nontrivial recurrent trajectories are dense in N, the open classes $\Gamma(\eta_l)$ and $\Gamma(\eta_r)$ contain respective points η'_l and η'_r that lie in the domain $\mathrm{Dom}(P)$ of the Poincaré mapping $P\colon C \to C$ and are such that the interval $\overline{\eta'_l, \eta'_r}$ contains points of the element η (it is possible that $\eta'_l = \eta_l$, $\eta'_r = \eta_r$).

The closed curve $S(\eta) = \overline{\eta'_l \eta'_r} \cup \overline{P(\eta'_l)P(\eta'_r)} \cup \overset{\frown}{\eta'_l P(\eta'_l)} \cup \overset{\frown}{\eta'_r P(\eta'_r)}$ is called a characteristic curve of the element $\eta \in \xi_N$ (Figure 3.5).

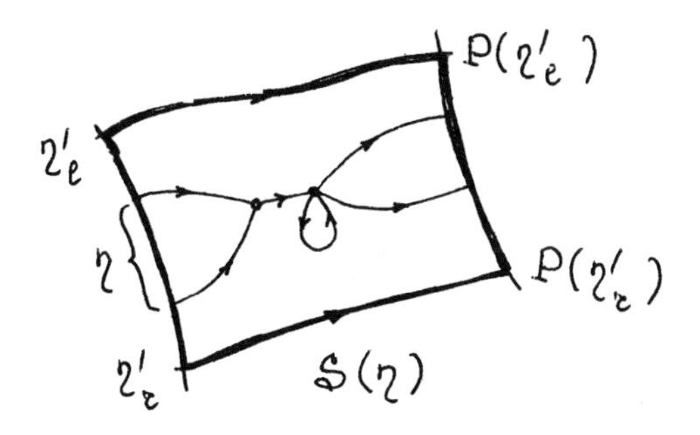

FIGURE 3.5

Let $\eta^{-k}, \dots, \eta^{-1}, \eta^0, \eta^{+1}, \dots, \eta^{+k}$ be a saddle set, and take an element $\eta^{2j} = [\eta_l^{2j}, \eta_r^{2j}]$ of it with even superscript together with points $\widetilde{\eta}_l^{2j} \in \Gamma(\eta_l^{2j})$ and $\widetilde{\eta}_r^{2j} \in \Gamma(\eta_r^{2j})$ belonging to the domain Dom(P).

The closed curve $S(\eta^{-k}, \dots, \eta^{-1}, \eta^0, \eta^{+1}, \dots, \eta^{+k})$ consisting of the segments $[\widetilde{\eta}_l^{2j}, \widetilde{\eta}_r^{2j}], [P(\widetilde{\eta}_l^{2j}), P(\widetilde{\eta}_r^{2j})] \subset C$ and the arcs $\overset{\frown}{\widetilde{\eta}_l^{2j} P(\widetilde{\eta}_l^{2j})}$ and $\overset{\frown}{\widetilde{\eta}_r^{2j} P(\widetilde{\eta}_r^{2j})}$, where η^{2j} runs through all the elements of the saddle set with even superscripts, is called a characteristic curve of the saddle set $\eta^{-k}, \dots, \eta^0, \dots, \eta^{+k}$ (Figure 3.6, for $k = 2$).

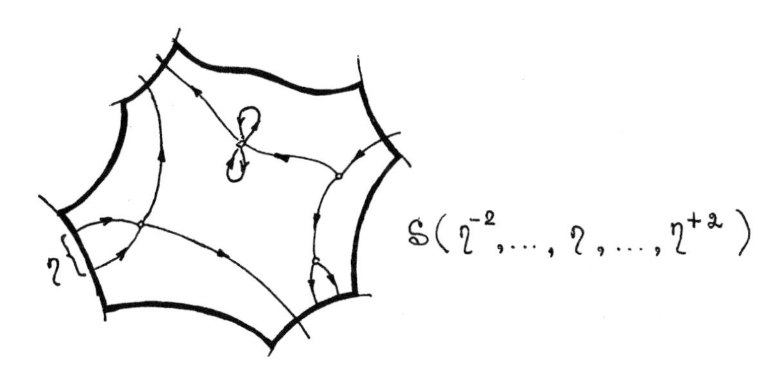

FIGURE 3.6

From the definition of Gutierrez equivalence classes it follows that a characteristic curve (of an element of type 1 or a saddle set) is determined up to a free homotopy.

DEFINITION. The family $S(N, C)$ of characteristic curves constructed for all elements of type 1 and saddle sets of the partition ξ_N is called the *characteristic family of curves of the quasiminimal set N*.

We impose additional restrictions on the characteristic curves. Let $\eta = [\eta_l, \eta_r]$ be an element of ξ_N of type 1. Since there are no one-sided contours in the quasiminimal set N (otherwise, the limit set of each nontrivial recurrent semitrajectory in N would coincide with a one-sided contour by Theorem 3.6 in Chapter 2), the element $[P_l(\eta_l), P_r(\eta_r)] \in \xi_N$ does not intersect $\eta = [\eta_l, \eta_r]$. Therefore, taking points $\eta_l' \in \Gamma(\eta_l)$ and $\eta_r' \in \Gamma(\eta_r)$ sufficiently close to the respective points η_l and η_r, we get a simple (that is, without selfintersections) characteristic $S(\eta)$.

If the elements in a saddle set $\eta^{-k}, \dots, \eta, \dots, \eta^{+k}$ are distinct with the exception of the extreme elements $\eta^{-k} = \eta^{+k}$, then a simple characteristic curve $S(\eta^{-k}, \dots, \eta, \dots, \eta^{+k})$ is constructed similarly.

In what follows we assume that characteristic curves constructed for elements of type 1 and characteristic curves of saddle sets are simple under the condition indicated above.

2.2. Periodic elements of a partition. In the notation of the preceding subsection we give the following

DEFINITION. An element $\eta \in \xi_N$ is said to be periodic if there exists a collection of elements $\eta = \eta_1, \eta_2, \dots, \eta_n = \eta$ such that:

a) if the element $\eta_i = [\eta_l^{(i)}, \eta_r^{(i)}]$ is of type 1, then $\eta_{i+1} = [P_l(\eta_l^{(i)}), P_r(\eta_r^{(i)})]$ for $i \in \{1, \dots, n-1\}$;

b) if a subfamily $\eta_{i_1}, \eta_{i_1+1}, \dots, \eta_{i_2}$ is in a saddle set $G = \{\eta^{-k}, \dots, \eta^0, \dots, \eta^{+k}\}$, $\eta_{i_1-1} \notin G$ for $i_1 > 1$, and $\eta_{i_2+1} \notin G$ for $i_2 < n$, then the first element η_{i_1} of the subfamily is an element of G with even superscript, while the last element η_{i_2} of the subfamily is an element of G with odd superscript (Figure 3.7);

c) except for the extreme elements $\eta_1 = \eta_n$, the elements of the collection $\eta_1, \dots, \eta_n$ are distinct.

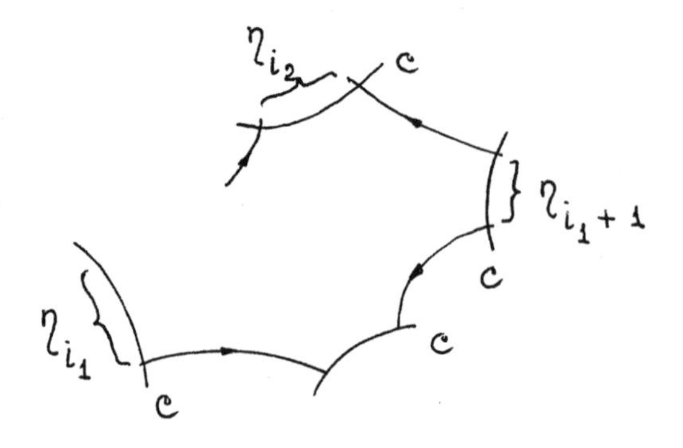

FIGURE 3.7

The elements $\eta_1 = \eta, \eta_2, \dots, \eta_n = \eta$ are called a *periodic chain* containing the periodic element η.

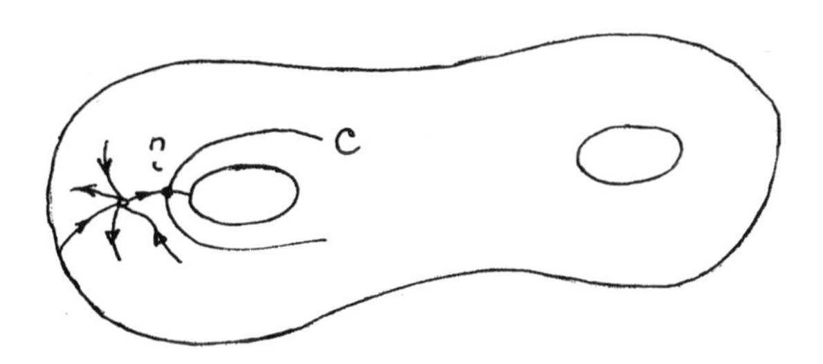

FIGURE 3.8

Figure 3.8 shows a periodic element η of the partition ξ_N on a closed transversal C of a flow f^t; f^t is transitive, and its unique quasiminimal set N coincides with the pretzel.

LEMMA 2.1. *On an orientable compact surface $\mathcal{M}$ let f^t be a flow with quasiminimal set N, and let C be a special contact-free cycle intersecting N (that is, either $C \cap N = C$ or $C \cap N$ is a Cantor set on C). Assume one of the following conditions:*

1) *there is a periodic element of the partition ξ_N;*

2) *the family $S(N, C)$ contains a characteristic curve that is not homotopic to zero.*

Then there exists a simple closed curve that is not homotopic to zero and does not intersect nontrivial recurrent semitrajectories of f^t.

PROOF. Suppose that 1) holds, and let $\mathrm{ch}(\eta) = \{\eta_1 = \eta, \eta_2, \dots, \eta_n = \eta\}$ be a periodic chain containing the periodic element $\eta \in \xi_N$.

We take an element $\eta_i \in \mathrm{ch}(\eta)$ and assume that $\eta_i = [\eta_l^i, \eta_r^i]$ is a one-point set; that is, $\eta_l^i = \eta_r^i = \eta^i$. We show that the two semitrajectories $l^-(\eta^i)$ and $l^+(\eta^i)$ are α- and ω-separatrices, respectively. Assume not: for definiteness assume that $l^+(\eta^i)$ is not an ω-separatrix. Since η_i is a one-point set, the curve C contains points arbitrarily close to η^i from both sides of η^i that belong to nontrivial recurrent trajectories in N. Therefore, all the points of the intersection $l^+(\eta^i) \cap C$ form one-point elements of ξ_N in view of the theorem on continuous dependence of trajectories on the initial conditions. Since a quasiminimal set does not contain closed trajectories nor one-point contours (Theorem 3.6), it follows that the one-point elements containing points in $l^+(\eta^i) \cap C$ are distinct. This contradicts the periodicity of the element η. The contradiction shows that for a one-point element $\eta_i = [\eta^i, \eta^i] \in \mathrm{ch}(\eta)$ the trajectory $l(\eta^i)$ is a separatrix joining equilibrium states.

We take an element $\eta_i \in \mathrm{ch}(\eta)$ of type 1. It follows from the preceding result that there is a simple arc $\gamma(\eta_i)$ that has endpoints in the respective intervals $[\eta_l^i, \eta_r^i]$ and $[P_l(\eta_l^i), P_r(\eta_r^i)]$ and does not intersect nontrivial recurrent semitrajectories (Figure 3.9).

An analogous arc $\gamma(\eta_{i_1}, \dots, \eta_{i_2})$ exists for a family $\{\eta_{i_1}, \eta_{i_1+1}, \dots, \eta_{i_2}\} \subset \mathrm{ch}(\eta)$ that satisfies the condition b) in the definition of periodicity of an element η.

Since η is periodic, the arcs $\gamma(\eta_i)$ constructed for the elements $\eta_i \in \mathrm{ch}(\eta)$ of type 1 and the arcs $\gamma(\eta_{i_1}, \dots, \eta_{i_2})$ constructed for all subfamilies $\{\eta_{i_1}, \dots, \eta_{i_2}\} \subset \mathrm{ch}(\eta)$ satisfying b) can be supplemented by segments of C so as to form a simple closed curve $\gamma(\eta_1, \dots, \eta_n)$ that does not intersect nontrivial recurrent semitrajectories.

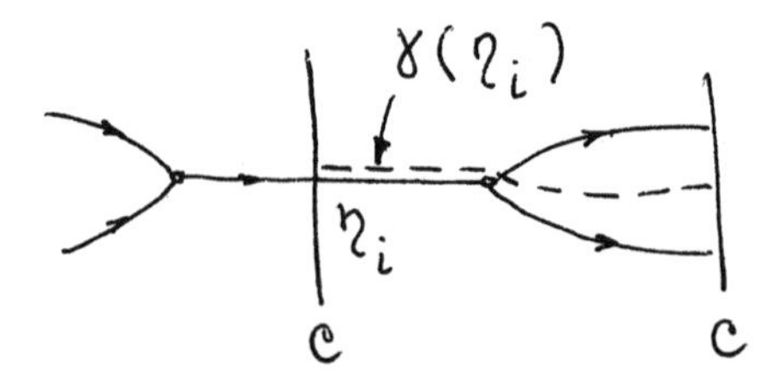

FIGURE 3.9

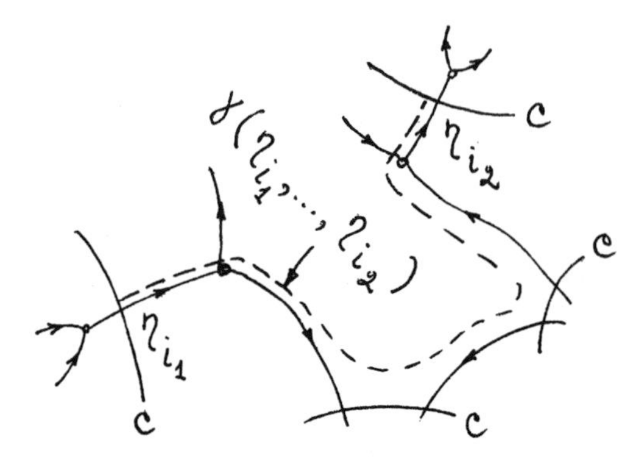

FIGURE 3.10

Let us orient the curve $\gamma(\eta_1, \ldots, \eta_n)$. It follows from the construction of the arcs $\gamma(\eta_i)$ and $\gamma(\eta_{i_1}, \ldots, \eta_{i_2})$ that the intersection index of the curve $\gamma(\eta_1, \ldots, \eta_n)$ with the contact-free cycle C is nonzero. Since C is not homotopic to zero (Lemma 2.3 in Chapter 2), it follows that $\gamma(\eta_1, \ldots, \eta_n)$ is also not homotopic to zero.

Suppose that the condition 2) holds, and let $S(\eta)$ be a characteristic curve that is constructed for an element $\eta \in \xi_N$ of type 1 and is not homotopic to zero. If the elements $\eta = [\eta_l, \eta_r]$ and $[P_l(\eta_l), P_r(\eta_r)]$ are one-point sets, then there is a simple closed curve $\widetilde{S}(\eta)$ made up of equilibrium states and separatrices joining equilibrium states that is homotopic to the characteristic curve $S(\eta)$ (Figure 3.11). Then $\widetilde{S}(\eta)$ is not homotopic to zero and does not intersect nontrivial recurrent semitrajectories.

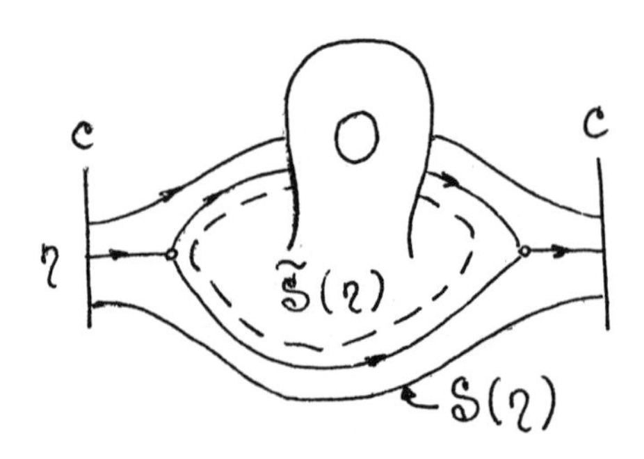

FIGURE 3.11

If at least one of the elements η and $[P(\iota(\eta_l), P_r(\eta_r)]$ is not a one-point set, then the characteristic curve $S(\eta)$ can be deformed "inside" the set $\mathcal{M} \setminus N$ to form a simple curve $\widetilde{S}(\eta)$ satisfying the assertion of the lemma (Figure 3.12). The required curve can be constructed similarly from a characteristic curve of a saddle set that is not homotopic to zero. □

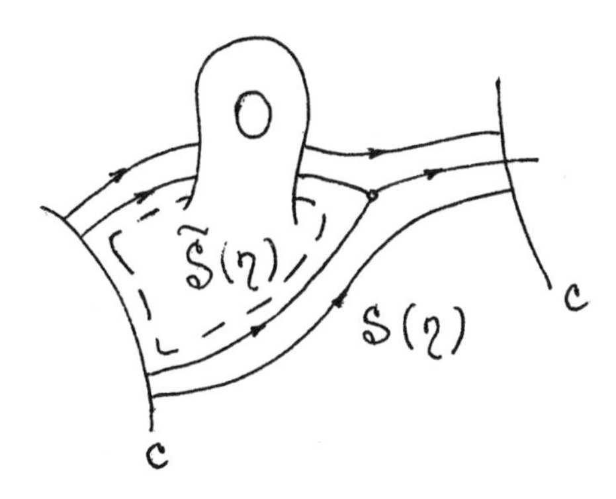

FIGURE 3.12

2.3. Criterion for a flow to be irreducible.

DEFINITION. A flow f^t on a two-dimensional manifold $\mathcal{M}$ is said to be *irreducible* if f^t has only one quasiminimal set, and any closed curve that is not homotopic to zero on $\mathcal{M}$ intersects at least one nontrivial recurrent semitrajectory of f^t.

Any highly transitive flow is irreducible. In the next section it will be shown that an irreducible flow on an orientable surface either is highly transitive or can be obtained from a highly transitive flow by means of a so-called blowing-up operation.

Suppose that a flow f^t on a closed orientable surface $\mathcal{M}$ has a quasiminimal set N (not unique in general). According to Lemma 4.1 in Chapter 2, there is a special contact-free cycle C_N satisfying the following conditions: 1) C_N intersects N and does not intersect quasiminimal sets different from N; 2) either $C_N \cap N = C_N$ or $C_N \cap N$ is a Cantor set.

In the given notation we formulate a criterion for a flow to be irreducible.

THEOREM 2.1. *Suppose that f^t is a flow with a quasiminimal set N on a closed orientable surface $\mathcal{M}^2$, and let C_N be a corresponding special contact-free cycle. Then f^t is irreducible if and only if all curves in the characteristic family $S(N, C_N)$ of curves of N are homotopic to zero, and the partition ξ_N on C_N does not have periodic elements.*

PROOF. NECESSITY. It follows from Lemma 2.1.

SUFFICIENCY. Suppose that all the curves in the characteristic family $S(N, C_N)$ are not homotopic to zero, and the partition ξ_N on C_N does not have periodic elements. Then each curve $\widehat{S} \in S(N, C_N)$ is simple and bounds a disk $\mathcal{D}(\widehat{S})$ on $\mathcal{M}^2$.

We show that $\mathcal{M} = C_N \cup \mathcal{D}(\widehat{S})$ is open and closed, where the union is over all curves $\widehat{S}$ in the family $S(N, C_N)$.

It follows from the construction of characteristic curves (see §2.1) that $N \subset \mathcal{M}$.

We prove that $\mathcal{M}$ is open. Since the disks $\mathcal{D}(\widehat{S})$ are open, it suffices to show that any point $m \in C_N$ is an interior point. Any point $m \in C_N$ belongs to some element $\eta \in \xi_N$, so there are two characteristic curves $S_1, S_2 \in S(N, C_N)$ containing m (Figure 3.13). It follows from the construction of characteristic curves that there exists a neighborhood $\mathcal{U}(m)$ of m such that $\mathcal{U}(m) \subset \mathcal{D}(S_1) \cup C_N \cup \mathcal{D}(S_2)$. Therefore, $\mathcal{M}$ is open.

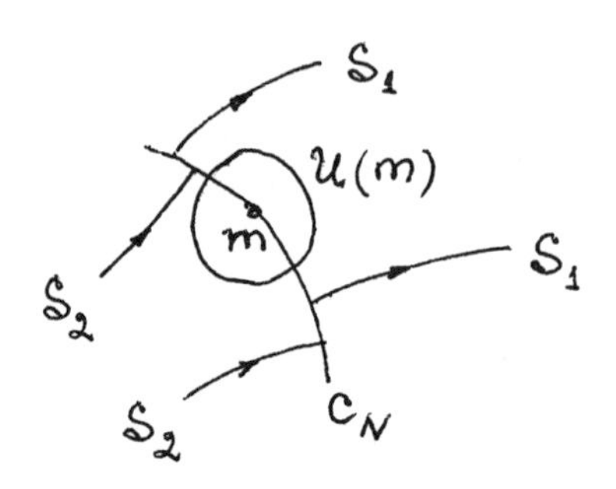

FIGURE 3.13

Let m be a limit point of a sequence $\{m_i\}_{i=1}^{\infty}$, $m_i \in \mathcal{M}$. It will be assumed that $m, m_i \notin C_N$ (otherwise there is nothing to prove), $m_i \in \mathcal{D}(\widehat{S}_i)$, and $\widehat{S}_i \in S(N, C_N)$. Since each disk $\mathcal{D}(\widehat{S}_i)$ is bounded by the curve $\widehat{S}_i$, m is a limit point for the points $\widetilde{m}_i \in \widehat{S}_i$, $i = 1, 2, \dots$. It follows from $m \notin C_N$ and the construction of characteristic curves that for sufficiently large i the points $\widetilde{m}_i$ lie on nontrivial recurrent trajectories of N, and hence $m \in N \subset \mathcal{M}$. It is proved that $\mathcal{M}$ is closed.

Since the surface $\mathcal{M}^2$ is connected, $\mathcal{M} = \mathcal{M}^2$.

We show that f^t has only one quasiminimal set N. Assume the contrary: let N_1 be a quasiminimal set different from N. Since N_1 does not intersect C_N and $\mathcal{M}^2 = C_N \cup \mathcal{D}(\widehat{S})$, it follows that N_1 lies in one of the disks $\mathcal{D}(\widehat{S})$, $\widehat{S} \in S(N, C_N)$, which contradicts Lemma 2.4 in Chapter 2.

Let γ be a closed curve that is not homotopic to zero, and assume that it does not intersect nontrivial recurrent semitrajectories. Then γ lies in a union of finitely many closed disks $\operatorname{cl}\mathcal{D}(\widehat{S}_{i_1}), \dots, \operatorname{cl}\mathcal{D}(\widehat{S}_{i_k})$ and can intersect their boundaries only in segments of C_N that are elements of ξ_N. Then from the elements of ξ_N that intersect γ we can form a periodic chain; that is, we can get a periodic element of ξ_N. The contradiction means that on $\mathcal{M}^2$ there are no closed curves that are not homotopic to zero and do not intersect nontrivial recurrent semitrajectories. This concludes the proof that f^t is irreducible. □

REMARK. In the preceding arguments the contact-free cycle can be replaced by a contact-free segment Σ with the additional condition that the endpoints of Σ do not lie on nontrivial recurrent semitrajectories. We introduce the partition ξ_Σ of Σ, and the concepts of characteristic curves and a periodic element. It is proved as above that if ξ_Σ does not have periodic elements and all the characteristic curves are homotopic to zero, then the flow is irreducible.

2.4. Decomposition of a flow into irreducible flows and flows without nontrivial recurrent semitrajectories. Let f^t be a flow on a two-dimensional manifold $\mathcal{M}$. We consider a compact C^0-submanifold $\widehat{\mathcal{M}} \subset \mathcal{M}$. At each point of the boundary $\partial\widehat{\mathcal{M}}$ we locate an equilibrium state, and we denote the resulting flow by f_1^t. If f^t is given by a system of differential equations, then by multiplying the right-hand side of the system by a function equal to 0 on $\partial\widehat{\mathcal{M}}$ and strictly positive on $\mathcal{M}\setminus\partial\widehat{\mathcal{M}}$ we get a system of differential equations determining the flow f_1^t. Note that f_1^t is not uniquely determined by f^t, but two such flows f_1^t and f_2^t are topologically orbitally equivalent (f_1^t and f_2^t can differ by phase velocities, but their trajectories coincide).

The submanifold $\widehat{\mathcal{M}}$ is invariant with respect to the flow f_1^t. Denote by $f^t|_{\widehat{\mathcal{M}}}$ the restriction $f_1^t|_{\widehat{\mathcal{M}}}$ of f_1^t to this invariant submanifold.

We regard $\widehat{\mathcal{M}}$ as an independent manifold with the flow $f_1^t|_{\widehat{\mathcal{M}}} = f^t|_{\widehat{\mathcal{M}}}$ given on it. Let us identify each component of $\partial\widehat{\mathcal{M}}$ with a point and denote the resulting closed manifold by $\mathcal{M}_*$. The flow $f^t|_{\widehat{\mathcal{M}}}$ passes into a flow f_*^t on $\mathcal{M}_*$, and each component of $\partial\widehat{\mathcal{M}}$ under the natural mapping $\widehat{\mathcal{M}} \to \mathcal{M}_*$ passes into an equilibrium state of f_*^t.

DEFINITION. The flow $f^t|_{\widehat{\mathcal{M}}}$ is said to be irreducible if the flow f_*^t on $\mathcal{M}_*$ is irreducible.

We say that the flow $f^t|_{\widehat{\mathcal{M}}}$ does not have nontrivial recurrent semitrajectories if f_*^t does not have nontrivial recurrent semitrajectories on $\mathcal{M}_*$.

THEOREM 2.2. *Let f^t be a flow on a closed orientable surface $\mathcal{M}$. Then on $\mathcal{M}$ there is a finite family $\mathcal{E}$ of simple closed curves $C_1, \dots, C_r$ that are not homotopic to zero, have union $\bigcup_{i=1}^r C_i$ disjoint from nontrivial recurrent semitrajectories of f^t, and are such that for the closure $\widehat{\mathcal{M}}$ of a component of the set $\mathcal{M}\setminus\bigcup_{i=1}^r C_i$ either the flow $f^t|_{\widehat{\mathcal{M}}}$ is irreducible or it does not have nontrivial recurrent semitrajectories.*

PROOF. If f^t does not have nontrivial recurrent semitrajectories, then the assertion of the theorem is obvious. Therefore, we assume that f^t has a quasiminimal set N_1. Let $C_{N_1} \overset{\text{def}}{=} C_1$ be a special contact-free cycle ($N_1 \cap C_1 \neq \emptyset$), let $\xi_{N_1} = \xi_1$ be the partition on C_1, and let $S(N_1, C_1)$ be the characteristic family of the quasiminimal set N_1.

By Lemma 2.1, for each periodic chain $\eta_1, \dots, \eta_n$ of ξ_1 we take a simple closed curve $\gamma(\eta_1, \dots, \eta_n)$ that is not homotopic to zero, and for each characteristic curve $S \in S(N_1, C_1)$ not homotopic to zero we take a simple curve $\widetilde{S}$ that is not homotopic to zero. Denote by $\mathcal{E}_1$ the family of all such curves $\gamma(\eta_1, \dots, \eta_n)$, $\widetilde{S}$. Since the number of saddle sets is finite, the number of periodic (distinct) chains is also finite. Similarly, we get that the number of characteristic curves constructed for saddle sets is finite. It follows from Lemma 2.8 in Chapter 2 that the number of characteristic curves $S(\eta)$ not homotopic to zero and constructed for elements $\eta \in \xi_1$ of type 1 is finite. Therefore, $\mathcal{E}_1$ is a finite family.

We remark that the curves in $\mathcal{E}_1$ intersect in general. It follows from the construction that by a slight perturbation outside the quasiminimal sets we can make the curves in $\mathcal{E}_1$ intersect at a finite (possibly zero) number of points or in a finite number of arcs.

According to Lemma 2.1, each curve in $\mathcal{E}_1$ does not intersect nontrivial recurrent semitrajectories.

Let us cut the surface $\mathcal{M}$ successively along the curves of the family $\mathcal{E}_1$. We take the closure $\widehat{\mathcal{M}_1}$ of the component of $\mathcal{M} \setminus \cup C_i$ ($C_i \in \mathcal{E}_1$) containing N_1. By the criterion for a flow to be irreducible (Theorem 2.1, and the remark after it), the flow $f^t|_{\widehat{\mathcal{M}_1}}$ is irreducible.

According to Theorem 4.1 in Chapter 2, f^t has finitely many quasiminimal sets. Therefore, there are only finitely many components of the set $\mathcal{M} \setminus \cup C_i$ ($C_i \in \mathcal{E}_1$) different from $\widehat{\mathcal{M}}_1$ and containing nontrivial recurrent semitrajectories. Continuing the process with these components, we get the required family of curves $C_1, \dots, C_r$ and the required decomposition of the original flow f^t into irreducible flows and flows without nontrivial recurrent semitrajectories. □

A family of curves $C_1, \dots, C_r$ satisfying Theorem 2.2 is said to be *reductive*, and each curve in this family is said to be a *reductive curve*.

REMARKS. 1) Gardiner [**83**] proved that a reductive family can be constructed from semitransversals, that is, curves consisting of finitely many contact-free segments and finitely many arcs of trajectories.

2) Theorem 2.2 is analogous to Theorem B in Gutierrez's paper, *Smoothing continuous flows and a converse of the Denjoy–Schwarz theorem*, An. Acad. Brasil. Ciênc. **51** (1979), 581–589.

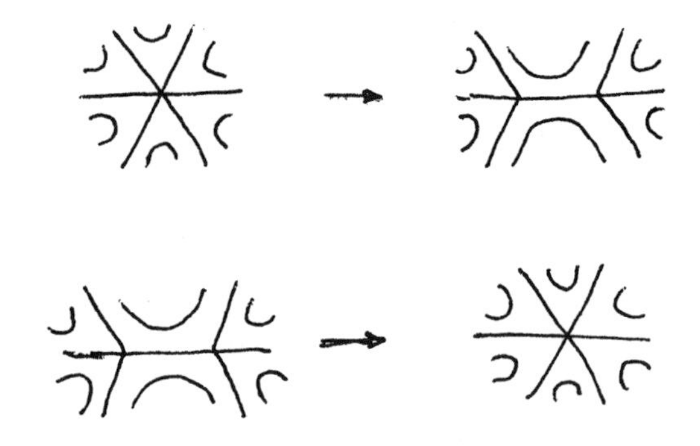

FIGURE 3.14

3) In his dissertation [**93**] Levitt obtained a result (Theorem III.4.1) analogous to Theorem 2.2 for arational foliations on a compact surface that have singularities only of saddle and thorn type (arational means that the foliation does not have leaves homeomorphic to a circle, nor leaves joining singularities). We formulate the result of Levitt. The transformations of a foliation in a neighborhood of a saddle pictured in Figure 3.14 are called Whitehead transformations. Foliations obtained from each other by Whitehead transformations are said to be Whitehead equivalent.

THEOREM [**93**]. *Let $\mathcal{F}$ be an arational foliation on a compact surface $\mathcal{M}$. Then there exist a foliation $\mathcal{F}_1$ Whitehead equivalent to $\mathcal{F}$ and a finite family $\{C_i\}_{i=1}^r$ of disjoint transversals of $\mathcal{F}_1$ such that no curve in the family $\{C_i\}_{i=1}^r$ intersects quasiminimal sets of $\mathcal{F}_1$, and there is at most one quasiminimal set in any component of the set $\mathcal{M} \setminus \bigcup_{i=1}^r C_i$.*

2.5. The Levitt decomposition. We denote by $\mathcal{F}_p^t$ the set of flows on a closed orientable surface $\mathcal{M}_p$ of genus $p \geq 2$ that have only saddles as equilibrium states and that do not have separatrices joining equilibrium states.

The set $\mathcal{F}_p^t$ contains all transitive flows on $\mathcal{M}_p$ with structurally stable saddles.

We consider a flow $f^t \in \mathfrak{F}_p^t$ and a subset $E \subset \mathcal{M}_p$ homeomorphic to a closed disk with two holes.

DEFINITION. We say that a flow $f^t \in \mathfrak{F}_p^t$ has *standard structure* on E if: 1) f^t has exactly one equilibrium state (a saddle) on E; 2) all the components of the boundary ∂E are contact-free cycles of f^t; 3) there are no closed trajectories of f^t on E (Figure 3.15).

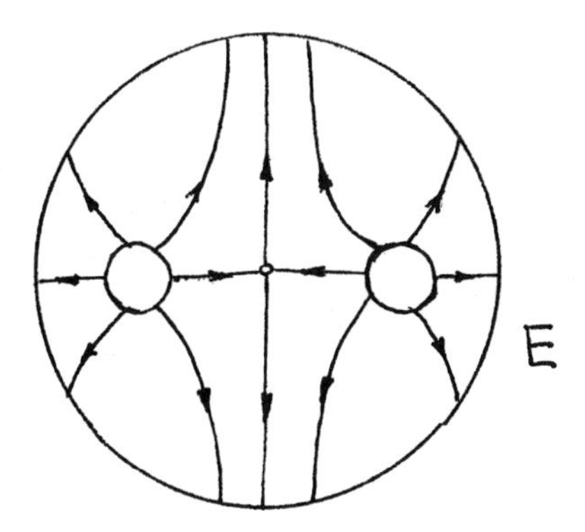

FIGURE 3.15

A flow $f^t \in \mathfrak{F}_p^t$ has an *almost standard structure* on E if the standard structure conditions 1) and 2) hold for f^t and instead of 3) it is assumed that E has closed trajectories of f^t that are homotopic to components of ∂E (Figure 3.16).

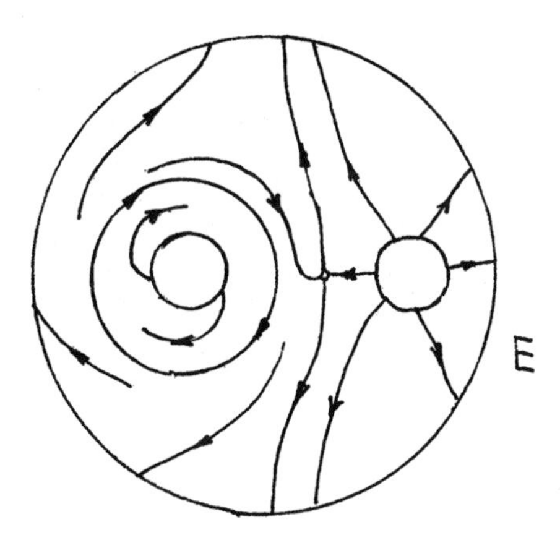

FIGURE 3.16

In other words, an almost standard structure is obtained from a standard structure by attaching to the components of ∂E annuli on which f^t has closed trajectories but not equilibrium states, with the boundaries of the annuli transversal to f^t.

THEOREM 2.3. *For any flow $f^t \in \mathfrak{F}_p^t$, $p \geq 2$, there exists a finite family $C_1, \ldots, C_k$ of contact-free cycles such that: 1) the closure of each component of the set $\mathcal{M}_p \setminus \bigcup_{i=1}^k C_i$ is homeomorphic to a disk E with two holes; 2) the curves*

$C_1, \dots, C_k$ do not intersect closed trajectories of f^t; 3) f^t has standard or almost standard structure on the closure of each component of $\mathcal{M}_p \setminus \bigcup_{i=1}^k C_i$.

PROOF. By the definition of the class $\mathcal{F}_p^t$, the equilibrium states of f^t are isolated. Therefore, f^t can have only finitely many equilibrium states on the closed surface $\mathcal{M}_p$. Since the index of each saddle is equal to -1, and the sum of all the indices of f^t is equal to $2-2p \le -2$ (Theorem 4.3 in Chapter 1), f^t has a nonzero finite number of saddles $S_1, \dots, S_l$.

For each saddle S_i we construct a closed neighborhood E_i homeomorphic to E on which f^t has standard structure.

We suppose that we have already constructed $j-1$ neighborhoods, where $1 \le j < l$, and we take the saddle S_j, which belongs to the component $\mathcal{M}_j$ of $\mathcal{M}_p \setminus \bigcup_{i=1}^{j-1} E_i$. The saddle S_j has two α-separatrices l_j and l'_j and two ω-separatrices. The boundary $\partial\mathcal{M}_j$ of $\mathcal{M}_j$ either is empty or consists of contact-free cycles that do not intersect closed trajectories of f^t.

Inside $\mathcal{M}_j$ we construct contact-free cycles C and C' with the following properties:

a) either $C = C'$ or $C \cap C' = \emptyset$;

b) C and C' do not intersect closed trajectories of f^t;

c) with increasing time the α-separatrix l_j (respectively, l'_j) intersects C (C') at a point m (m'), and the semitrajectory $l_j^-(m)$ (respectively, $l_j'^{(-)}(m')$) is disjoint from C' (C).

Obviously, there are contact-free cycles C and C' satisfying the conditions a)–c) if l_j and l'_j intersect $\partial\mathcal{M}_j$ (recall that there are no closed trajectories of f^t in some neighborhood of $\partial\mathcal{M}_j$).

Assume that l_j is disjoint from $\partial\mathcal{M}_j$. It follows from the definition of the class $\mathcal{F}_p^t$ of flows and Theorem 3.6 in Chapter 2 that the ω-limit set of l_j is a closed trajectory, a one-sided contour, or a quasiminimal set. In all cases there exists a contact-free segment Σ_j intersecting l_j at more than one point. Therefore, by Lemma 1.2 in Chapter 2, there is a contact-free cycle C intersecting l_j. Since $\omega(l_j) \subset \operatorname{int}\mathcal{M}_j$, we can construct a cycle C lying in $\operatorname{int}\mathcal{M}_j$ according to the proof of Lemma 1.2 in Chapter 2. Further, since l_j is a nonclosed trajectory, there is a contact-free segment Σ_j disjoint from the closed trajectories of f^t. Therefore, there exists a contact-free cycle C disjoint from the closed trajectories (Figure 3.17 shows the cycle C in the case when $\omega(l_j)$ consists of a closed trajectory).

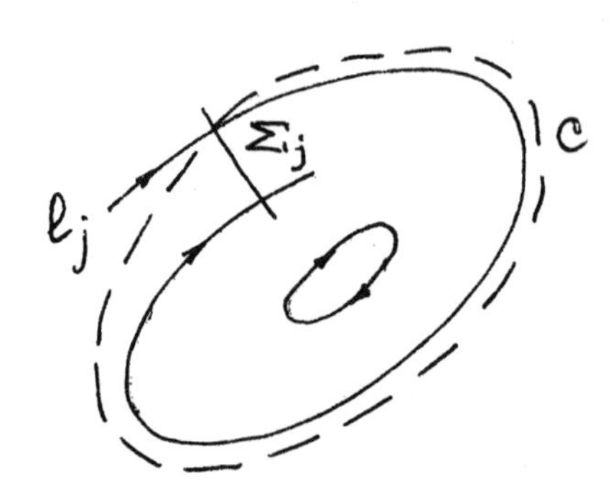

FIGURE 3.17

If the α-separatrix l'_j intersects C, then we set $C' = C$. If l'_j does not intersect C but does intersect $\partial \mathcal{M}_j$, then we take C' to be a closed transversal in a sufficiently small neighborhood of the component of $\partial \mathcal{M}_j$ intersecting l'_j. If l'_j intersects neither C nor $\partial \mathcal{M}_j$, then the construction of the cycle C' is analogous to that of the cycle C for l_j. As a result we obtain cycles C and C' ($C \cap C' = \emptyset$) satisfying the conditions a)–c).

We proceed to the construction of the closed neighborhood E_j. Let $C = C'$. By the definition of the class $\mathcal{F}_p^t$, there are four hyperbolic sectors in a neighborhood of the curve $l_j^-(m) \cup S_j \cup l_j'^{(-)}(m')$. The points m and m' lie on a single contact-free cycle C. Therefore, since $\mathcal{M}_p$ is orientable, there exist in a neighborhood of the union $C \cup l_j^-(m) \cup S_j \cup l_j'^{(-)}(m')$ two contact-free cycles C_1^j and C_2^j which together with C bound a domain $\mathring{E}_j$ on $\mathcal{M}_p$ containing S_j (Figure 3.18). It is not hard to see that the closed neighborhood $E_j = \mathrm{cl}(\mathring{E}_j)$ is homeomorphic to a disk with two holes, and the flow f^t has standard structure on E_j.

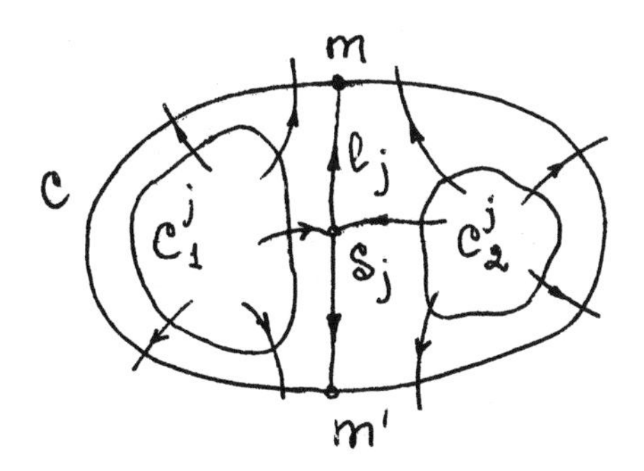

FIGURE 3.18

Let $C \neq C'$. As above, there exists in a neighborhood of $C \cup l_j^-(m) \cup S_j \cup l_j'^{(-)}(m') \cup C'$ a contact-free cycle C_j which together with C and C' bounds a domain $\mathring{E}_j$ on $\mathcal{M}_p$, and f^t has standard structure on $E_j = \mathrm{cl}(\mathring{E}_j)$.

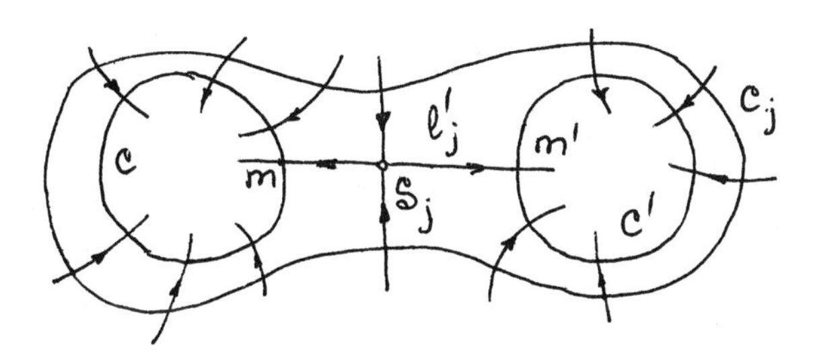

FIGURE 3.19

Continuing this process, we construct for each saddle S_i ($i = 1, \dots, l$) a closed neighborhood E_i on which f^t has standard structure.

We take the closure K of a component of the set $\mathcal{M}_p \setminus \bigcup_{i=1}^{l} E_i$. By construction, each component of the boundary ∂K of K is a closed transversal of f^t. According to the definition of $\mathcal{F}_p^t$, there are no equilibrium states interior to K. Then K is homeomorphic to an annulus by Corollary 4.5 in Chapter 1. □

It follows from Theorem 2.3 that if a flow $f^t \in \mathcal{F}_p^t$ does not have closed trajectories, then there exists a family of contact-free cycles that partition $\mathcal{M}_p$ into submanifolds with standard structure. The flow f^t can be represented as pictured in Figure 3.20.

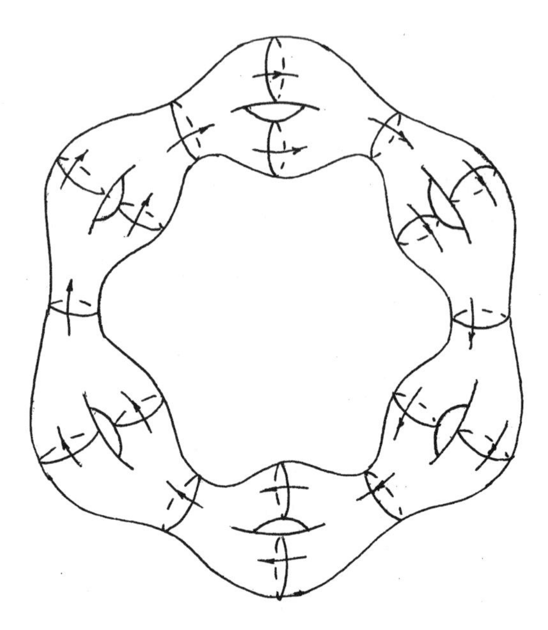

FIGURE 3.20

Remark 1. The canonical decomposition. In [**91**] Levitt generalized Theorem 2.3 to foliations on surfaces with boundary. We sketch this generalization.

Denote by $\mathcal{M}_{p,b}$ an orientable compact surface of genus $p \geq 0$ with $b \geq 0$ boundary components, and by $\mathcal{F}_{p,b}$ the class of orientable foliations on $\mathcal{M}_{p,b}$ that have singularities only of saddle type and are transversal to the boundary $\partial\mathcal{M}_{p,b}$. Besides this we require that each foliation in $\mathcal{F}_{p,b}$ not have leaves joining singularities and that any compact leaf intersect at least one closed transversal of the foliation.

Since we are considering orientable foliations, for each foliation $\mathcal{F} \in \mathcal{F}_{p,b}$ the components of $\partial\mathcal{M}_{p,b}$ can be broken up into "incoming" and "outgoing" components. Denote by $\mathcal{F}_{p,r,s}$ the class of foliations $\mathcal{F} \in \mathcal{F}_{p,b}$ such that $\mathcal{F}$ has $r \geq 0$ incoming components of $\partial\mathcal{M}_{p,b}$ and $s \geq 0$ outgoing components (obviously, $r+s=b$).

For an arbitrary triple (p, r, s) with $2p+r+s \geq 0$ Levitt proposed two models $\mathcal{M}_{p,r,s}$ and $\mathcal{M}'_{p,r,s}$, which he called canonical, of foliations belonging to $\mathcal{F}_{p,r,s}$.

It is proved in [**91**] that any foliation $\mathcal{F} \in \mathcal{F}_{p,r,s}$ (where $2p+r+s \geq 0$) is topologically equivalent either to the foliation $\mathcal{M}_{p,r,s}$ or to the foliation $\mathcal{M}'_{p,r,s}$, according to the following conditions: 1) any semileaf of $\mathcal{F}$ intersects $\partial\mathcal{M}_{p,b}$; 2) there exists a semileaf of $\mathcal{F}$ disjoint from $\partial\mathcal{M}_{p,b}$.

An analogous canonical representation of orientable foliations on nonorientable compact surfaces was obtained in V. Nordon's paper, *Description canonique de champs de vecteur sur une surface*, Ann. Inst. Fourier **32** (1982), no. 4, 151–156.

Remark 2. The center of a flow. The *center* $\Omega(f^t)$ of a flow f^t is defined to be the closure of the points lying on recurrent semitrajectories of f^t.

One method of determining the center of f^t goes back to Birkhoff. Let N be an invariant set of a flow f^t on a manifold $\mathcal{M}$. We denote by $f^t|_N$ the restriction of f^t to N, that is, the one-parameter group of homeomorphisms of N induced by f^t. The definition of the nonwandering set $NW(f^t|_N)$ of the flow $f^t|_N$ is analogous to §1.5 in Chapter 2. Let $\Omega_1 = NW(f^t), \dots, \Omega_{i+1} = NW(f^t|_{\Omega_i}), \dots, \Omega_\omega = \bigcap_{i<\omega} \Omega_i$ (where ω is the ordinal type or ordinal number of the set of natural numbers), $\Omega_{\omega+1} = NW(f^t|_{\Omega_\omega}), \dots$.

There exists a smallest ordinal number α for which $\Omega_\alpha = \Omega_{\alpha+1} = \dots$. It can be shown that Ω_α is the center of the flow f^t (there is a proof of this in [**61**]). The ordinal number α is called the *depth of the center*.

The center of a flow f^t is the largest closed invariant set $N \subset \mathcal{M}$ such that the flow $f^t|_N$ does not have wandering trajectories. The depth of the center of a flow is one of the basic topological invariants characterizing the topological structure of a flow, and it shows the number of steps needed to reach the center in the transfinite process described above.

In a paper of A. J. Schwartz and E. S. Thomas (*The depth of the center of 2-manifolds*, Proc. Sympos. Pure Math., vol. 14, Amer. Math. Soc., Providence, RI, 1970, pp. 253–264) it is proved that the depth of the center of any flow on an orientable compact surface does not exceed 2. On a nonorientable compact surface the depth of the center of any flow does not exceed 3 (see E. S. Thomas, *Flows on nonorientable 2-manifolds*, J. Differential Equations **7** (1970), 448–453). The estimates are sharp in the orientable and nonorientable cases. (We mention that both the cited papers treat flows for a collection of surfaces including compact surfaces.)

§3. The structure of an irreducible flow

A transitive flow is irreducible. We recall that a flow f^t on $\mathcal{M}$ is said to be transitive if there is a semitrajectory of f^t that is dense in $\mathcal{M}$. Obviously, this semitrajectory is a locally dense nontrivial recurrent semitrajectory, and the two-dimensional manifold $\mathcal{M}$ is itself a quasiminimal set.

The main result in this section is a proof that an irreducible flow on an orientable closed surface either is highly transitive or is obtained from a highly transitive flow by means of a blowing-up operation.

3.1. Blowing-down and blowing-up operations.

Let f^t be a flow on a manifold $\mathcal{M}$.

DEFINITION. A *Poisson pencil* is defined to be either a single nontrivial recurrent trajectory, or, in the case of a transitive flow f^t, a collection of equilibrium states (saddles) joined by separatrices, together with all the separatrices of these equilibrium states (Figure 3.21).

THEOREM 3.1. *Let f^t be an irreducible flow on a closed orientable surface $\mathcal{M}$ with finitely many equilibrium states, and suppose that each equilibrium state has finitely many separatrices (possibly none) tending to it. Then there exist a highly transitive flow g^t on $\mathcal{M}$ and a continuous mapping $h\colon \mathcal{M} \to \mathcal{M}$ homotopic to the identity and with the following properties*:

1) *if S is a Poisson pencil of the flow g^t, then $h^{-1}(S)$ is an invariant set of f^t*;

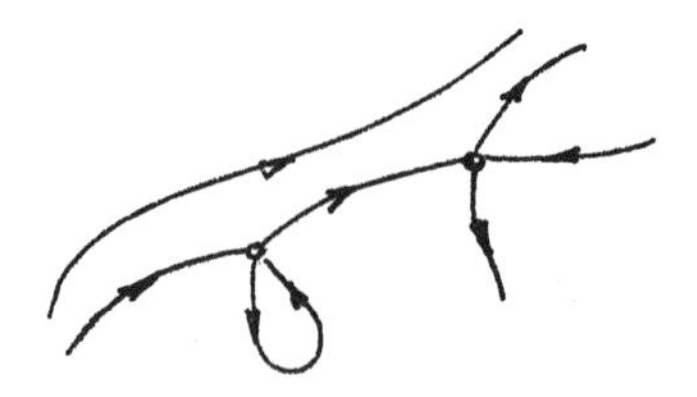

FIGURE 3.21

2) *if L is a nontrivial recurrent trajectory of g^t, then $h^{-1}(L)$ contains at most two nontrivial recurrent trajectories and at least one nontrivial recurrent semitrajectory of f^t, and, furthermore,*

a) *if $h^{-1}(L)$ contains two nontrivial recurrent trajectories, then they both lie on the accessible (from within) boundary of $h^{-1}(L)$,*[1]

b) *if $h^{-1}(L)$ contains a single trajectory l, then l is a nontrivial recurrent trajectory, and the restriction $h|_l: l \to L$ is a homeomorphism onto its image;*

3) *any P^+ (P^-) nontrivial recurrent trajectory of f^t is mapped by h homeomorphically onto its image, which is a union of finitely many saddles (possibly none), finitely many separatrices (possibly none) joining these saddles, and a single P^+ (P^-) nontrivial recurrent trajectory of g^t;*

4) *if $m, h(m) \in \mathcal{M}$ are regular points of the respective flows f^t and g^t lying on nontrivial recurrent semitrajectories of these flows, then h preserves the direction in time in some neighborhood of m;*

5) *if N is the quasiminimal set of f^t (by assumption, f^t has exactly one quasiminimal set), then $h(N) = \mathcal{M}$;*

6) *for any point $m \in \mathcal{M}$ the set $h^{-1}(m)$ is arcwise connected and contractible;*

7) *let $h = \theta \circ h_1$, where the mappings θ and h_1 satisfy the conditions* 1)–6) *with the corresponding transitive flows g_θ^t and g_1^t. Then θ is a homeomorphism realizing a topological equivalence between the flows g^t and g_1^t.*

PROOF. The flow g^t is obtained from f^t by "removing" (blowing down) the domains lying in the complement of the quasiminimal set N.

We first show that if f^t is a highly transitive flow, then we can set $h = \mathrm{id}$ (that is, it is impossible to "blow down" any further). For this we prove the property 7). Assume that θ is not a homeomorphism, and that h, θ, and h_1 satisfy the conditions 1)–6). This assumption and the transitivity of the flow g^t imply the existence of a point $m \in \mathcal{M}$ lying on a nontrivial recurrent semitrajectory l of g^t such that $\theta^{-1}(m)$ contains at least two points. According to 1) and 2), the set $\theta^{-1}(l)$ is invariant and contains a nontrivial recurrent semitrajectory L of the flow g_1^t. If $\theta^{-1}(l)$ contains only one nontrivial recurrent semitrajectory, then $L = \theta^{-1}(l)$, and $\theta|_L$ is a homeomorphism onto its image (the property 3)), which contradicts our assumption. If $\theta^{-1}(l)$ contains two nontrivial recurrent semitrajectories, then by 2) both of them lie on the accessible (from within) boundary of $\theta^{-1}(l)$, and hence $\theta^{-1}(l)$ contains interior points. According to 2), the invariant set $\operatorname{int}\theta^{-1}(l)$ (the set of interior points) does not contain nontrivial recurrent semitrajectories, and this contradicts the transitivity of the flow g_1^t. The assertion 7) is proved.

[1]The accessible (from within) boundary δK of the set K is defined to be the subset of the boundary ∂K such that for any point $x \in \delta K \subset \partial K$ there is an arc λ with x as one endpoint and all its remaining points interior to K.

It follows from 7) that if f^t is transitive and h satisfies the conditions 1)–6), then h realizes a topological equivalence between f^t and g^t. Obviously, $h = \mathrm{id}$ satisfies 1)–6).

The flow f^t will be assumed to be nontransitive in what follows. For such a flow we construct h as the composition of two quotient mappings.

Denote by $\widetilde{\mathfrak{D}}$ the family of simply connected domains bounded by separatrices going from an equilibrium state to an equilibrium state (Figure 3.22). By our assumption that there are finitely many equilibrium states and separatrices, the family $\widetilde{\mathfrak{D}}$ consists of finitely many domains.

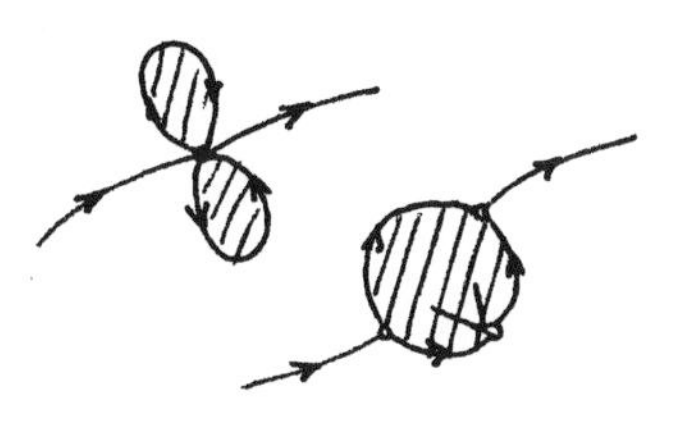

FIGURE 3.22

We take a quotient of $\mathcal{M}$, regarding all the points in the closure of each domain $\mathfrak{D} \in \widetilde{\mathfrak{D}}$ as equivalent. Then the resulting quotient manifold $\mathcal{M}_1$ is homeomorphic to $\mathcal{M}$, and the natural projection $\tau\colon \mathcal{M} \to \mathcal{M}_1$ is a continuous mapping homotopic to the identity.

The regular trajectories of f^t induce on $\mathcal{M}_1$ via the mapping τ a family of curves that satisfies the Whitney theorem (Theorem 2.3 in Chapter 1); therefore, this family can be imbedded in a flow f_1^t. The points $\tau(\mathfrak{D})$ $(\mathfrak{D} \in \widetilde{\mathfrak{D}})$ and the images of the equilibrium states of f^t under τ are equilibrium states of the flow f_1^t. By construction, f_1^t is irreducible and has exactly one quasiminimal set N_1. We remark that each nontrivial recurrent semitrajectory of f^t is mapped by τ homeomorphically into a nontrivial recurrent semitrajectory of f_1^t, and, conversely, for any nontrivial recurrent semitrajectory l_1 of f_1^t the complete inverse image $\tau^{-1}(l_1)$ is a nontrivial recurrent semitrajectory of f^t. It can be assumed without loss of generality that τ preserves the direction in time on such semitrajectories.

According to Lemma 4.1 in Chapter 2, there exists a contact-free cycle C for the flow f_1^t such that either $C \cap N_1 = C$ or $C \cap N_1$ is a Cantor set. Let $P\colon C \to C$ be the Poincaré mapping, and ξ the corresponding partition on the cycle C (see §4.3 in Chapter 2).

We consider an element $\eta = [\eta_l, \eta_r] \in \xi$ of type 1. Let $\overline{P}(\eta)$ stand for the element $[P_l(\eta_l), P_r(\eta_r)] \in \xi$. Since f_1^t is irreducible, $\eta \neq \overline{P}(\eta)$ by virtue of Theorem 2.1, and the characteristic curve $S(\eta) \in S(N_1, C)$ is a simple closed curve.

If $\eta_l \in \mathrm{Dom}(P)$, then we denote by $\Lambda(\eta_l, P_l(\eta_l))$ the arc of the trajectory $l(\eta_l)$ between η_l and $P_l(\eta_l)$. If $\eta_l \notin \mathrm{Dom}(P)$, then η_l is the right endpoint of a component of the domain $\mathrm{Dom}(P)$. It follows from Lemma 3.7 in Chapter 2 that there is a sequence of separatrices $l^+(\eta_l) = l_1, \dots, l_k$ such that the separatrix l_i $(1 < i \leq k)$ is a Bendixson extension of l_{i-1} to the left, and the separatrices $l_1, \dots, l_{k-1}$ are disjoint from C, but l_k intersects C first at the point $P_l(\eta_l)$ as time increases. In this case we denote by $\Lambda(\eta_l, P_l(\eta_l))$ the union of the separatrices $l^+(\eta_l), l_2, \dots, l_{k-1}, l^-(P_l(\eta_l))$ and the equilibrium states to which these separatrices tend.

The curve $\Lambda(\eta_r, P_r(\eta_r))$ is defined similarly (Figure 3.23).

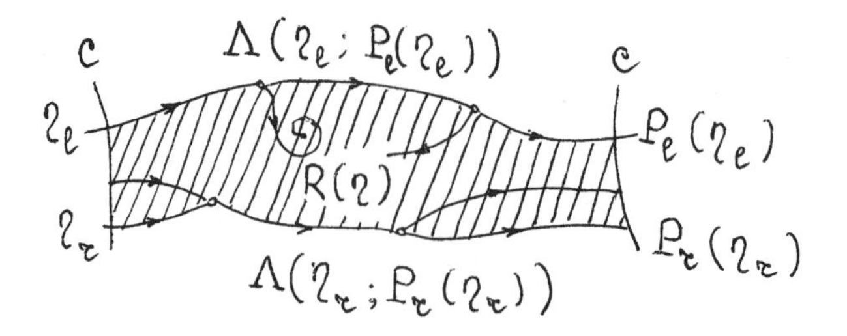

FIGURE 3.23

The closed curve $\widetilde{S}(\eta) = [\eta_l, \eta_r] \cup \Lambda(\eta_l, P_l(\eta_l)) \cup [P_l(\eta_l), P_r(\eta_r)] \cup \Lambda(\eta_r, P_r(\eta_r))$ is homotopic to the characteristic curve $S(\eta)$. Since $S(\eta)$ bounds a disk $\mathcal{D}(\eta)$ on $\mathcal{M}_1$ and $\widetilde{S}(\eta) \subset \mathcal{D}(\eta)$, $\widetilde{S}(\eta)$ bounds a closed set $R(\eta) \subset \mathcal{D}(\eta)$ (Figure 3.23). We remark that $R(\eta)$ is not homeomorphic to a closed disk in general. In particular, for a continuum of one-point elements η the set $R(\eta)$ is the arc $\overset{\frown}{\eta_l P(\eta_l)}$. The possible types of the sets $R(\eta)$ are pictured in Figure 3.24.

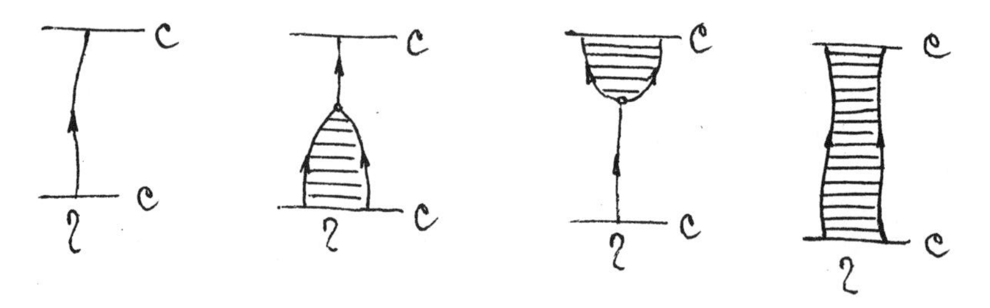

FIGURE 3.24

The subsets of $R(\eta)$ homeomorphic to a closed disk are broken up into segments I_α in such a way that they form a foliation, and each segment I_α joins points on the curves $\Lambda(\eta_l, P_l(\eta_l))$ and $\Lambda(\eta_r, P_r(\eta_r))$ (Figure 3.24).

We realize an analogous construction for the saddle set $\eta^{-k}, \dots, \eta^0, \dots, \eta^{+k}$. Since f_1^t is irreducible, the characteristic curve $S(\eta^{-k}, \dots, \eta^0, \dots, \eta^{+k})$ is homotopic to zero. As above, we define a closed curve $\widetilde{S}(\eta^{-k}, \dots, \eta^0, \dots, \eta^{+k})$ that is homotopic to the characteristic curve $S(\eta^{-k}, \dots, \eta^{+k})$ and that bounds a closed set $R(\eta^{-k}, \dots, \eta^0, \dots, \eta^{+k})$ (Figure 3.25). The subsets of $R(\eta^{-k}, \dots, \eta^0, \dots, \eta^{+k})$ homeomorphic to a closed disk are broken up into segments I_α forming a foliation with one singularity; namely, if the saddle set $\eta^{-k}, \dots, \eta^{+k}$ has index $1-k$, then the foliation of segments I_α has a single saddle singularity $O(\eta^{-k} \dots, \eta^{+k})$ (of index $1-k$) with $2k$ separatrices (Figure 3.25).

We take a quotient of the manifold $\mathcal{M}_1$, regarding the points lying on each I_α as equivalent. Since the segments I_α form a foliation on simply connected subsets homeomorphic to a disk, the resulting quotient manifold $\mathcal{M}_2$ is homeomorphic to $\mathcal{M}_1$, and hence to the original manifold $\mathcal{M}$. Denote by $\tau_1\colon \mathcal{M}_1 \to \mathcal{M}_2$ the natural projection.

The subsets of $R(\eta)$ and $R(\eta^{-k}, \dots, \eta^0, \dots, \eta^{+k})$ (where η runs through all the elements of type 1, and the family $\eta^{-k}, \dots, \eta^0, \dots, \eta^{+k}$ runs through all saddle sets

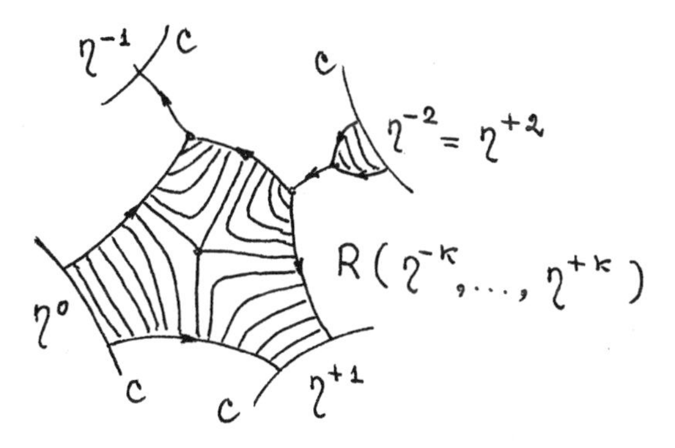

FIGURE 3.25

of the partition ξ) homeomorphic to a closed disk are transformed under the action of τ_1 into arcs $\mathcal{I}(\eta)$ and $\mathcal{I}(\eta^{-k}, \ldots, \eta^{+k})$. Since the closed intervals $[\eta_l, \eta_r] \subset C$ corresponding to elements of ξ with more than one point belong to the family $\{I_\alpha\}$, the arcs $\mathcal{I}(\eta)$ and $\mathcal{I}(\eta^{-k}, \ldots, \eta^{+k})$ can intersect each other only at the endpoints. Therefore, the curves formed by these arcs and the images of the nontrivial recurrent semitrajectories of f_1^t under τ_1 determine a foliation $\mathcal{F}$ on $\mathcal{M}_2$ with singularity set $\mathfrak{s}(\mathcal{F})$. We remark that $\mathcal{F}$ has only saddle singularities at the points $\tau_1(O(\eta^{-k}, \ldots, \eta^{+k}))$, where $\{\eta^{-k}, \ldots, \eta^{+k}\}$ runs through all the saddle sets of ξ.

By construction, the family of regular leaves of $\mathcal{F}$ (that is, the leaves lying in $\mathcal{M}_2 \setminus \mathfrak{s}(\mathcal{F})$) satisfy Whitney's theorem (Theorem 2.3 in Chapter 1); therefore, this family can be imbedded in a flow f_2^t.

Since f_1^t is irreducible, the complement $\mathcal{M}_1 \setminus N_1$ is covered by the sets $R(\eta)$ and $R(\eta^{-k}, \ldots, \eta^{+k})$, and thus through each point of $\mathcal{M}_1 \setminus N_1$ there passes some segment I_α in the foliation $\{I_\alpha\}$ of segments constructed above. Each segment I_α is mapped by τ_1 into a point. Since the endpoints of any I_α belong to N_1, it follows that $\tau_1(N_1) = \mathcal{M}_2$, and the flow f_2^t is transitive.

By construction, τ_1 maps the nontrivial recurrent semitrajectories of f_1^t homeomorphically onto their images. Let $\tau_1(m_1) = m_2$, where $m_1 \in \mathcal{M}_1$ and $m_2 \in \mathcal{M}_2$ are regular points of the respective flows f_1^t and f_2^t, and suppose that m_1 lies on a nontrivial recurrent semitrajectory $l_1^{()}$ of f_1^t. It can be assumed that in a neighborhood of m_1 on $l_1^{()}$ the mapping τ_1 preserves the direction in time (otherwise, we replace the motion in time for the flow f_2^t by its opposite motion). Since the foliation $\mathcal{F}$ is orientable, τ_1 then preserves the direction in time for any regular points m_1 and m_2, $\tau_1(m_1) = m_2$, with the above properties.

By construction, τ_1 is a continuous mapping homotopic to the identity. We show that the transformation $h = \tau_1 \circ \tau$ and the flow $g^t \overset{\text{def}}{=} f_2^t$ satisfy the theorem. Indeed, 1), 3), 4), 5), and 6) follow immediately from the definitions of τ and τ_1. We prove 2). Let L be a nontrivial recurrent trajectory of g^t, and assume that a nontrivial recurrent semitrajectory $l^{()}$ of f^t passes through an interior point m of the set $h^{-1}(L)$. Let Σ be a contact-free segment containing m and lying in $\operatorname{int}[h^{-1}(L)]$. In view of Lemma 1.2 in Chapter 2 and the invariance of $\operatorname{int}[h^{-1}(L)]$ we can construct a contact-free cycle C^* intersecting $l^{()}$ and lying in $h^{-1}(L)$. According to Lemma 2.3 in Chapter 2, C^* is not homotopic to zero, which contradicts the fact that h is homotopic to the identity. This and the fact that any nontrivial recurrent semitrajectory of f_1^t intersects the contact-free cycle C gives us that if $h^{-1}(L)$ contains interior points, then any nontrivial recurrent semitrajectory of f^t in the set $h^{-1}(L)$ lies on the accessible (from within) boundary of $h^{-1}(L)$.

Suppose that $h^{-1}(L)$ contains more than two nontrivial recurrent trajectories. Then $\tau_1^{-1}(L)$ contains at least three nontrivial recurrent trajectories l_1, l_2, and l_3 of f_1^t. Since l_1, l_2, and l_3 are nontrivial recurrent trajectories, each of them is a component of the accessible (from within) boundary of $\tau_1^{-1}(L)$. By the construction of τ_1, the set $\tau_1^{-1}(L)$ then contains saddle sets of the partition ξ, and this contradicts the fact that L is a trajectory that is not a separatrix. Thus, $h^{-1}(L)$ contains at most two nontrivial recurrent trajectories.

Suppose that $h^{-1}(L)$ contains one trajectory l of f^t. Since $h^{-1}(L)$ is invariant, $l = h^{-1}(L)$. According to 5), $h(N) = \mathcal{M}_2$, and hence $l \subset N$. Since $h^{-1}(L)$ contains only one trajectory, l is not a separatrix of an equilibrium state. It follows from Theorem 3.4 in Chapter 2 that l is a nontrivial recurrent trajectory. □

DEFINITION. A mapping h satisfying 1)–7) in Theorem 3.1 is called a *blowing-down operation* (or a blowing down) of the flow f^t to the flow g^t. If g^t is obtained from f^t by a blowing-down operation, then we say that f^t is obtained from g^t by a *blowing-up operation.*

For an irreducible flow f^t the construction of a blowing-down operation and a highly transitive flow g^t satisfying Theorem 3.1 is not unique. The connection between different blowing-down operations and highly transitive flows for a fixed flow f^t is given in the following theorem of Gardiner [**83**].

THEOREM 3.2. *Let f^t be an irreducible flow on a closed orientable surface $\mathcal{M}$, and let $h_1, h_2\colon \mathcal{M} \to \mathcal{M}$ be blowing-down operations of f^t to highly transitive flows g_1^t and g_2^t, respectively. Then g_1^t and g_2^t are topologically equivalent by a homeomorphism $\nu\colon \mathcal{M} \to \mathcal{M}$ (carrying a trajectory of g_1^t into a trajectory of g_2^t) such that for any point $m \in \mathcal{M}$ the points $\nu \circ h_1(m)$ and $h_2(m)$ belong to a single Poisson pencil of g_2^t.*

3.2. Irreducible flows on the torus.

LEMMA 3.1. *Any flow f^t on the torus T^2 having a nontrivial recurrent semi-trajectory is irreducible.*

PROOF. By assumption, f^t has a quasiminimal set N. Since the genus of the torus is 1, it follows from Theorem 4.2 in Chapter 2 that N is the unique quasiminimal set of f^t.

Let γ be a closed curve on T^2 that is not homotopic to zero. We show that γ intersects any nontrivial recurrent trajectory l of f^t. Assume the contrary. Then there exists a simple closed curve $\widetilde{\gamma} \subset T^2$ that is not homotopic to zero and disjoint from l. The set $T^2 \setminus \widetilde{\gamma}$ is homeomorphic to an annulus K, and $l \subset K$, which contradicts Lemma 2.4 in Chapter 2. □

Recall that a Denjoy flow is defined to be a flow f^t without equilibrium states on T^2 that has a limit set of Cantor type.

LEMMA 3.2. *A flow f^t without equilibrium states on T^2 is a Denjoy flow if and only if f^t is an irreducible nontransitive flow.*

PROOF. Let f^t be a Denjoy flow. According to the catalogue of limit sets (Theorem 3.6 in Chapter 2), f^t has a quasiminimal set N.

We show that f^t does not have closed trajectories. For if it had a closed trajectory homotopic to zero, then it would have an equilibrium state by virtue of Corollary 4.3 in Chapter 1. Further, by Lemma 3.1, f^t is irreducible, and hence does not have closed trajectories which are not homotopic to zero.

Since f^t does not have equilibrium states nor closed trajectories, the ω- and α-limit sets of any trajectory of f^t coincide with N. Consequently, N is a limit set of Cantor type and $N \neq T^2$. This implies that f^t is not transitive.

Let f^t be an irreducible nontransitive flow, and let N be the unique quasiminimal set of f^t. Then f^t does not have closed trajectories. This and the absence of equilibrium states imply that N is the unique minimal set of f^t. The nontransitivity gives us that $N \neq T^2$. Therefore, N is locally homeomorphic to the direct product of a closed bounded interval and the Cantor set in view of Theorem 3.7 in Chapter 2; that is, N is a limit set of Cantor type. □

COROLLARY 3.1. *Let f^t be a Denjoy flow. Then:*

a) *f^t does not have closed trajectories;*

b) *f^t has a unique quasiminimal set, which coincides with the unique minimal set N of f^t;*

c) *N is nowhere dense and locally homeomorphic to the direct product of a closed bounded interval and the Cantor set.*

DEFINITION. A flow f^t on a manifold $\mathcal{M}$ is said to be *minimal* if $\mathcal{M}$ is a minimal set of f^t.

It follows from the definition and Lemma 1.7 in Chapter 2 that any semitrajectory or trajectory of a minimal flow is dense in the manifold. Obviously, a minimal flow is transitive.

According to Theorem 3.7 in Chapter 2, the torus is the only orientable closed two-dimensional manifold on which there is a minimal flow. It can be shown that minimal flows do not exist on nonorientable closed surfaces.

LEMMA 3.3. *Suppose that f^t is a Denjoy flow. Then there exists a blowing-down operation $h\colon T^2 \to T^2$ of f^t to a minimal flow g^t on T^2 with the following properties:*

1) *h carries each trajectory of f^t into a trajectory of g^t with preservation of the direction in time;*

2) *h maps any nontrivial recurrent trajectory of f^t homeomorphically onto its image (which is a nontrivial recurrent trajectory of g^t);*

3) *$h(N) = T^2$, where N is the unique minimal (quasiminimal) set of f^t;*

4) *if w is a component of the set $T^2 \setminus N$, then*

a) *w is simply connected,*

b) *the accessible (from within) boundary of w consists of precisely two trajectories l_1 and l_2 belonging to N,*

c) *$h(w \cup l_1 \cup l_2)$ is a trajectory of g^t;*

5) *if $\mathring{N} \subset N$ consists of the trajectories not lying on the accessible (from within) boundary of any component of the set $T^2 \setminus N$, then the restriction $h|_{\mathring{N}}$ of h to $\mathring{N}$ is a homeomorpism of $\mathring{N}$ onto its image.*

PROOF. The simple connectedness of the component w follows from the irreducibility of f^t (Lemma 3.2). The remaining properties follow from Theorem 3.1. □

REMARK. In [**20**] there is a proof of Lemma 3.3 independent of Theorem 3.1, and the mapping h is constructed explicitly ([**20**], published in 1976, and Theorem 3.1 was proved in [**83**], published in 1985).

LEMMA 3.4. 1) *Any minimal flow on T^2 is topologically equivalent to the suspension over a minimal homeomorphism of the circle S^1.*

2) *Any transitive flow on T^2 with finitely many equilibrium states is obtained from a minimal flow by adjoining a certain number of impassable grains (in particular, the equilibrium states of a transitive flow on T^2 are impassable grains).*

3) *A transitive flow without equilibrium states on T^2 is minimal.*

PROOF. 1) By Lemma 2.3 in Chapter 2, there exists a contact-free cycle C for f^t that is not homotopic to zero. According to Lemma 2.4 in the same chapter, any semitrajectory of a minimal flow f^t intersects C; that is, C is a global section of the flow. Therefore, the Poincaré mapping $P\colon C \to C$ is a homeomorphism. Since the flow is minimal, the homeomorphism P is minimal (Lemma 2.1 in Chapter 1). The fact that C is not homotopic to zero implies that $T^2 \setminus C$ is homeomorphic to an open annulus. The trajectories of the flow pass from one boundary component of the annulus to the other with increasing time because there are no equilibrium states (Figure 3.26). Therefore, f^t is topologically equivalent to the suspension over the homeomorphism $P\colon C \to C$.

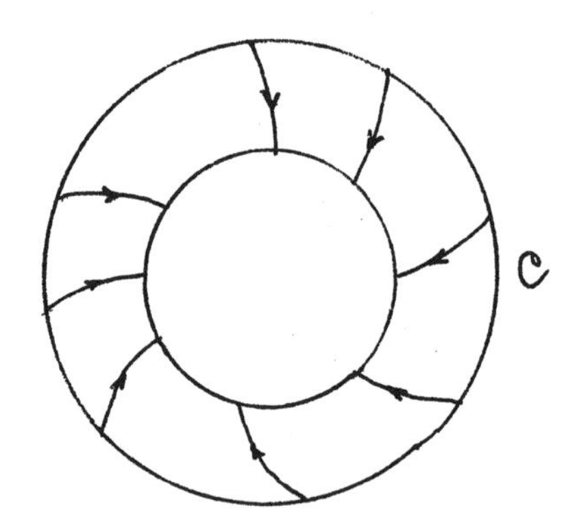

FIGURE 3.26

We prove 2). It follows from the transitivity of f^t that its equilibrium states are topological saddles (Corollary 3.1 in Chapter 2). Since the sum of the indices of all the equilibrium states is equal to the Euler characteristic of the surface (the Euler characteristic of the torus is equal to 0) according to Theorem 4.3 in Chapter 1, and since any topological saddle has nonpositive index, the index of each equilibrium state of f^t is equal to zero. Consequently, each equilibrium state is a topological saddle with two hyperbolic sectors (Theorem 4.1 in Chapter 1), that is, is an impassable grain (Figure 3.27).

To conclude the proof it remains to show that a transitive flow g^t on T^2 without equilibrium states is minimal. Indeed, since g^t is irreducible and does not have

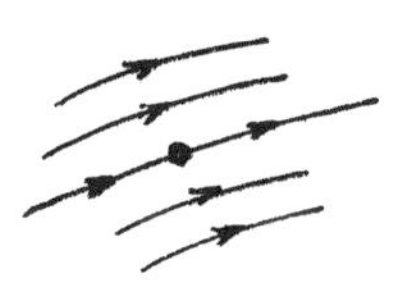

FIGURE 3.27

equilibrium states, it does not have closed trajectories. It follows from Theorem 3.7 in Chapter 2 (the catalogue of minimal sets) that T^2 is a minimal set of g^t. □

§4. Flows without nontrivial recurrent trajectories

In this section we consider flows not having nontrivial recurrent trajectories (and hence semitrajectories; see Cherry's theorem in Chapter 2). While topological invariants connected with the asymptotic behavior of recurrent semitrajectories are effective (see Chapter 6) in the investigation of the topological structure of a flow having nontrivial recurrent trajectories (in particular, of an irreducible flow), in the investigation of flows without nontrivial recurrent trajectories a large role is played by trajectories separating the manifold into domains in which the trajectories have the same behavior (the meaning of the concept of "the same" behavior is revealed in Lemma 4.1). These domains are called cells, and the trajectories separating the manifold into cells are called singular trajectories.

The definition of singular trajectories is connected with the class of flows being considered. For flows on the two-dimensional sphere the singular trajectories, which determine the qualitative structure of the flow, were found by Leontovich and Maĭer ([**50**], [**51**]). In these papers the concept of a singular trajectory is based on the concept of an orbitally unstable trajectory ([**3**], [**4**]). Another approach for determining the singular trajectories, based on the concept of topological equivalence, was presented in [**29**].

4.1. Singular trajectories. We define singular trajectories of flows on surfaces by explicitly listing such trajectories.

DEFINITION. Let f^t be a flow without nontrivial recurrent trajectories on a closed surface $\mathcal{M}$. The *singular trajectories* of f^t are the following types of trajectories:

1) equilibrium states;

2) separatrices of equilibrium states;

3) limit cycles;[2]

4) closed trajectories l such that for any neighborhood $\mathcal{U}$ of l there are both closed trajectories and nonclosed trajectories in $\mathcal{U} \setminus \{l\}$ (a foliated pie);

5) closed trajectories l such that for any neighborhood $\mathcal{U} \supset l$ there exists a neighborhood $V \subset \mathcal{U}$ of l homeomorphic to a Möbius band and filled by closed trajectories.

The family of singular trajectories of a flow f^t will be denoted by $E(f^t)$.

REMARK. Maĭer [**55**] defined singular trajectories of flows having nontrivial recurrent trajectories on orientable surfaces. For such flows a singular trajectory

[2]A limit cycle is defined to be an isolated closed trajectory of a flow, that is, a trajectory with a neighborhood in which there are no other closed trajectories.

is defined to be either one of the trajectories 1)–4) above, or a nontrivial recurrent trajectory l satisfying the following condition: for any point $m \in l$ there exists a neighborhood $\mathcal{U}$ of m such that the component of $l \cap \mathcal{U}$ (an arc of l) containing m partitions $\mathcal{U}$ into two half-neighborhoods, one of which does not contain points belonging to nontrivial recurrent trajectories. We remark that orientable surfaces do not have singular trajectories of type 5).

The singular trajectories on nonorientable surfaces were determined by Aranson [**8**].

The next lemma shows that the nonsingular trajectories (that is, trajectories that are not singular) form an open set, and the limit behavior of nearby nonsingular trajectories is the same in a certain sense.

LEMMA 4.1. *Let f^t be a flow with isolated equilibrium states on an oriented closed surface $\mathcal{M}$, and let $l(m)$, $m \in \mathcal{M}$, be a nonsingular trajectory of f^t. Then there exists a neighborhood $\mathcal{U}$ of m with the following properties:*

1) all the trajectories $l(\widetilde{m})$, $\widetilde{m} \in \mathcal{U}$, are nonsingular;

2) if $l(m)$ is closed, then all the $l(\widetilde{m})$, $\widetilde{m} \in \mathcal{U}$, are closed, and together with $l(m)$ they bound an annular domain;

3) if $l(m)$ is nonclosed, then all the trajectories $l(\widetilde{m})$, $\widetilde{m} \in \mathcal{U}$, are nonclosed, and

$$\omega(l^+(m)) = \omega(l^+(\widetilde{m})), \qquad \alpha(l^-(m)) = \alpha(l^-(\widetilde{m})).$$

PROOF. If $l(m)$ is a closed trajectory, then in view of its nonsingularity it cannot be a limit for nonclosed semitrajectories and limit cycles. Therefore, all the trajectories in some neighborhood of $l(m)$ are closed (and hence nonsingular), and together with $l(m)$ they bound an annular domain because $\mathcal{M}$ is orientable.

Suppose that $l(m)$ is nonclosed. Then $\omega(l^+(m))$ is a single equilibrium state m_0, a single closed trajectory l_0, or a single one-sided contour (recall that in this section we are considering flows without nontrivial recurrent trajectories). Let us analyze all these cases.

Suppose that $\omega(l^+(m)) = m_0$ is an equilibrium state, and that the neighborhood $\mathcal{U}(m_0)$ of m_0 does not contain equilibrium states other than m_0. We assume that for any neighborhood $\mathcal{U}(m)$ of m there is a point $\widetilde{m} \in \mathcal{U}(m)$ with $\omega(l^+(\widetilde{m})) \neq m_0$. Then the semitrajectory $l^+(\widetilde{m})$ leaves $\mathcal{U}(m_0)$ as time increases, and hence $l^+(m)$ can be extended with respect to $\mathcal{U}(m_0)$. According to Lemma 3.2 in Chapter 2, $l^+(m)$ is an ω-separatrix, which is a contradiction. Therefore, there exists a neighborhood $\mathcal{U}$ of m such that $\omega(l^+(\widetilde{m})) = \omega(l^+(m)) = m_0$ for all points $\widetilde{m} \in \mathcal{U}$. This implies that the trajectories $l(\widetilde{m})$, $\widetilde{m} \in \mathcal{U}$, are nonclosed. Since $l^+(m)$ is not extendible with respect to $\mathcal{U}(m_0)$, the semitrajectories $l^+(\widetilde{m})$, $\widetilde{m} \in \mathcal{U}$, are not extendible with respect to $\mathcal{U}(m_0)$ for a sufficiently small neighborhood $\mathcal{U}$.

Let $\omega(l^+(m)) = l_0$ be a closed trajectory. By Corollary 1.3 in Chapter 2, there is a neighborhood $\mathcal{U}$ of m such that $\omega(l^+(\widetilde{m})) = l_0$ for all $\widetilde{m} \in \mathcal{U}$. It can be assumed without loss of generality that $l_0 \cap \mathcal{U} = \emptyset$. Then all the trajectories $l(\widetilde{m})$, $\widetilde{m} \in \mathcal{U}$, are nonclosed. The case when $\omega(l^+(m))$ is a one-sided contour is handled in a completely analogous way.

The preceding arguments are repeated for the α-limit set of the semitrajectory $l^-(m)$.

If $\omega(l^\pm(m))$ $(\alpha(l^\pm(m)))$ is an equilibrium state, then the trajectories $l(\widetilde{m})$, $\widetilde{m} \in \mathcal{U}$, are nonsingular because the semitrajectories $l^\pm(\widetilde{m})$, $\widetilde{m} \in \mathcal{U}$, are not extendable.

If $\omega(l^{\pm}(m))$ $(\alpha(l^{\pm}(m)))$ is a closed trajectory or a one-sided contour, then the trajectories $l(\widetilde{m})$, $\widetilde{m} \in \mathcal{U}$ are nonsingular because they are nonclosed. □

COROLLARY 4.1. *The union of the singular trajectories of a flow f^t with isolated equilibrium states on an orientable closed surface forms a closed invariant set. The ω- (α-) limit set of any semitrajectory of f^t consists of singular trajectories.*

The proof is left to the reader as an exercise.

4.2. Cells. Denote by $\widetilde{E}(f^t)$ the union of the singular trajectories of f^t.

DEFINITION. For a flow f^t on a manifold $\mathcal{M}$, a component of $\mathcal{M} \setminus \widetilde{E}(f^t)$ is called a *cell* of f^t.

LEMMA 4.2. *Suppose that f^t is a flow with a finite family $E(f^t)$ of singular trajectories on an orientable closed surface $\mathcal{M}$. Then f^t has finitely many cells.*

PROOF. Let l be a singular trajectory of f^t that is not an equilibrium state. We take a point $m \in l$ and show that only finitely many cells intersect some neighborhood $\mathcal{U}(m)$ of m.

Assume first that l is not a limit for any singular trajectory of f^t. Since the boundary of any cell consists of singular trajectories, it follows from the theorem on continuous dependence of trajectories on the initial conditions that l lies on the boundary of at most two cells.

Suppose now that l can be a limit for singular trajectories. Since f^t does not have nontrivial recurrent trajectories, all the singular trajectories tending to l cannot be limit trajectories for other trajectories (Theorem 2.2 in Chapter II), and hence such singular trajectories lie on the boundary of at most two cells. The finiteness of the family $E(f^t)$ gives us that only finitely many cells intersect some neighborhood $\mathcal{U}(m)$. It follows from the theorem on continuous dependence of trajectories on the initial conditions and the finiteness of $E(f^t)$ that l belongs to the boundary of finitely many cells.

The boundary of each cell contains at least one regular singular trajectory. The finiteness of the number of cells of f^t follows from this, the compactness of $\mathcal{M}$, and the finiteness of the family $E(f^t)$. □

4.3. Topology of cells.

LEMMA 4.3. *On an orientable closed surface $\mathcal{M}$ let f^t be a flow with isolated equilibrium states and let R be a cell of f^t. Then the trajectories in R are either all closed or all nonclosed.*

PROOF. This follows from Lemma 4.1 and the connectedness of R. □

We now investigate the topological structure of cells filled by closed or by nonclosed trajectories.

THEOREM 4.1. *Suppose that f^t is a flow on an orientable closed surface $\mathcal{M}$, and let R be a cell of the flow that is filled by closed trajectories. Then:*

1) *R is homeomorphic either to an open annulus or to the torus T^2 (in the latter case f^t is a rational winding on the torus);*

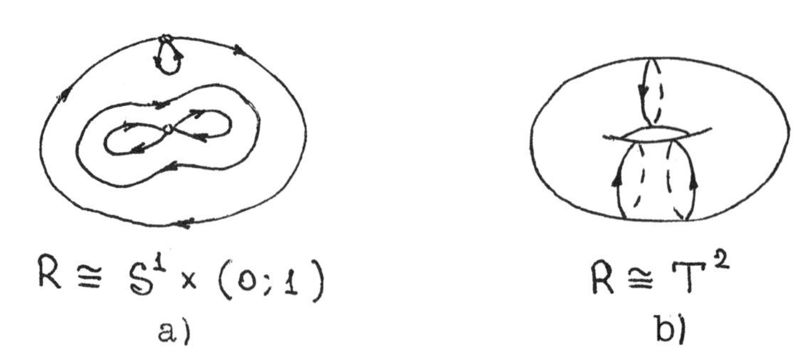

FIGURE 3.28

2) *any two closed trajectories in R bound a domain homeomorphic to an open annulus and lying in R (Figure 3.28).*

PROOF. Let l_1 and l_2 be closed trajectories in R. Assume that l_1 and l_2 do not bound an annular domain, and take an arc d_{12} with endpoints $d_1 \in l_1$ and $d_2 \in l_2$ that lies completely in R. We partition the interior points of d_{12} into two classes $\mathcal{D}_1$ and $\mathcal{D}_2$ by setting $\mathcal{D}_i = \{$the points in d_{12} through which there pass trajectories bounding an annular domain together with $l_i\}$, $i = 1, 2$. According to Lemma 4.1, $\mathcal{D}_i \neq \emptyset$ $(i = 1, 2)$, and $\mathcal{D}_1 \cap \mathcal{D}_2 = \emptyset$ by our assumption. It follows from the connectedness of the arc d_{12} that there is a point $m \in d_{12}$ that does not lie in the union $\mathcal{D}_1 \cup \mathcal{D}_2$ but does have points of $\mathcal{D}_1 \cup \mathcal{D}_2$ in any of its neighborhoods. This contradicts Lemma 4.1, 2). Thus, any two closed trajectories in R bound an annular domain in R.

Two cases are possible: 1) $\partial R = \emptyset$; 2) $\partial R \neq \emptyset$. Since $\mathcal{M}$ is connected, we get $\mathcal{M} = R$ in case 1), and hence all the trajectories of f^t are closed. It follows from Theorem 4.3 in Chapter 1 that the Euler characteristic of $\mathcal{M}$ is equal to zero, and thus $\mathcal{M} = R = T^2$.

Since $\mathcal{M}$ is orientable, in case 2) R contains two sequences $\{m_i\}_1^\infty$ and $\{m_i'\}_1^\infty$ of points tending to the boundary ∂R and such that the open annular domains K_i bounded by the closed trajectories $l(m_i)$ and $l(m_i')$ form an expanding sequence of sets $K_1 \subset \ldots \subset K_i \subset \ldots$. Since $m_i, m_i' \to \partial R$ $(i \to \infty)$, it follows that $R = \cup K_i$. This implies that R is homeomorphic to an open annulus. □

THEOREM 4.2. *Let f^t be a flow with finitely many singular trajectories on an orientable closed surface $\mathcal{M}$, and let R be a cell of the flow that is filled by nonclosed trajectories. Then:*

1) *R is homeomorphic either to an open disk or to an open annulus (Figure 3.29);*

2) *∂R has at most two connected components;*

3) *all the trajectories in R have the same ω- and α-limit sets;*

4) *the ω- (α-) limit set of any trajectory in R belongs to ∂R, and, moreover, each component of ∂R contains points in the ω- or α-limit sets of trajectories in R.*

PROOF. The connectedness of R and Lemma 4.1, 3) imply 3).

According to Corollary 3.1, the ω- (α-) limit set of any nonclosed trajectory $l \subset R$ consists of singular trajectories. Consequently, $\omega(l) \cup \alpha(l) \subset \partial R$.

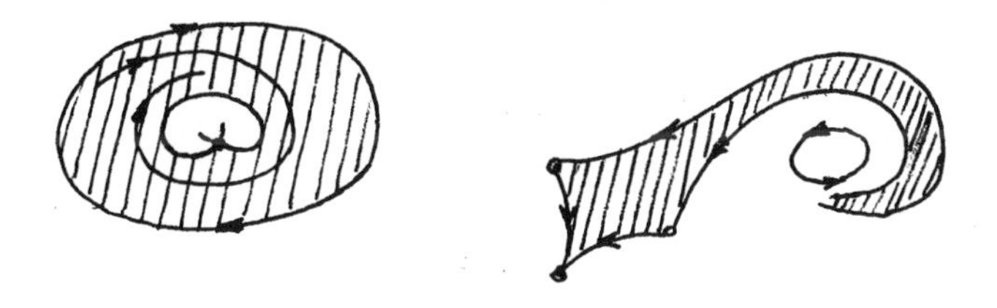

FIGURE 3.29

We show that each component R^* of the boundary ∂R contains limit points of trajectories in R. Note that since the set of singular trajectories is invariant (Corollary 4.1), R^* is a connected invariant set. We assume that R^* does not contain limit points of any trajectory in R.

Suppose that R^* consists solely of equilibrium states. Since R^* is connected, and each equilibrium state is isolated, R^* is a single equilibrium state m_0. Any neighborhood of m_0 intersects nonclosed trajectories in R. This and the connectedness of R give us that there is no closed trajectory of f^t lying arbitrarily close to m_0 and enclosing m_0 inside itself. According to the Bendixson theorem on equilibrium states (Theorem 3.3 in Chapter 2), there are trajectories of f^t tending to m_0. By the fact that m_0 is an isolated equilibrium state and by our assumption that m_0 is not a limit point of any trajectory in R, there exist a neighborhood $\mathcal{U}(m_0)$ not containing equilibrium states other than m_0 and a hyperbolic (saddle) sector S in $\mathcal{U}(m_0)$ that is bounded by separatrices l_1^+ and l_2^- and intersects the cell R (Figure 3.30). Since S contains points in R arbitrarily close to m_0, the separatrices l_1^+ and l_2^- belong to ∂R. This contradicts the fact that R^* consists only of the single equilibrium state m_0.

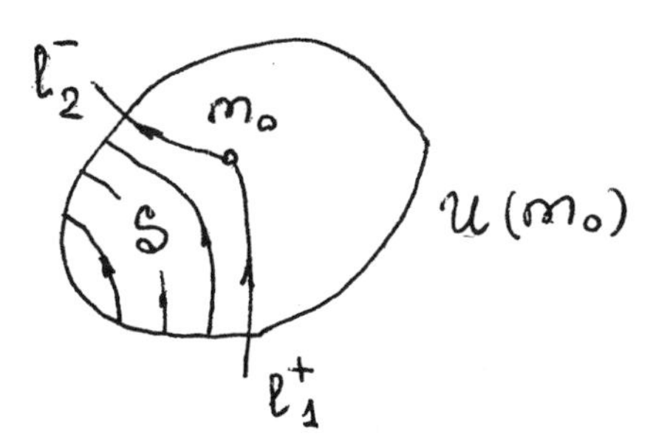

FIGURE 3.30

Suppose that the component R^* contains a regular point m_*. Since f^t does not have nontrivial recurrent trajectories, the singular trajectory $l(m_*)$ is either a closed trajectory or a separatrix of some equilibrium state. If $l(m_*)$ is a closed trajectory, then some neighborhood of $l(m_*)$ does not contain closed trajectories other than $l(m_*)$ because there are finitely many singular trajectories. It follows that the trajectory $l(m_*) \subset R^*$ belongs to the limit set of the trajectories in the cell R.

We consider the case when $l(m_*)$ is a separatrix. If $l(m_*)$ belongs to a one-sided contour K, then it can be shown as above that R^* contains limit points for the trajectories in R. If $l(m_*)$ does not belong to a one-sided contour, then finitely many Bendixson extensions of it from the side from which the trajectories in R approach m_* lead to a separatrix l_* of an equilibrium state O_*, with l_* either not Bendixson extendible (beyond O_*) or tending to a closed trajectory l_0. In this case either the equilibrium state O_* is in R^*, or the closed trajectory $l_0 \subset R^*$ belongs to the limit set of the trajectories in R.

Thus, each component R^* of ∂R contains limit points of (all) the trajectories in R. From this, the connectedness of the ω- (α-) limit set of any trajectory (Lemma 1.5 in Chapter 2) and the fact that all the trajectories in R have the same ω- and α-limit sets, it follows that ∂R consists of at most two components.

We proceed to the proof of 1). Two cases are possible:

a) any contact-free segment intersects each trajectory in R at most at one point;

b) There exist a contact-free segment and a trajectory in R that intersect at more than one point.

Let us consider the case a). Since all the trajectories in R are nonclosed, the ω- (α-) limit set of each trajectory in R consists of a single equilibrium state in this case. Therefore, any two sufficiently close trajectories in R bound a domain in R that is homeomorphic to the open strip $(0, 1) \times \mathbb{R}^1$, and hence homeomorphic to an open disk (note that the indicated transformation $(0,1) \times \mathbb{R}^1 \to R$ is a homeomorphism of the strip $(0,1) \times \mathbb{R}^1$ onto its image in the topology induced by this mapping; in the topology of R induced by the topology of $\mathcal{M}$ the transformation $(0,1)\times\mathbb{R}^1 \to R$ is not a homeomorphism in general). Since R is connected, any two trajectories in R bound a domain in R homeomorphic to an open disk. In this case it can be shown as in the proof of Theorem 4.1 that R is homeomorphic to an open disk.

We consider the case b). By Lemma 1.2 in Chapter 2, there exists a contact-free cycle C intersecting trajectories in the cell R. Two subcases are possible:

b1) C lies entirely in R;

b2) C intersects R in an arc $\mathcal{J}$.

In the subcases b1) we denote by R_1 (respectively, R_2) the set of trajectories of R intersecting (respectively, not intersecting) the cycle C. By the compactness of C and the theorem on continuous dependence of trajectories on the initial conditions, both the sets R_1 and R_2 are open. By construction, $R_1 \neq \emptyset$. If $R_2 = \emptyset$, then we get a contradiction to the connectedness of R. Therefore, $R_2 = \emptyset$, that is, any trajectory in R intersects C.

We show that in the subcase b1) each trajectory in R intersects C exactly once. Suppose not. Then a Poincaré mapping $P\colon C \to C$ is defined on C with nonempty domain $\mathrm{Dom}(P)$. If $\mathrm{Dom}(P) \neq C$, then by Lemma 3.7 in Chapter 2, a separatrix of some equilibrium state passes through an endpoint of a component of $\mathrm{Dom}(P)$, and this contradicts the fact that R does not contain singular trajectories. Therefore, $\mathrm{Dom}(P) = C$, and the restriction of the flow f^t to R is topologically equivalent to a suspension over the circle. Then $R = T^2$, and f^t is a flow on T^2 without equilibrium states and closed trajectories. According to Theorem 3.6 in Chapter 2, such a flow has nontrivial recurrent trajectories, which is impossible. The contradiction shows that each trajectory in R intersects C at exactly one point. This implies that the cell R is homeomorphic to an open annulus.

In the subcase b2) it again follows from the connectedness of R that the arc $\mathcal{J}$ intersects any trajectory in R. If $\mathcal{J}$ intersects each trajectory at exactly one point, then R is homeomorphic to the open strip $(0,1) \times \mathbb{R}^1$, and hence to an open disk. If $\mathcal{J}$ intersects some trajectory in R at more than point, then by Lemma 1.2 in Chapter 2, there exists a contact-free cycle in R; that is, the subcase b2) reduces to the subcase b1) treated above, and the cell R is homeomorphic to an open disk. □

4.4. Structure of a flow in cells. Let R be a cell of the flow f^t. Since the set R is invariant, the restriction $f^t|_R$ of f^t to R can be regarded as a flow on an open submanifold. There are only four types of flows $f^t|_R$; that is, $f^t|_R$ is topologically equivalent to one of four standard flows ([**1**], [**100**]). This subsection is devoted to a description of these standard flows. The main result is one of the key results in solving the problem of topological equivalence of flows not having nontrivial recurrent trajectories (Chapter 6).

Let $\mathbb{R}^2$ be the Euclidean plane with Cartesian coordinates x and y. Denote by Π the open strip bounded by the lines $y = 0$ and $y = 1$.

DEFINITION. The flow f^t_Π on Π given by the system $\dot{x} = 1$, $\dot{y} = 0$ is called a *parallel flow on an open strip* (Figure 3.31). Any flow on an open manifold or submanifold that is topologically equivalent to f^t_Π will also be called a parallel flow on an open strip.

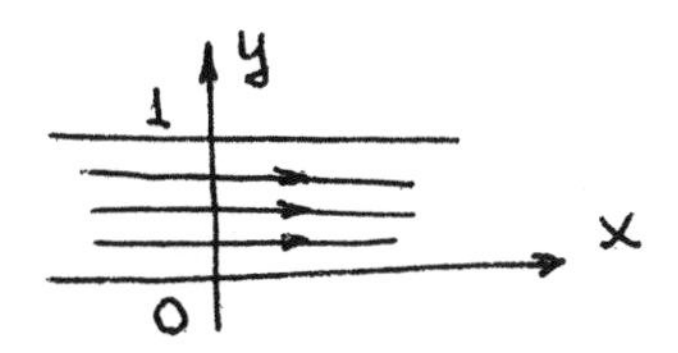

FIGURE 3.31

We introduce a polar coordinate system (r, φ) on $\mathbb{R}^2$, and denote by O the pole $(r = 0)$. Obviously, the set $K = \mathbb{R}^2 \setminus O$ is homeomorphic to the open annulus $(0, 1) \times S^1$.

DEFINITION. The flow f^t_{K1} on K given by the system $\dot{r} = 0$, $\dot{\varphi} = 1$ is called *a parallel flow on an open annulus* (Figure 3.32, a)).

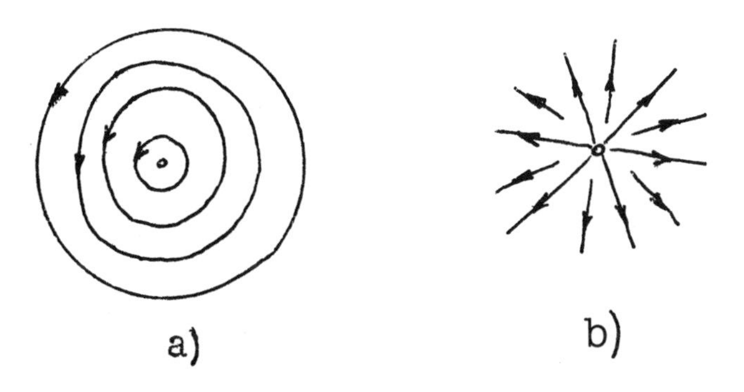

FIGURE 3.32

The flow f^t_{K2} on K given by the system $\dot{r} = r$, $\dot{\varphi} = 0$ is called *a spiral flow on an open annulus* (Figure 3.32, b)).

Any flow on an open manifold or submanifold that is topologically equivalent to the flow f^t_{K1} (respectively, f^t_{K2}) will also be called a parallel (respectively, spiral) flow on an open annulus.

We recall that a rational winding on the torus T^2 is defined to be a flow on T^2 for which a covering flow on $\mathbb{R}^2$ with respect to the universal covering $\mathbb{R}^2 \to \mathbb{R}^2/\mathbb{Z}^2 \cong T^2$ has the form $\dot{x} = 1$, $\dot{y} = p/q$, where p/q is a rational number (see §2.3.2 in Chapter 1).

The next result follows directly from the proofs of Theorems 4.1 and 4.2.

THEOREM 4.3. *Let f^t be a flow with finitely many singular trajectories on an orientable closed surface $\mathcal{M}$, and let R be a cell of f^t. Then the restriction $f^t|_R$ of the flow to R is a parallel flow on an open strip, or a spiral flow on an open annulus, or a parallel flow on an open annulus, or a rational winding on the torus (in the last case $R = \mathcal{M} = T^2$).*

4.5. Smooth models. We recall that the accessible (from within) boundary δR of a set R is defined to be the subset of points x in ∂R such that there is an arc λ with x as one endpoint and all other points interior to R.

For example, let R be the cell pictured in Figure 3.33. The trajectories l_1 and l_2 (separatrices) belong to the boundary ∂R of R, as does the limit cycle l_0. But l_0 does not belong to the accessible (from within) boundary δR. The trajectories l_1 and l_2 do belong to δR.

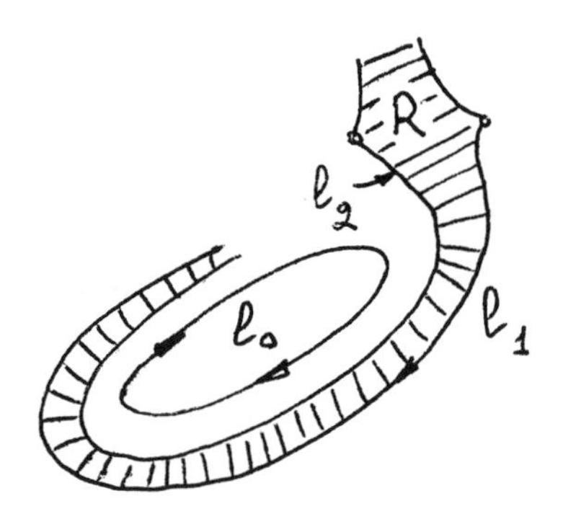

FIGURE 3.33

We consider a flow f^t not having nontrivial recurrent trajectories on a closed orientable surface $\mathcal{M}$. Let R be a cell of f^t. It follows from the theorem on continuous dependence of trajectories on the initial conditions that δR consists of whole trajectories. We remove the equilibrium states from δR (if there are any) and denote the resulting set by $\delta\check{R}$. Then the set $R \cup \delta\check{R}$ is invariant and consists solely of regular trajectories of f^t.

This subsection is devoted to the study of the topological structure of the flow $f^t|_{R\cup\delta\check{R}}$ (that is, the restriction of f^t to the invariant set $R \cup \delta\check{R}$). For cells R with nonempty boundary it is shown that there are four standard constructions enabling us to construct, from a given flow f^t and a given cell R, a C^∞-flow topologically equivalent to $f^t|_{R\cup\delta\check{R}}$. In other words, any flow $f^t|_{R\cup\delta\check{R}}$ is topologically equivalent (and even topologically orbitally equivalent) to a certain C^∞-flow, which is called a smooth model for the flow $f^t|_{R\cup\delta\check{R}}$. This result is given in Chapter 7 in the proof of the fact that any continuous flow with finitely many singular trajectories on a compact orientable surface is topologically equivalent to some C^∞-flow, that is, is smoothable (Neumann's theorem). The collection of all smooth models breaks up into four classes corresponding to the four constructions that reduce to these models. We proceed to a description of the constructions and the corresponding families of smooth models. Our presentation follows [**100**].

1) A flow $\check{f}_0^t$ on a strip. As in the preceding subsection, we use Cartesian and polar coordinate systems in the Euclidean plane $\mathbb{R}^2$.

On the line $y = i$ $(i = 0, 1)$ we take a set P_i that is empty, or coincides with the whole line $y = i$, or is a family of isolated points. There exists a C^∞-function $\varphi\colon \overline{\Pi} \to [0,1]$ equal to 0 on $P_0 \cup P_1$ and strictly positive on the set $\check{\Pi} \overset{\text{def}}{=} \overline{\Pi} \setminus P_0 \cup P_1$ (where $\overline{\Pi}$ is the closed strip on $\mathbb{R}^2$ bounded by the lines $y = 0$ and $y = 1$) [**47**]. Then the flow f_0^t on $\overline{\Pi}$ given by the system $\dot{x} = \varphi(x, y)$, $\dot{y} = 0$, $(x, y) \in \overline{\Pi}$, is smooth of class C^∞ (Figure 3.34). The restriction of f_0^t to $\Pi = \operatorname{int} \overline{\Pi}$ is a parallel flow on an open strip.

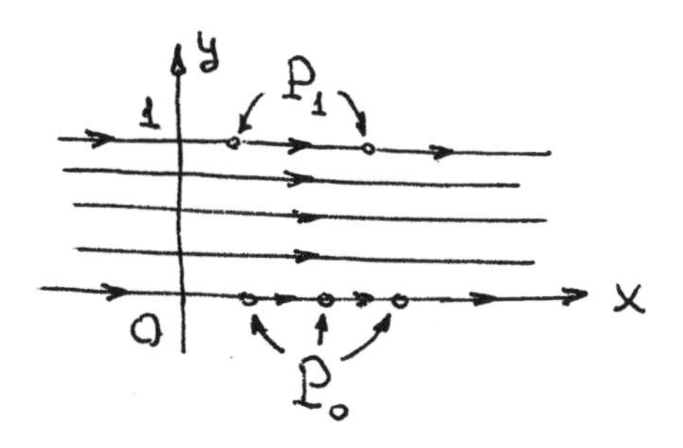

FIGURE 3.34

The restriction of f_0^t to $\check{\Pi}$ is denoted by $\check{f}_0^t$. Then $\check{f}_0^t$ has only nonclosed regular trajectories and is a C^∞-flow.

2) A parallel flow $\check{f}_1^t$ on an annulus. Let $K = \{(r, \varphi) : 1 \le r \le 2\}$ be the closed annulus bounded by the circles $\mathfrak{I}_1 = \{r = 1\}$ and $\mathfrak{I}_2 = \{r = 2\}$ (where (r, φ) are polar coordinates on $\mathbb{R}^2$). On the circle $\mathfrak{I}_i$ $(i = 1, 2)$ we take a set P_{i+1} that coincides with $\mathfrak{I}_i$, or is empty, or is a family of isolated points. There exists a C^∞-function $\psi\colon \overline{K} \to [0,1]$ equal to 0 on $P_2 \cup P_3$ and taking positive values on the set $\check{K} \overset{\text{def}}{=} \overline{K} \setminus P_2 \cup P_3$ [**47**]. Then the flow f_1^t on $\overline{K}$ given by the system $\dot{r} = 0$, $\dot{\varphi} = \psi(r, \varphi)$, $(r, \varphi) \in \overline{K}$, is smooth of class C^∞ (Figure 3.35). The restriction of f_1^t to $\operatorname{int} \overline{K}$ is a parallel flow on an open annulus.

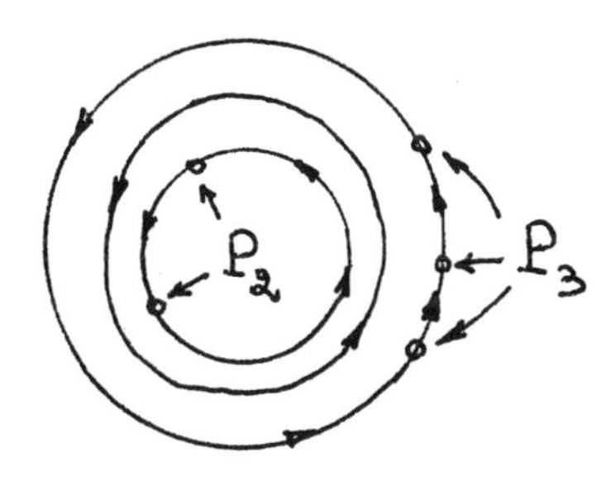

FIGURE 3.35

Let $\check{f}_1^t$ denote the restriction of f_1^t to $\check{K}$. By construction, $\check{f}_1^t$ is a C^∞-flow.

3) Spiral flows $\check{f}_2^t$ and $\check{f}_3^t$ on an annulus. We use the notation of case 2). On the closed annulus $\overline{K}$ there are only two topological types of flows without equilibrium states and without closed trajectories other than $\mathcal{J}_1, \mathcal{J}_2 \subset \partial\overline{K}$ (by the definition of a flow, the boundary $\partial\overline{K}$ of the annulus $\overline{K}$ is an invariant set of any flow on $\overline{K}$; therefore, if a flow on $\overline{K}$ does not have equilibrium states, then both the boundary components $\mathcal{J}_1, \mathcal{J}_2 \subset \partial\overline{K}$ are closed trajectories). Any such flow on $\overline{K}$ is topologically equivalent to one of the flows in Figure 3.36. One topological class contains flows for which the motions along the trajectories $\mathcal{J}_1$ and $\mathcal{J}_2$ with increasing time are realized either both clockwise or both counterclockwise (Figure 3.36, a)). The second topological class consists of flows for which the motions along $\mathcal{J}_1$ and $\mathcal{J}_2$ with increasing time are in opposite directions (Figure 3.36, b)). A flow in the first (second) topological class will be indicated by a "plus" ("minus") sign.

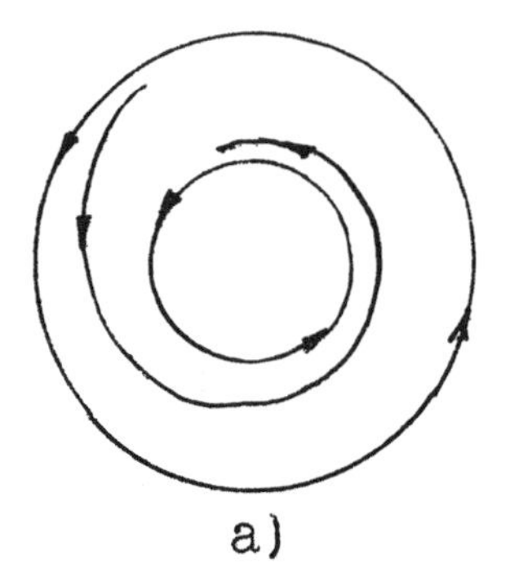

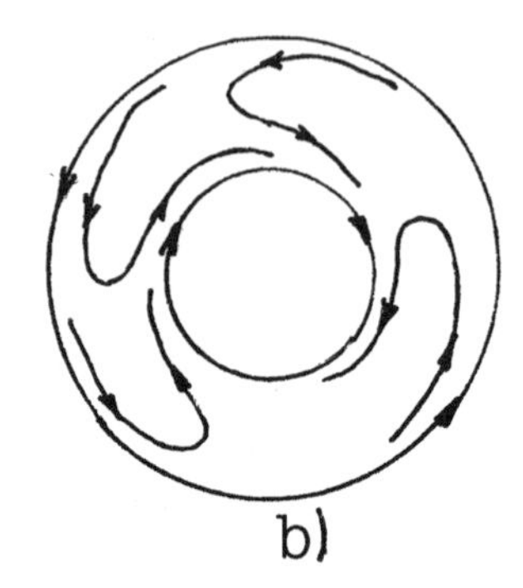

FIGURE 3.36

a) A spiral flow $\check{f}_2^t$ of "plus" type. Let the sets P_2 and P_3 and the function $\psi(r,\varphi)$ be the same as in case 2), and let the C^∞-function $v\colon \overline{K} \to [0,1]$ be given by

$$v(r,\varphi) = \begin{cases} e^{\frac{1}{(r-1)(r-2)}}, & 1 < r < 2 \\ 0, & r = 1,\ r = 2. \end{cases}$$

Then the flow f_2^t on $\overline{K}$ given by the system $\dot{r} = v(r,\varphi)$, $\dot{\varphi} = \psi(r,\varphi)$, $(r,\varphi) \in \overline{K}$, is smooth of class C^∞. The restriction of f_2^t to $\operatorname{int}\overline{K}$ is a spiral flow on an open annulus.

The restriction of f_2^t to $\check{K} = \overline{K} \setminus P_2 \cup P_3$ is denoted by $\check{f}_2^t$. By construction, $\check{f}_2^t$ is a C^∞-flow, and is called a spiral flow of "plus" type on an annulus.

b) A spiral flow $\check{f}_3^t$ of "minus" type. We use the preceding notation, and we take a C^∞-function $\psi_1\colon \overline{K} \to [0,1]$ that is equal to 0 on $P_2 \cup P_3$ and on the closed neighborhood $1.5 - \varepsilon \le r \le 1.5 + \varepsilon$ $(0 < \varepsilon < 0.125)$ of the circle $r = 1.5$. At the remaining points of $\overline{K}$ the function ψ_1 takes strictly positive values. Then the function

$$\psi_2(r,\varphi) = \begin{cases} -\psi_1(r,\varphi), & 1 \le r \le 1.5 \\ \psi_1(r,\varphi), & 1.5 \le r \le 2 \end{cases}$$

is smooth of class C^∞, and the flow f_3^t on $\overline{K}$ given by the system $\dot{r} = v(r,\varphi)$, $\dot{\varphi} = \psi_2(r,\varphi)$, $(r,\varphi) \in \overline{K}$, is a C^∞-flow. The restriction of f_3^t to $\operatorname{int}\overline{K}$ is also a spiral flow on an open annulus.

Let $\check{f}_3^t$ be the restriction of f_3^t to $\check{K}$. By construction, $\check{f}_3^t$ is a C^∞-flow, and is called a spiral flow of "minus" type on an annulus.

The flows $\check{f}_0^t$, $\check{f}_1^t$, $\check{f}_2^t$, and $\check{f}_3^t$ constructed according to the above four constructions are called model flows or smooth models.

LEMMA 4.4. *Suppose that f^t is a flow with finitely many singular trajectories on an orientable closed surface $\mathcal{M}$, and let R be a cell of f^t with nonempty boundary ∂R. Then the restriction $f^t|_{R\cup\delta\check{R}}$ of f^t to the invariant set $R \cup \delta\check{R}$ (where δR is the accessible (from within) boundary of R) is topologically equivalent (and even topologically orbitally equivalent) to one of the model C^∞- flows $\check{f}_i^t$, $i = 0, 1, 2, 3$.*

PROOF. Since $\partial R \neq \emptyset$, Theorem 4.3 gives us that the restriction $f^t|_R$ of f^t to R is a parallel flow on an open strip, or a spiral flow on an open annulus, or a parallel flow on an open annulus. Let us consider the case when $f^t|_R$ is a parallel flow on an open annulus. We let R_i ($i = 0,\ 1$) be the line $y = i$, and we use the notation in §4.5, 1).

Let f_0^t be the flow given by the system $\dot{x} = 1$, $\dot{y} = 0$ on the open strip Π bounded by the lines R_0 and R_1 (Figure 3.31). We take a trajectory L of f_0^t that coincides as a set with the line $y = y_0$, $0 < y_0 < 1$. Define $\Pi^+ = \{(x, y) \in \Pi : y_0 \leq y < 1\}$ and $\Pi^- = \{(x, y) \in \Pi : 0 < y \leq y_0\}$.

Let $\widetilde{L}$ be a trajectory in the cell R. By Theorem 4.3, $\widetilde{L}$ separates R into two disjoint invariant subsets R_0^+ and R_0^-, where R_0^+ (respectively, R_0^-) denotes the subset locally lying to the left (right) of $\widetilde{L}$ upon moving along $\widetilde{L}$ in the positive direction (Figure 3.37). We set $R^+ = R_0^+ \cup \widetilde{L}$ and $R^- = R_0^- \cup \widetilde{L}$. Denote by δR^+ the accessible (from within) boundary of R^+, with $\widetilde{L}$ removed, and let $\delta\check{R}^+$ be the family of regular trajectories in δR^+. We introduce the analogous concepts and notation δR^- and $\delta\check{R}^-$ for the set R^-.

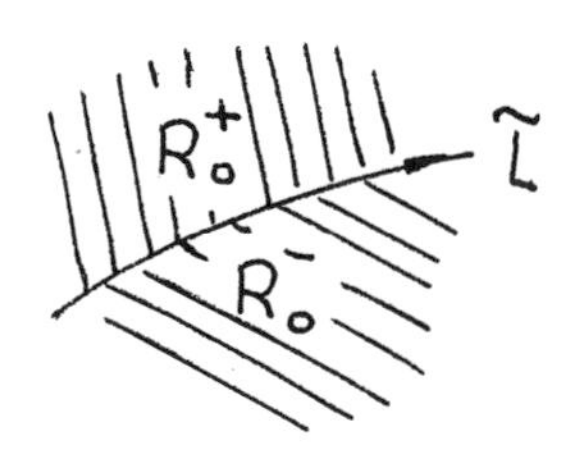

FIGURE 3.37

If $\delta\check{R} = \delta\check{R}^+ \cup \delta\check{R}^- = \emptyset$, then the flow $f^t|_{R\cup\delta\check{R}} = f^t|_R$ is topologically equivalent to the parallel flow f_0^t on the open strip Π. The flow f_0^t is the model C^∞-flow $\check{f}_0^t$ with $P_0 = R_0$ and $P_1 = R_1$.

Suppose that $\delta\check{R} \neq \emptyset$; say $\delta\check{R}^+ \neq \emptyset$. By the structure of the flow $f^t|_R$ (Theorem 4.3), there exists a contact-free segment Σ with endpoints on $\delta\check{R}^+$ and $\widetilde{L}$ that lies in $R^+ \cup \delta\check{R}^+$ and is such that each trajectory of the invariant set R_0^+ intersects Σ at exactly one point (Figure 3.38, a)).

The trajectories in the cell R induce a positive direction $\mathcal{J}$ on δR^+ (Figure 3.38, a)). Let $\dots, l_{-1}, l_0, l_1, \dots$ be regular trajectories on δR^+ (that is, trajectories in $\delta\check{R}^+$), written in the order in which they occur on δR^+ upon moving in the positive direction $\mathcal{J}$. Here we require that one of the endpoints of Σ belong to l_0. In the notation of §4.5, 1) we take a set $P_1 \subset R_1$ such that the number of regular

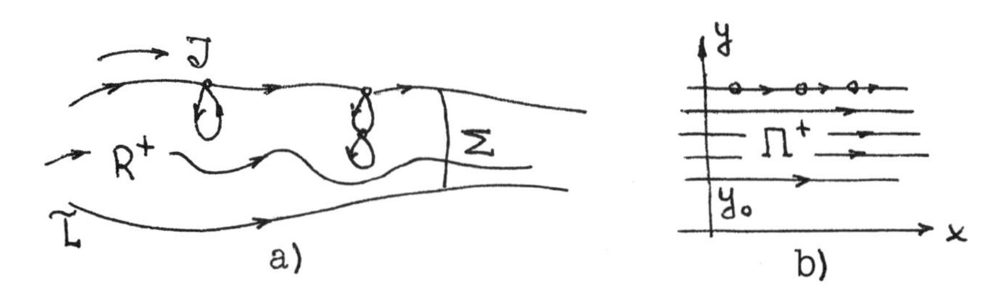

FIGURE 3.38

trajectories of the flow $\check{f}_0^t$, constructed as in §4.5, 1), on the line R_1 is equal to the number of regular trajectories of the flow f^t on δR^+, that is, to the number of trajectories of the family $\dots, l_{-1}, l_0, l_1, \dots$. Let $\dots, L_{-1}, L_0, L_1, \dots$ be the regular trajectories of the flow $\check{f}_0^t$ on R_1, written out in the order in which they occur upon moving along R_1 in the positive direction of the axis Ox. Then there exists a one-to-one correspondence $\tau\colon \{\dots, l_{-1}, l_0, l_1, \dots\} \to \{\dots, L_{-1}, L_0, L_1, \dots\}$ preserving the order of the trajectories. It can be assumed without loss of generality that $\tau(l_0) = L_0$, and that L_0 intersects the axis Oy.

The segment $\Sigma_0 = \{(x, y) \in \mathbb{R}^2 : x = 0,\ y_0 \le y \le 1\}$ is a contact-free segment of the flow $\check{f}_0^t$. Let $h\colon \Sigma \to \Sigma_0$ be an arbitrary homeomorphism satisfying the conditions $h(l_0 \cap \Sigma) = L_0 \cap \Sigma_0$ and $h(\widetilde{L} \cap \Sigma) = L \cap \Sigma_0$. Since each trajectory of the invariant set Π^+ of $\check{f}_0^t$ intersects Σ_0 at exactly one point (Figure 3.38, b)), the homeomorphism h can be extended with the help of τ to a homeomorphism $H^+\colon R^+ \cup \delta\check{R}^+ \to \Pi^+ \cup \delta\check{\Pi}^+$ realizing a topological equivalence (and even a topological orbital equivalence) of the flows $f^t|_{R^+ \cup \delta\check{R}^+}$ and $\check{f}_0^t|_{\Pi^+ \cup \delta\check{\Pi}^+}$. Similarly, we construct a homeomorphism $H^-\colon R^- \cup \delta\check{R}^- \to \Pi^- \cup \delta\check{\Pi}^-$ realizing a topological orbital equivalence of the flows $f^t|_{R^- \cup \delta\check{R}^-}$ and $\check{f}_0^t|_{\Pi^- \cup \delta\check{\Pi}^-}$ and coinciding with H^+ on $\widetilde{L}$. The homeomorphisms H^+ and H^- thus form a homeomorphism $H\colon R \cup \delta\check{R} \to \Pi \cup \delta\check{\Pi}$ realizing a topological orbital equivalence of the flows $f^t|_{R \cup \delta\check{R}}$ and $\check{f}_0^t$.

The cases when $f^t|_R$ is topologically equivalent to parallel or spiral flows on an open annulus can be handled according to the scheme presented above. We leave it as an exercise for the reader to fill in the details of the proof. □

We remark that $\delta\check{R}$ cannot be replaced by δR in the assertion of Lemma 4.4 because the number of equilibrium states has nothing to do with the number of regular trajectories of an arbitrary flow f^t on δR in general, in contrast to model flows (see Figure 3.38, a), b)).

4.6. Morse–Smale flows. The Morse–Smale flows occupy a special position among flows without nontrivial recurrent trajectories. For example, in the class of flows on closed orientable surfaces the Morse–Smale flows form a dense open set. Before giving the precise definition of a Morse–Smale flow we introduce the concept of a hyperbolic equilibrium state and the concept of a hyperbolic closed trajectory.

Let f^t be a C^r-flow ($r \ge 1$) on a surface $\mathcal{M}$, and let m_0 be an equilibrium state of f^t. Recall that f_t denotes the shift of the points of $\mathcal{M}$ along trajectories by the time t. Then f_t is a C^r-diffeomorphism of $\mathcal{M}$ onto itself, and m_0 is a fixed point of f_t for any $t \in \mathbb{R}$.

Denote by $T_{m_0}\mathcal{M}$ the tangent space of $\mathcal{M}$ at the point m_0, a real two-dimensional vector space. The derivative $\mathcal{D}f_t(m_0)$ of the diffeomorphism f_t at m_0 is a linear transformation $T_{m_0}\mathcal{M} \to T_{m_0}\mathcal{M}$, which is given by the Jacobi function matrix (of partial derivatives) in a chart containing m_0 [**76**].

DEFINITION. An equilibrium state m_0 of a C^r-flow f^t $(r \geq 1)$ is said to be *hyperbolic* if the eigenvalues of the linear transformation $\mathcal{D}f_1(m_0)$ are not equal to 1 in modulus.

We give an equivalent definition of a hyperbolic equilibrium state. Suppose that in a chart $\mathcal{U}$ containing m_0 and in a coordinate system $(x, y)\colon \mathcal{U} \to \mathbb{R}^2$ the flow f^t is given by the system

$$\dot{x} = P(x, y), \qquad \dot{y} = Q(x, y),$$

and let (x_0, y_0) be the coordinates of m_0 in $\mathcal{U}$. Then the functions $P(x, y)$ and $Q(x, y)$ are smooth of class C^r $(r \geq 1)$, and $P(x_0, y_0) = Q(x_0, y_0) = 0$.

An equilibrium state is hyperbolic if and only if

$$\Delta = \begin{vmatrix} P'_x(x_0, y_0) & P'_y(x_0, y_0) \\ Q'_x(x_0, y_0) & Q'_y(x_0, y_0) \end{vmatrix} \neq 0$$

and $\sigma = P'_x(x_0, y_0) + Q'_y(x_0, y_0) \neq 0$ for $\Delta > 0$; that is, the roots of the equation

$$\lambda^2 - \sigma\lambda + \Delta = 0$$

are not purely imaginary.

Both definitions of hyperbolicity are independent of the coordinate system in a neighborhood of m_0.

If an equilibrium state is hyperbolic, then it is isolated. What is more, a hyperbolic equilibrium state is locally topologically equivalent either to a node (stable or unstable), or to a saddle [**3**] (Figure 3.39).

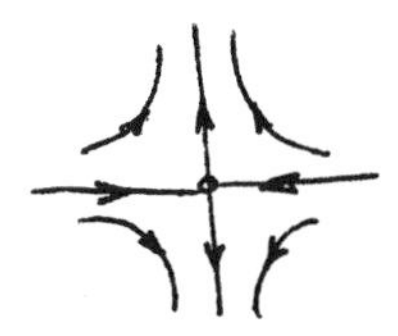

FIGURE 3.39

Let l_0 be a closed trajectory of a C^r-flow f^t of period $\tau > 0$, and let $m \in l_0$ be an arbitrary point.

DEFINITION. A closed trajectory l_0 of period $\tau > 0$ for a C^r-flow f^t $(r \geq 1)$ is said to be *hyperbolic* if exactly one eigenvalue of the linear transformation $\mathcal{D}f_\tau(m)$ is equal to 1, and the modulus of the second is different from 1 (consequently, the second eigenvalue is a real number with modulus different from 1).

By the group property $f_{t_1+t_2} = f_{t_1} \circ f_{t_2}$, the definition of hyperbolicity of a closed trajectory l_0 does not depend on the choice of the point $m \in l_0$.

We give an equivalent definition of hyperbolicity of a closed trajectory. Let Σ be a contact-free segment containing a point m, and let $P\colon \Sigma \to \Sigma$ be the Poincaré mapping for the trajectory l_0. According to Lemma 1.4 in Chapter 2, P is a C^r-diffeomorphism ($r \geq 1$) in its domain. Clearly, $m \in \mathrm{Dom}(P)$.

A closed trajectory l_0 is hyperbolic if and only if $|P'(m)| \neq 1$.

If a closed trajectory l_0 is hyperbolic, then it is isolated in the set of closed trajectories of f^t; that is, l_0 is a limit cycle. For $|P'(m)| < 1$ the limit cycle l_0 is stable (Figure 3.40, a)), while for $|P'(m)| > 1$ it is unstable (Figure 3.40, b)).

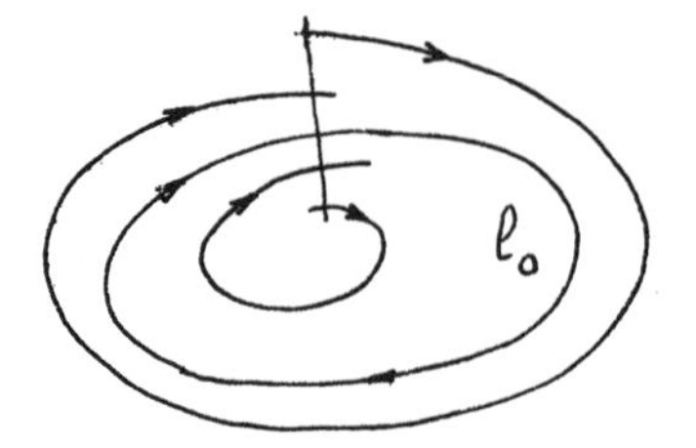

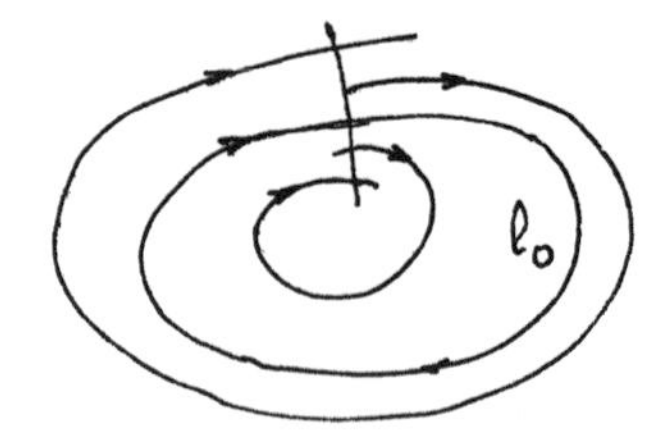

FIGURE 3.40

DEFINITION. A C^r-flow f^t ($r \geq 1$) on a surface $\mathcal{M}$ is called a *Morse–Smale flow* if the following conditions hold:

1) f^t has finitely many equilibrium states and finitely many closed trajectories, and they are all hyperbolic;

2) the limit set of any semitrajectory of f^t consists either of a single equilibrium state or of a single closed trajectory;

3) there are no separatrices going from a saddle to a saddle (in particular, no separatrix loops).

It follows from the condition 2) that in Morse–Smale flows there are no non-trivial recurrent trajectories.

The simplest example of a Morse–Smale flow on the sphere is the flow in §2.1 of Chapter 1, of "North-South Pole" type.

In Figure 3.41 we picture the simplest Morse–Smale flow with four equilibrium states on the torus (one stable node, one unstable node, and two saddles).

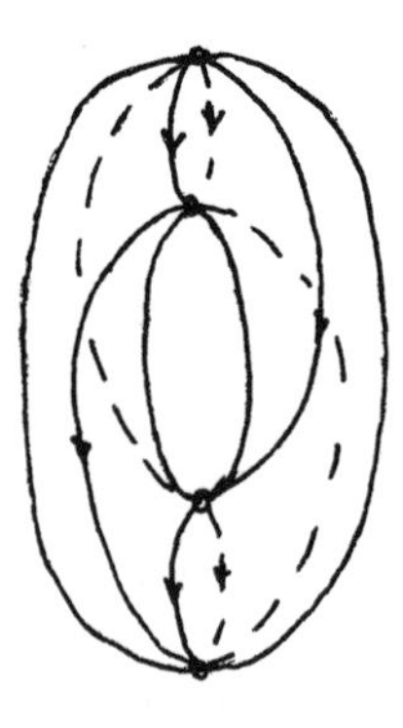

FIGURE 3.41

As shown in §2.3 of Chapter 1, a pretzel (a closed orientable surface of genus two) can be obtained from an octagon after an appropriate identification of its sides (Figure 3.42, a)). Using this, we picture in Figure 3.42, b) a Morse–Smale flow on

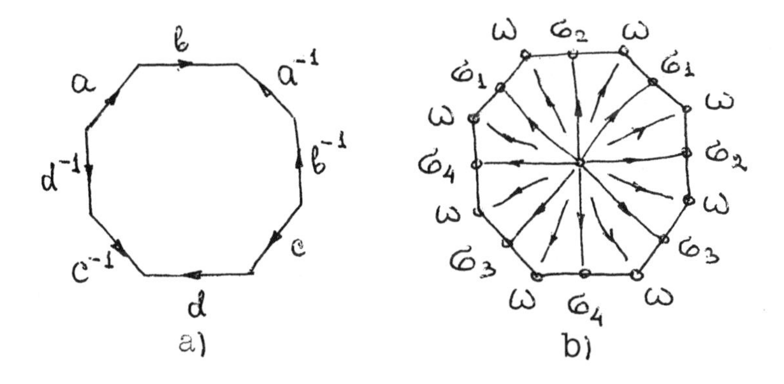

FIGURE 3.42

a pretzel. The flow has one stable node ω, one unstable node (at the center of the octagon), and four saddles σ_1, σ_2, σ_3, and σ_4.

DEFINITION. A Morse–Smale flow is said to be *polar* if it does not have closed trajectories, and the family of equilibrium states contains exactly one stable node and exactly one unstable node (the rest of the equilibrium states, if there are any, are saddles).

All the above examples of Morse–Smale flows are polar. There are other examples of Morse–Smale flows in the books [**3**], [**64**], and [**74**].

4.7. Cells of Morse–Smale flows. The next result is an immediate consequence of the definition of a Morse–Smale flow and of Theorems 4.1–4.3.

LEMMA 4.5. *Let f^t be a Morse–Smale flow on an orientable closed surface $\mathcal{M}$, and let R be a cell of f^t. Then:*

1) *R consists of nonclosed trajectories;*

2) *as $t \to \infty$ all the trajectories in R tend either to a single equilibrium state (a stable node or focus) or to a single limit cycle, and the same is also true as $t \to -\infty$;*

3) *if R is homeomorphic to an open disk, then its accessible (from within) boundary contains a saddle and at least two separatrices.*

§5. The space of flows

In this section flows on a fixed surface $\mathcal{M}$ are regarded as points of a set. In this set of flows we introduce a metric, which in general depends on the differentiable structure (and hence the metric) on $\mathcal{M}$, and which turns the set of flows into a metric topological space. This enables us to investigate various classes of flows (in particular, Morse–Smale flows) according to their situation in the space of flows. Using the metric introduced, we define important concepts such as structural stability, degree of structural instability, and others.

5.1. The metric in the space of flows. Suppose that a compact surface $\mathcal{M}$ has the structure of a two-dimensional differentiable manifold. This means that there is a covering $\mathcal{E}(\mathcal{M})$ of $\mathcal{M}$ by domains (charts) homeomorphic to an open disk such that each domain $\mathcal{U} \in \mathcal{E}(\mathcal{M})$ has a coordinate system $(x, y)\colon \mathcal{U} \to \mathbb{R}^2$, and the transition from one set of coordinates to another in overlapping domains is realized by means of analytic functions with nonzero Jacobian.

Let f^t and g^t be C^{r+0}-flows on $\mathcal{M}$ $(r \geq 0)$. Then in each chart $\mathcal{U} \in \mathcal{E}(\mathcal{M})$ with coordinate system $(x, y)\colon \mathcal{U} \to \mathbb{R}^2$ the flows f^t and g^t are given by respective systems of differential equations

$$\dot{x} = P(x, y), \qquad \dot{y} = Q(x, y)$$
$$\dot{x} = \widetilde{P}(x, y), \qquad \dot{y} = \widetilde{Q}(x, y)$$

where $P(x, y)$, $Q(x, y)$, $\widetilde{P}(x, y)$, and $\widetilde{Q}(x, y)$ are smooth functions of class C^r.

For an integer $0 \leq k \leq r$ we define the distance $\rho_k(f^t, g^t)$ between the flows f^t and g^t to be the maximum of the following quantities over all domains $\mathcal{U} \in \mathcal{E}(\mathcal{M})$:

$$\max_{\mathcal{U}} |P - \widetilde{P}|, \qquad \max_{\mathcal{U}} |Q - \widetilde{Q}|$$
$$\max_{\mathcal{U}} \left| \frac{\partial^s P}{\partial x^i \partial y^j} - \frac{\partial^s \widetilde{P}}{\partial x^i \partial y^j} \right|, \qquad \max_{\mathcal{U}} \left| \frac{\partial^s Q}{\partial x^i \partial y^j} - \frac{\partial^s \widetilde{Q}}{\partial x^i \partial y^j} \right|,$$

where $i + j = s \leq k$.

The distance introduced determines a metric ρ_k on the set of C^r-flows, $0 \leq k \leq r$.

Let $\mathfrak{X}_r^k(\mathcal{M})$ be the metric space of C^r-flows on $\mathcal{M}$, equipped with the metric ρ_k. For $k = r$ we let $\mathfrak{X}^r(\mathcal{M}) \overset{\text{def}}{=} \mathfrak{X}_r^r(m)$.

5.2. The concepts of structural stability and the degree of structural instability. On a compact surface $\mathcal{M}$ we fix a metric ρ compatible with the differentiable structure on $\mathcal{M}$.

DEFINITION. A homeomorphism $\varphi\colon \mathcal{M} \to \mathcal{M}$ is called an *ε-homeomorphism* $(\varepsilon > 0)$ if $\rho(m, \varphi(m)) < \varepsilon$ for any point $m \in \mathcal{M}$.

Let f^t and g^t be C^r-flows $(r \geq 1)$ on $\mathcal{M}$, and let $1 \leq k \leq r$.

DEFINITION. The two flows f^t and g^t are said to be *δ-close* $(\delta > 0)$ in the space $\mathfrak{X}_r^k(\mathcal{M})$ if $\rho_k(f^t, g^t) < \delta$.

DEFINITION. A C^r-flow f^t is said to be *structurally stable* in the space $\mathfrak{X}_r^k(\mathcal{M})$ if for any $\varepsilon > 0$ there is a $\delta > 0$ such that each flow g^t that is δ-close to f^t in $\mathfrak{X}_r^k(\mathcal{M})$ is topologically equivalent to f^t by an ε-homeomorphism.

Thus, structural stability of a flow means that any small perturbation of this flow results in a flow topologically equivalent to the original flow, with the homeomorphism implementing the topological equivalence close to the identity in the C^0-topology.

The concept of structural stability was first introduced in 1937 by Andronov and Pontryagin [**5**]. They considered dynamical systems in a planar domain bounded by a contact-free cycle, and they defined δ-closeness in the space $\mathfrak{X}_r^1$, $r \geq 1$. Anosov therefore proposed using the name Andronov–Pontryagin structural stability for the above definition of structural stability [**6**].

If in the definition of structural stability it is not required that the homeomorphism implementing a topological equivalence of f^t and g^t be an ε-homeomorphism, then we get the definition of *weak structural stability* of a flow f^t (or structural stability in the Peixoto sense [**6**]), which was introduced in [**103**].

We proceed to the concept of the degree of structural instability.

DEFINITION. A C^r-flow f^t $(r \geq 1)$ on a compact surface $\mathcal{M}$ is called a flow *of the first degree of structural instability* in the space $\mathfrak{X}_r^k(\mathcal{M})$, $1 \leq k \leq r$, if it is not structurally stable and if for any $\varepsilon > 0$ there is a $\delta > 0$ such that each structurally unstable flow g^t that is δ-close to f^t in $\mathfrak{X}_r^k(\mathcal{M})$ is topologically equivalent to f^t by an ε-homeomorphism.

Thus, flows of the first degree of structural instability are flows that are structurally unstable but are "structurally stable in the set of structurally unstable flows" (that is, are relatively structurally stable). These flows play a large role in the theory of bifurcation of flows on surfaces, since the simplest bifurcations of structurally stable flows pass through them.

Similarly, flows of the jth degree of structural instability $(j \geq 2)$ are defined to be flows that are not structurally stable and are not flows of the 1st, $\dots, j-1$st degrees of structural instability, but are relatively structurally stable in the set of flows remaining after removal from $\mathfrak{X}_r^k(\mathcal{M})$ of the structurally stable flows and the flows of the 1st, $\dots, j-1$st degrees of structural instability.

The degrees of structural instability establish a hierarchy in the space of flows according to the degree of sensitivity to perturbations.

5.3. The space of structurally stable flows. This subsection has the nature of a survey.

THEOREM 5.1. *A C^r-flow f^t $(r \geq 1)$ on a closed orientable surface $\mathcal{M}$ is structurally stable in the space $\mathfrak{X}_r^k(\mathcal{M})$ for any $1 \leq k \leq r$ if and only if f^t is a Morse–Smale flow.*

THEOREM 5.2. *On a closed orientable surface $\mathcal{M}$ the set of Morse–Smale C^r-flows $(r \geq 1)$ is open and dense in the space $\mathfrak{X}_r^k(\mathcal{M})$ for any $1 \leq k \leq r$.*

For the sphere $S^2 = \mathcal{M}$ Theorems 5.1 and 5.2 follow from [**5**]. They are proved in [**103**] for the general case. There are proofs of these theorems in the books [**64**] and [**74**].

The structural stability of the flow in Theorems 5.1 and 5.2 can be understood both in the Andronov–Pontryagin sense and in the Peixoto sense.

As follows from [**8**], [**64**], [**65**], [**85**], and [**103**], Theorems 5.1 and 5.2 are valid for closed nonorientable surfaces of genus $p = 1$ (the projective plane), $p = 2$ (the Klein bottle), and $p = 3$ (the torus with a Möbius cap attached). Further, by the C^1-closing lemma [**69**], both the theorems hold in the space $\mathfrak{X}^1(\mathcal{M})$, both for orientable and for nonorientable closed surfaces $\mathcal{M}$. As for the remaining possibilities of generalizing Theorems 5.1 and 5.2 for two-dimensional surfaces, there are several open questions here. It is not known (at present) whether a structurally stable flow $f^t \in \mathfrak{X}_r^k(\mathcal{M})$, $1 < k \leq r$, necessarily belongs to the class of Morse–Smale flows for nonorientable closed surfaces of genus $p \geq 4$ (this assertion is valid in the other direction). The question of whether the set of structurally stable C^r-flows $(r \geq 1)$ is dense in $\mathfrak{X}_r^k(\mathcal{M})$ $(1 < k \leq r)$ also remains open for a nonorientable closed surface of genus $p \geq 4$. The main difficulty in the investigation of these questions is the existence of nonorientable nontrivial recurrent trajectories. We describe this result in greater detail.

Let f^t be a flow with a nontrivial recurrent trajectory l on a closed nonorientable surface $\mathcal{M}$. This trajectory is called a *nonorientable nontrivial recurrent trajectory* if for any point $m \in l$ and any contact-free segment Σ passing through m the semitrajectories $l^+(m)$ and $l^-(m)$ both have the following property: there exists a Σ-arc $\overset{\frown}{ab}$ of the semitrajectory $l^{()}(m)$ $(a, b \in l^{()}(m) \cap \Sigma)$ such that the simple closed curve $\overset{\frown}{ab} \cup \overline{ab}$ (where $\overline{ab} \subset \Sigma$) is a one-sided curve (that is, has a neighborhood homeomorphic to an open Möbius band (Figure 3.43).

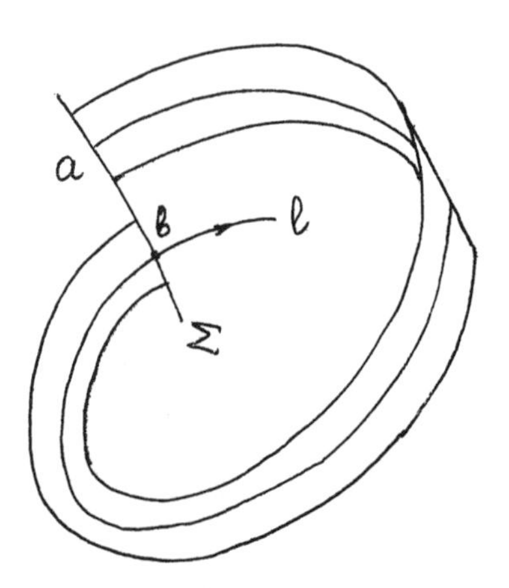

FIGURE 3.43

In the paper, *Smooth nonorientable nontrivial recurrence on two-manifolds* (J. Differential Equations **29** (1978), 338–395), Gutierrez showed that on any nonorientable compact surface of genus $p \geq 4$ there exists a smooth flow with a nonorientable nontrivial recurrent trajectory. The existence of such trajectories prevents extending Peixoto's proofs of Theorems 5.1 and 5.2 to the case of a nonorientable closed surface $\mathcal{M}$ of genus $p \geq 4$ and the space $\mathfrak{X}_r^k(\mathcal{M})$, $1 < k \leq r$. We remark that the flow with a nonorientable nontrivial recurrent trajectory constructed by Gutierrez in the above paper can be approximated by Morse–Smale flows in the C^∞-topology.

Recall that on nonorientable closed surfaces of genera $p = 1$ and $p = 2$ there are no flows with nontrivial recurrent trajectories (§2.2 in Chapter 2). Such flows exist on a nonorientable closed surface of genus $p = 3$, but as shown in [**85**], they do not have nonorientable nontrivial recurrent trajectories. Therefore, the proofs of Theorems 5.1 and 5.2 extend to the case of a nonorientable closed surface of genus 1, 2, or 3.

In [**79**] there is a treatment of flows on the torus T^2 given by polynomial vector fields. Let $\widehat{\pi}\colon \mathbb{R}^2 \to T^2$ be the universal covering of the form $\widehat{\pi}(x, y) = (e^{ix}, e^{iy})$. Recall that the expression

$$M_k(x,y) = \sum_{j=0}^{k} \Bigg[\sum_{m+n=j} (a_{mn} \cos mx \cos ny + b_{mn} \sin mx \cos ny + c_{mn} \cos mx \sin ny + d_{mn} \sin mx \sin ny) \Bigg]$$

is called a trigonometric polynomial of degree k.

A vector field $\vec{V}$ on T^2 is a polynomial vector field of degree k if for a covering vector field on $\mathbb{R}^2$ (with respect to the covering $\widehat{\pi}$) both components are trigonometric polynomials of degree k.

Denote by A_k the set of polynomial vector fields of degree k that determine a flow on the torus with at least one equilibrium state.

It is shown in [**79**] that the set of Morse–Smale vector fields (that is, the fields giving Morse–Smale flows) is open and dense in A_k for all $k \in \mathbb{N}$.

We dwell on a paper by Gutierrez and de Melo, *The connected components of Morse–Smale vector fields on two-manifolds* (Lecture Notes in Math., vol. 597, Springer-Verlag, Berlin, 1977, pp. 230–251). Here the set of Morse–Smale vector fields on an orientable compact surface is regarded as a topological space, and the arcwise connected components of this space are investigated.

Denote by $\Sigma^r(\mathcal{M})$ the space of Morse–Smale vector fields of class C^r $(r \geq 1)$ on an orientable compact surface $\mathcal{M}$.

DEFINITION. Two vector fields $X, Y \in \Sigma^r(\mathcal{M})$ are said to be *isotopically equivalent* if there exists a continuous mapping $F\colon [0,1] \to \Sigma^r(\mathcal{M})$ such that $F(0) = X$ and $F(1) = Y$.

In the paper under consideration there are necessary and sufficient conditions for Morse–Smale vector fields in $\Sigma^r(\mathcal{M})$ to be isotopically equivalent. We illustrate the results in the simplest case of polar Morse–Smale vector fields.

Let $X \in \Sigma^r(\mathcal{M})$ be a polar Morse–Smale vector field which determines a flow $f^t \in \mathfrak{X}^r(\mathcal{M})$, $r \geq 1$. By the definition of a polar vector field, f^t has exactly one unstable node $u(X)$. For each saddle S of the flow f^t two of its ω-separatrices combine with $u(X)$ and S to form a simple closed curve called a *stable cycle* of the vector field X. Two different stable cycles of X intersect only at the point $u(X)$.

It is proved in the paper that two polar vector fields $X, Y \in \Sigma^r(\mathcal{M})$ are isotopically equivalent if and only if there exists a bijection (a one-to-one correspondence) between the stable cycles of X and the stable cycles of Y such that stable cycles corresponding under this bijection are homotopic.

Similar assertions are proved for gradient-like Morse–Smale vector fields also in the general case.

5.4. Flows of the first degree of structural instability. We first describe the structurally unstable equilibrium states and closed trajectories that are possible for a flow f^t of the first degree of structural instability on an orientable surface $\mathcal{M}$.

It can be assumed without loss of generality that the equilibrium state m_0 or the closed trajectory l of f^t under consideration belongs to a single chart $\mathcal{U} \in \mathcal{E}(\mathcal{M})$ with coordinate system $(x, y)\colon \mathcal{U} \to \mathbb{R}^2$ in which the flow f^t of the first degree of structural instability is given by the system

$$\dot{x} = P(x, y), \qquad \dot{y} = Q(x, y). \tag{5.1}$$

It is shown in [**3**] that the concept of the first degree of structural instability is meaningful in the space $\mathfrak{X}^r(\mathcal{M})$ of flows, $r \geq 3$. Therefore, it will be assumed that $f^t \in \mathfrak{X}^r(\mathcal{M})$, $r \geq 3$, and hence the functions $P(x, y)$ and $Q(x, y)$ are smooth of class C^r, $r \geq 3$.

Let $(0, 0)$ be the coordinates of the equilibrium state $m_0 \in \mathcal{U} \subset \mathcal{M}$, that is, $P(0, 0) = Q(0, 0) = 0$. We assume that

$$\Delta = (P'_x Q'_y - Q'_x P'_y)|_{(0,0)} = 0,$$
$$\sigma = (P'_x + Q'_y)|_{(0,0)} \neq 0.$$

It is known [**3**] that with the help of a nonsingular linear change of the coordinates (x, y) we can pass to new coordinates in which the right-hand sides of the system (5.1) have the form (we denote the new coordinates again by x, y)

$$P(x,y) = P_2(x,y) + \varphi(x,y), \qquad Q(x,y) = y + Q_2(x,y) + \psi(x,y),$$

where $P_2(x,y)$ and $Q_2(x,y)$ are homogeneous second-degree polynomials in x and y, and the functions φ and ψ and their derivatives through second order are equal to zero at $x = y = 0$.

The equilibrium state $(0,0)$ of the system (5.1) is called a *saddle-node of multiplicity two* if $P_2(1,0) \neq 0$ (Figure 3.44).

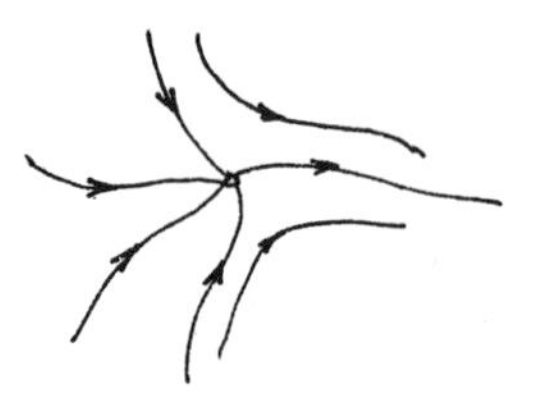

FIGURE 3.44

Let us consider the equilibrium state $(0,0)$ of (5.1) when

$$\Delta(0,0) > 0, \qquad \sigma(0,0) = 0.$$

Then ([**3**], [**4**]) the system (5.1) can be reduced by a change of coordinates to the form

$$\dot{x} = y + \varphi(x,y), \qquad \dot{y} = -x + \psi(x,y) \tag{5.2}$$

where φ and ψ and their first-order derivatives are equal to zero at $x = y = 0$. After passage to polar coordinates (ρ, θ) the system (5.2) reduces to the equation

$$\frac{d\rho}{d\theta} = R(\rho, \theta),$$

where $R(0,\theta) \equiv 0$.

We expand the function $R(\rho,\theta)$ at the point $\rho = 0$ according to the Taylor formula: $R(\rho,\theta) = R_1(\theta)\rho + R_2(\theta)\rho^2 + R_3(\theta)\rho^3 + \dots$. It is known ([**2**]) that the first nonzero quantity $\alpha_i = \int_0^{2\pi} R_i(\theta)\,d\theta$, $i = 1, 2, \dots$, has odd index. Since $\sigma = 0$, it follows that $\alpha_1 = 0$, and hence $\alpha_2 = 0$.

The equilibrium state $(0,0)$ of the system (5.1) is called a *compound focus of multiplicity* 1 if $\Delta(0,0) > 0$, $\sigma(0,0) = 0$, and $\alpha_3 \neq 0$.

We proceed to a description of structurally unstable closed trajectories.

Let l_0 be a closed trajectory of the flow f^t. Through a point $m \in l_0$ we draw a contact-free segment Σ and introduce on it a regular parameter $S\colon [-1,1] \to \Sigma$ with $S(0) = m$. The Poincaré mapping $P\colon \Sigma \to \Sigma$ is defined in a neighborhood of m. We expand the function $\varphi(n) = S^{-1} \circ P \circ S(n)$, $n \in [-1,1]$, in a neighborhood of the point $n = 0$: $\varphi(n) = h_1 n + h_2 n^2 + \dots$ (obviously, $\varphi(0) = 0$).

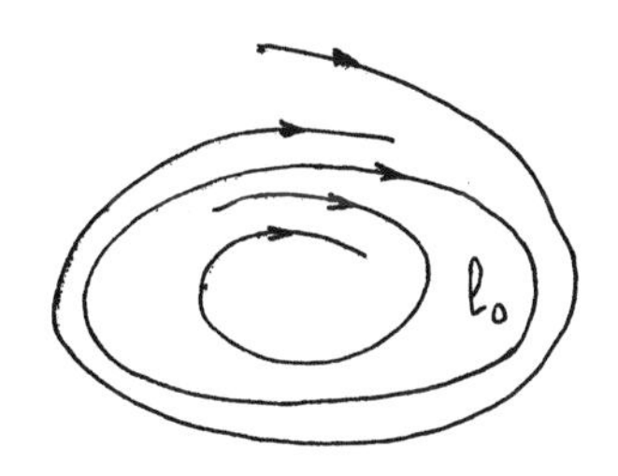

FIGURE 3.45

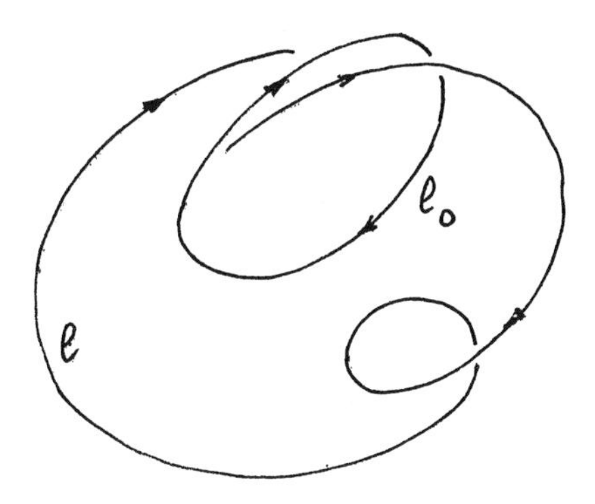

FIGURE 3.46

The closed trajectory l_0 is called a *double limit cycle* if $h_1 = 1$ and $h_2 \neq 0$ (Figure 3.45). This definition does not depend on the choice of the point $m \in l_0$ and the segment Σ.

The trajectory l is said to be *doubly asymptotic* to the limit cycle l_0 if $\omega(l) = \alpha(l) = l_0$ (Figure 3.46).

Obviously, in this case the limit cycle l_0 is semistable (from one side the trajectories wind off of the cycle l_0, and from the other side they wind onto l_0) and nonhyperbolic.

DEFINITION. A trajectory l of a flow f^t on an orientable surface $\mathcal{M}$ is called a *structurally unstable singular trajectory* if it is one of the following trajectories: 1) a nonhyperbolic equilibrium state; 2) a nonhyperbolic closed trajectory; 3) a separatrix going from an equilibrium state to an equilibrium state.

It follows from Theorem 5.1 that a structurally unstable flow f^t on an orientable closed surface $\mathcal{M}$ has a structurally unstable singular trajectory or a nontrivial recurrent trajectory.

THEOREM 5.3 [**7**]. *A C^r-flow f^t $(r \geq 3)$ on a closed orientable surface $\mathcal{M}$ is a flow of the first degree of structural instability in the space $\mathfrak{X}_r^k(\mathcal{M})$, $3 \leq k \leq r$, if and only if the following conditions hold:*

1) *f^t does not have nontrivial recurrent trajectories.*

2) *f^t does not have trajectories doubly asymptotic to a double limit cycle.*

3) *f^t has one and only one structurally unstable singular trajectory, which is one of the following types:*

a) *a compound focus of multiplicity* 1;

b) *a saddle-node of multiplicity* 2;

c) *a double limit cycle*;

d) *a separatrix going from a saddle to a saddle, and if it returns to the initial saddle, then the saddle value is nonzero.*

4) *the separatrices of saddles and of a saddle-node satisfy the following conditions*:

a) *a separatrix cannot twist onto a separatrix forming a loop (or twist off of this separatrix)*;

b) *a separatrix cannot twist onto (twist off of) a double limit cycle if there is another separatrix twisting off of it (twisting onto it)*;

c) *the separatrices of a saddle-node cannot go to a saddle (that is, they are not two-sided separatrices) and cannot belong to the boundary of a node sector of a saddle-node that forms a loop.*

There is a proof in [**7**], but the reader of this paper should first see [**2**] and [**3**].

5.5. On denseness of flows of the first degree of structural instability in the space of structurally unstable flows. According to Theorem 5.2, the set of structurally stable C^r-flows ($r \geq 1$) on a closed orientable surface $\mathcal{M}$ is dense in the space $\mathfrak{X}_r^k(\mathcal{M})$ of C^r-flows, $1 \leq k \leq r$. It is natural to suppose that the set of flows of the first degree of structural instability (that is, relatively structurally stable flows) is dense in the set of structurally unstable flows. However, as the next theorem shows, the situation here is more complicated and depends on the genus of the surface.

Denote by $\mathfrak{N}^r(\mathcal{M})$ the set of structurally unstable C^r-flows ($r \geq 1$) on an orientable closed surface $\mathcal{M}$. We shall regard $\mathfrak{N}^r(\mathcal{M})$ as a topological space with the topology induced by the space $\mathfrak{X}^r(\mathcal{M})$.

Let $\widetilde{\mathfrak{X}}_{1,r}(\mathcal{M})$ be the set of C^r-flows ($r \geq 3$) of the first degree of structural instability on an orientable closed surface $\mathcal{M}$.

THEOREM 5.4. 1) *The set $\widetilde{\mathfrak{X}}_{1,r}(\mathcal{M})$ of flows of the first degree of structural instability is open in the space $\mathfrak{N}^r(\mathcal{M})$ of structurally unstable flows.*

2) *The space $\widetilde{\mathfrak{X}}_{1,r}(S^2)$ of flows of the first degree of structural instability on the sphere S^2 is dense in the space $\mathfrak{N}^r(S^2)$, $r \geq 3$.*

3) *The set $\widetilde{\mathfrak{X}}_{1,r}(\mathcal{M}_p)$ of flows of the first degree of structural instability on a closed orientable surface $\mathcal{M}_p$ of genus $p \geq 2$ is not dense in the space $\mathfrak{X}^r(\mathcal{M}_p)$, $r \geq 3$.*

4) *If $\mathcal{M}_1 = T^2$ is the torus (an orientable closed surface of genus $p = 1$), then:*

a) *in the space of C^r-flows on T^2 without equilibrium states the set of flows of the first degree of structural instability is dense in the space of structurally unstable flows*;

b) *in the space of C^r-flows on T^2 with equilibrium states the set of flows of the first degree of structural instability is not dense in the space of structurally unstable flows, $r \geq 3$.*

There is a proof of Theorem 5.4 in [**12**] and [**13**].

CHAPTER 4

Local Structure of Dynamical Systems

The so-called method of normal forms, which goes back to Poincaré, is a powerful instrument for studying the local structure of dynamical systems. The essence of this method is to choose a coordinate system in a neighborhood of the set of most interest to us in which the system has as simple a form as possible (a "normal form"). The dynamics of a normal form is usually easy to study by virtue of its simplicity.

The rectification theorem (see §2.7.2 in Chapter 1) is the simplest example of this approach. The system

$$\dot{\xi} = 1, \quad \dot{\eta} = 0 \qquad (x = (\xi, \eta) \in \mathbb{R}^2)$$

is taken as the normal form of a dynamical system in a neighborhood of a regular point. The trajectories of this system are straight lines. The rectification theorem thereby completely solves the problem of the local structure of trajectories in a neighborhood of a regular point.

The structure of a dynamical system in a neighborhood of a rest point is considerably more complicated in general. The simplest system in a neighborhood of the point $x = 0$ serving as a rest point is a linear system

$$\dot{x} = \Lambda x.$$

Therefore, it is natural first to study the possibility of reducing a system to linear normal form in a neighborhood of a rest point.

In connection with local behavior the general concept of structural stability leads to the definition of a *hyperbolic singular point*. For a system of differential equations

$$\dot{x} = Ax + o(x)$$

of class C^1 the point $x = 0$ is said to be hyperbolic if the spectrum of the linear part A does not intersect the imaginary axis.

A C^1-diffeomorphism

$$F(x) = \Lambda x + o(x)$$

is said to be hyperbolic at the point $x = 0$ if the spectrum of Λ does not intersect the unit circle.

We recall that diffeomorphisms F and G are said to be $C^{r,\alpha}$-*conjugate*[1] in a neighborhood of a common fixed point x_0, or *locally* $C^{r,\alpha}$-*conjugate at* x_0 if there exist a neighborhood $V \ni x_0$ and a $C^{r,\alpha}$- diffeomorphism

$$\Phi\colon (V, x_0) \to (W, x_0)$$

[1]In this chapter the notation $C^{r,\alpha}$ assumes that $r \geq 0$ is an integer and $\alpha \in [0,1]$. This is the space of mappings in C^r with highest derivatives in $\operatorname{Lip}\alpha$. By definition, $C^{r,0} = C^r$.

(with a different neighborhood W in general) such that

$$(\Phi F \Phi^{-1})(x) = G(x)$$

for all x sufficiently close to x_0. We also say that F and G are *locally conjugate in the class* $C^{r,\alpha}$.

If $r = 0$, then Φ must be a homeomorphism. A *homeomorphism of class* $C^{0,\alpha}$ is understood to be a homeomorphism Φ that is in $\operatorname{Lip}\alpha$ along with its inverse Φ^{-1}.

The concept of local conjugacy carries over in the obvious way to flows: it is required that

$$\Phi \circ F^t \circ \Phi^{-1}(x) = G^t(x)$$

for small t and for x sufficiently close to x_0. In the classification of flows, especially the topological classification, it is natural to consider equivalence and orbital equivalence (see the definitions in §1 of Chapter 3). However, it should be noted that, from the point of view of the methods and results to be discussed in this chapter, conjugacy and equivalence of flows do not differ significantly, at least in dimension ≥ 2.

§1. Dynamical systems on the line

1.1. Linearization of a diffeomorphism. Let

$$F\colon (\mathbb{R}^1, 0) \to (\mathbb{R}^1, 0)$$

be a diffeomorphism of class $C^{r,\alpha}$ with the fixed point $x_0 = 0$, that is, $F(0) = 0$, and let

$$\lambda = F'(0).$$

The general definition of a hyperbolic fixed point reduces here to the inequality

$$|\lambda| \neq 1.$$

THEOREM 1.1. *If $r + \alpha > 1$ and if a diffeomorphism F of the line of class $C^{r,\alpha}$ is hyperbolic at the point $x_0 = 0$, then it is locally conjugate in the class $C^{r,\alpha}$ to a linear diffeomorphism. A diffeomorphism F of class C^1 is locally conjugate in the class $C^{0,\beta}$ to a linear diffeomorphism for any $\beta < 1$.*

We give straightway an example showing that the statement of the theorem is sharp: the inequality $\beta < 1$ cannot be replaced by $\beta = 1$ in general.

Let us consider the local diffeomorphism

$$F(x) = \lambda x + \frac{\lambda x}{\ln |x|}, \qquad F(0) = 0.$$

We show that it is not locally conjugate to a linear diffeomorphism in the class $C^{0,1}$. Indeed, suppose that $\Phi \in C^{0,1}$, $\Phi(0) = 0$, and

$$\Phi(Fx) = \lambda \Phi(x)$$

for small x. Then

$$\frac{\lambda^{-n}\Phi(F^n x)}{\Phi(x)} = 1 \qquad (n = 0, 1, \dots).$$

Since $\Phi, \Phi^{-1} \in \operatorname{Lip} 1$ (according to the definition of a homeomorphism of class $C^{0,1}$), it follows that

$$c_2|x| \leq \Phi(x) \leq c_1|x|$$

for some constants $c_i > 0$. Therefore,

$$1 \geq \frac{c_2}{c_1} \frac{|\lambda^{-n} F^n x|}{|x|} = \frac{c_2}{c_1} \prod_{k=1}^{n} \frac{\lambda^{-1}|F(F^{k-1}x)|}{|F^{k-1}x|} = \frac{c_2}{c_1} \prod_{k=1}^{n} \left(1 + \frac{1}{\ln |F^{k-1}x|}\right).$$

To get a contradiction it suffices to verify that

$$\sum_{k=1}^{\infty} \frac{1}{\ln |F^{k-1}x|} = -\infty. \tag{1.1}$$

Indeed, if $\delta > 0$ is sufficiently small, then

$$|F^k x| \geq \lambda^k (1-\varepsilon)^k |x| \qquad (|x| \leq \delta, \quad k = 0, 1, \dots).$$

Consequently,

$$\frac{1}{\ln |F^k x|} \leq \frac{1}{k \ln \lambda (1-\varepsilon) + \ln |x|},$$

which yields (1.1).

1.2. Lemmas on functional equations. The proofs of the conjugacy theorems reduce to proofs that the equation

$$\Phi(Fx) = G(\Phi(x)), \qquad \Phi(x_0) = x_0, \tag{1.2}$$

is solvable in a neighborhood of x_0 with the additional condition that Φ be a local diffeomorphism of class $C^{r,\alpha}$. For $r \geq 1$ the last condition is in turn equivalent to the derivative $\Phi'(x_0)$ being nonsingular. If $r = 0$, then an additional check that Φ be a local homeomorphism is required.

We look for a reducing diffeomorphism Φ in the form

$$\Phi(x) = x + \varphi(x), \qquad \varphi(0) = 0.$$

Then for φ the equation (1.2) takes the form

$$\varphi(Fx) = G(x + \varphi(x)) - F(x).$$

If we set

$$F(x) = \lambda x + f(x), \qquad G(x) = \lambda x + g(x)$$

here, where

$$f(0) = g(0) = 0, \qquad f'(0) = g'(0) = 0,$$

then we get that

$$\varphi(Fx) - \lambda\varphi(x) = g(x + \varphi(x)) - f(x). \tag{1.3}$$

If G is linear, then $g = 0$, and we arrive at the linear equation

$$\varphi(Fx) - \lambda\varphi(x) = -f(x). \tag{1.4}$$

We prove two lemmas, needed in the proof of Theorem 1.1, about the solvability of the equations (1.3) and (1.4).

LEMMA 1.1. *Let $|\lambda| < 1$ and $r + \alpha > 1$. The equation (1.4) has a solution $\varphi \in C^{r,\alpha}$ with $\varphi(0) = \varphi'(0) = 0$ in a neighborhood of the origin. Any two such solutions coincide in some neighborhood of the point $x = 0$.*

PROOF. Let us write (1.4) in the form

$$\varphi(x) = \lambda^{-1}\varphi(Fx) + \lambda^{-1}f(x). \tag{1.5}$$

We let

$$(T\varphi)(x) = \lambda^{-1}\varphi(Fx) + \lambda^{-1}f(x)$$

and choose $\delta > 0$ such that

$$|F(x)| \leq \delta \qquad (|x| \leq \delta).$$

The operator T acts in the space $L_0^{r,\alpha}(\delta)$ of all functions φ of class $C^{r,\alpha}$ defined on $[-\delta, \delta]$ and such that

$$\varphi(0) = \varphi'(0) = 0.$$

By making δ smaller if necessary, we can ensure that

$$q \equiv |\lambda^{-1}| \cdot |F'(x)|^{1+\theta} < 1 \qquad (|x| \leq \delta),$$

where

$$\theta = \theta(\alpha) = \begin{cases} \alpha & \text{if } \alpha > 0, \\ 1 & \text{if } \alpha = 0. \end{cases}$$

Let $c_1 > 0$. Then for $\varphi \in L_0^{r,\alpha}(\delta)$ with

$$|\varphi'(x)| \leq c_1|x|^\theta \qquad (|x| \leq \delta) \tag{1.6}$$

we have that

$$|(T\varphi)'(x)| \leq (qc_1 + b_1)|x|^\theta \qquad (|x| \leq \delta),$$

where

$$b_1 = \max_{|x|\leq\delta} \frac{|\lambda^{-1}f'(x)|}{|x|^\theta}.$$

Fix c_1 such that

$$c_1 > \frac{b_1}{1-q}.$$

Then it follows from the inequality (1.6) that

$$|(T\varphi)'(x)| \leq c_1|x|^\theta \qquad (|x| \leq \delta). \tag{1.7}$$

Next, let $\bar{c} = \{c_j\}_{j=1}^r$ be a sequence of positive numbers. Denote by $S_0(\bar{c})$ the subset of functions $\varphi \in L_0^{r,\alpha}(\delta)$ for which (1.6) holds along with the inequalities

$$|\varphi^{(j)}(x)| \leq c_j \qquad (j \leq r).$$

If $\varphi \in S_0(\bar{c})$, then for $j \geq 2$

$$|(T\varphi)^{(j)}(x)| \leq qc_j + \sum_{i=1}^{j-1} a_{ij}c_i + b_j,$$

where

$$b_j = \max_{|x|\leq\delta} |\lambda^{-1}f^{(j)}(x)|,$$

and the coefficients a_{ij} are determined by the derivatives of F. We choose numbers c_j so that

$$c_j > \frac{1}{1-q}\left(\sum_{i=1}^{j-1} a_{ij}c_i + b_j\right) \qquad (j \geq 2).$$

Then by (1.7) the set $S_0(\overline{c})$ is invariant under T. If $r = \infty$, then $S_0(\overline{c})$ is a compact convex set. By the fixed point principle, equation (1.5) has a solution $\varphi \in S_0(\overline{c})$.

Suppose now that $r < \infty$, and denote by $S(\overline{c}, N)$ the compact convex set of functions $\varphi \in S_0(\overline{c})$ such that

$$|\varphi^{(r)}(x') - \varphi^{(r)}(x'')| \le N\omega(|x' - x''|) \qquad (|x'|, |x''| \le \delta),$$

where $\omega(u)$ is the modulus of continuity of the function[2] $f^{(r)}(x)$ ($\omega(u) = |u|^\alpha$ if $\alpha > 0$). There is a number $N = N(\overline{c})$ large enough that the compact set $S(\overline{c}, N)$ is invariant under T. Consequently, (1.5) has a solution $\varphi \in S(\overline{c}, N)$.

Suppose now that $\widetilde{\varphi}$ is another solution, and let

$$M = \max_{|x| \le \delta} \frac{|\varphi'(x) - \widetilde{\varphi}'(x)|}{|x|^\alpha}.$$

Then by the equation,

$$M \le qM.$$

Since $q < 1$, it follows that $M = 0$. The proof of the lemma is complete.

Let us now consider (1.3). We regard it on the whole line and assume accordingly that $f(x)$ and $g(x)$ are functions bounded on the line. Moreover,

$$|f'(x)| \le \varepsilon, \qquad |g'(x)| \le \varepsilon. \tag{1.8}$$

LEMMA 1.2. *Suppose that $|\lambda| < 1$, $r = 1$, and $\alpha = 0$. For each $\beta < 1$ there is a number $\varepsilon = \varepsilon(\beta, \lambda)$ such that (1.8) implies the existence of a bounded solution $\varphi \in \operatorname{Lip}\beta$ of (1.3) on the whole line, unique in the class of all bounded functions on the line.*

PROOF. We rewrite (1.3) in the form

$$\varphi(x) = \lambda\varphi(F^{-1}x) + g(F^{-1}x + \varphi(F^{-1}x)) - f(F^{-1}x). \tag{1.9}$$

If (1.8) holds, then

$$F^{-1}x = \lambda^{-1}x + \widetilde{f}(x), \qquad |\widetilde{f}'(x)| \le \varepsilon \cdot \frac{|\lambda|^{-1}}{|\lambda| - \varepsilon}.$$

Let $\beta < 1$ be fixed. Denote by $T\varphi$ the operator on the right-hand side of (1.9), and by $S(c, N)$ the compact convex set of functions $\varphi\colon \mathbb{R}^1 \to \mathbb{R}^1$ satisfying the estimates

$$|\varphi(x)| \le c, \qquad |\varphi(x') - \varphi(x'')| \le N \cdot |x' - x''|^\beta.$$

Since

$$|\lambda| \cdot |\lambda^{-1}|^\beta < 1,$$

for sufficiently small ε there exist numbers c and N large enough that $S(c, N)$ is invariant under T. Consequently, (1.9) has a solution $\varphi \in S(c, N)$.

Suppose now that $\widetilde{\varphi}$ is another bounded solution, and let

$$M = \sup_x |\varphi(x) - \widetilde{\varphi}(x)|.$$

---

[2]Recall that the modulus of continuity $\omega_h(u)$ of a function $h(\xi)$ is defined by $\omega_h(u) = \max_{|\xi' - \xi''| \le u} |h(\xi') - h(\xi'')|$.

Then in view of the equation,

$$M \leq (|\lambda| + \varepsilon)M.$$

Consequently, $M = 0$, and the lemma is proved.

By analogous arguments we can prove the solvability of a somewhat more complicated equation. Namely,

$$\varphi(x) = \mu\varphi(F(x, \varphi(x))) + g(x, \varphi(x)), \tag{1.10}$$

where

$$F \colon \mathbb{R}^1 \times \mathbb{R}^1 \to \mathbb{R}^1, \qquad g \colon \mathbb{R}^1 \times \mathbb{R}^1 \to \mathbb{R}^1$$

are given $C^{r,\alpha}$-functions. Let

$$\lambda = \frac{\partial F}{\partial x}(0,0), \qquad f(x,y) = F(x,y) - \lambda x.$$

Under the assumptions

$$f(0) = g(0) = 0, \qquad f'(0) = g'(0) = 0$$

we have the following result.

LEMMA 1.3. *Suppose that* $|\lambda| < 1$ *and the integer* $l \leq r$ *is such that*

$$|\mu| \cdot |\lambda|^l < 1, \qquad g(x,0) = O(x^l) \qquad (x \to 0).$$

Then (1.10) *has in a neighborhood of the origin a solution* $\varphi \in C^{r,\alpha}$ *satisfying the condition*

$$\varphi(x) = O(x^l) \qquad (x \to 0).$$

Any two solutions with this condition coincide in some neighborhood of the point $x = 0$.

By analogy with Lemma 1.2 we now consider the equation (1.10) on the whole line. Suppose that $r \geq 0$, $\alpha > 0$, and f and g are bounded everywhere in the plane together with all derivatives. It is assumed that the rth-order derivatives satisfy the condition $\operatorname{Lip}\alpha$ uniformly in the plane. Also, let

$$\left|\frac{\partial f}{\partial x}\right| \leq \varepsilon, \quad \left|\frac{\partial f}{\partial y}\right| \leq \varepsilon, \quad \left|\frac{\partial g}{\partial x}\right| \leq \varepsilon, \quad \left|\frac{\partial g}{\partial y}\right| \leq \varepsilon. \tag{1.11}$$

LEMMA 1.4. *Suppose that* $|\lambda| \geq 1$, $|\mu| < 1$, *and*

$$|\mu| \cdot |\lambda|^{r+\alpha} < 1.$$

For each $\beta < \alpha$ *there is an* $\varepsilon = \varepsilon(\beta)$ *such that the conditions* (1.11) *imply the existence of a solution* $\varphi \in C^{r,\beta}$ *of* (1.10) *that is bounded together with its derivatives, and* φ *is unique in the class of all bounded functions on the line.*

Although the proof uses a topological fixed point principle, in all the compact sets arising we can introduce a metric such that our operators are contractions, and then we can employ the Banach contraction mapping theorem in a metric space.

1.3. Proof of Theorem 1.1. In the case $r + \alpha > 1$ it suffices for us to establish the existence of a solution $\varphi \in C^{r,\alpha}$ of (1.4) such that $\varphi(0) = \varphi'(0) = 0$. It can be assumed that $|\lambda| < 1$, and the assertion we need is contained in Lemma 1.1.

We now consider the case $r = 1$. The necessity of verifying also that $\Phi(x) = x + \varphi(x)$ is a homeomorphism leads to a certain complication.

Since we are interested in local conjugacy, we can assume that

$$|f'(x)| \leq \varepsilon \qquad (x \in \mathbb{R}^1),$$

where ε is sufficiently small. Let us consider the two equations

$$\varphi(Fx) - \lambda\varphi(x) = -f(x),$$
$$\psi(\lambda x) - \lambda\psi(x) = f(x + \psi(x))$$

on the line. By Lemma 1.2, each equation has a bounded solution of class $C^{0,\beta}$ on the line.

Let

$$\Phi(x) = x + \varphi(x), \qquad \Psi(x) = x + \psi(x).$$

Then

$$\Phi(Fx) = \lambda\Phi(x), \qquad \Psi(\lambda x) = F(\Psi(x)).$$

Consequently,

$$\Psi \circ \Phi \circ F = F \circ \Psi \circ \Phi,$$

and

$$(\Phi \circ \Psi)(\lambda x) = \lambda(\Phi \circ \Psi)(x).$$

Let

$$H(x) = (\Psi \circ \Phi)(x).$$

Then

$$H(x) = x + h(x),$$

where

$$h(x) = \varphi(x) + \psi(x + \varphi(x))$$

is a bounded function on the line such that

$$h(Fx) - \lambda h(x) = f(x + h(x)) - f(x) \qquad (x \in \mathbb{R}^1).$$

By the uniqueness of the solution, $h = 0$. Similarly,

$$(\Phi \circ \Psi)(x) = x + \widetilde{h}(x),$$

where $\widetilde{h}$ is a bounded solution of the equation

$$\widetilde{h}(\lambda x) = \lambda\widetilde{h}(x)$$

on the line; that is, $\widetilde{h} = 0$. Thus,

$$\Psi \circ \Phi = \mathrm{id}, \qquad \Phi \circ \Psi = \mathrm{id}.$$

Consequently, Φ and Ψ are mutually inverse homeomorphisms of class $C^{0,\beta}$.

REMARK. In reality we have proved global conjugacy to the linear part of a diffeomorphism $F(x) = \lambda x + f(x)$ under the condition

$$|f'(x)| \leq \varepsilon,$$

where ε is sufficiently small.

1.4. Flows on the line. The local orbital classification of one-dimensional flows is empty of content: all curves coincide locally with $\mathbb{R}^1$.

However, conjugacy of flows has more meaning, and the results here are completely analogous to the corresponding results on diffeomorphisms.

We consider the following equation on the line:

$$\dot{x} = \nu x + f(x), \qquad \nu \neq 0, \quad f(0) = f'(0) = 0. \tag{1.12}$$

Let $f \in C^{r,\alpha}$. Then corresponding to (1.12) the local flow

$$F^t(x) = e^{\nu t} x + f_t(x), \qquad f_t(0) = f_t'(0) = 0,$$

is also of class $C^{r,\alpha}$.

THEOREM 1.2. *If $r + \alpha > 1$, then the flow F^t is conjugate in the class $C^{r,\alpha}$ to the linear flow $e^{\nu t}$. If $r = 1$, then F^t is locally conjugate in the class $C^{0,\beta}$ to a linear flow for any $\beta < 1$.*

As in the case of diffeomorphisms, the assertion of the theorem is sharp.

EXAMPLE. Suppose that $\nu < 0$ in (1.12), and

$$f(x) = \int_0^x \frac{1}{\ln|t|}\, dt, \qquad |x| \leq \tfrac{1}{2}.$$

This equation is of class C^1. We show that it cannot be linearized in this class. Indeed, suppose that a transformation $x \mapsto \Phi(x)$ of class C^1 carries the equation into the linear equation

$$\dot{x} = \nu x.$$

From the equation

$$\Phi(F^t x) = e^{\nu t}\Phi(x) \qquad (t \in \mathbb{R}_+),$$

where F^t is the flow of the equation under consideration, we get that

$$e^{-\nu t}\frac{dF^t(x)}{dx} = \frac{\Phi'(x)}{\Phi'(F^t x)} \to \frac{\Phi'(x)}{\Phi'(0)}, \qquad t \to +\infty. \tag{1.13}$$

On the other hand, we get from the formula

$$F^t(x) = e^{\nu t} x + \int_0^t e^{\nu(t-s)} f(F^s x)\, ds$$

after differentiation with respect to x that

$$e^{-\nu t}\frac{dF^t(x)}{dx} = 1 + \int_0^t \frac{e^{-\nu s}}{\ln|F^s x|} \cdot \frac{dF^s(x)}{dx}\, ds.$$

By (1.13), the left-hand side of this equality has a limit as $t \to +\infty$, while the integral on the right-hand side diverges, since

$$\ln|F^s x| \sim \nu s.$$

PROOF OF THE THEOREM. We can assume that the function f is defined everywhere on the line and is bounded, and that

$$|f'(x)| \leq \varepsilon,$$

where ε is sufficiently small. Then the field (1.12) has a global flow

$$F^t(x) = e^{\nu t}x + f_t(x),$$

where $f_t(x)$ is bounded for each $t \in \mathbb{R}^1$. It can be assumed that

$$|f_t'(x)| \leq \varepsilon \qquad (|t| \leq 1,\ x \in \mathbb{R}^1).$$

By Theorem 1.1, the diffeomorphism $F \equiv F^t|_{t=1}$ is conjugate to a linear diffeomorphism, in the class $C^{r,\alpha}$ for $r + \alpha > 1$ and in the class $C^{0,\beta}$ for $r = 1$. Further, the conjugating diffeomorphism Φ satisfies the conditions $\Phi(0) = 0$ and $\Phi'(0) = 1$ in the case $r + \alpha > 1$, while in the case $r = 1$ it can be assumed that $\Phi(x) = x + \varphi(x)$, where φ is bounded on the line. Let us consider the flow

$$\widetilde{F}^t(x) = \Phi F^t \Phi^{-1}(x).$$

This flow is either of class $C^{r,\alpha}$ (for $r + \alpha > 1$), or of class $C^{0,\beta}$ with $\beta < 1$. Since $\widetilde{F}^1(x) = e^{\nu}x$, the flow $\widetilde{F}^t$ commutes with the linear mapping:

$$\widetilde{F}^t(e^{\nu}x) = e^{\nu}\widetilde{F}^t(x).$$

Let

$$\Phi_t(x) = e^{-\nu t}\widetilde{F}^t(x) \qquad (|t| \leq 1).$$

From the commutation condition we get that

$$\Phi_t(e^{\nu}x) = e^{\nu}\Phi_t(x).$$

If we now set

$$\Phi_t(x) = x + \varphi_t(x),$$

then

$$\varphi_t(e^{\nu}x) = e^{\nu}\varphi_t(x).$$

Further, if $r + \alpha > 1$, then $\varphi(0) = \varphi'(0) = 0$, that is, $\varphi = 0$ (the uniqueness in Lemma 1.1). But if $r = 1$, then φ is bounded on the line, and hence $\varphi = 0$ (the uniqueness in Lemma 1.2). Thus,

$$e^{-\nu t}\widetilde{F}^t(x) = x,$$

from which

$$\Phi F^t \Phi^{-1}(x) = e^{\nu t}x.$$

The theorem is proved.

In passing from diffeomorphisms to flows we can use another device due to Sternberg. Namely, as above let $\Phi \in C^{r,\alpha}$ be a linearizing diffeomorphism for $F \equiv F^t|_{t=1}$. Then the family of diffeomorphisms

$$\Psi_t = e^{-\nu t} \circ \Phi \circ F^t$$

is 1-periodic in t. Indeed, $\Psi_0 = \Phi$, and

$$\Psi_{t+1} = e^{-\nu t}e^{-\nu} \circ \Phi \circ F \circ F^t = e^{-\nu t} \circ \Phi \circ F^t = \Psi_t.$$

Let

$$H(x) = \int_0^1 \Psi_s(x)\, ds.$$

Then by periodicity,

$$e^{\nu t} H(x) = \int_{-t}^{-t+1} \Psi_s(F^t x)\, ds = H(F^t x).$$

If $r + \alpha > 1$, then H is a local diffeomorphism because $H'(0) = 1$. Consequently, H conjugates the flows $e^{\nu t}$ and F^t. In the case $\Phi \in C^{0,\beta}$ it must also be checked that Φ is a homeomorphism (see [**74**]).

§2. Topological linearization on the plane

The well-known Grobman–Hartman theorem asserts that a C^1-diffeomorphism is locally topologically conjugate to a linear diffeomorphism at a hyperbolic fixed point $x_0 \in \mathbb{R}^n$. We refine this theorem in the spirit of §1.1.

2.1. Formulation of the theorem. Let

$$F(x) = \Lambda x + f(x), \qquad f(0) = 0,\ f'(0) = 0,$$

be a diffeomorphism of the plane that is hyperbolic at the point $x = 0$. Denote by λ_1 and λ_2 the eigenvalues of Λ. We define a number $\tau = \tau(\Lambda)$. If the spectrum of Λ is not separated by the unit circle (the case of a node), that is, either $\max(|\lambda_1|, |\lambda_2|) < 1$ or $\min(|\lambda_1|, |\lambda_2|) > 1$, then we set

$$\tau(\Lambda) = \min\left(\frac{\ln|\lambda_1|}{\ln|\lambda_2|}, \frac{\ln|\lambda_2|}{\ln|\lambda_1|}\right).$$

If the numbers λ_1 and λ_2 lie on different sides of the unit circle (the case of a saddle), then $\tau(\Lambda) = 1$.

Obviously, $\tau(\Lambda) \leq 1$. In addition to the case of a saddle, equality also holds in the case when the eigenvalues have the same modulus: $|\lambda_1| = |\lambda_2|$.

THEOREM 2.1. *For any $\alpha < \tau(\Lambda)$ the diffeomorphism F is locally conjugate in the class $C^{0,\alpha}$ to a linear diffeomorphism.*

The theorem carries over to the multidimensional situation. We need only refine the definition of the number $\tau(\Lambda)$. Namely, let $\lambda_1, \dots, \lambda_l$ be the eigenvalues lying interior to the unit disk, and let $\mu_1, \dots, \mu_s$ be those exterior to it. Then the proper definition is

$$\tau(\Lambda) = \min\left(\min_{i,j} \frac{\ln|\lambda_i|}{\ln|\lambda_j|}, \min_{i,j} \frac{\ln|\mu_i|}{\ln|\mu_j|}\right).$$

With this definition the formulation (and proof) of the theorem carries over immediately to the multidimensional case.

2.2. Proof of the theorem. We must solve the equation (1.2) in which $x \in \mathbb{R}^2$, and

$$F(x) = \Lambda x + f(x), \qquad G(x) = \Lambda x + g(x)$$

are diffeomorphisms of the plane such that

$$f(0) = g(0) = 0, \qquad f'(0) = g'(0) = 0.$$

In this notation the equation for the transformation

$$\Phi(x) = x + \varphi(x), \qquad \varphi(0) = 0,$$

reduces to the form (1.3):

$$\varphi(Fx) - \Lambda\varphi(x) = g(x + \varphi(x)) - f(x). \tag{2.1}$$

We can assume that f and g are bounded in the whole plane.

LEMMA 2.1. *For each $\alpha \in [0, \tau(\Lambda))$ there is an $\varepsilon = \varepsilon(\alpha, \Lambda)$ such that the conditions*

$$\|f'(x)\| \leq \varepsilon, \qquad \|g'(x)\| \leq \varepsilon$$

imply the existence of a solution $\varphi \in C^{0,\alpha}$ that is bounded in the whole plane, unique in the class of bounded mappings.

PROOF. In the case of a node we assume that $|\lambda_i| < 1$. We rewrite (2.1) in the form

$$\varphi(x) = \Lambda\varphi(F^{-1}x) + g(F^{-1}x + \varphi(F^{-1}x)) - f(F^{-1}x). \tag{2.2}$$

Under the assumption that

$$|\lambda_1| < 1 < |\lambda_2|$$

we rewrite (2.1) in the case of a saddle as the following system in coordinates:

$$\begin{aligned} \varphi^1(x) &= \lambda_1\varphi^1(F^{-1}x) + g^1(F^{-1}x + \varphi(F^{-1}x)) - f^1(F^{-1}x), \\ \varphi^2(x) &= \lambda_2^{-1}\varphi^2(Fx) - \lambda_2^{-1}g^2(x + \varphi(x)) + \lambda_2^{-1}f^2(x). \end{aligned} \tag{2.3}$$

Here $\varphi = (\varphi^1, \varphi^2)$, $g = (g^1, g^2)$, and $f = (f^1, f^2)$. The proof of existence of a bounded solution of class $C^{0,\alpha}$ for the equations (2.2) and (2.3) is carried out by the scheme of the proof of Lemma 1.2 in §1.2 with the definition of $\tau(\Lambda)$ taken into account. The proof of uniqueness is analogous.

Theorem 2.1 can now be derived from the lemma as in the one-dimensional case.

§3. Invariant curves of local diffeomorphisms

We consider the following linear operator in the plane:

$$\Lambda x = (\lambda_1\xi, \lambda_2\eta) \qquad (x = (\xi, \eta) \in \mathbb{R}^1).$$

The coordinate lines

$$l_1 = \{x = (\xi, 0)\}, \qquad l_2 = \{x = (0, \eta)\}$$

are invariant under Λ. Let

$$F(x) = (\lambda_1\xi + f^1(x), \lambda_2\eta + f^2(x)), \qquad F(0) = 0,$$

be a diffeomorphism with linear approximation Λ. If F is locally conjugate to Λ by the conjugating diffeomorphism Φ,

$$\Phi F \Phi^{-1} = \Lambda,$$

then the curves

$$M_i = \Phi^{-1}(l_i) \qquad (i = 1, 2)$$

are invariant under F. The smoothness class of these curves is not less than that of Φ. In particular, the diffeomorphism F has invariant curves of class $C^{0,\alpha}$ by Theorem 2.1.

According to the well-known (multidimensional) Hadamard–Perron theorem, a diffeomorphism of class C^r has at a saddle point invariant manifolds of class C^r whose tangents coincide with l_1 and l_2, respectively. Here we prove this theorem for the plane and consider also the case of a node.

These invariant curves are described by equations

$$M_1 = \{x \mid \eta = \gamma_1(\xi)\}, \qquad M_2 = \{x \mid \xi = \gamma_2(\eta)\},$$

where the γ_i are smooth functions such that

$$\gamma_i(0) = 0, \qquad \gamma_i'(0) = 0.$$

The smoothness class of the curves M_i is the same as that of the functions γ_i.

3.1. Invariant curves of a node. Suppose that

$$|\lambda_1| < |\lambda_2| < 1.$$

THEOREM 3.1. *The diffeomorphism $F \in C^{r,\alpha}$ has a unique local invariant manifold M_1 of class $C^{r,\alpha}$, and it also has an invariant manifold M_2 of class $C^{q,\beta}$ for*

$$q + \beta < \min\left(r + \alpha, \frac{\ln|\lambda_1|}{\ln|\lambda_2|}\right).$$

As we see, M_2 has lower smoothness class than the original diffeomorphism in general. Moreover, M_2 is not unique.

EXAMPLE 3.1. Let

$$F(x) = (\lambda^2\xi + \tfrac{1}{2}\lambda^2\eta^2, \lambda\eta), \qquad 0 < \lambda < 1.$$

This diffeomorphism is analytic. Our theorem ensures the existence of a curve M_2 of class $C^{1,\beta}$ for any $\beta < 1$. We show that F does not have an invariant curve M_2 of class $C^{1,1}$. Indeed, assume that

$$M_2 = \{\xi = \gamma(\eta)\}, \qquad \gamma(0) = \gamma'(0) = 0,$$

is an invariant curve. Then the function γ satisfies the equation

$$\lambda^2\gamma(\eta) + \tfrac{1}{2}\lambda^2\eta^2 = \gamma(\lambda\eta).$$

If $\gamma \in C^{1,1}$, then the function

$$\delta(\eta) = \frac{1}{\eta}\gamma'(\eta)$$

is bounded in a neighborhood of zero and satisfies the equation

$$\delta(\lambda\eta) - \delta(\eta) = 1.$$

However, iterating the last equality, we get that

$$\delta(\lambda^n\eta) - \delta(\eta) = n,$$

which contradicts the boundedness.

The uniqueness of the manifold M_2 is violated even for a linear mapping. Indeed, let $\lambda_i > 0$. The curve

$$\xi = |\eta|^\tau \gamma\left(\frac{\ln|\eta|}{\ln\lambda_1}\right), \qquad \tau = \frac{\ln\lambda_1}{\ln\lambda_2},$$

where γ is any smooth 1-periodic function, is invariant under Λ. If $\gamma \in C^r$, $r \geq [\tau]$, then the smoothness class of such a curve is not less than $[\tau]$ ($\tau - 1$, if τ is an odd integer).

In any case if $|\lambda_1| \neq |\lambda_2|$, then in view of Theorem 3.1 the diffeomorphism F has smooth invariant curves M_1 and M_2. Let $|\lambda_1| < |\lambda_2|$ and let

$$\Phi(\xi,\eta) = (\xi - \gamma_2(\eta), \eta - \gamma_1(\xi)).$$

Then Φ is a local diffeomorphism of class $C^{q,\beta}$, where

$$q + \beta < \min\left(r + \alpha, \frac{\ln|\lambda_1|}{\ln|\lambda_2|}\right).$$

Thus, we have

COROLLARY 3.1. *Let $|\lambda_1| \neq |\lambda_2|$. Then the diffeomorphism F is locally conjugate in the class $C^{q,\beta}$ to a diffeomorphism for which the lines l_i are invariant.*

PROOF OF THEOREM 3.1. The function γ_1 in the definition of the curve M_1 must satisfy the equation

$$\gamma_1(\xi) = \lambda_2^{-1}\gamma_1(\lambda_1\xi + f^1(\xi,\gamma_1(\xi))) - \lambda_2^{-1} f^2(\xi,\gamma_1(\xi)). \tag{3.1}$$

Since $|\lambda_2^{-1}| \cdot |\lambda_1| < 1$, it follows from Lemma 1.3 in §1.2 that the equation has a unique local solution $\gamma_1(\xi) = o(\xi)$ of class $C^{r,\alpha}$.

To prove the second part of the theorem it suffices to establish the existence of the curve M_2 for the inverse diffeomorphism

$$F^{-1}(\xi,\eta) = (\lambda_1^{-1}\xi + \widetilde{f}^1(x), \lambda_2^{-1}\eta + \widetilde{f}^2(x)).$$

Then for the function γ_2 corresponding to M_2 we get the equation

$$\widetilde{\gamma}_2(\eta) = \lambda_1\widetilde{\gamma}_2(\lambda_2^{-1}\eta + \widetilde{f}^2(\widetilde{\gamma}_2(\eta),\eta)) - \lambda_1\widetilde{f}^1(\widetilde{\gamma}_2(\eta),\eta).$$

Here

$$|\lambda_2^{-1}| > 1, \qquad |\lambda_1| \cdot |\lambda_2^{-1}|^{q+\beta} < 1.$$

By Lemma 1.4, this equation has a local solution $\widetilde{\gamma}_2 \in C^{q,\beta}$. The theorem is proved.

3.2. Invariant curves of a saddle point. Here the situation is better. Let

$$|\lambda_1| < 1 < |\lambda_2|.$$

We have

THEOREM 3.2. *At a saddle point the diffeomorphism F has unique local invariant curves M_1 and M_2 of class $C^{r,\alpha}$.*

COROLLARY 3.2. *At a saddle point a diffeomorphism of class $C^{r,\alpha}$ is locally conjugate in $C^{r,\alpha}$ to a diffeomorphism for which the axes l_1 and l_2 are invariant.*

PROOF OF THEOREM 3.2. The function γ_1 satisfies the equation (3.1), which, by Lemma 1.3, has a unique local solution $\gamma_1(\xi) = O(\xi)$ of class $C^{r,\alpha}$. Similarly, passing to the inverse F^{-1}, we get the existence of a unique curve M_2 of class $C^{r,\alpha}$.

§4. C^1-linearization on the plane

First, suppose that λ_i are the eigenvalues of the linear approximation $\Lambda = F'(0)$ of a local diffeomorphism

$$F(x) = \Lambda x + f(x) \qquad (x \in \mathbb{R}^2)$$

of class $C^{1,\alpha}$. Recall that $\tau = \tau(\Lambda) = 1$ if $x = 0$ is a saddle point, and

$$\tau(\Lambda) = \min\left(\frac{\ln|\lambda_1|}{\ln|\lambda_2|}, \frac{\ln|\lambda_2|}{\ln|\lambda_1|}\right)$$

if it is a node.

THEOREM 4.1. *Let $\alpha > 1 - \tau(\Lambda)$. Then F is locally conjugate in the class C^1 to a linear diffeomorphism.*

PROOF. We must prove that the equation

$$\varphi(Fx) = \Lambda\varphi(x) - f(x), \qquad f = (f^1, f^2),$$

has a local C^1-solution φ with $\varphi(0) = 0$ and $\varphi'(0) = 0$. For definiteness, assume that

$$|\lambda_1| \leq |\lambda_2|.$$

In the case of a node ($|\lambda_2| < 1$) and under the condition

$$1 + \alpha > \frac{\ln|\lambda_1|}{\ln|\lambda_2|}$$

let us rewrite our equation in the form

$$\varphi(x) = \Lambda^{-1}\varphi(Fx) + \Lambda^{-1}f(x).$$

Then the operator

$$(T\varphi)(x) = \Lambda^{-1}\varphi(Fx), \qquad \varphi(0) = 0, \quad \varphi'(0) = 0,$$

is a contraction with respect to the norm

$$\|\varphi\| = \sup \frac{\|\varphi'(x)\|}{\|x\|^\alpha}, \qquad \|x\| \leq \delta,$$

for some small δ, and our equation has the required solution φ. Now assume that either $|\lambda_2| < 1$ and

$$1 + \alpha \leq \frac{\ln|\lambda_1|}{\ln|\lambda_2|},$$

or $|\lambda_1| < 1 < |\lambda_2|$. Then, as follows from Corollaries 3.1 and 3.2, we can assume that the coordinate axes l_i are invariant under F. If we write F in coordinates,

$$F(x) = (\lambda_1\xi + f^1(x), \lambda_2\eta + f^2(x)),$$

then the invariance means that

$$f^1(\xi, 0) = f^2(0, \eta) = 0.$$

Denote by $\mathfrak{L}$ the space of C^1-mappings

$$h\colon \mathbb{R}^2 \to \mathbb{R}^2, \qquad h(0) = 0,\ h'(0) = 0.$$

We have the following direct sum decomposition into subspaces:

$$\mathfrak{L} = \mathfrak{L}_1 + \mathfrak{L}_2 + \mathfrak{L}_3,$$

where

$$\begin{aligned}\mathfrak{L}_1 &= \{h \in \mathfrak{L} \mid \frac{\partial h}{\partial \eta} = 0\},\\ \mathfrak{L}_2 &= \{h \in \mathfrak{L} \mid \frac{\partial h}{\partial \xi} = 0\},\\ \mathfrak{L}_3 &= \{h \in \mathfrak{L} \mid h(0,\eta) = h(\xi,0) = 0\}.\end{aligned}$$

In other words, the mappings in $\mathfrak{L}_1$ do not depend on the second argument, and the mappings in $\mathfrak{L}_2$ do not depend on the first. Onto each of the subspaces $\mathfrak{L}_i$ there is a natural projection

$$P_i\colon \mathfrak{L} \to \mathfrak{L}_i \qquad (i = 1, 2, 3).$$

Namely,

$$(P_1 h)(x) = h(\xi, 0), \quad (P_2 h)(x) = h(0,\eta), \quad (P_3 h)(x) = h(x) - h(\xi, 0) - h(0, \eta).$$

Let

$$(T\varphi)(x) = \varphi(Fx) - \Lambda\varphi(x).$$

The subspaces

$$\mathfrak{L}_2 + \mathfrak{L}_3 \subset \mathfrak{L}, \qquad \mathfrak{L}_3 \subset \mathfrak{L}$$

are invariant under T. To prove that the equation

$$T\varphi = -f, \qquad \varphi \in \mathfrak{L},$$

is solvable it suffices to prove that each of the following equations is solvable:

$$\begin{aligned}P_1 T\delta &= -P_1 f, &\quad \delta \in \mathfrak{L}_1,\\ P_2 T\psi &= -P_2(f + T\delta), &\quad \psi \in \mathfrak{L}_2,\\ P_3 T\varphi &= -P_3(f + T\delta + T\psi), &\quad \varphi \in \mathfrak{L}_3.\end{aligned} \tag{4.1}$$

Indeed, since the subspaces $\mathfrak{L}_2 + \mathfrak{L}_3$ and $\mathfrak{L}_3$ are invariant, we have the equalities

$$\begin{aligned}P_1 Th &= P_1 T(\delta + \psi + \varphi),\\ P_2 T(\delta + \psi) &= P_2 T(\delta + \psi + \varphi).\end{aligned}$$

Consequently,

$$\begin{aligned}T(\delta + \psi + \varphi) &= (P_1 + P_2 + P_3) T(\delta + \psi + \varphi)\\ &= P_1 T(\delta) + P_2 T(\delta + \psi) + P_3 T(\delta + \psi + \varphi) = -f.\end{aligned}$$

Thus, the proof has been reduced to the solvability of the three equations (4.1). The first two are in a single variable. Here we can invoke results in §1.2.

We consider the last of the equations (4.1). It has the form

$$\varphi(Fx) - \Lambda\varphi(x) = \gamma(x), \qquad \gamma \in \mathfrak{L}_3, \tag{4.2}$$

where γ belongs to the class $C^{1,\alpha}$. We must prove the existence of a solution $\varphi \in \mathfrak{L}_3$ of class C^1. In coordinates, the equation (4.2) reduces to a pair of equations of the form

$$h(Fx) - \mu h(x) = \theta(x), \qquad \theta(0,\eta) = \theta(\xi, 0) = 0 \tag{4.3}$$

with respect to a function $h(\xi, \eta)$ of class C^1 satisfying the condition

$$h(0,\eta) = h(\xi, 0) = 0.$$

Further, the coefficient μ in (4.3) satisfies one of the conditions

$$|\mu| > \max(|\lambda_1| \cdot |\lambda_2|^\alpha, |\lambda_1|^\alpha \cdot |\lambda_2|), \tag{4.4}$$

$$|\mu| < \min(|\lambda_1| \cdot |\lambda_2|^\alpha, |\lambda_1|^\alpha \cdot |\lambda_2|). \tag{4.5}$$

LEMMA 4.1. *Suppose that in the equation* (4.3) *the function* θ *belongs to the class* $C^{1,\alpha}$, *and the number* μ *satisfies one of the conditions* (4.4) *or* (4.5). *Then the equation has a local solution of class* C^1.

PROOF. Assume the condition (4.4). We write (4.3) in the form

$$h(x) = \mu^{-1} h(Fx) - \mu^{-1}\theta(x). \tag{4.6}$$

We can assume that the function θ satisfies the condition

$$\sup_{\xi,\eta} \frac{1}{|\eta|^\alpha} \cdot \left|\frac{\partial \theta}{\partial \xi}\right| < \infty, \qquad \sup_{\xi,\eta} \frac{1}{|\xi|^\alpha} \left|\frac{\partial \theta}{\partial \eta}\right| < \infty,$$

and the derivative $f'(x)$ is sufficiently small on the whole plane:

$$\|f'(x)\| \le \varepsilon.$$

We consider the space $\mathfrak{L}_3'$ of all C^1-functions $h\colon \mathbb{R}^2 \to \mathbb{R}^2$ satisfying the conditions

$$h(0,\eta) = h(\xi,0) = 0,$$

$$\|h\| \equiv \max\left(\sup_x \frac{1}{|\eta|^\alpha}\left|\frac{\partial h}{\partial \xi}\right|, \sup_x \frac{1}{|\xi|^\alpha}\left|\frac{\partial h}{\partial \eta}\right|\right) < \infty.$$

The operator on the right-hand of (4.6) is a contraction in this norm; therefore, the equation has a solution $h \in \mathfrak{L}_3'$.

Suppose now that the second condition on μ is satisfied, that is, (4.5). Then we rewrite (4.3) in the form

$$h(x) = \mu h(F^{-1}x) + \theta(F^{-1}x),$$

and the rest of the argument is analogous.

The lemma is proved, and with it the theorem on C^1-linearization.

§5. Formal transformations

To investigate the local structure of a dynamical system it is sometimes required to reduce it to "normal form" with the help of a transformation of a higher class of smoothness than C^1. However, in the linearization of a system obstacles arise already in the class C^2, for example, connected with the "formal" solvability of the equation (1.3). For instance, as we saw in §3, the diffeomorphism

$$F(x) = (\lambda^2\xi + \tfrac{1}{2}\lambda^2\eta^2, \lambda\xi), \qquad 0 < \lambda < 1,$$

not only cannot be linearized in C^2, but does not even have a local invariant C^2-manifold M_2. This necessitates looking for other "normal forms", different from linear forms, to which a diffeomorphism can be reduced by a C^r-transformation with $r \geq 2$.

5.1. Formal mappings. We recall that a *formal mapping*

$$P\colon \mathbb{R}^n \to \mathbb{R}^m$$

at the point $x_0 = 0$ is defined to be a formal series

$$P = \sum_{k=0}^{\infty} P^{(k)},$$

where $P^{(k)}\colon \mathbb{R}^n \to \mathbb{R}^m$ is a homogeneous polynomial mapping of degree k. In particular,

$$P^{(0)}(x) = \text{const} \in \mathbb{R}^m$$

is the *free term* of the mapping P. In the coordinates

$$x = (\xi^1, \dots, \xi^n) \in \mathbb{R}^n$$

the formal mapping can be written as a series

$$P(x) = \left\{ \sum_{k=0}^{\infty} \sum_{|I|=k} P_I^j \cdot x^I \right\}_{j=1}^{m}.$$

Here $I = (I_1, \dots, I_n)$ is an integer multi-index, and

$$|I| = I_1 + \cdots + I_n, \qquad x^I \equiv (\xi^1)^{I_1} \cdots (\xi^n)^{I_n}.$$

The set of all formal mappings will be denoted by $\mathbb{R}[n, m]$. This set is a linear space over $\mathbb{R}$ with respect to the obvious operations.

The space $\mathbb{R}[n, 1]$ has an obvious structure of an algebra over $\mathbb{R}$, and the space $\mathbb{R}[n, m]$ has the structure of an $\mathbb{R}[n, 1]$-module. Finally, if $P \in \mathbb{R}[n, m]$ is a mapping without free term ($P^{(0)} = 0$), and $\Phi \in \mathbb{R}[m, l]$ is arbitrary, then the *composition*

$$\Phi \circ P \in \mathbb{R}[n, l]$$

is defined (*substitution of a series in a series*).

The space $\mathbb{R}_0[n, n]$ with zero free term is a semigroup with respect to composition. The subgroup of invertible mappings consists of the elements

$$P(x) = Tx + \sum_{k \geq 2} P^{(k)}(x)$$

with nonsingular linear part:

$$\det T \neq 0.$$

We call such mappings *formal diffeomorphisms.*

In coordinates we can define the partial derivative

$$\frac{\partial P}{\partial \xi^i} = \sum_k \frac{\partial P^{(k)}}{\partial \xi^i}, \qquad P \in \mathbb{R}[n, m],$$

and the *Jacobi matrix* (over the ring $\mathbb{R}[n, 1]$)

$$P'(x) = \left(\frac{\partial P^j}{\partial \xi^i}\right).$$

With each C^r-mapping $F\colon \mathbb{R}^n \to \mathbb{R}^m$ we can associate the formal mapping

$$\beta(F) = \sum_{k=0}^{r} F^{(k)},$$

where

$$F^{(k)}(x) = \sum_{|I|=k} \frac{1}{I_1! \cdots I_n!} \frac{\partial^k F}{\partial x^I}(0) \cdot x^I.$$

For $r < \infty$ the mapping $\beta(F)$ is called *the Taylor series segment*, and for $r = \infty$ *the Taylor series* of the mapping F.

According to a well-known lemma of Borel (see, for example, [**27**]), for $r = \infty$ the correspondence

$$F \mapsto \beta(F) \tag{5.1}$$

is surjective: for each formal mapping $P \in \mathbb{R}[n, m]$ there is a C^∞-mapping $F\colon \mathbb{R}^n \to \mathbb{R}^m$ such that

$$\beta(F) = P.$$

The correspondence (5.1) is clearly not injective.

5.2. Conjugacy of formal mappings. Two formal mappings $F, G \in \mathbb{R}_0[n, n]$ are said to be *conjugate* if there exists a formal diffeomorphism $\Phi \in \mathbb{R}_0[n, n]$ such that

$$\Phi F \Phi^{-1} = G. \tag{5.2}$$

Let

$$F(x) = \Lambda x + \cdots \in \mathbb{R}_0[n, n].$$

We determine the simplest possible form to which this can be reduced by conjugation. Let

$$\Phi(x) = x + \varphi(x),$$

where $\varphi \in \mathbb{R}_0[n, n]$ is a mapping with zero linear part:

$$\varphi(x) = \sum_{k \geq 2} \varphi^{(k)}(x).$$

Then we get from (5.2)

$$\varphi^{(2)}(\Lambda x) - \Lambda \varphi^{(2)}(x) = G^{(2)}(x) - F^{(2)}(x),$$

and for $k \geq 3$ the recursion relation

$$\varphi^{(k)}(\Lambda x) - \Lambda\varphi^{(k)}(x) = G^{(k)}(x) - R_k(\varphi^{(2)} \dots \varphi^{(k-1)})(x),$$

in which R_k depends only on "lower" terms. We consider the subspace

$$\mathbb{R}^{(k)}[n,n] \subset \mathbb{R}[n,n]$$

of kth-degree homogeneous polynomial mappings $h\colon \mathbb{R}^n \to \mathbb{R}^n$. Acting in this space is the operator

$$\Lambda_k h = h \circ \Lambda - \Lambda h.$$

Denote by $\lambda_1, \dots, \lambda_n$ the eigenvalues of the operator Λ. Then the spectrum of Λ_k consists of all possible numbers of the form

$$\{\lambda_1^{p_1} \cdots \lambda_n^{p_n} - \lambda_j\}, \qquad p_1 + \cdots + p_n = k, \quad p_i \in \mathbb{Z}_+.$$

Therefore, if

$$\lambda_j \neq \lambda_1^{p_1} \cdots \lambda_n^{p_n} \qquad (p_1 + \cdots + p_n = k,\ p_i \in \mathbb{Z}_+),$$

then Λ_k is a nonsingular operator. The equality

$$\lambda_j = \lambda_1^{p_1} \cdots \lambda_n^{p_n},$$

in which $p_1 + \cdots + p_n = k$, is called a *resonance relation* (or *resonance*) of order k. If for some $k \geq 2$ the operator Λ_k does not have any resonances of order k, then F can be reduced by conjugation to a mapping G such that $G^{(k)} = 0$. Indeed, let $\varphi^{(l)} = 0$ $(2 \leq l \leq k-1)$. Then the recursion formula for $\varphi^{(k)}$ takes the form

$$\Lambda_k \varphi^{(k)} = G^{(k)} - F^{(k)}.$$

Using the surjectivity of Λ_k, we choose $\varphi^{(k)}$ so that

$$-F^{(k)} = \Lambda_k \varphi^{(k)}.$$

Then $G^{(k)} = 0$, and the formal mapping

$$\Phi(x) = x + \varphi^{(k)}(x)$$

reduces F to the necessary form.

Further, if for some $r \geq 2$ the operator Λ does not have any resonances of orders $2 \leq l \leq r$, then F can be reduced by conjugation to the form

$$G(x) = \Lambda x + \sum_{k \geq r+1} G^{(k)}(x).$$

In particular, the mapping F can be linearized in the absence of resonances of all orders $r \geq 2$:

$$(\Phi F \Phi^{-1})(x) = \Lambda x.$$

In general, linearization is impossible in the presence of resonances, but it is possible to determine a certain normal form.

We consider what resonances can be realized at a hyperbolic singular point in the plane.

Let $\operatorname{spec}\Lambda = \{\lambda_1, \lambda_2\}$. If the numbers λ_1 and λ_2 are not separated by the unit circle, then a unique resonance is possible. It has the form

$$\lambda_1 = \lambda_2^p$$

for some integer $p \geq 2$. Assume that

$$|\lambda_1| < 1 < |\lambda_2|.$$

If

$$\lambda_1 = \lambda_1^p \lambda_2^q$$

for some $p, q \in \mathbb{Z}_+$, then $p \geq 2$ and $q \geq 1$. Therefore, if there is at least one resonance, then

$$\lambda_1^l \lambda_2^s = 1$$

for some integers $l, s \geq 1$. It can be assumed that the numbers l and s are relatively prime. Then all resonances have the form

$$\lambda_1 = (\lambda_1^l \lambda_2^s)^k \cdot \lambda_1, \qquad \lambda_2 = (\lambda_1^l \lambda_2^s)^k \cdot \lambda_2.$$

Corresponding to these possibilities we consider the following three types of formal mappings on the plane:

$$\text{I. } G(x) = \Lambda x;$$

$$\text{II. } G(x) = (\lambda^p \xi + \eta^p, \lambda\eta);$$

$$\text{III. } G(x) = \begin{cases} \lambda_1 \xi + \xi \cdot \sum_{j=1}^{\infty} c_j (\xi^l \eta^s)^j, \\ \lambda_2 \eta + \eta \cdot \sum_{j=1}^{\infty} d_j (\xi^l \eta^s)^j. \end{cases}$$

THEOREM 5.1. *If the eigenvalues of the linear operator $\Lambda \colon \mathbb{R}^2 \to \mathbb{R}^2$ are different from* 1 *in modulus, then the formal diffeomorphism*

$$F(x) = \Lambda x + \dots$$

can be reduced to a diffeomorphism of one of the types I–III *by conjugation.*

This is the well-known theorem, going back to Dulac, on reduction to a so-called *resonance normal form* (see, for example, [**27**]).

5.3. Formal vector fields and flows. A formal mapping $P \in \mathbb{R}_0[n, n]$ can be interpreted as a *formal vector field* giving a *formal system*

$$\dot{x} = P(x). \tag{5.3}$$

According to this interpretation, the space $\mathbb{R}_0[n, n]$ can be regarded as a Lie algebra $\mathfrak{L}[n]$ with the commutation operation

$$[P, Q](x) = P'(x)Q(x) - Q'(x)P(x).$$

It is natural to regard this algebra, in turn, as the Lie algebra of the (Lie) group $\mathfrak{G}[n]$ of formal diffeomorphisms. The existence of an *exponential mapping*

$$\exp \mathfrak{L}[n] \to \mathfrak{G}[n]$$

is ensured by a "formal Cauchy theorem", which asserts that to each field $P(x)$ there corresponds a one-parameter subgroup $F^t \in \mathfrak{G}[n]$ such that

$$\frac{d}{dt} F^t(x) = P(F^t x).$$

The subgroup F^t will be called a *formal flow*.

The formal transformation

$$x \mapsto \Phi(x)$$

carries the system (5.3) into the system

$$\dot{x} = \widetilde{P}(x), \qquad \widetilde{P}(x) = (\Phi'(x))^{-1}P(\Phi(x)).$$

Two systems connected by such a transformation will be said to be *conjugate*. Supplementing coordinate transformations by multiplication by a "formal function" $h \in \mathbb{R}[n,1]$ with nonzero free term, we arrive at the concept of (formal) *equivalence*, or *orbital equivalence* of formal vector fields.

Let

$$P(x) = Ax + \cdots \in \mathfrak{L}[n],$$

and let

$$\operatorname{spec} A = \{\nu_i\}_{i=1}^{n}.$$

The equalities

$$\nu_j = \sum_{i=1}^{n} p_i \nu_i, \qquad \sum p_i = k \geq 2,$$

serve as analogues of resonance relations in connection with the system (5.3). If such equalities are absent for all $k \geq 2$, then the system is conjugate to a linear system.

By analogy with formal diffeomorphisms, either the resonances

$$\nu_1 = p\nu_2, \qquad \operatorname{Re}\nu_1,\ \operatorname{Re}\nu_2 > 0,$$

or the resonances

$$\nu_1 = (kl+1)\nu_1 + ks\nu_2, \quad \nu_2 = kl\nu_1 + (ks+1)\nu_2 \qquad (k = 1, 2, \dots)$$

are realized for hyperbolic vector fields on the plane, where $l, s \in \mathbb{Z}_+$ are relatively prime, and

$$l\nu_1 + s\nu_2 = 0.$$

On the plane we consider the following three types of formal vector fields:

$$\mathrm{I}'.\ Q(x) = Ax;$$

$$\mathrm{II}'.\ Q(x) = (p\nu\xi + \eta^p, \nu\eta);$$

$$\mathrm{III}'.\ Q(x) = \begin{cases} \nu_1\xi + \xi \cdot \sum_{j=1}^{\infty} c_j(\xi^l\eta^s)^j, \\ \nu_2\eta + \eta \cdot \sum_{j=1}^{\infty} d_j(\xi^l\eta^s)^j, \end{cases} \qquad \text{where } l\nu_1 + s\nu_2 = 0.$$

THEOREM 5.2. *A formal hyperbolic vector field on the plane is conjugate to a field of one of the types* I′–III′.

COROLLARY. *A formal hyperbolic vector field on the plane is orbitally equivalent to a linear field, or to a field*

$$Q(x) = (p\xi + \eta^p, \eta), \qquad p \in \mathbb{Z}_+,$$

or to a field of the form

$$Q(x) = (\nu\xi + \xi \cdot \sum_{j=1}^{\infty} b_j(\xi^l\eta^s)^j, \eta), \qquad \nu = -\frac{l}{s}. \tag{5.4}$$

§6. Smooth normal forms

Assume that the operator Λ is reduced to Jordan normal form. Then the mappings I–III in §5.2 satisfy the relation

$$G(x) = \Lambda x + g(x), \qquad g(\Lambda^* x) = \Lambda^* g(x), \tag{6.1}$$

where Λ^* is the adjoint operator, and the series g begins with quadratic terms. It is obvious that the equality

$$g(\Lambda x) = \Lambda g(x) \tag{6.2}$$

also holds, but this is connected with the specific nature of two-dimensional hyperbolic operators: any (not necessarily hyperbolic) formal diffeomorphism on $\mathbb{R}^n$ can be reduced by conjugation to the form (6.1) (see, for example, [**31**]), but it is impossible to reduce to the form (6.2) in general.

Conversely, if G satisfies (6.1), then it has the form I–III. Correspondingly, we call a local C^∞-diffeomorphism G with the form (6.1) a *normal form*. In the case of a node, such a diffeomorphism is always a polynomial diffeomorphism. In the case of a saddle point, this is no longer so, not even in the absence of resonances.

EXAMPLE. Let $|\lambda_1| < 1 < |\lambda_2|$. There exist $\alpha, \beta > 0$ such that

$$|\lambda_1|^\alpha \cdot |\lambda_2|^\beta = 1.$$

Let

$$h(\xi, \eta) = \begin{cases} \xi \exp\left(-\frac{1}{|\xi|^\alpha \cdot |\eta|^\beta}\right) & (\xi \cdot \eta \neq 0), \\ 0 & (\xi \cdot \eta = 0). \end{cases}$$

Then the C^∞-diffeomorphism

$$G(x) = (\lambda_1 \xi + h(\xi, \eta), \lambda_2 \eta)$$

is a normal form by our definition. It turns out, however, that the transition to C^∞-normal forms does not introduce new invariants: *if two (hyperbolic) normal forms have the same (or conjugate) formal Taylor series, then they are locally conjugate in the class* C^∞. This assertion is contained in the well-known multidimensional theorem of Sternberg and Chern, which will be presented in §6.4.

6.1. Normal forms with flat residual. Let

$$F(x) = \Lambda x + f(x), \qquad f(0) = f'(0) = 0,$$

be a C^r-diffeomorphism of the plane that is hyperbolic at zero.

THEOREM 6.1. *There exists a local C^∞-diffeomorphism $\Phi\colon (\mathbb{R}^2, 0) \to (\mathbb{R}^2, 0)$ reducing F to the form*

$$(\Phi F \Phi^{-1})(x) = G(x) + h(x),$$

where G is a normal form, and the residual $h \in C^r$ *has all derivatives equal to zero at the point $x = 0$:*

$$\frac{\partial^{p+q} h}{\partial \xi^p \partial \eta^q}(0) = 0 \qquad (p + q \leq r). \tag{6.3}$$

A mapping h with the condition (6.3) is said to be *r-flat* at $x = 0$, or simply flat in the case $r = \infty$.

Subsequent normalization of the mapping F now consists in "killing" the flat residuals h.

PROOF OF THEOREM 6.1. Let $r < \infty$. We consider a formal mapping $\widehat{\Phi}$ reducing the Taylor series segment $\beta(F)$ to the normal form $\widehat{G}$. For the mappings $\widehat{\Phi}$ and $\widehat{G}$ we throw out all terms of orders $\geq r+1$, and denote by Φ and G the resulting polynomial mappings. Then the difference

$$h = \Phi F \Phi^{-1} - G$$

is r-flat at zero, and G is a normal form.

Suppose now that $r = \infty$. Following Borel's lemma, we "extend" the formal diffeomorphisms $\widehat{G}$ and $\widehat{\Phi}$ to local diffeomorphisms $\widetilde{G}$ and $\widetilde{\Phi}$. Then the difference

$$h = \widetilde{\Phi} F \widetilde{\Phi}^{-1} - \widetilde{G}$$

is flat at zero. However, $\widetilde{G}$ can fail to be a normal form in general. To finish the proof we need to "correct" $\widetilde{G}$ so that (6.1) holds. With this goal we set

$$G(x) = \widetilde{G}(x) + \tau(x),$$

where τ is an unknown flat residual. The relation (6.1) gives for τ the equation

$$\tau(\Lambda x) - \Lambda \tau(x) = \Lambda \widetilde{G}(x) - \widetilde{G}(\Lambda x).$$

Here the right-hand side is flat at zero. The assertion that such equations are solvable in flat mappings τ is part of the proof of the Sternberg–Chern theorem (see §6.4).

6.2. Smooth normal forms of a node. Let $F \in C^{r,\alpha}$ be a diffeomorphism of the form

$$F(x) = G(x) + h(x),$$

where G is a normal form, and h is r-flat. We assume here that

$$|\lambda_1| \leq |\lambda_2| < 1.$$

Let

$$\tau = r\frac{\ln|\lambda_2|}{\ln|\lambda_1|} = r\tau(\Lambda).$$

THEOREM 6.2. *If $\alpha > 1 - \tau$, then F is locally conjugate in the class C^r to G.*

In the case of a node there are only the normal forms I–II, so we get the

COROLLARY. *If $r\tau(\Lambda) > 1$, then a C^r-diffeomorphism F can be reduced by a local C^r-transformation either to a linear diffeomorphism or to the normal form* II *with some $p < r$. If $r\tau(\Lambda) = 1$ and $\alpha > 0$, then a $C^{r,\alpha}$-diffeomorphism F can be reduced by a local C^r-transformation either to a linear diffeomorphism or to the normal form* II *with $p = r$. If $r\tau(\Lambda) < 1$ and $\alpha > 1 - r\tau(\Lambda)$, then F is locally conjugate in the class C^r to a linear diffeomorphism.*

We shall see later (§6.5) that the inequality for α in the case $r\tau(\Lambda) \leq 1$ cannot be replaced by the equality $\alpha = 0$. In this connection it is interesting to note that the condition $r\tau(\Lambda) > 1$ is equivalent to the absence of "nonintegral" resonances

$$|\lambda_1| \neq |\lambda_1|^\beta \cdot |\lambda_2|^\gamma \qquad (\beta + \gamma = r,\ \beta, \gamma \geq 0).$$

At the same time, the inequality $r\tau(\Lambda) \le 1$ means that

$$|\lambda_1| = |\lambda_1|^\beta \cdot |\lambda_2|^{r-\beta}, \qquad \beta = 1 + (r-1)\frac{\tau(\Lambda)}{\tau(\Lambda)-1} \ge 0.$$

PROOF OF THEOREM 6.2. Let

$$G(x) = \Lambda x + g(x), \qquad g(0) = g'(0) = 0.$$

Actually, either $g = 0$ or

$$g(\xi, \eta) = (\eta^p, 0).$$

Suppose first that $r\tau(\Lambda) \ge 1$. As before, we look for a reducing diffeomorphism in the form

$$\Phi(x) = x + \varphi(x),$$

where $\varphi \in C^r$ is r-flat. The equation for φ is

$$\varphi(Fx) - \Lambda\varphi(x) - [g(x+\varphi(x)) - g(x)] = -h(x).$$

We rewrite it in the form

$$\varphi(x) = \Lambda^{-1}\varphi(Fx) - \Lambda^{-1}[g(x+\varphi(x)) - g(x)] + \Lambda^{-1}h(x). \tag{6.4}$$

Let $L_0^r(\delta)$ be the space of all mappings $\varphi\colon \mathbb{R}^2 \to \mathbb{R}^2$ defined in the δ-neighborhood of the origin and r-flat at zero. If $\delta > 0$ is such that

$$\|F(x)\| \le \delta \qquad (\|x\| \le \delta),$$

then the operator $T\varphi$ on the right-hand side of (6.4) acts in the space $L_0^r(\delta)$.

We shall denote by $\varphi^{(k)}(x)$ the k-th order derivative of φ, that is, the collection of all partial derivatives

$$\varphi^{(k)}(x) = \left\{ \frac{\partial^k \varphi(x)}{\partial\xi^p \partial\eta^q}, \quad p+q=k \right\}.$$

For a fixed $x \in \mathbb{R}^2$ this derivative can be interpreted as an element of the space of tensors of corresponding rank. Assume next that $r < \infty$. Then

$$(T\varphi)^{(r)}(x) = A(x)\varphi^{(r)}(Fx) - \Lambda^{-1}g'(x+\varphi(x))\varphi^{(r)}(x) + R[\varphi(x), \dots, \varphi^{(r-1)}(x)],$$

where

$$A(x) = \Lambda^{-1} \otimes (F'(x))^{\otimes r}.$$

The set of eigenvalues of the operator $A(0)$ coincides with the set of all numbers of the form

$$\{\lambda_1^{-1}\lambda_1^p\lambda_2^{r-p}\}, \quad \{\lambda_2^{-1}\lambda_1^q\lambda_2^{r-q}\} \qquad (p, q \in \mathbb{Z}_+).$$

It follows from the inequalities

$$|\lambda_2|^r \le |\lambda_1| \le |\lambda_2| < 1$$

that all these numbers have modulus ≤ 1. Therefore, we can introduce a norm in $\mathbb{R}^2$ such that for sufficiently small $\delta > 0$

$$\|A(x)\| \cdot \|F'(x)\|^\alpha \le q < 1 \qquad (\|x\| \le \delta). \tag{6.5}$$

In $L_0^r(\delta)$ we consider the subspace of mappings φ such that

$$\|\varphi\| \equiv \sup_x \frac{\|\varphi^{(r)}(x)\|}{\|x\|^\alpha} < \infty.$$

It follows from (6.5) that for sufficiently small δ there is a ball

$$S(D) = \{\varphi \in L_0^r(\delta) \mid \|\varphi\| \leq D\}$$

in which T is a contraction (recall that $g(x) = (c\eta^p, 0))$. The equation (6.4) has a solution $\varphi \in S(D)$.

It also follows from (6.5) that any two solutions φ and $\widetilde{\varphi}$ satisfying the relations

$$\varphi(x) = O(\|x\|^{r+\alpha}), \qquad \widetilde{\varphi}(x) = O(\|x\|^{r+\alpha})$$

coincide in some neighborhood of zero.

Now let $r = \infty$. We fix a number r_0 such that

$$\|\Lambda^{-1}\| \cdot \|F'(x)\|^k \leq q < 1 \qquad (k \geq r_0, \ \|x\| \leq \delta). \tag{6.6}$$

For a sequence $\overline{c} = \{c_j\}_{j=r_0}^{\infty}$ we denote by $S(\overline{c}) \subset L_0^{\infty}(\delta)$ the compact convex set consisting of all C^∞-mappings satisfying

$$\max_{r \leq k} \|\varphi^{(r)}(x)\| \leq c_k \qquad (\|c\| \leq \delta, \ k = r_0, r_0 + 1, \ldots).$$

By (6.6) the sequence $\overline{c}$ can be chosen so that the compact set $S(\overline{c})$ is invariant with respect to T. The equation (6.4) has a solution $\varphi \in S(\overline{c})$.

Suppose now that $r\tau(\Lambda) < 1$. Then G is linear, and the proof of the theorem is almost a repetition of the arguments in §2 relating to the case $r = 1$.

Namely, if

$$r + \alpha > \frac{\ln|\lambda_1|}{\ln|\lambda_2|},$$

then, as we have seen, the operator T is a contraction with respect to the norm

$$\|\varphi\| = \sup_x \frac{\|\varphi^{(r)}(x)\|}{\|x\|^\alpha}, \qquad \varphi \in L_0^r(\delta).$$

In the case

$$r + \alpha \leq \frac{\ln|\lambda_1|}{\ln|\lambda_2|},$$

it can be assumed that the axes l_1 and l_2 are invariant with respect to F. Under this condition we should prove the existence of a C^r-solution of the equation

$$\varphi(Fx) - \Lambda\varphi(x) = h(x),$$

with an r-flat part $h \in C^{r,\alpha}$. Repeating the arguments for the case $r = 1$, we reduce the proof to a proof that the three equations in (4.1) are solvable, with the difference that the solution must be in C^r. The first two equations are in a single variable, and we again can use the results in §1.2. The third equation in (4.1) can be written in the form

$$\varphi(x) = \Lambda^{-1}\varphi(Fx) + \theta(x),$$

where $\theta \in \mathfrak{L}_3$ is a mapping of class $C^{r,\alpha}$. By the condition on the number α, to get a contraction we should give the norm by the formula

$$\|\varphi\| = \max\left(\sup_x \frac{1}{|\xi|^\alpha}\left\|\frac{\partial^r\varphi}{\partial\eta^r}\right\|, \sup_{k\geq 1}\left\|\frac{\partial^r\varphi}{\partial\xi^k\partial\eta^{r-k}}\right\|\right). \tag{6.7}$$

This concludes the proof of the theorem.

6.3. Smooth normalization in a neighborhood of a saddle point. Here there is a more essential lowering of the smoothness class of the reducing diffeomorphism than in the case of a node.

As before, let

$$F(x) = G(x) + h(x) \in C^{r,\alpha}, \qquad r < \infty, \tag{6.8}$$

where G is a normal form and $h \in C^{r,\alpha}$ is an r-flat residue. We assume that

$$|\lambda_1| < 1 < |\lambda_2|, \qquad \lambda_i \in \operatorname{spec} F'(0).$$

THEOREM 6.3. *For any $\alpha > 0$ the diffeomorphism F is locally conjugate in the class C^p to the normal form G, where $p = [r/2]$. The condition $\alpha > 0$ can be replaced by the condition $\alpha = 0$ provided that it is required in addition that*

$$|\lambda_i| \neq |\lambda_1|^p \cdot |\lambda_2|^p \qquad (i \neq 1, 2).$$

PROOF. This is based on an elaboration of the methods used in Theorem 6.2. We consider the following collection of linear spaces:

$$\mathfrak{L}_i = \{h \mid h(x) = \xi^{i-1}\gamma(\eta)\} \qquad (i = 1, \dots, p+1),$$

where $\gamma\colon (\mathbb{R}^1, 0) \to (\mathbb{R}^2, 0)$ is a mapping of class $C^{r-i+1,\alpha}$ that is $(r-i+1)$-flat at the point $\eta = 0$. Further,

$$\mathfrak{L}_i = \{h \mid h(x) = \eta^{i-p-2}\delta(\xi)\} \qquad (i = p+2, \dots, 2p+2),$$

where $\delta\colon (\mathbb{R}^1, 0) \to (\mathbb{R}^2, 0)$ is a mapping of class $C^{r-i+p+2,\alpha}$ that is also $(r-i+p+2)$-flat at the point $\xi = 0$. Finally, we denote by $\mathfrak{L}_{2p+3}$ the space of all C^p-mappings

$$h\colon (\mathbb{R}^2, 0) \to (\mathbb{R}^2, 0)$$

that satisfy the conditions

$$\text{a)} \quad \left.\frac{\partial^{i+j} h}{\partial \xi^i \partial \eta^j}\right|_{\xi=0} = \left.\frac{\partial^{i+j} h}{\partial \xi^i \partial \eta^j}\right|_{\eta=0} = 0 \qquad (i + j \leq p)$$

$$\text{b)} \quad \left|\frac{\partial^p h(x)}{\partial \xi^i \partial \eta^{p-i}}\right| \leq c|\xi|^{p-i+\alpha} \cdot |\eta|^i \qquad (0 \leq i \leq p).$$

Let

$$\widetilde{L}_0^{p,\alpha} = \sum_{i=1}^{2p+3} \mathfrak{L}_i.$$

As before, let $L_0^{k,\alpha}$ be the space of all $C^{k,\alpha}$-mappings that are k-flat at zero. We have the inclusions

$$L_0^{r,\alpha} \subset \widetilde{L}_0^{p,\alpha} \subset L_0^p$$

(recall that $p = [r/2]$).

The natural projections

$$N_i\colon \widetilde{L}_0^{p,\alpha} \to \mathfrak{L}_i$$

act in the space $\widetilde{L}_0^{p,\alpha}$. Namely,

$$(N_i h)(x) = \frac{1}{(i-1)!}\xi^{i-1}\frac{\partial^{i-1}h}{\partial\xi^{i-1}}\bigg|_{\eta=0} \qquad (i \le p+1)$$

$$(N_i h)(x) = \frac{1}{(i-p-2)!}\eta^{i-1}\frac{\partial^{i-p-2}h}{\partial\eta^{i-p-2}}\bigg|_{\eta=0} \qquad (p+2 \le i \le 2p+2)$$

$$(N_{2p+3}h)(x) = h(x) - \sum_{i=1}^{2p+2}(N_i h)(x).$$

To prove the theorem we must prove the existence of a p-flat solution of class C^p for the equation

$$\varphi(Fx) - \Lambda\varphi(x) - [g(x+\varphi(x)) - g(x)] = h(x), \tag{6.9}$$

where $h = g - f$.

It can be assumed from the start that the coordinate axes l_1 and l_2 are invariant under F and that

$$|\lambda_\nu| \ne |\lambda_1|^j \cdot |\lambda_2|^{r-j+\alpha} \qquad (0 \le j \le p,\ \nu = 1,2),$$
$$|\lambda_\nu| \ne |\lambda_1|^{r-j+\alpha} \cdot |\lambda_2|^j \qquad (0 \le j \le p,\ \nu = 1,2),$$
$$|\lambda_\nu| \ne |\lambda_1|^{p+\alpha} \cdot |\lambda_2|^p \qquad (\nu = 1,2).$$

Under these conditions we prove the inclusion

$$T(\widetilde{L}_0^{p,\alpha}) \supset L_0^{r,\alpha},$$

where T is the operator on the left-hand side of (6.9). The theorem will thereby be proved.

Observe first that it suffices to prove the existence of solutions $\varphi_i \in \mathfrak{L}_i$ of the equations

$$\begin{aligned} N_1 T\varphi_1 &= h_1 \\ N_i T(\varphi_1 + \cdots + \varphi_i) &= h_i \qquad (2 \le i \le 2p+3), \end{aligned} \tag{6.10}$$

where $h_i = N_i h$. Indeed, let

$$\varphi = \varphi_1 + \cdots + \varphi_{2p+3}.$$

It follows from the definitions of the spaces $\mathfrak{L}_i$ and the projections N_i that

$$N_i T(\varphi_1 + \cdots + \varphi_i) = N_i T(\varphi) \qquad (1 \le i \le 2p+3).$$

Consequently,

$$\begin{aligned} h &= N_1 T\varphi_1 + N_2 T(\varphi_1 + \varphi_2) + \cdots + N_{2p+3}T(\varphi) \\ &= N_1 T(\varphi) + N_2 T(\varphi) + \cdots + N_{2p+3}T(\varphi) = T(\varphi), \end{aligned}$$

which is what was required.

Proceeding to the proof that the equations in (6.10) are solvable, we note first of all that the equations with indices $i = 1$, $i = p+2$, and $i = 2p+3$ are nonlinear in φ_i, while the remaining equations are linear. More precisely, let

$$\varphi_i(\xi,\eta) = \xi^{i-1}\gamma_i(\eta) \qquad (i \le p+1),$$
$$\varphi_i(\xi,\eta) = \eta^{i-p-2}\delta_{i-p-1}(\xi) \qquad (p+2 \le i \le 2p+2).$$

Then for $i = 1$ we get the equation

$$\gamma_1(\lambda_2\eta + f^2(0,\eta)) - \Lambda\gamma_1(\eta) - [g(\gamma_1^1(\eta), \eta + \gamma_1^2(\eta)) - g(0,\eta)] = h(0,\eta). \tag{6.11}$$

Further, for $2 \le i \le p+1$

$$\begin{aligned}(\lambda_1 + f^1(0,\eta))^{i-1}\gamma_i(\lambda_2\eta + f^2(0,\eta)) - \Lambda\gamma_i(\eta)& \\ - g^1(\gamma_1^1(\eta), \eta + \gamma_1^2(\eta))\gamma_i(\eta) &= \widetilde{h}_i(\eta),\end{aligned} \tag{6.12}$$

where

$$\widetilde{h}_i(\eta) = h_i(\eta) - N_i[g(x + \varphi_1 + \cdots + \varphi_{i-1}) - g(x)].$$

Similarly, for $\delta_1(\xi)$ we have the equation

$$\delta_1(\lambda_1\xi + f^1(\xi,0)) - \Lambda\delta_1(\xi) - [g(\xi + \delta_1^1(\xi), \delta_1^2(\xi)) - g(\xi,0)] = h(\xi,0), \tag{6.13}$$

and for $p+3 \le i \le 2p+2$ the linear equations

$$\begin{aligned}(\lambda_2 + f^2(\xi,0))^{i-p-2}\delta_{i-p-1}(\lambda_1\xi + f^1(\xi,0)) - \Lambda\delta_{i-p-1}(\xi)& \\ - g^1(\xi + \delta_1^1(\xi), \delta_1^2(\xi))\delta_{i-p-1}(\xi) &= \widetilde{h}_i(\xi),\end{aligned} \tag{6.14}$$

where

$$\widetilde{h}_i(\xi) = h_i(\xi) - N_i[g(x + \varphi_1 + \cdots + \varphi_{i-1}) - g(x)].$$

Finally, for $\varphi = \varphi_{2p+3}$ we again have a nonlinear equation

$$\begin{aligned}\varphi(Fx) - \Lambda\varphi(x) - g(x + \varphi_1 + \cdots + \varphi_{2p+2} + \varphi)& \\ + g(x + \varphi_1 + \cdots + \varphi_{2p+3}) &= \widetilde{h}_{2p+3}(x),\end{aligned} \tag{6.15}$$

where

$$\widetilde{h}_{2p+3}(x) = h_{2p+3}(x) - N_i[g(x + \varphi_1 + \cdots + \varphi_{2p+2}) - g(x)].$$

The proof that (6.11)–(6.15) are solvable uses the same functional methods as above. For instance, let us consider (6.15).

Since we are interested in local solutions, we can assume that the mappings f and g have compact support. We consider the Banach space $\mathfrak{B}$ of all C^p-mappings

$$\varphi\colon \mathbb{R}^2 \to \mathbb{R}^2, \qquad \varphi \in \mathfrak{L}_{2p+3},$$

that are p-flat at zero, with the norm

$$\|\varphi\| \equiv \sup_x \max_i \frac{1}{|\xi|^{p-i+\alpha}|\eta|^i}\left\|\frac{\partial^p\varphi(x)}{\partial\xi^i\partial\eta^{p-i}}\right\| < \infty.$$

It can be assumed that $\widetilde{h}_{2p+3} \in \mathfrak{B}$.

Suppose now that

$$\|f'(x)\| \le \varepsilon, \quad \|g'(x)\| \le \varepsilon \qquad (x \in \mathbb{R}^2).$$

LEMMA 6.1. *There is an $\varepsilon = \varepsilon(\Lambda)$ such that the equation* (6.15) *has a unique solution $\varphi \in \mathfrak{B}$ defined in the whole plane.*

PROOF. For definiteness assume that

$$|\lambda_1| < |\lambda_1|^{p+\alpha} \cdot |\lambda_2|^p < |\lambda_2|.$$

We write out the equation as the system

$$\varphi^1(x) = \lambda_1 \varphi^1(F^{-1}x) + g^1(F^{-1}x + \theta(F^{-1}x) + \varphi(F^{-1}x)) \\ - g^1(F^{-1}x + \theta(F^{-1}x)) + \widetilde{h}^1_{2p+3}(x),$$

$$\varphi^2(x) = \lambda_2^{-1}\varphi^2(Fx) - \lambda_2^{-1}g^2(x + \theta(x) + \varphi(x)) + \lambda_2^{-1}g^2(x + \theta(x)) - \lambda_1^{-1}\widetilde{h}_{2p+3}(x),$$

where

$$\theta = \sum_{i=1}^{2p+2} \varphi_i.$$

We write this system, in turn, in the operator form

$$\varphi(x) = (T\varphi)(x) + \gamma(x), \tag{6.16}$$

where

$$\gamma(x) = (\widetilde{h}^1_{2p+3}(x), -\lambda_2^{-1}\widetilde{h}^2_{2p+3}(x)).$$

If $\varphi \in \mathfrak{B}$, then the quantities

$$D_l(\varphi) \equiv \max_{i\le l} \sup_x \frac{1}{|\xi|^{p-i+\alpha} \cdot |\eta|^{p+i-l}} \left\| \frac{\partial^l \varphi(x)}{\partial \xi^i \partial \eta^{l-i}} \right\|$$

are finite for $l = 0, \dots, p$. If ε is sufficiently small, then

$$D_l(T\varphi) \le qD_l(\varphi) + R_l(D_0(\varphi), \dots, D_{l-1}(\varphi)),$$

where $q < 1$, and R_l is a function determined by f, g, and θ. In particular,

$$R_0 = D_0(\gamma).$$

Let

$$c_0 = \frac{D_0(\gamma)}{1-q},$$

and let the numbers $c_1, \dots, c_p$ be chosen by induction so that

$$c_l = \frac{R_l(c_0, \dots, c_{l-1})}{1-q} \qquad (1 \le l \le p).$$

Then the ball

$$S(\overline{c}) = \{\varphi \in \mathfrak{B} \mid D_l(\varphi) \le c_l,\ 0 \le l \le p\}$$

is invariant under T. We choose a metric in this ball so that T is a contraction. With this purpose we observe that if $\varphi, \psi \in S(\overline{c})$, then

$$D_l(T\varphi - T\psi) \le qD_l(\varphi + \psi) + b_l \cdot \max_{j \le l-1} D_j(\varphi - \psi), \tag{6.17}$$

where $b_l = b_l(\overline{c})$. In particular, $b_0 = 0$.

Let $b = \max_{l\ge 1} b_l$. We fix a number $\widetilde{q} \in (q, 1)$ and define the numbers

$$a_p = 1, \qquad a_i = \frac{b}{\widetilde{q} - q} \sum_{j \ge i+1} a_j \qquad (0 \le i \le p-1).$$

For $\varphi, \psi \in S(\bar{c})$ let

$$d(\varphi, \psi) = \sum_{l=0}^{p} a_l D_l(\varphi - \psi).$$

Then by the choice of the numbers a_l we get from (6.17) that

$$d(T\varphi, T\psi) \leq \tilde{q} d(\varphi, \psi).$$

Consequently, (6.16) has a solution $\varphi \in S(\bar{c})$.

To prove uniqueness in the whole of $\mathfrak{B}$ we set

$$M = D_0(\varphi - \psi),$$

where $\widetilde{\varphi}$ is another solution. Then by the equation,

$$M \leq qM, \qquad q < 1,$$

that is, $M = 0$.

The existence (and uniqueness in a suitable class) of solutions of the equations (6.11)–(6.14) is proved similarly.

This proves Theorem 6.3.

6.4. The Sternberg–Chern theorem. We recall that this is a theorem about conjugacy of C^∞-diffeomorphisms.

THEOREM 6.4. *If the formal Taylor series of C^∞-diffeomorphisms F and G at a hyperbolic fixed point are conjugate, then F and G are locally conjugate in the class C^∞.*

This theorem has already been proved for the case of a node in §6.2. It remains only for us to treat the case of a saddle point. For this we must prove the existence of a local C^∞-solution φ of (6.9) such that $\varphi(0) = \varphi'(0) = 0$.

We can assume from the start that the right-hand side $h = f - g$ of (6.9) is flat at zero. Beginning with this, we prove the existence of a local solution $\varphi \in C^\infty$ that is flat at zero.

As before, we assume that the coordinate axes are invariant with respect to F.

We note first that each mapping $h \in C^\infty$ that is flat at zero can be written as a sum

$$h(x) = h_+(x) + h_-(x), \qquad h_\pm \in C^\infty,$$

where

$$\left.\frac{\partial^{i+j} h_+(x)}{\partial \xi^i \partial \eta^j}\right|_{\xi=0} = 0, \qquad \left.\frac{\partial^{i+j} h_-(x)}{\partial \xi^i \partial \eta^j}\right|_{\eta=0} = 0.$$

To prove that there is such a decomposition we construct a mapping h_+ such that

$$\left.\frac{\partial^{i+j} h_+(x)}{\partial \xi^i \partial \eta^j}\right|_{\xi=0} = 0, \qquad \left.\frac{\partial^{i+j} h_+(x)}{\partial \xi^i \partial \eta^j}\right|_{\eta=0} = \left.\frac{\partial^{i+j} h(x)}{\partial \xi^i \partial \eta^j}\right|_{\eta=0}$$

for all i and j. The mapping can be constructed in this way by generalizing the proof of Borel's lemma to the case when the derivatives of all orders are given on the "coordinate cross" $l_1 \cup l_2$. However, we can invoke the general Whitney extension theorem (see [**59**]). After constructing the term h_+ it remains to set

$$h_-(x) = h(x) - h_+(x).$$

Next, to prove that (6.9) is solvable it now suffices to prove that the following two equations are solvable:

$$\begin{aligned} \varphi_+(Fx) - \Lambda\varphi_+(x) - [g(x+\varphi_+(x)) - g(x)] &= h_+(x), \\ \varphi_-(Fx) - \Lambda\varphi_-(x) - g(x+\varphi_+(x)+\varphi_-(x)) + g(x+\varphi_+(x)) &= h_-(x). \end{aligned} \tag{6.18}$$

Then the sum $\varphi = \varphi_+ + \varphi_-$ is the desired solution.

As before, assume that the mappings f, g, and $h_\pm$ are given in the whole plane and are bounded, together with their derivatives of all orders. Also as before, it is assumed that

$$\|f'(x)\| \le \varepsilon, \qquad \|g'(x)\| \le \varepsilon.$$

LEMMA 6.2. *There is an $\varepsilon = \varepsilon(\Lambda)$ such that each of the equations in* (6.18) *has in the plane a unique solution satisfying for all $r = 0, 1, \dots,$ and $\nu = 0, 1, \dots$ the conditions*

$$\|\varphi_+^{(r)}(x)\| \le c_{r\nu}^+ |\xi|^\nu$$

and

$$\|\varphi_-^{(r)}(x)\| \le c_{r\nu}^- |\eta|^\nu,$$

respectively.

PROOF. Let us consider the first of our pair of equations. We write it in the form

$$\varphi_+(x) = \Lambda^{-1}\varphi_+(Fx) - \Lambda^{-1}[g(x+\varphi_+(x)) - g(x)] - \Lambda^{-1}h_+(x).$$

For each $r = 0, 1, \dots$ we choose a number $\nu(r)$ large enough that

$$\|\Lambda^{-1}\| \cdot \|\Lambda\|^r \cdot |\lambda_1|^\nu \le q < 1 \qquad (\nu \ge \nu(r)).$$

Let

$$D_{r\nu}(\varphi) = \sup_x \frac{\|\varphi_+^{(r)}(x)\|}{|\xi|^\nu}.$$

In this case if ε is sufficiently small, then for $\nu \ge \nu(r)$

$$D_{r,\nu}(T\varphi_+) \le \widetilde{q} D_{r,\nu}(\varphi_+) + R_{r,\nu}(D_{r-1,\nu}(\varphi_+) \dots D_{0,\nu}(\varphi_+)).$$

Here, as before, T denotes the operator on the right-hand side of the equation, and $R_{r,\nu}$ is a function determined by the given mappings. In particular,

$$R_{0,\nu} = D_{0,\nu}(-\Lambda^{-1}h_+).$$

The factor $\widetilde{q}$ is less than 1.

Suppose now that

$$\overline{c} = \{c_{r,\nu}\}_{\nu \ge \nu(r)}$$

is the infinite collection of numbers determined by

$$\begin{aligned} c_{0,\nu} &= D_{0,\nu}(-\Lambda^{-1}h_+) \cdot \frac{1}{1-\widetilde{q}}, \\ c_{r,\nu} &= \frac{1}{1-\widetilde{q}} R_{r,\nu}(c_{r-1,\nu}, \dots, c_{0,\nu}) \qquad (r \ge 1). \end{aligned}$$

Then the compact convex set

$$S(\overline{c}) = \{\varphi \mid D_{r,\nu}(\varphi) \le c_{r,\nu}\} \qquad (r = 0, 1, \dots, \ \nu \ge \nu(r))$$

is invariant under T. Consequently, T has a fixed point φ_+ in $S(\overline{c})$.

We prove uniqueness. Let $\widetilde{\varphi}_+$ be another solution satisfying our conditions. We set

$$M = D_{0,\nu_0}(\varphi - \widetilde{\varphi}), \qquad \nu_0 = \nu(0).$$

Then by the equation,

$$M \leq \widetilde{q}M, \qquad \widetilde{q} < 1,$$

that is, $M = 0$.

The assertions of the lemma for the second of the equations in (6.18) are proved similarly (just make the substitution $x \mapsto F^{-1}x$).

The lemma is proved, and with it the Sternberg-Chern theorem.

We remark that a metric in which T is a contraction can be introduced in the compact set $S(\bar{c})$.

We remark also that Lemma 6.2 also finishes the proof of Theorem 6.1 for the case C^∞.

6.5. The smoothness class as an obstacle to smooth normalization. By Theorem 1.1, a diffeomorphism of the line of class C^r, $r \geq 2$, can be linearized in the same smoothness class at a hyperbolic fixed point. The example after the formulation of the theorem shows that this is no longer so for $r = 1$. Thus, there exist one-dimensional C^1-diffeomorphisms that are not smoothly conjugate to a diffeomorphism of higher smoothness class (not even $C^{1,\alpha}$ for an $\alpha > 0$) in a neighborhood of a hyperbolic fixed point. The global variant of this effect is considerably stronger: in §5 of Chapter 5 we shall see that for any $r < \infty$ there exist structurally stable C^r-diffeomorphisms on the circle that are not smoothly (not even in the class C^1) conjugate to C^{r+1}-diffeomorphisms. In some sense the smoothness class is an invariant of smooth conjugacy.

Even in the local situation there is a greater diversity on the plane. In the case of $\mathbb{R}^2$ we now show that *for any $r < \infty$ there exist C^r-diffeomorphisms that are not conjugate in the class C^r to C^{r+1}-diffeomorphisms in a neighborhood of a hyperbolic fixed point.*

To construct an appropriate example we set

$$h(\eta) = \frac{\eta^r}{\ln|\eta|} \quad (\eta \neq 0), \qquad h(0) = 0.$$

Then $h \in C^r$ is an r-flat function. Let us consider the diffeomorphism

$$H(x) = (\lambda^r(\xi + h(\eta)), \lambda\eta), \qquad 0 < \lambda < 1.$$

We show that H not only cannot be linearized in the class C^r but cannot even have an invariant curve M_2 of class C^r. Thus, it follows from Theorem 6.2 (the case $r\tau(\Lambda) = 1$ of the corollary) that H is not locally conjugate in C^r to a $C^{r,\alpha}$-diffeomorphism for $\alpha > 0$.

An invariant curve

$$M_2 = \{\xi = \gamma(\eta)\}$$

must satisfy the equation

$$\lambda^{-r}\gamma(\lambda\eta) - \gamma(\eta) = h(\eta).$$

Assume now that $\gamma \in C^r$. Then $\gamma(\eta) = O(\eta^r)$, and the function

$$\delta(\eta) = \frac{1}{\eta^r}\gamma(\eta)$$

must have limit equal to $\gamma^{(r)}(0)$ as $\eta \to 0$. At the same time,

$$\delta(\lambda\eta) - \delta(\eta) = \frac{1}{\ln|\eta|}.$$

Integrating, we get that

$$\delta(\lambda^n\eta) - \delta(\eta) = \sum_{s=0}^{n-1} \frac{1}{s\ln\lambda + \ln|\eta|}.$$

The left-hand side of the last equality is bounded, while the right-hand side, being a sum of the terms of a harmonic series, is not. Consequently, $\gamma \notin C^r$.

More subtle examples are given in [**108**]. In particular, there are examples showing that the estimate of the smoothness class for normalization of a saddle cannot be improved.

§7. Local normal forms of two-dimensional flows

The results on normal forms for diffeomorphisms carry over to flows.

7.1. Topological and C^1-linearization. Let us consider the equation

$$\dot{x} = Ax + h(x), \qquad h(0) = h'(0) = 0, \tag{7.1}$$

where $x \in \mathbb{R}^2$, $h \in C^1$, and $A\colon \mathbb{R}^2 \to \mathbb{R}^2$ is a linear operator whose spectrum does not intersect the imaginary axis. By analogy with §2.1, we define the number $\theta = \theta(A)$. Namely, if the spectrum of A is not separated by the imaginary axis (the case of a node), that is, if either

$$\max(\operatorname{Re}\nu_1, \operatorname{Re}\nu_2) < 0$$

or

$$\min(\operatorname{Re}\nu_1, \operatorname{Re}\nu_2) > 0$$

(here $\nu_i \in \operatorname{spec} A$), then

$$\theta(A) = \min\left(\frac{\operatorname{Re}\nu_1}{\operatorname{Re}\nu_2}, \frac{\operatorname{Re}\nu_2}{\operatorname{Re}\nu_1}\right).$$

But if the numbers ν_1 and ν_2 lie on different sides of the imaginary axis, then

$$\theta(A) = 1.$$

THEOREM 7.1. *For any $\alpha < \theta(A)$ the system* (7.1) *is conjugate in the class $C^{0,\alpha}$ to a linear system.*

The proof goes according to the same scheme as in the one-dimensional case with the use of Lemma 2.1.

THEOREM 7.2. *Let $h \in C^{1,\alpha}$, $\alpha > 1-\theta(A)$. Then the system* (7.1) *is conjugate in the class C^1 to a linear system.*

Since a normal form is linear, and the smoothness class is ≥ 1, we can make direct use of the Sternberg device mentioned in §1.4 for the proof. Namely, let $\Phi \in C^1$ be a local diffeomorphism linearizing the mapping $F = F^t|_{t=1}$, where F^t

is the flow of the system (7.1). Such a diffeomorphism Φ exists in view of Theorem 4.1. Then the local diffeomorphism

$$\Psi(x) = \int_0^1 e^{-tA}\Phi(F^t x)\, dt$$

linearizes the field (7.1).

7.2. Invariant curves of a flow. As in the case of diffeomorphisms, invariant curves of the system (7.1) whose tangents coincide with the coordinate axes are described by equations

$$M_1 = \{x \mid \eta = \gamma_1(\xi)\}, \qquad M_2 = \{x \mid \xi = \gamma_2(\eta)\},$$

where the γ_i are smooth functions with $\gamma_i(0) = \gamma_i'(0) = 0$. The smoothness class of the curves coincides with the smoothness class of the function γ_i. We assume that $r + \alpha > 1$ and that $\nu_1 < \nu_2 < 0$ in the case of a node.

THEOREM 7.3. *In the case of a node, a system (7.1) of class $C^{r,\alpha}$ has a unique invariant manifold M_1 of class $C^{r,\alpha}$, as well as an invariant manifold M_2 of class $C^{q,\beta}$ for*

$$q + \beta < \min\left(r + \alpha, \frac{\nu_1}{\nu_2}\right).$$

In the case of a saddle, the system has unique local invariant curves M_1 and M_2 of class $C^{r,\alpha}$.

PROOF. As before, let F^t be the flow of the system, and let $F = F^t|_{t=1}$. Let M_i be the invariant curves for F; their existence is ensured by Theorem 3.1. Let

$$M_i^t = F^t(M_i) \qquad (i = 1, 2).$$

For each t the curves M_i^t are invariant under F, and the coordinate axes l_i are tangent at zero. Therefore, $M_i^t = M_i$ in the cases when Theorem 3.1 ensures uniqueness; that is, the M_i are invariant for the system (7.1). It thus remains to prove the existence of M_2 in the case of a node, when there is certainly not uniqueness. We indicate one method with use of the uniqueness in Lemma 1.4. Recall that in our notation $\nu_1 < \nu_2 < 0$. The function γ_2 giving the curve M_2 satisfies the differential equation

$$\gamma_2'(\eta)(\nu_2\eta + h^2(\gamma_2(\eta), \eta)) = \nu_1\gamma_2(\eta) + h^1(\gamma_2(\eta), \eta).$$

Here, as before, $x = (\xi, \eta) \in \mathbb{R}^2$, and $h = (h^1, h^2)$. Passing to the flow, we get the equation

$$\gamma_2(\eta) = e^{-\nu_1 t}\gamma_2(e^{\nu_2 t}\eta + f_t^2(\gamma_2(\eta), \eta)) - e^{-\nu_1 t} f_t^1(\gamma_2(\eta), \eta), \tag{7.2}$$

where

$$F^t(x) = (e^{\nu_1 t}\xi + f_t^1(x), e^{\nu_2 t}\eta + f_t^2(x)) \tag{7.3}$$

is the flow of the system. The equation (7.2) must be satisfied for small t. We extend the residual h to the whole plane in such a way that it is bounded together with all its derivatives, and the first-order derivatives of the functions f_t^1 and f_t^2 are sufficiently small (see Lemma 1.4) over the whole plane (with respect to x) for $|t| \leq 1$. By Lemma 1.4, (7.2) then has a solution $\gamma_2 \in C^{q,\beta}$ for $t = -1$ that is defined on the whole axis and unique in the class of bounded functions. For a fixed t we consider the curve

$$F^t(\gamma_2(\eta), \eta) = \{\xi = \gamma_2^t(\eta)\}.$$

This curve is invariant with respect to the diffeomorphism $F^{-1} = F^t|_{t=-1}$. Consequently, γ_2^t satisfies the equation

$$\gamma_2^t(\eta) = e^{\nu_1}\gamma_2^t(e^{-\nu_2}\eta + f_{-1}^2(\gamma_2^t(\eta), \eta)) - e^{\nu_1} f_{-1}^1(\gamma_2^t(\eta), \eta).$$

From its definition, γ_2^t is bounded. By uniqueness,

$$\gamma_2^t(\eta) = \gamma_2(\eta) \qquad (|t| \leq 1).$$

Consequently, $M_2 = \{\xi = \gamma_2(\eta)\}$ is the desired curve.

7.3. Smooth normal forms. The normal forms I′–III′ in §5.3 satisfy the relation

$$G(x) = Ax + g(x), \qquad g'(x)A^*x = A^*g(x). \tag{7.4}$$

By analogy with the discrete case, each C^∞-flow satisfying (7.4) will be called a *normal form*. As in the case of a diffeomorphism, the equation (7.1) can be reduced by a C^∞-transformation to the form

$$\dot{x} = G(x) + h(x), \tag{7.5}$$

where G is a normal form, and $h \in C^{r,\alpha}$ is an r-flat residual.

THEOREM 7.4. *Suppose that the origin of coordinates is a node for the system (7.5). If $\alpha > 1 - r\theta(A)$, then the system (7.5) is locally conjugate in the class C^r to the normal form*

$$\dot{x} = G(x).$$

PROOF. Assume first that

$$r\theta(A) \geq 1.$$

Again passing to the flows F^t and G^t and invoking Theorem 6.2, we get a C^r-diffeomorphism

$$\Phi(x) = x + \varphi(x), \qquad \varphi(x) = O(\|x\|^{r+\alpha}), \tag{7.6}$$

which conjugates F^t and G^t for $t = 1$. Let

$$\Phi_s = G^{-s} \circ \Phi \circ F^s.$$

For each s the diffeomorphism Φ_s also has the form (7.6) and conjugates the diffeomorphisms F^t and G^t for $t = 1$. However, as noted in the proof of Theorem 6.2, a local diffeomorphism with the indicated properties is unique. Consequently,

$$\Phi_s = \Phi.$$

If $r\theta(A) < 1$, then the normal form is linear, and we can use the Sternberg device of passing from diffeomorphisms to flows.

Finally, there is also an analogue of Theorem 6.3 for a saddle point. We formulate it also for $r = \infty$, so that the formulation will include the Sternberg–Chern theorem for flows.

THEOREM 7.5. *Suppose that either $r = \infty$ or that $r < \infty$ and $\alpha > 0$, and assume that the origin of coordinates is a saddle singular point for the system (7.5) of class $C^{r,\alpha}$. Then (7.5) is conjugate in the class C^p to a normal form, $p = [r/2]$. In the case $r < \infty$ it is possible to take $\alpha = 0$ if it is required in addition that*

$$\nu_j \neq p(\nu_1 + \nu_2) \qquad (j \neq 1, 2).$$

PROOF. It suffices to prove the existence of a local C^p-solution φ of

$$\varphi'(x)(Ax + g(x) + h(x)) - A\varphi(x) - [g(x + \varphi) - g(x)] = h(x)$$

that is p-flat at the origin. We can assume that the coordinate axes are invariant for the original system. Letting $\widetilde{T}\varphi$ be the operator on the left-hand side of this equation, we reduce solvability of the equation

$$\widetilde{T}\varphi = h$$

to solvability of the equations

$$\begin{aligned} N_1\widetilde{T}\varphi_1 &= h_1 \\ N_i\widetilde{T}(\varphi_1 + \cdots + \varphi_i) &= h_i \qquad (2 \leq i \leq 2p + 3) \end{aligned} \tag{7.7}$$

with respect to the mappings $\varphi_i \in \mathfrak{L}_i$ (see the proof of Theorem 6.3).

The proof of solvability of (7.7) can be carried out independently, by "functional" methods. However, it is also possible to employ the arguments already used for diffeomorphisms, with uniqueness of the solutions taken into account. For instance, let us consider (7.7) for $i = 2p + 3$.

We write it in the form

$$\varphi'(x)(Ax + g(x) + h(x)) = A\varphi(x) + g(x + \theta(x) + \varphi(x)) - g(x + \varphi(x)) + \gamma(x),$$

where $\gamma \in \mathfrak{L}_{2p+3}$, and

$$\theta(x) = \varphi_1(x) + \cdots + \varphi_{2p+2}(x).$$

We consider the system

$$\begin{aligned} \dot{x} &= Ax + g(x) + h(x), \\ \dot{y} &= Ay + g(x + \theta(x) + y) - g(x + \theta(x)) + \gamma(x). \end{aligned}$$

Its flow has the form

$$V^t(x, y) = (F^t(x), H^t(x, y)).$$

Further,

$$H^t(x, 0) = \int_0^t e^{(t-s)A}\gamma(F^s x)\, ds \in \mathfrak{L}_{2p+3}.$$

Moreover, we have the formula

$$H^{t+s}(x, y) = H^t(F^s x, H^s(x, y)). \tag{7.8}$$

It suffices to prove the existence of a p-flat C^p-mapping φ such that

$$\varphi(F^t x) = H^t(x, \varphi(x)) \tag{7.9}$$

for small t. We can assume that the mappings g, h, and γ are extended to the whole plane in such a way that for small t the residuals

$$f_t(x) = F^t x - e^{At}x,$$
$$g_t(x,y) = H^t(x,y) - e^{tA}y = \int_0^t e^{(t-s)A}[g(F^s x + \theta(F^s x) + y) - g(F^s x + \theta(F^s x)]\, ds$$

satisfy the conditions of Lemma 6.1 for the equation (7.9) with $t = 1$,

$$\varphi(Fx) = e^A \varphi(x) + \int_0^1 e^{(1-s)A} g(F^s x + \theta(F^s x) + \varphi(x))\, ds$$
$$- \int_0^1 e^{(1-s)A} g(F^s x + \theta(F^s x))\, ds + H^1(x,0).$$

Unlike in Lemma 6.1, the nonlinearities in φ are not essential to the equation here: if

$$\|f'(x)\| \le \varepsilon, \qquad \|g'(x)\| \le \varepsilon$$

and ε is sufficiently small, then the last equation has a unique solution $\varphi \in \mathfrak{B}$ defined for all $x \in \mathbb{R}^2$. We now let

$$\varphi_t(x) = H^t(F^{-t}x, \varphi(F^{-t}x)).$$

Then $\varphi_t \in \mathfrak{B}$ and, moreover,

$$\varphi_t(Fx) = H^t(F^{1-t}x, \varphi(F^{1-t}x)) = H^t(F^{1-t}x, H^1(F^{-t}x, \varphi(F^{-t}x)))$$
$$= H^{t+1}(F^{-t}x, \varphi(F^{-t}x)) = H^1(x, H^t(F^{-t}x, \varphi(F^{-t}x)) = H^1(x, \varphi_t(x))).$$

This long chain of formulas uses the flow condition (7.8) and the equality

$$\varphi(Fx) = H^1(x, \varphi(x)) \qquad (F = F^t|_{t=1},\ H^1 = H^t|_{t=1}),$$

which means that φ is a solution of (7.9) for $t = 1$. Finally, we get that

$$\varphi_t(Fx) = H^1(x, \varphi_t(x)).$$

Since $\varphi_t \in \mathfrak{B}$, it follows from uniqueness that $\varphi_t = \varphi$ for all small t. Consequently, φ satisfies (7.9) for small t, which is what was required.

The case $r = \infty$ (the Sternberg–Chern theorem) can be treated similarly, with the uniqueness in Lemma 6.2 taken into account.

7.4. The correspondence mapping at a saddle point. Suppose that the origin of coordinates is a saddle singular point for the system (7.1). We fix numbers $\varepsilon > 0$ and $\delta > 0$ and consider the two "sections"

$$\Sigma_1 = \{x = (\varepsilon, \eta)\}, \qquad \Sigma_2 = \{x = (\xi, \delta)\}.$$

Let us take a point $(\varepsilon, \eta) \in \Sigma_1$ with $\eta > 0$. The integral curve of our system passing through the point (ε, η) intersects the surface element Σ_2 at some point

$$x' = (f(\eta), \delta).$$

The smooth function f arising in this way is called the *Poincaré mapping* or *correspondence mapping* of the system (7.1) at the point $x = 0$.

EXAMPLE. Suppose that the system (7.1) is linear:

$$\dot{\xi} = \nu_1 \xi, \qquad \dot{\eta} = \nu_2 \eta, \qquad \nu_1 < 0 < \nu_2.$$

Its flow has force

$$F^t(\xi, \eta) = (e^{\nu_1 t}\xi, e^{\nu_2 t}\eta).$$

The curve $F^t(\xi, \eta)$ intersects the surface element Σ_2 at the time

$$t_0 = t_0(\eta) = \frac{1}{\nu_2} \ln \frac{\delta}{\eta}.$$

Therefore,

$$f(\eta) = \eta^\nu \cdot \varepsilon \cdot \delta^{-\nu}, \qquad \nu = -\nu_1/\nu_2.$$

Obviously, the function f depends on the choice of ε and δ. However, in different dynamical problems (see, for example, [**48**], and also §2 in Chapter 3) it is only the power behavior of f and its derivatives as $\eta \to 0$ that is important. This behavior is independent of the choice of ε and δ and the local coordinate system. Sometimes the coordinate system is required to have sufficiently high (≥ 2) smoothness. Nevertheless, to study the behavior of the correspondence mapping we can use a normal form (which may be nonlinear).

If the original function is sufficiently smooth, then it can be reduced by a C^2-transformation to a linear form, or to the normal form

$$\dot{\xi} = -\xi + c\xi^2\eta, \qquad \dot{\eta} = \eta,$$

or to the normal form

$$\dot{\xi} = -2\xi + c\xi^3\eta, \qquad \dot{\eta} = \eta,$$

or to the normal form

$$\dot{\xi} = -\tfrac{1}{2}\xi + c\xi\eta^3, \qquad \dot{\eta} = \eta.$$

The behavior of the Poincaré mapping for these nonlinear normal forms was studied in [**48**], and will be used by us in §2 of Chapter 7.

§8. Normal forms in a neighborhood of an equilibrium state (survey and comments)

We present a summary of the preceding sections.

1. In a neighborhood of a hyperbolic fixed point any (C^1-) diffeomorphism (flow) is conjugate in the class $C^{0,\alpha}$ to a linear diffeomorphism (flow) for any $\alpha < \tau(\Lambda)$ (see Theorem 2.1). This assertion refines the Grobman–Hartman theorem ensuring topological linearization, and carries over to the multidimensional case for an appropriate definition of $\tau(\Lambda)$.

2. In a neighborhood of a hyperbolic point any diffeomorphism (flow) in $\mathbb{R}^2$ of class $C^{1,\alpha}$ with $\alpha > 1 - \tau(\Lambda)$ is conjugate in the class C^1 to a linear diffeomorphism (flow). This is a refinement of the theorem of Hartman [**74**]: in a neighborhood of a hyperbolic point any C^2-diffeomorphism of the plane is conjugate in the class C^1 to a linear diffeomorphism. The Hartman theorem carries over to the multidimensional case only for a node (see [**74**]). However, already in $\mathbb{R}^3$ there are examples (see [**74**]) of analytic local diffeomorphisms that cannot be smoothly linearized in a neighborhood of a saddle. A condition is given in [**31**] that ensures

smooth linearization. Namely, suppose that the eigenvalues $\lambda_1, \dots, \lambda_n$ of a linear transformation Λ satisfy the condition

$$|\lambda_j| \neq |\lambda_i| \cdot |\lambda_k| \qquad (|\lambda_i| < 1 < |\lambda_k|).$$

Then the C^2-diffeomorphism $F(x) = \Lambda x + \dots$ is locally conjugate in the class C^1 to a linear diffeomorphism.

3. A diffeomorphism of the plane $\mathbb{R}^2$ of class $C^{r,\alpha}$, $\alpha > 1 - r\tau(\Lambda)$, in a neighborhood of a node can be reduced by a C^r-transformation either to linear form or to the normal form

$$G(x) = (\lambda^p \xi + \eta^p, \lambda\eta), \qquad p < r.$$

Consequently, the system of equations can be reduced either to linear form or to the form

$$\dot{\xi} = p\xi + \eta^p, \qquad \dot{\eta} = \eta, \qquad p \in \mathbb{Z}_+,\ p \geq 2.$$

For $r < \infty$ and $\alpha > 0$ the diffeomorphism can be reduced in a neighborhood of a saddle point by a C^p-transformation with $p = [r/2]$ to the polynomial normal form

$$G(x) = \begin{cases} \lambda_1 \xi + \xi \sum_{j=1}^m c_j (\xi^l \eta^s)^j, \\ \lambda_2 \eta + \eta \sum_{j=1}^m d_j (\xi^l \eta^s)^j. \end{cases}$$

This normal form corresponds to the resonance

$$\lambda_1^l \lambda_2^s = 1,$$

where l and s are relatively prime, and

$$m \leq \frac{r-1}{l+s}.$$

In particular, if $l + s > r - 1$, that is, there are only "distant" resonances, then the normal form is linear.

Sharp estimates are given in [**108**] (with appropriate counterexamples) for the smoothness class of the reducing transformation in dependence on the value of the ratio $\tau = \left|\dfrac{\ln|\lambda_2|}{\ln|\lambda_2|}\right|$. However, the estimate $[r/2]$ cannot be improved in the set of all linear approximations. For linear normal forms there is a multidimensional analogue of this estimate: a C^r-diffeomorphism

$$F(x) = \Lambda x + o(\|x\|^r), \qquad x \in \mathbb{R}^n,$$

that is hyperbolic at the origin is locally conjugate in the class C^p to a linear diffeomorphism, $p = [\frac{r-1}{n}]$ (see [**31**]).

4. In general a C^∞-diffeomorphism cannot be reduced by a C^∞-transformation to a polynomial mapping in a neighborhood of a hyperbolic point. However, this can be done by a transformation of any finite smoothness class. For example, by Theorem 6.3, a C^∞-diffeomorphism can be reduced in a neighborhood of a saddle point by a C^k-transformation to the form

$$G(\xi, \eta) = \begin{cases} \lambda_1 \xi + \xi \cdot \sum_{j=1}^m c_j (\xi^l \eta^s)^j, \\ \lambda_2 \eta + \eta \cdot \sum_{j=1}^m d_j (\xi^l \eta^s)^j, \end{cases}$$

where

$$m \leq \frac{2k-1}{l+s}, \qquad \lambda_1^l \lambda_2^s = 1, \tag{8.1}$$

with $l, s \in \mathbb{Z}_+$ relatively prime. However, the degrees of the polynomials increase in general as the smoothness of the reducing transformation increases.

The following question arises in this connection. Fix numbers $d, k \in \mathbb{Z}_+$. What conditions must be satisfied by a local diffeomorphism

$$F\colon (\mathbb{R}^n, 0) \to (\mathbb{R}^n, 0)$$

in order to be reducible by a C^k-transformation to a polynomial of degree $\leq d$?

In view of the Sternberg–Chern theorem the answer must be formulated for hyperbolic C^∞-diffeomorphisms in terms of the properties of the coefficients of the normal form of the Taylor series $\beta(F)$. For instance, in the presence of the resonance (8.1) C^k-linearization is possible if in the normal form III we have the condition

$$|d_j| + |c_j| \neq 0 \implies j > \frac{2k-1}{l+s}. \tag{8.2}$$

The case $d = 1$, that is, the possibility of C^k-linearization, seems most interesting. It has been thoroughly studied by Samovol [**72**] (see also [**113**]) for arbitrary values of k and arbitrary dimension n.

5. Nonhyperbolic equilibrium states are also of interest in a whole series of problems, in particular, for bifurcation theory. Here the formal theory of normal forms becomes richer in content (see [**31**]). In the one-dimensional case all C^∞-diffeomorphisms can be reduced in a neighborhood of a fixed point by a local C^∞-transformation to one of the normal forms

$$G(x) = \pm x + x^{sl+1} + \alpha x^{2sl+1}, \qquad l \in \mathbb{Z}_+,$$

up to a set of infinite codimension, where $s = 1$ or 2, depending on the sign of the first term. This normal form was found by Takens in [**109**], where an orbital normal form was determined for "almost all" systems on the plane. For C^∞-diffeomorphisms and vector fields in $\mathbb{R}^n$ satisfying the so-called single-resonance condition (a multidimensional analogue of a resonance of the form (8.1)) a classification and normal forms were obtained in [**34**] and [**35**].

6. Finally, analytic dynamical systems (real and complex) are a classical subject of investigation. The first results are due to Poincaré and laid the foundation of the theory of normal forms. He proved the theorem on analytic linearization of a node in the absence of resonances. Then Dulac proved the possibility of analytic reduction to normal form in a neighborhood of a node. At a singular point that is not a node, analytic normalization is hindered by so-called small denominators: the differences $\lambda_j - \lambda_1^{p_1} \dots \lambda_n^{p_n}$, where the λ_i are the eigenvalues of the linear approximation, appear in the denominators of the formal transformations and can interfere with convergence when they are "pathologically" small. In the absence of resonances small denominators are the only obstacle to linearization. However, finding precise conditions ensuring convergence is a difficult and substantive problem. The first step was made by Siegel. He proved that a one-dimensional complex-analytic diffeomorphism

$$F(z) = \lambda z + f(z)$$

can be linearized by a locally analytic transformation under the arithmetic condition

$$|\lambda^q - 1| \geq c q^{-\nu}, \qquad q \in \mathbb{Z}_+. \tag{8.3}$$

with some $c, \nu > 0$. The condition (8.3) holds for almost all numbers λ with respect to Lebesgue measure. This theorem was carried over by Siegel [**45**] to the multidimensional case. However, all the fundamental difficulties already appear in the one-dimensional problem. Siegel used the direct majorant method, and that involved a considerable quantity of technical details. Essential progress became possible with the emergence of the KAM method. By using this method Bryuno [**38**] replaced the condition (8.3) by a weaker condition. Recently Yoccoz [**112**] determined that the Bryuno condition is necessary: if this condition fails to hold, then there is a residual f such that the diffeomorphism F cannot be analytically linearized.

In the problem of analytic normalization in the presence of resonances other obstacles arise in addition to small denominators. This problem has been thoroughly studied by Bryuno [**38**]. He determined conditions on a formal normal form which together with "nonsmallness" of the denominators ensure an analytic normalization. These conditions are very stringent. For example, even the one-dimensional complex-analytic mappings

$$F(z) = z + o(z), \qquad z \in \mathbb{C},$$

cannot as a rule be reduced to their formal normal form (which here has the form

$$G(z) = z + z^q + \alpha z^{2q+1}, \qquad q \geq 2,$$

with some $q \in \mathbb{Z}_+$ and $\alpha \in \mathbb{C}$). Moreover, the analytic classification of such mappings leads to functional invariants (the Écalle–Voronin moduli; see [**41**]).

CHAPTER 5

Transformations of the Circle

The investigation of flows on two-dimensional manifolds leads in a natural way to the study of transformations of the circle, because a flow with a closed transversal induces on it a Poincaré mapping (a transformation of the closed transversal into itself). Many properties of a flow can be extracted from the properties of the Poincaré mapping on a global section of the flow (that is, a closed transversal that intersects any regular trajectory of the flow).

The results in this chapter are used in the next chapter in studying topological equivalence problems and the topological classification of transitive flows and Denjoy flows on the torus, as well as in Chapter 7 in studying the interrelation between smoothness properties and topological properties of flows.

The study of transformations of the circle is also of independent interest. Several directions of investigation have a finished form and serve as prototypes of corresponding results in the multidimensional theory of dynamical systems.

We present rigorous results mainly for transformations of the circle of degree 1 (in particular, for homeomorphisms and diffeomorphisms) with nondecreasing covering mapping. However, for completeness we mention (in the form of remarks) results of interest in our view on transformations of the circle with other properties.

§1. The Poincaré rotation number

In 1885 Poincaré [**68**] introduced for homeomorphisms of the circle the concept of the rotation number, which has turned out to be very fruitful in the theory of dynamical systems. The meaning of the rotation number of a transformation of the circle is that for a particular point of the circle the rotation number gives an asymptotic indication (that is, in the limit) of the average rotation of the point along the circle under the action of the transformation. For a large class of transformations the rotation number does not depend on the point (in particular, for homeomorphisms) and characterizes the transformation itself.

In this section we introduce the rotation number for a transformation of the circle of degree 1 with a monotonically nondecreasing covering mapping, and we prove some properties of the rotation number.

1.1. Definitions and notation. Denote by $P(\mathbb{R})$ the set of transformations $\overline{f}\colon \mathbb{R} \dashrightarrow \mathbb{R}$ satisfying the conditions:

1) $\overline{f}(x+1) = \overline{f}(x) + 1$, $x \in \mathbb{R}$;

2) $\overline{f}$ is monotonically nondecreasing, that is, $\overline{f}(x_1) \leq \overline{f}(x_2)$ if $x_1 < x_2$.

These conditions immediately yield the properties:

3) the transformation $\overline{f}^k \overset{\text{def}}{=} f \circ \ldots \circ f$ is monotonically nondecreasing for $k \in \mathbb{N}$;

4) $\overline{f}^k(x+r) = \overline{f}^k(x) + r$, $x \in \mathbb{R}$, $k \in \mathbb{N}$, $r \in \mathbb{Z}$.

LEMMA 1.1. *Let $\overline{f} \in P(\mathbb{R})$, and suppose that $r < \overline{f}^k(x_0) - x_0 < r+1$ for some $k \in \mathbb{N}$, $r \in \mathbb{Z}$, and $x_0 \in \mathbb{R}$. If $\overline{f}^k(x) - x \notin \mathbb{Z}$ for all $x \in \mathbb{R}^1$, then $r < \overline{f}^k(x) - x < r+1$ for all $x \in \mathbb{R}$.*

PROOF. Let $\mathcal{D} \stackrel{\text{def}}{=} \{x \in \mathbb{R} : r < \overline{f}^k(x) - x\}$. By the condition 1) in the definition of the set $P(\mathbb{R})$, $x_0 + n \in \mathcal{D}$ for all $n \in \mathbb{Z}$.

Let $x_1 \in \mathcal{D}$. We show that $[x_1, \infty) \in \mathcal{D}$. Indeed, since $r < \overline{f}^k(x_1) - x_1$, the condition 3) gives us that $x + r \leq \overline{f}^k(x_1) \leq \overline{f}^k(x)$ for all $x_1 \leq x \leq \overline{f}^k(x_1) - r$. Since $\overline{f}^k(x) - x \notin \mathbb{Z}$, we have the strict inequality $r < \overline{f}^k(x) - x$, that is, $[x_1, \overline{f}^k(x_1) - r] \subset \mathcal{D}$.

Let $\{x_n\}_1^\infty$ be a monotonically increasing sequence of points $x_n \in \mathcal{D}$ converging to a point $x_\infty \in \mathbb{R}$. If we assume that $x_\infty \notin \mathcal{D}$, that is, $r > \overline{f}^k(x_\infty) - x_\infty$, then we get that $\overline{f}^k(x_n) \leq \overline{f}^k(x_\infty) < x_n + r$ for the points x_n with $\overline{f}^k(x_\infty) - r < x_n < x_\infty$, and this contradicts the inclusion $x_n \in \mathcal{D}$. Consequently, $x_\infty \in \mathcal{D}$, and hence $[x_1, \infty) \subset \mathcal{D}$.

The equality $\mathcal{D} = \mathbb{R}$ follows from this and the inclusion $x_0 + n \in \mathcal{D}$, $n \in \mathbb{Z}$; that is, $r < \overline{f}^k(x) - x$ for all $x \in \mathbb{R}$. It can be proved similarly that $\overline{f}^k(x) - x < r + 1$ for all $x \in \mathbb{R}$. $\square$

THEOREM 1.1. *For any transformation $\overline{f} \in P(\mathbb{R})$ the limit*

$$\lim_{n\to\infty} \frac{\overline{f}^n(x)}{n} \stackrel{\text{def}}{=} \alpha \tag{1.1}$$

exists and does not depend on the choice of the point $x \in \mathbb{R}$. If $\overline{f}^k(x_0) = x_0 + r$ for some $x_0 \in \mathbb{R}$, $k \in \mathbb{N}$, and $r \in \mathbb{Z}$, then the number α is rational and equal to r/k.

PROOF. We first assume that the limit (1.1) exists for some point x_0, and we take an arbitrary point $x \in \mathbb{R}$. Let $s \in \mathbb{Z}$ be such that $x_0 + s \leq x < x_0 + s + 1$. By the properties 3) and 4), $\overline{f}^n(x_0 + s) = \overline{f}^n(x_0) + s \leq \overline{f}^n(x) \leq \overline{f}^n(x_0 + s + 1) = \overline{f}^n(x_0) + s + 1$. From this,

$$\left| \frac{\overline{f}^n(x) - \overline{f}^n(x_0)}{n} \right| \leq \frac{|s| + 1}{n} \to 0 \qquad \text{as } n \to \infty,$$

and hence the limit (1.1) exists for all $x \in \mathbb{R}$ and does not depend on x.

We now prove that the limit (1.1) exists for some point.

Let $\overline{f}^k(x_0) = x_0 + r$ for some $x_0 \in \mathbb{R}$, $k \in \mathbb{N}$, and $r \in \mathbb{Z}$. Any positive integer n can be written in the form $n = qk + s$, where $0 \leq s < k$. Then by the property 4) and the equality

$$\overline{f}^{qk}(x_0) = \underbrace{\overline{f}^k \circ \ldots \circ \overline{f}^k}_{q}(x_0) = x_0 + qr,$$

we have that $\overline{f}^n(x_0) = \overline{f}^s \circ \overline{f}^{qk}(x_0) = \overline{f}^s(x_0 + qr) = \overline{f}^s(x_0) + qr$, which implies that

$$\frac{\overline{f}^n(x_0)}{n} = \frac{\overline{f}^s(x_0) + qr}{n} = \frac{\overline{f}^s(x_0)}{n} + \frac{qr}{qk + s} = \frac{\overline{f}^s(x_0)}{n} + \frac{r}{k + s/q} \to \frac{r}{k} \quad (q, n \to \infty).$$

Thus, in this case the limit (1.1) exists and is equal to $r/k \in Q$.

Assume now that $\overline{f}^k(x) - x \notin \mathbb{Z}$ for all $x \in \mathbb{R}$ and $k \in \mathbb{N}$. Then for fixed $x \in \mathbb{R}$ and $k \in \mathbb{N}$ there exists a number $q_k(x) \in \mathbb{Z}$ such that $q_k(x) < \overline{f}^k(x) - x < q_k(x) + 1$. According to Lemma 1.1, $q_k(x) \overset{\text{def}}{=} q$ does not depend on the point $x \in \mathbb{R}$, that is,

$$q < \overline{f}^k(x) - x < q + 1 \tag{1.2}$$

for any $x \in \mathbb{R}$. We apply (1.2) to the points $x = 0, \overline{f}^k(0), \dots, \overline{f}^{k(n-1)}(0)$:

$$\begin{aligned} &q < \overline{f}^k(0) < q + 1 \\ &q < \overline{f}^{2k}(0) - \overline{f}^k(0) < q + 1 \\ &\qquad \vdots \\ &q < \overline{f}^{nk}(0) - \overline{f}^{k(n-1)}(0) < q + 1. \end{aligned}$$

Adding these inequalities, we get that

$$qn < \overline{f}^{nk}(0) < (q+1)n,$$

or

$$\frac{q}{n} < \frac{\overline{f}^{nk}(0)}{nk} < \frac{q}{k} + \frac{1}{k}. \tag{1.3}$$

In view of (1.2) for $x = 0$, we have that

$$\frac{q}{k} < \frac{\overline{f}^{k}(0)}{k} < \frac{q}{k} + \frac{1}{k}.$$

This and (1.3) give us that

$$\left| \frac{\overline{f}^{nk}(0)}{nk} - \frac{\overline{f}^{k}(0)}{k} \right| < \frac{1}{k}.$$

Since k and n are arbitrary, the above computations can be repeated with k and n interchanged. Then we get that

$$\left| \frac{\overline{f}^{n}(0)}{n} - \frac{\overline{f}^{nk}(0)}{nk} \right| < \frac{1}{n}.$$

Combining the last two inequalities, we get that

$$\left| \frac{\overline{f}^{n}(0)}{n} - \frac{\overline{f}^{k}(0)}{k} \right| < \frac{1}{n} + \frac{1}{k},$$

that is, the sequence $\{\overline{f}^n(0)/n\}_1^\infty$ satisfies the Cauchy condition. Therefore, the limit (1.1) exists for $x = 0$, and hence for all $x \in \mathbb{R}$. □

DEFINITION. Let $\overline{f} \in P(\mathbb{R})$. The number

$$\lim_{n \to \infty} \frac{\overline{f}^n(x)}{n} \overset{\text{def}}{=} \operatorname{rot}(\overline{f}), \qquad x \in \mathbb{R},$$

is called the *rotation number* of the transformation $\overline{f}$.

LEMMA 1.2. *For $\overline{f} \in P(\mathbb{R})$ and for fixed $k \in \mathbb{N}$ and $r \in \mathbb{Z}$ the transformation $\overline{f}_1(x) = \overline{f}^k(x) + r$ belongs to the set $P(\mathbb{R})$ and has rotation number* $\operatorname{rot}(\overline{f}_1) = \operatorname{rot}(\overline{f}^k + r) = k \operatorname{rot}(\overline{f}) + r$.

PROOF.

$$\operatorname{rot}(\overline{f}_1) = \lim_{n\to\infty} \frac{\overline{f}_1^n(x)}{n} = \lim_{n\to\infty} \frac{\overline{f}^{nk}(x) + nr}{n}$$
$$= k \lim_{n\to\infty} \frac{\overline{f}^{nk}(x)}{nk} + r = k \operatorname{rot}(\overline{f}) + r. \qquad \square$$

LEMMA 1.3. *Let $\overline{f} \in P(\mathbb{R})$.*

1) If $r \le \overline{f}^k(x) - x \le r+1$ for some $k \in \mathbb{N}$, $r \in \mathbb{Z}$, and $x \in \mathbb{R}$, then

$$\frac{r}{k} \le \operatorname{rot}(\overline{f}) \le \frac{r+1}{k}.$$

2) If $\overline{f}$ is continuous and $r/k < \operatorname{rot}(\overline{f}) < (r+1)/k$, then there exists a number $\eta > 0$ such that the inequalities

$$r + \eta < \overline{f}^k(x) - x < r + 1 - \eta$$

hold for all $x \in \mathbb{R}$.

PROOF. 1) Since the transformation $\overline{f}^k$ is nondecreasing and since $\overline{f}^k(x+r) = \overline{f}^k(x)+r$ for $x \in \mathbb{R}$ and $r \in \mathbb{Z}$, it follows from the inequalities $r+x \le \overline{f}^k(x) \le r+1+x$ that $nr + x \le \overline{f}^{nk}(x) \le n(r+1) + x$. Consequently,

$$\frac{r}{k} \le \lim_{n\to\infty} \frac{nr+x}{nk} \le \lim_{n\to\infty} \frac{\overline{f}^{nk}(x)}{nk} = \operatorname{rot}(\overline{f}) \le \lim_{n\to\infty} \frac{n(r+1)+x}{nk} = \frac{r+1}{k}.$$

We prove 2). According to the (proven) assertion 1), $r \le \overline{f}^k(x) - x$ for all $x \in \mathbb{R}$. It follows from Theorem 1.1 that if $r = \overline{f}^k(x) - x$ for some $x \in \mathbb{R}$, then $\operatorname{rot}(\overline{f}) = r/k$. Therefore, $r < \overline{f}^k(x) - x$ for all $x \in \mathbb{R}$.

The equality $\overline{f}^k(x+1) = \overline{f}^k(x) + 1$ implies that the function $\overline{f}^k(x) - x$ is periodic with period 1. Consequently, there exists a number $\eta_1 > 0$ such that $r + \eta_1 < \overline{f}^k(x) - x$ for all $x \in \mathbb{R}$.

It can be proved similarly that there is an $\eta_2 > 0$ such that $\overline{f}^k(x) - x < r+1-\eta_2$. The required η is $\min\{\eta_1, \eta_2\}$. $\square$

REMARK. The assertion 2) in Lemma 1.3 is not true in general for discontinuous transformations in the class $P(\mathbb{R})$ (Figure 5.1). This assertion is used in what follows for proving that the rotation number depends continuously on a parameter. It can be shown that under the additive introduction of a parameter (see Theorem 1.2) the rotation number is a discontinuous function in general for a discontinuous transformation in $P(\mathbb{R})$.

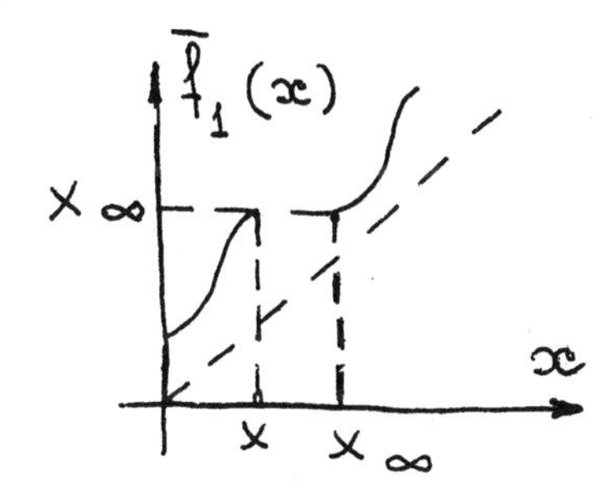

FIGURE 5.1

The group $\mathbb{Z}$ of integers can be identified in a natural way with the group of mappings of $\mathbb{R}$ of the form $x \mapsto x+n$, $x \in \mathbb{R}$, $n \in \mathbb{Z}$. The quotient space $\mathbb{R}/\mathbb{Z}$ is homeomorphic to the circle S^1, which can be represented as the closed interval $[0,1]$ with endpoints identified. The natural projection $\pi\colon \mathbb{R} \to \mathbb{R}/\mathbb{Z} \cong S^1$ is a universal covering.

By the condition 1) in the definition of the set $P(\mathbb{R})$, each transformation $\overline{f} \in P(\mathbb{R})$ is a covering for some transformation $f\colon S^1 \to S^1$, that is,

$$f \circ \pi = \pi \circ \overline{f}.$$

Denote by $P(S^1)$ the set of transformations of the circle that have covering transformations in $P(\mathbb{R})$.

LEMMA 1.4. *Let $f \in P(S^1)$.*

1) *If $\overline{f} \in P(\mathbb{R})$ is a covering transformation for f, then so is $\overline{f}+n$, $n \in \mathbb{Z}$.*

2) *If $\overline{f}_1, \overline{f}_2 \in P(\mathbb{R})$ are covering transformations for f, then there exists a $k \in \mathbb{Z}$ such that $\overline{f}_2(x) = \overline{f}_1(x) + k$ for all $x \in \mathbb{R}$.*

PROOF. The assertion 1) follows from the definition of the covering π because $\pi \circ \overline{f} = \pi \circ (\overline{f}+n)$.

We prove 2). By assumption, for each $x \in \mathbb{R}$ there exists an integer $k(x)$ such that $\overline{f}_2(x) - \overline{f}_1(x) = k(x)$. Since $\overline{f}_i(x+1) = \overline{f}_i(x)+1$ and since $\overline{f}_i$ is monotonically nondecreasing ($i = 1, 2$), $|\overline{f}_i(x_1) - \overline{f}_i(x_2)| \leq 1$ for sufficiently close points $x_1, x_2 \in \mathbb{R}$. This implies that the function $k(x)$ is locally constant. Then $k(x) = k = \text{const}$ because $k(x)$ is in $\mathbb{Z}$. □

DEFINITION. Let $f \in P(S^1)$. The number

$$\lim_{n\to\infty} \frac{\overline{f}^n(x)}{n} \pmod 1 \stackrel{\text{def}}{=} \operatorname{rot}(f), \qquad x \in \mathbb{R},$$

where $\overline{f} \in P(\mathbb{R})$ is a covering transformation for f, is called the *rotation number* of the transformation f.

By Lemma 1.2 and 1.4, the definition of the rotation number of a transformation $f \in P(S^1)$ does not depend on the choice of the covering $\overline{f} \in P(\mathbb{R})$.

Remark 1. The rotation set of a continuous transformation of degree 1. We recall that a transformation $f\colon S^1 \to S^1$ is a transformation of degree k if there exists a covering transformation $\overline{f}\colon \mathbb{R} \to \mathbb{R}$ for f such that $\overline{f}(x+1) =$

$\overline{f}(x) + k$. Thus, all the transformations in the set $P(S^1)$ are transformations of degree 1.

Let $\overline{f}\colon \mathbb{R} \to \mathbb{R}$ be a covering for some transformation $f\colon S^1 \to S^1$ of degree 1. If the condition 2) in the definition of $P(\mathbb{R})$ does not hold for $\overline{f}$, that is, $\overline{f}$ is not monotonically nondecreasing, then the limit (1.1) can fail to exist in general or can depend on the point x. Newhouse, Palis, and Takens (*Stable families of dynamical systems.* I: *Diffeomorphisms*, preprint, I.M.P.A., Pio, Brasil, 1978) proposed a generalization of the concept of the rotation number to continuous mappings (not necessarily one-to-one) of the circle of degree 1. For a given $x \in \mathbb{R}$ the rotation number is defined to be the limit superior of the sequence $\{\overline{f}^n(x)/n\}_1^\infty$:

$$\operatorname{rot}(\overline{f}, x) = \varlimsup_{n\to\infty} \frac{\overline{f}^n(x)}{n}$$

(the largest limit of the convergent subsequences). The *rotation set* of $\overline{f}$ is $\operatorname{rot}(\overline{f}) = \{\operatorname{rot}(\overline{f}, x) : x \in \mathbb{R}\}$. Since a covering of a transformation f is determined up to an integer addend, the rotation set of f, defined to be $\operatorname{rot}(f) = \operatorname{rot}(\overline{f}) \pmod 1$, does not depend on the covering $\overline{f}$.

The following two lemmas were proved by R. Ito in the paper, *Rotation sets are closed* (Math. Proc. Cambridge Philos. Soc. **89** (1981), no. 1, 107–111).

LEMMA. *The rotation set* $\operatorname{rot}(\overline{f})$ *is bounded.*

PROOF. Since $\overline{f}(x)$ is a covering for a continuous transformation of the circle of degree 1, $\overline{f}(x) - x$ is a periodic continuous function of period 1. Therefore, setting $L \stackrel{\text{def}}{=} \max |\overline{f}(x) - x|$, we have that

$$|\overline{f}^n(x) - x| \le \left| \sum_{i=1}^n [\overline{f}^i(x) - \overline{f}^{i-1}(x)] \right| \le \sum_{i=1}^n \left| \overline{f}(\overline{f}^{i-1}(x)) - \overline{f}^{i-1}(x) \right| \le nL,$$

and from this

$$-L \le \varlimsup_{n\to\infty} \frac{\overline{f}^n(x) - x}{n} = \operatorname{rot}(\overline{f}, x) \le L. \qquad \square$$

LEMMA. *If* $\lambda, \mu \in \operatorname{rot}(\overline{f})$ *and* $\lambda \le p/q \le \mu$, *where* $p \in \mathbb{Z}$ *and* $q \in \mathbb{N}$, *then there exists an* $x \in \mathbb{R}$ *such that* $\overline{f}^q(x) = x + p$ *(that is, the transformation* $f\colon S^1 \to S^1$ *has a periodic point* $\pi(x)$ *of period* q *with rotation number* $p/q \pmod 1$*).*

PROOF. Assume the contrary. Then since $\overline{f}^q(x) - x$ is continuous, either $\overline{f}^q(x) - x < p$ for all $x \in \mathbb{R}$ or $\overline{f}^q(x) - x > p$ for all $x \in \mathbb{R}$. For definiteness assume the former. Since $\overline{f}^q(x) - x$ is periodic and continuous, there exists an $\varepsilon > 0$ such that $\overline{f}^q(x) - x < p - \varepsilon$. Then $\overline{f}^{nq}(x) - x < n(p - \varepsilon)$, and hence $\operatorname{rot}(\overline{f}, x) \le (p - \varepsilon)/q < p/q$ for all $x \in \mathbb{R}$, which contradicts our assumption. $\square$

With the help of these lemmas, Ito proved the following theorem.

THEOREM (R. Ito). *Let* $\overline{f}\colon \mathbb{R} \to \mathbb{R}$ *be a covering for a continuous transformation of the circle of degree* 1. *Then the rotation set of* $\overline{f}$ *is either a point or a closed bounded interval.*

In the paper, *Rotation intervals of endomorphisms of the circle* (Ergodic Theory Dynamical Systems **4** (1984), 493–498), Bamon, Malta, Pacifico, and Takens defined the rotation set $\widetilde{\mathrm{rot}}(\overline{f}, x)$ of a point x with respect to a continuous transformation $\overline{f}: \mathbb{R} \to \mathbb{R}$ covering a transformation of the circle of degree 1. The set $\widetilde{\mathrm{rot}}(\overline{f}, x)$ was defined to be the union of all the limit points of the sequence $\{\overline{f}^n(x)/n\}_1^\infty$. In that paper it was proved that $\widetilde{\mathrm{rot}}(\overline{f}, x)$ is a closed subinterval of the interval $\mathrm{rot}(\overline{f})$ for any $x \in \mathbb{R}$, and for a given $[\alpha, \beta] \subseteq \mathrm{rot}(\overline{f})$ ($\alpha \leq \beta$) there is a point $x \in \mathbb{R}$ such that $\widetilde{\mathrm{rot}}(\overline{f}, x) = [\alpha, \beta]$.

Remark 2. The rotation set of a topological Markov chain. Following a paper of V. M. Alekseev on symbolic dynamics (Eleventh Mathematical School, Kiev, 1976), we consider a topological Markov chain (TMC) (L, Π, σ) with a Hausdorff state space L, a transition matrix Π that is a closed subset of $L \times L$, and the left-shift mapping σ on the space Ω_Π of two-sided infinite sequences $\{x_n\}_{-\infty}^\infty$, $x_n \in L$, such that $(x_n, x_{n+1}) \in \Pi$ for all $n \in \mathbb{Z}$.

We represent the state space L as the union $L_1 \cup L_2$ of two disjoint subsets L_1 and L_2. Denote by χ_2 the indicator function of the subset $L_2 \subset L$: $\chi_2(x) = 1$ for $x \in L_2$ and $\chi_2(x) = 0$ for $x \notin L_2$. It will be assumed that $L_2 \neq \emptyset$.

The concepts of the rotation number of a point $\underline{x} = \{x_n\}_{-\infty}^\infty \in \Omega_\Pi$ and the rotation set of a TMC were introduced in the following way in [**57**].

The number $\mathrm{rot}(\sigma, \underline{x}) = \overline{\lim}_{n\to\infty} \frac{1}{n} \sum_{k=0}^{n-1} \chi_2(x_k)$ is called the rotation number of a point $\underline{x} = \{x_k\}_{-\infty}^\infty \in \Omega_\Pi$ with respect to the partition $L = L_1 \cup L_2$. The set $\mathrm{rot}(\sigma, \Omega_\Pi) = \{\mathrm{rot}(\sigma, \underline{x}) : \underline{x} \in \Omega_\Pi\}$ is called the rotation set of the TMC (L, Π, σ) with respect to the partition $L = L_1 \cup L_2$.

THEOREM [**57**]. *Let (L, Π, σ) be an irreducible finite TMC, and let $L = L_1 \cup L_2$ be a fixed partition of the state space. Then:*

1) *the rotation set $\mathrm{rot}(\sigma, \Omega_\Pi)$ with respect to the partition $L = L_1 \cup L_2$ is either a point or a closed bounded interval;*

2) *if the transformation $\sigma|_{\Omega_\Pi}$ is topologically mixing, then $\mathrm{rot}(\sigma, \Omega_\Pi)$ is a closed interval with rational endpoints that are the rotation numbers of two periodic points.*

Remark 3. The rotation set of a mapping of Lorenz type. The results in the preceding remark can be applied to Lorenz-type mappings of an interval [**58**]. Denote by $\mathfrak{F}_L$ the set of mappings $f: I \to I$ of $I = [0, 1]$ having a single point $c = c(f)$ of discontinuity and satisfying the following conditions:

1) f is continuous and monotonically increasing on $[0, c)$ and $(c, 1]$;

2) the set $\mathcal{D} = \bigcup_{n\geq 0} f^{-n}(c)$ is dense in I;

3) $\lim_{x\downarrow c} f(x) = 0$ and $\lim_{x\uparrow c} f(x) = 1$ (Figure 5.2).

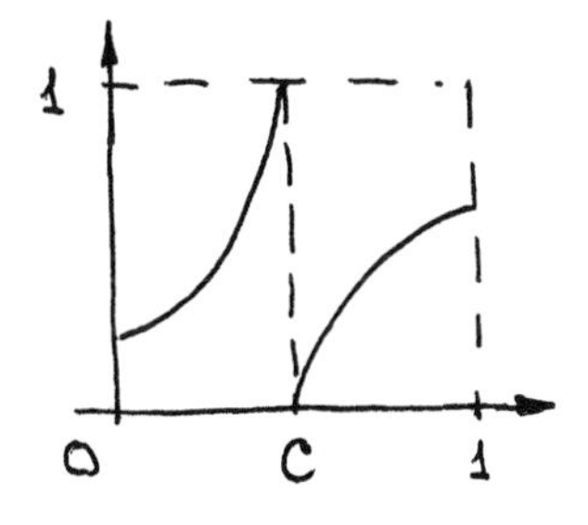

FIGURE 5.2

A symbolic model of a mapping $f \in \mathfrak{F}_L$ can be constructed as follows. Let us associate with an arbitrary point $x \in I \setminus \mathfrak{D}$ the sequence $k(x) = \{k_n(x)\}_{n=0}^{\infty}$ of symbols ± 1 with

$$k_n(x) = \begin{cases} -1, & f^n(x) \in [0, c), \\ +1, & f^n(x) \in (c, 1]. \end{cases}$$

We get a mapping k from $I \setminus \mathfrak{D}$ to a one-sided Bernoulli scheme Ω_2^+ in two symbols (with the product topology). Denote by Σ_f^+ the closure of $k(I \setminus \mathfrak{D})$ in Ω_2^+, and by σ_+ the one-sided shift. The symbolic model of f is defined to be the TMC (L, Π, σ), where $L = \{+1, -1\}$ and $\Omega_\Pi = \Sigma_f^+$. The rotation set $\operatorname{rot}(f)$ of a mapping $f \in \mathfrak{F}_L$ is defined to be the rotation set of the corresponding TMC with respect to the partition $L = L_1 \cup L_2$, where $L_1 = \{-1\}$ and $L_2 = \{+1\}$.

According to [**57**], $\operatorname{rot}(f)$ is either a point or a closed bounded interval.

THEOREM [**58**]. *If $\operatorname{rot}(f)$, $f \in \mathfrak{F}_L$, is an interval, then the topological entropy of f is strictly positive, and there exists a $k \in \mathbb{N}$ such that for any $n \geq k$ the mapping f has a periodic point with smallest period n.*

We remark that if we identify the endpoints of the interval I, then on the resulting circle a Lorenz-type mapping induces a transformation of the circle for which it is also possible to introduce the rotation set in terms of a covering transformation [**57**].

1.2. Invariance of the rotation number.

DEFINITION. A mapping $\overline{f} \in P(\mathbb{R})$ is *semiconjugate* to a mapping $\overline{f}_0 \in P(\mathbb{R})$ if there exists a continuous mapping $\overline{h} \in P(\mathbb{R})$ such that

$$\overline{f} \circ \overline{h} = \overline{h} \circ \overline{f}_0. \tag{1.4}$$

We say that $\overline{f}$ is semiconjugate to $\overline{f}_0$ by means of $\overline{h}$.

Semiconjugacy of an $f \in P(S^1)$ and an $f_0 \in P(S^1)$ by means of a continuous $h \in P(S^1)$ is defined similarly:

$$f \circ h = h \circ f_0. \tag{1.5}$$

LEMMA 1.5. a) *If $\overline{f}$ is semiconjugate to $\overline{f}_0$ by means of $\overline{h}$, where $\overline{f}, \overline{f}_0, \overline{h} \in P(\mathbb{R})$, then*

$$\operatorname{rot}(\overline{f}) = \operatorname{rot}(\overline{f}_0).$$

b) *If f is semiconjugate to f_0 by means of h, where $f, f_0, h \in P(S^1)$, then*

$$\operatorname{rot}(f) = \operatorname{rot}(f_0).$$

PROOF. a) By virtue of (1.4), $\overline{f}^n \circ \overline{h} = \overline{h} \circ \overline{f}_0^n$, $n \in \mathbb{N}$. Therefore,

$$\operatorname{rot}(\overline{f}) = \lim_{n\to\infty} \frac{\overline{f}^n(\overline{h}(x))}{n} = \lim_{n\to\infty} \frac{\overline{h} \circ \overline{f}_0^n(x)}{n}.$$

The function $\overline{h}(x) - x \overset{\text{def}}{=} \overline{\varphi}(x)$ is periodic and continuous, so $\sup_{x\in\mathbb{R}} |\overline{\varphi}(x)| \leq M < +\infty$. Consequently,

$$\operatorname{rot}(\overline{f}) = \lim_{n\to\infty} \frac{\overline{h} \circ \overline{f}_0^n(x)}{n} = \lim_{n\to\infty} \frac{\overline{f}_0^n(x) + \overline{\varphi}(\overline{f}_0^n(x))}{n} = \lim_{n\to\infty} \frac{\overline{f}_0^n(x)}{n} = \operatorname{rot}(\overline{f}_0).$$

Let us prove b). Take coverings $\overline{f}$, $\overline{f}_0$, and $\overline{h}$ for f, f_0, and h, respectively. By (1.5), $\overline{f} \circ \overline{h} = \overline{h} \circ \overline{f}_0 + k = \overline{h} \circ (\overline{f}_0 + k)$ for some $k \in \mathbb{Z}$. From this,

$$\operatorname{rot}(\overline{f}) = \operatorname{rot}(\overline{f}_0 + k) = \operatorname{rot}(\overline{f}_0) + k,$$

and hence $\operatorname{rot}(f) = \operatorname{rot}(f_0)$. □

DEFINITION. Mappings $\overline{f}, \overline{g} \in P(\mathbb{R})$ are *conjugate* if there exists a homeomorphism $\overline{h} \in P(\mathbb{R})$ such that $\overline{f} \circ \overline{h} = \overline{h} \circ \overline{g}$.

DEFINITION. Mappings $f, g \in P(S^1)$ are *conjugate* if there exists a homeomorphism $h \in P(S^1)$ such that $f \circ h = h \circ g$.

The next result follows immediately from Lemma 1.5.

COROLLARY 1.1. a) *If the transformations* $\overline{f}, \overline{g} \in P(\mathbb{R})$ *are conjugate, then* $\operatorname{rot}(\overline{f}) = \operatorname{rot}(\overline{g})$.

b) *If the transformations* $f, g \in P(S^1)$ *are conjugate, then* $\operatorname{rot}(f) = \operatorname{rot}(g)$.

1.3. Continuous dependence of the rotation number on a parameter. If a transformation $\overline{f}$ belongs to the set $P(\mathbb{R})$, then for any fixed number $t \in \mathbb{R}$ the transformation $\overline{f}_t(x) \overset{\text{def}}{=} \overline{f}(x) + t$ also belongs to $P(\mathbb{R})$.

THEOREM 1.2. *If* $\overline{f} \in P(\mathbb{R})$ *is continuous, then the function*

$$r(t) = \operatorname{rot}(\overline{f}_t) = \operatorname{rot}[\overline{f}(x) + t]$$

is continuous and monotonically nondecreasing, and $r(\mathbb{R}) = \mathbb{R}$.

PROOF. We prove that $r(t)$ is continuous at a point t_0. For a given $\varepsilon > 0$ we choose a natural number $k > 2/\varepsilon$ and an integer r such that

$$\frac{r}{k} < \operatorname{rot}(\overline{f}_0) < \frac{r+1}{k},$$

where $\overline{f}_0 = \overline{f} + t_0$. By Lemma 1.3, $r + \eta < \overline{f}_0^k(x) - x < r + 1 - \eta$ for all $x \in \mathbb{R}$ and some $\eta > 0$.

Then there exists a $\delta > 0$ such that for $|t_0 - t| < \delta$

$$r \le \overline{f}_t^k(x_0) - x_0 \le r + 1.$$

According to Lemma 1.3, 1), we have $r/k \le \operatorname{rot}(\overline{f}_t) \le (r+1)/k$. Therefore, $|\operatorname{rot}(\overline{f}_0) - \operatorname{rot}(\overline{f}_t)| \le 2/k < \varepsilon$.

Since $\overline{f}$ is nondecreasing, it follows that $\overline{f}_{t_1}^n \le \overline{f}_{t_2}^n$ for $t_1 < t_2$ and $n \in \mathbb{N}$. From this,

$$\operatorname{rot}(\overline{f}_{t_1}) = \lim_{n\to\infty} \frac{\overline{f}_{t_1}^n(x)}{n} \le \lim_{n\to\infty} \frac{\overline{f}_{t_2}^n(x)}{n} = \operatorname{rot}(\overline{f}_{t_2}).$$

Consequently, $r(t)$ is a nondecreasing function.

It follows from Lemma 1.2 that $r(\mathbb{R}) = \mathbb{R}$. □

In the set $\operatorname{Homeo}_+(S^1)$ of orientation-preserving homeomorphisms of the circle we introduce the metric

$$\rho_0(f, g) = \max_{S^1} |f(x) - g(x)|,$$

$f, g \in \operatorname{Homeo}_+(S^1)$.

The rotation number $\operatorname{rot}(f)$, considered only for $f \in \operatorname{Homeo}_+(S^1)$, is a real-valued function rot with domain $\operatorname{Homeo}_+(S^1)$. The next theorem shows that this function is continuous.

THEOREM 1.3. *The function* $\operatorname{rot}\colon \operatorname{Homeo}_+(S^1) \to [0,1)$ *is continuous at each point* $f \in \operatorname{Homeo}_+(S^1)$.

PROOF. This is analogous to the proof of continuity of the function $r(t)$ in Theorem 1.2. We omit it (and leave it to the reader as an exercise).

1.4. The rotation number of a homeomorphism of the circle. In this subsection we establish a connection between an arithmetic property of the rotation number (namely, its rationality or irrationality) and a dynamical property of a homeomorphism of the circle with that rotation number (namely, the presence or absence of a periodic orbit).

LEMMA 1.6. *Suppose that the transformation* $f \in P(S^1)$ *is a homeomorphism. Then* $\operatorname{rot}(f)$ *is rational if and only if* f *has a periodic orbit, that is, there exist a point* $x_0 \in S^1$ *and a number* $k \in \mathbb{N}$ *such that* $f^k(x_0) = x_0$.

PROOF. Let $\overline{f} \in P(\mathbb{R})$ be a covering transformation for f. Then the presence of a periodic point for f is equivalent to the existence of a point $\overline{x}_0 \in \mathbb{R}$ and numbers $k \in \mathbb{N}$ and $r \in \mathbb{Z}$ such that $\overline{f}^k(\overline{x}_0) = \overline{x}_0 + r$.

According to Theorem 1.1, if f has a periodic orbit, then $\operatorname{rot}(f)$ is rational.

We prove that the rationality of $\operatorname{rot}(f)$ implies the existence of a periodic orbit. Take a covering $\overline{f}$ such that $\overline{f}(0) \in [0,1)$ (by Lemma 1.4, there is such a covering). We first consider the case $\operatorname{rot}(\overline{f}) = 0$ and show that $\overline{f}$ has a fixed point. Assume that $\overline{f}(x) \neq x$ for all $x \in \mathbb{R}$. Since $\overline{f}$ is a homeomorphism, it can be assumed for definiteness that $\overline{f}(x) > x$ for all $x \in \mathbb{R}$. Then $0 < \overline{f}(0) < \cdots < \overline{f}^n(0) < \dots$, that is, the sequence $\{\overline{f}^n(0)\}$ is monotonically increasing.

We show that $\overline{f}^n(0) < 1$ for all $n \in \mathbb{N}$. Indeed, if $\overline{f}^s(0) \geq 1$ for some $s \in \mathbb{N}$, then

$$\overline{f}^{2s}(0) = \overline{f}^s(\overline{f}^s(0)) \geq \overline{f}^s(1) = \overline{f}^s(0) + 1 \geq 2.$$

Similarly, $\overline{f}^{ns}(0) \geq n$ for $n \in \mathbb{N}$. Then

$$\operatorname{rot}(\overline{f}) = \lim_{n\to\infty} \overline{f}^{ns}(0)/ns \geq 1/s \neq 0,$$

which contradicts the assumption that $\operatorname{rot}(\overline{f}) = 0$. Therefore, $\overline{f}^n(0) < 1$ for all $n \in \mathbb{N}$, that is, the sequence $\{\overline{f}^n(0)\}$ is bounded.

Consequently, $\lim_{n\to\infty} \overline{f}^n(0) \overset{\text{def}}{=} \overline{x}_0$ exists, and $\overline{f}(\overline{x}_0) = \lim_{n\to\infty} \overline{f}(\overline{f}^n(0)) = \lim_{n\to\infty} \overline{f}^{n+1}(0) = \overline{x}_0$.

Suppose now that $\operatorname{rot}(\overline{f}) = r/k$. Then $\operatorname{rot}[\overline{f}^k - r] = k\operatorname{rot}(\overline{f}) - r = 0$ in view of Lemma 1.2. By what has been proved, there exists a point $\overline{x}_0$ such that $\overline{f}^k(\overline{x}_0) - r = \overline{x}_0$. Then $\pi(\overline{x}_0) \in S^1$ is a fixed point of the homeomorphism f^k. $\square$

COROLLARY 1.2. *A homeomorphism* $f \in P(S^1)$ *does not have periodic points if and only if the rotation number* $\operatorname{rot}(f)$ *is irrational.*

We remark that a transformation $f \in P(S^1)$ that is not a homeomorphism can fail to have periodic points, even if the rotation number $\operatorname{rot}(f)$ is rational. However, if f has a periodic point, then $\operatorname{rot}(f)$ is rational in view of Theorem 1.1.

§2. Transformations with irrational rotation number

In this section we consider transformations of the circle of degree 1 with irrational rotation number. According to Theorem 1.1, these transformations do not have periodic orbits. Therefore, such transformations arise in a natural way in the investigation of flows without closed trajectories.

2.1. Transformations semiconjugate to a rotation.

LEMMA 2.1. *Let* $\overline{f} \in P(\mathbb{R})$, *and suppose that* $\operatorname{rot}(\overline{f}) \in \mathbb{R} \setminus \mathbb{Q}$. *Then* $\overline{f}^{n_1}(x) + m_1 < \overline{f}^{n_2}(x) + m_2$ *if and only if* $n_1 \operatorname{rot}(\overline{f}) + m_1 < n_2 \operatorname{rot}(\overline{f}) + m_2$, *where* $n_1, n_2 \in \mathbb{N}$, $m_1, m_2 \in \mathbb{Z}$, *and* $x \in \mathbb{R}$.

PROOF. Let $\overline{f}^{n_1}(x_0) + m_1 < \overline{f}^{n_2}(x_0) + m_2$ for some $x_0 \in \mathbb{R}$. For definiteness assume that $n_1 \le n_2$. Then $y_0 + m_1 < \overline{f}^{n_2 - n_1}(y_0) + m_2$, where $y_0 = \overline{f}^{n_1}(x_0)$. By Lemma 1.1, $y + m_1 < \overline{f}^{n_2 - n_1}(y) + m_2$ for all $y \in \mathbb{R}$, so $\overline{f}^{n_1}(x) + m_1 < \overline{f}^{n_2}(x) + m_2$ for all $x \in \mathbb{R}$.

It follows from the definition of the rotation number that if $\overline{g}_1(x) \le \overline{g}_2(x)$ for all $x \in \mathbb{R}$, then $\operatorname{rot}(\overline{g}_1) \le \operatorname{rot}(\overline{g}_2)$. Using this property, we get that $\operatorname{rot}(\overline{f}^{n_1} + m_1) = n_1 \operatorname{rot}(\overline{f}) + m_1 \le \operatorname{rot}(\overline{f}^{n_2} + m_2) = n_2 \operatorname{rot}(\overline{f}) + m_2$. Since $\operatorname{rot}(\overline{f})$ is irrational, the last inequality must be strict, that is, $n_1 \operatorname{rot}(\overline{f}) + m_1 < n_2 \operatorname{rot}(\overline{f}) + m_2$. □

Denote by $\overline{R}_\alpha \colon \mathbb{R} \to \mathbb{R}$ a mapping of the form $x \mapsto x + \alpha$, $x \in \mathbb{R}$, where α is a fixed number. Obviously, $\overline{R}_\alpha \in P(\mathbb{R})$. Since $\overline{R}_\alpha^n(x) = x + n\alpha$, $n \in \mathbb{Z}$, it follows that $\operatorname{rot}(\overline{R}_\alpha) = \alpha$.

THEOREM 2.1. *Let* $\overline{f} \in P(\mathbb{R})$ *and* $\operatorname{rot}(\overline{f}) = \alpha \in \mathbb{R}/\mathbb{Q}$.

1) *If there exists a point* $x_0 \in \mathbb{R}$ *such that the set* $A = \{\overline{f}^n(x_0) + m : n \in \mathbb{N}, m \in \mathbb{Z}\}$ *is dense in* $\mathbb{R}$, *then* $\overline{f}$ *is a homeomorphism conjugate to the transformation* $\overline{R}_\alpha$.

2) *If* $\overline{h}_1$ *and* $\overline{h}_2$ *are two homeomorphisms* $\mathbb{R} \to \mathbb{R}$ *realizing a conjugacy between* $\overline{f}$ *and* $\overline{R}_\alpha$, *then there exists a* $\beta \in \mathbb{R}$ *such that* $\overline{h}_1 = \overline{R}_\beta \circ \overline{h}_2$.

PROOF. We define a mapping $\overline{h}_0$ of A into the set $B = \{n\alpha + m : n \in \mathbb{N}, m \in \mathbb{Z}\}$ according to the rule $\overline{h}_0(\overline{f}^n(x_0) + m) = n\alpha + m$. Since α is irrational, the mapping $\overline{h}_0$ is one-to-one. According to Lemma 2.1, $\overline{h}_0$ is monotonically increasing. This implies that $\overline{h}_0$ is continuous.

Since α is irrational, B is dense in $\mathbb{R}$.

Let $\widetilde{x} \in \mathbb{R}$ be arbitrary. Since $\operatorname{cl}(A) = \mathbb{R}$, there exists a sequence $\{x_n\}_1^\infty$ of points $x_n \in A$ converging to $\widetilde{x}$. It follows from the monotonicity of $\overline{h}_0$ that the sequence $\{\overline{h}_0(x_n)\}_1^\infty$ has at least one accumulation point $\widetilde{y} \in \mathbb{R}$. The equality $\operatorname{cl}(B) = \mathbb{R}$ implies that the accumulation point of $\{\overline{h}_0(x_n)\}_1^\infty$ is unique. Indeed, otherwise there would be two points $y_1, y_2 \in B$ between two accumulation points of $\{\overline{h}_0(x_n)\}_1^\infty$ (Figure 5.3). Since $\overline{h}_0$ is monotone, there exist two subsequences of $\{x_n\}_1^\infty$ that do not intersect the interval on $\mathbb{R}$ between $\overline{h}_0^{-1}(y_1)$ and $\overline{h}_0^{-1}(y_2)$ (Figure 5.3). This contradicts the uniqueness of the limit of the sequence $\{x_n\}_1^\infty$.

It can be shown similarly that $\widetilde{y}$ does not depend on the choice of the sequence $\{x_n\}_1^\infty$ converging to $\widetilde{x}$.

Let $\overline{h}(\widetilde{x}) = \widetilde{y}$. The mapping $\overline{h}$ is an extension of $\overline{h}_0$. Since $\operatorname{cl}(A) = \operatorname{cl}(B) = \mathbb{R}$, $\overline{h}$ is a mapping of $\mathbb{R}$ onto $\mathbb{R}$. Since $\overline{h}_0$ is monotone, so is $\overline{h}$. The equality

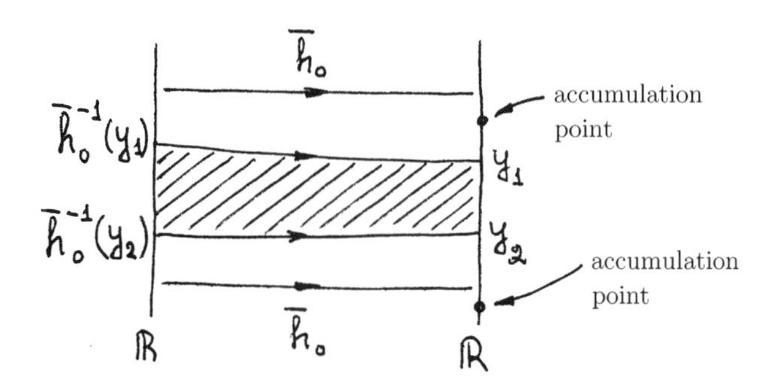

FIGURE 5.3

$\mathrm{cl}(B) = \mathbb{R}$ implies that $\overline{h}$ is one-to-one. This and monotonicity give us that $\overline{h}$ is a homeomorphism.

Let $x \in A$. Then $x = \overline{f}^n(x_0) + m$ for some $n \in \mathbb{N}$ and $m \in \mathbb{Z}$. We have that

$$\begin{aligned}\overline{h}_0 \circ \overline{f}(x) &= \overline{h}_0[\overline{f}^{n+1}(x_0) + m] = (n+1)\alpha + m \\ &= n\alpha + m + \alpha = \overline{h}_0(x) + \alpha = \overline{R}_\alpha \circ \overline{h}_0(x).\end{aligned}$$

By continuity we get that $\overline{h} \circ \overline{f}(x) = \overline{R}_\alpha \circ \overline{h}(x)$ for any $x \in \mathbb{R}$, that is, $\overline{f}$ is conjugate to $\overline{R}_\alpha$ by the homeomorphism $\overline{h}$ (we leave it to the reader as an exercise to verify the equality $\overline{h}(x+1) = \overline{h}(x) + 1$). This implies that $\overline{f}$ is a homeomorphism.

We prove 2). Suppose that the homeomorphisms $\overline{h}_1$ and $\overline{h}_2$ realize a conjugacy between $\overline{f}$ and $\overline{R}_\alpha$, and let $\beta = \overline{h}_1(x_0) - \overline{h}_2(x_0)$. Then $\overline{h}_1(x_0) = \overline{R}_\beta \circ \overline{h}_2(x_0)$. For any point $x = \overline{f}^n(x_0) + m \in A$ we have that

$$\begin{aligned}\overline{h}_1(x) &= \overline{h}_1[\overline{f}^n(x_0) + m] = \overline{R}_\alpha^n \circ \overline{h}_1(x_0) + m = \overline{R}_\alpha^n \circ \overline{R}_\beta \circ \overline{h}_2(x_0) + m \\ &= \overline{R}_\beta \circ \overline{R}_\alpha^n \circ \overline{h}_2(x_0) + m = \overline{R}_\beta \circ \overline{h}_2[\overline{f}^n(x_0) + m] = \overline{R}_\beta \circ \overline{h}_2(x).\end{aligned}$$

By continuity, $\overline{h}_1(x) = \overline{R}_\beta \circ \overline{h}_2(x)$ for any $x \in \mathrm{cl}(A) = \mathbb{R}$. □

THEOREM 2.2. *Suppose that $\overline{f} \in P(\mathbb{R})$, $\mathrm{rot}(\overline{f}) = \alpha \in \mathbb{R} \setminus \mathbb{Q}$, and there exists a point $x_0 \in \mathbb{R}$ such that the set $A = \{\overline{f}^n(x_0) + m : n \in \mathbb{N},\ m \in \mathbb{Z}\}$ is not dense in $\mathbb{R}$. In this case:*

1) $\overline{f}$ is semiconjugate to $\overline{R}_\alpha$ by means of a continuous mapping $\overline{h} \in P(\mathbb{R})$ that is not a homeomorphism;

2) if $\overline{h}_1$ and $\overline{h}_2$ are two continuous transformations realizing a semiconjugacy between $\overline{f}$ and $\overline{R}_\alpha$, then there exists a $\beta \in \mathbb{R}$ such that $\overline{h}_1 = \overline{R}_\beta \circ \overline{h}_2$ (consequently, any continuous transformation semiconjugating $\overline{f}$ and $\overline{R}_\alpha$ is not a homeomorphism);

3) if $\overline{h} \in P(\mathbb{R})$ realizes a semiconjugacy between $\overline{f}$ and $\overline{R}_\alpha$, and $\chi(\overline{f}, \overline{h}) \overset{\text{def}}{=} \{x \in \mathbb{R} : h^{-1}(x)$ contains more than one point$\}$, then

a) the inverse image $\overline{h}^{-1}(x)$ is a closed interval for any $x \in \chi(\overline{f}, \overline{h})$,

b) the set $\Omega(\overline{f}) \overset{\text{def}}{=} \mathbb{R} \setminus \cup \operatorname{int} \overline{h}^{-1}(x)$, $x \in \chi(\overline{f}, \overline{h})$, is a nonempty closed perfect set that does not depend on the semiconjugating transformation $\overline{h}$,

c) $\overline{h}[\Omega(\overline{f})] = \mathbb{R}$.

PROOF. Following the proof of Theorem 2.1, we define a monotonically increasing one-to-one continuous mapping $\overline{h}_0\colon A \to B = \{n\alpha + m : n \in \mathbb{N},\, m \in \mathbb{Z}\}$ by $\overline{h}_0[\overline{f}^n(x_0) + m] = n\alpha + m$. As in the proof of Theorem 2.1, $\overline{h}_0$ extends to a continuous monotonically nondecreasing mapping (which we again denote by $\overline{h}_0$) of the set $\mathrm{cl}(A)$ into $\mathrm{cl}(B) = \mathbb{R}$. Since the set A is not dense by assumption, the complement $\mathbb{R} \setminus \mathrm{cl}(A)$ contains at least one interval $I = (a, b)$ with endpoints $a, b \in \mathrm{cl}(A)$.

We show that $\overline{h}_0(a) = \overline{h}_0(b)$. Assume not. Since B is dense in $\mathbb{R}$, the interval $(\overline{h}_0(a), \overline{h}_0(b))$ contains a point $y_0 \in B$. We get that $\overline{h}_0^{-1}(y_0) \in (a, b)$ because the mapping $\overline{h}_0|_A\colon A \to B$ is one-to-one and monotone, and this contradicts the equality $A \cap (a, b) = \emptyset$.

For any component (a, b) of the set $\mathbb{R} \setminus \mathrm{cl}(A)$ we let $\overline{h}(x) = \overline{h}_0(a) = \overline{h}(b)$, where $x \in (a, b)$. By the foregoing, $\overline{h}$ is a monotonically nondecreasing continuous mapping of $\mathbb{R}$ onto $\mathbb{R}$ that coincides with $\overline{h}_0$ on $\mathrm{cl}(A)$. It follows from the definition of $\overline{h}$ that $\overline{h}$ is not a homeomorphism.

As in the proof of Theorem 2.1 we can show that $\overline{h}_0 \circ \overline{f}|_{\mathrm{cl}(A)} = \overline{R}_\alpha \circ \overline{h}_0|_{\mathrm{cl}(A)}$. The definition of $\overline{h}$ gives us that $\overline{h} \circ \overline{f} = \overline{R}_\alpha \circ \overline{h}$ on the whole line $\mathbb{R}$, that is, $\overline{h}$ realizes a semiconjugacy between $\overline{f}$ and $\overline{R}_\alpha$.

For $x = \overline{f}^n(x_0) + m \in A$ we have that

$$\overline{h}(x+1) = \overline{h}_0[\overline{f}^n(x_0) + m + 1] = n\alpha + m + 1 = \overline{h}_0(x) + 1 = \overline{h}(x) + 1.$$

This and the definition of the extension of $\overline{h}_0$ to the transformation $\overline{h}$ implies that $\overline{h}(x+1) = \overline{h}(x) + 1$ for all $x \in \mathbb{R}$, which proves 1).

The proof of 2) is completely analogous to that of 2) in Theorem 2.1.

Let us prove 3). The monotonicity of $\overline{h}$ implies 3a). Since the union $\cup \operatorname{int} \overline{h}^{-1}(x)$, $x \in \chi(\overline{f}, \overline{h})$, is an open set, the set $\Omega(\overline{f})$ is closed. By 3a), any point $x \in \Omega(\overline{f})$ cannot be simultaneously an endpoint of two intervals $\overline{h}^{-1}(y_1)$ and $\overline{h}^{-1}(y_2)$ with $y_1, y_2 \in \chi(\overline{f}, \overline{h})$; therefore, $\Omega(\overline{f})$ is perfect and nonempty.

The assertion 3c) follows immediately from the construction of $\overline{h}$ and the property 3a).

The fact that $\Omega(\overline{f})$ is independent of the semiconjugating transformation $\overline{h}$ follows from 2). □

COROLLARY 2.1. *Assume the conditions of Theorem 2.2, and let $\overline{f}$ be a homeomorphism. Then (in the notation of Theorem 2.2):*

1) *the set $\chi(\overline{f}, \overline{h})$ is invariant under $\overline{R}_\alpha$;*
2) *$\Omega(\overline{f})$ is a Cantor set and is invariant under $\overline{f}$;*
3) *$\overline{h}[\Omega(\overline{f})] = \mathbb{R}$.*

PROOF. The homeomorphism $\overline{f}$ carries an interval into an interval. From this, the equality $\overline{h} \circ \overline{f} = \overline{R}_\alpha \circ \overline{h}$, and the assertion 3a) in Theorem 2.2 we get the invariance of $\chi(\overline{f}, \overline{h})$ under $\overline{R}_\alpha$ and the invariance of the set $\cup \overline{h}^{-1}(x)$, $x \in \chi(\overline{f}, \overline{h})$, under $\overline{f}$. Since $\overline{f}$ is a homeomorphism, it follows that $\Omega(\overline{f}) = \mathbb{R} \setminus \cup \operatorname{int} \overline{h}^{-1}(x)$, $x \in \chi(\overline{f}, \overline{h})$, is also invariant under $\overline{f}$.

Since α is irrational and $\chi(\overline{f}, \overline{h})$ is invariant under $\overline{R}_\alpha$, $\chi(\overline{f}, \overline{h})$ is dense in $\mathbb{R}$. Therefore, between any two points in $\mathbb{R}$ there are points carried under the action of $\overline{h}$ into $\chi(\overline{f}, \overline{h})$. This implies that $\Omega(\overline{f})$ is nowhere dense, and hence a Cantor set. □

The transformation $\overline{R}_\alpha\colon x \mapsto x+\alpha$ is a covering for the mapping $R_\alpha\colon S^1 \to S^1$ given by $x \mapsto x+\alpha \pmod 1$, which is called a *rotation* or *shift of the circle.* It is not hard to see that $R_{\alpha+n} = R_\alpha$ for $n \in \mathbb{Z}$, and $\operatorname{rot}(R_\alpha) = \alpha \pmod 1$.

For a transformation $f\colon S^1 \to S^1$ the set $\{f^n(x_0) : n \text{ a nonnegative integer}\} \stackrel{\text{def}}{=} O^+(x_0)$ is called the *positive semi-orbit* of the point $x_0 \in S^1$.

The next two lemmas follow directly from Theorems 2.1 and 2.2.

LEMMA 2.2. *Suppose that $f \in P(S^1)$ has an irrational rotation number α, and that f has a dense positive semi-orbit. In this case:*

1) *f is a homeomorphism that is conjugate to the rotation R_α;*

2) *if h_1 and h_2 are homeomorphisms realizing a conjugacy between f and R_α, then there exists a number β such that $h_1 = R_\beta \circ h_2$.*

LEMMA 2.3. *Suppose that $f \in P(S^1)$ $(\operatorname{rot}(f) \in \mathbb{R}\setminus\mathbb{Q})$ has a positive semi-orbit that is not dense. In this case:*

1) *f is semiconjugate to R_α by means of a continuous mapping $h \in P(S^1)$ that is not a homeomorphism;*

2) *if h_1 and h_2 are continuous transformations realizing a semiconjugacy between f and R_α, then there exists a number β such that $h_1 = R_\beta \circ h_2$ (consequently, any continuous transformation semiconjugating f and R_α is not a homeomorphism);*

3) *if $h \in P(S^1)$ realizes a semiconjugacy between f and R_α and $\chi(f,h) \stackrel{\text{def}}{=} \{x \in S^1 : h^{-1}(x)$ contains more than one point$\}$, then*

a) *for any $x \in \chi(f,h)$ the inverse image $h^{-1}(x)$ is a closed interval,*

b) *the set $\Omega(f) \stackrel{\text{def}}{=} S^1 \setminus \bigcup \operatorname{int} h^{-1}(x)$, $x \in \chi(f,h)$, is a nonempty closed perfect set independent of the semiconjugating transformation h,*

c) *$h[\Omega(f)] = S^1$.*

COROLLARY 2.2. *Assume the conditions of Lemma 2.3, and let f be a homeomorphism. Then (in the notation of Lemma 2.3):*

1) *$\chi(f,h)$ is invariant under R_α;*

2) *$\Omega(f)$ is a Cantor set and is invariant under f;*

3) *$h[\Omega(f)] = S^1$.*

2.2. A criterion for being conjugate to a rotation.

DEFINITION. A transformation $f \in P(S^1)$ is said to be *transitive* if it has a dense positive semi-orbit on the circle.

EXAMPLE. A rotation $R_\alpha\colon S^1 \to S^1$ with irrational α is a transitive transformation. Moreover, the orbit of any point $x \in S^1$ with respect to R_α is dense in S^1; that is, R_α is a minimal transformation (see Example 1 in §2.3.2 of Chapter 1 for a proof).

LEMMA 2.4. *If a transformation $f \in P(S^1)$ with irrational rotation number $\operatorname{rot}(f) = \alpha$ is transitive, then f is a minimal homeomorphism that is conjugate to R_α.*

PROOF. This is immediate from Lemma 2.2 and the minimality of the rotation R_α. □

Denote by $\operatorname{Homeo}_+(S^1) = \operatorname{Diff}^\circ(S^1)$ the set of orientation-preserving homeomorphisms of the circle S^1. It is obvious that $P(S^1) \supset \operatorname{Homeo}_+(S^1)$.

For $f \in \mathrm{Homeo}_+(S^1)$ and for a point $x_0 \in S^1$ the set $O^-(x_0) = \{f^{-n}(x_0), n \geq 0\}$ is called the *negative semi-orbit of the point* x_0.

We recall that the family $\{f^{n_i}\}_{i=1}^{\infty}$ is uniformly continuous if for any $\varepsilon > 0$ there exists a $\delta > 0$ independent of the n_i such that if $x_1, x_2 \in S^1$ are δ-close, then $f^{n_i}(x_1)$ and $f^{n_i}(x_2)$ are ε-close.

The following criterion for being conjugate to a rotation holds for homeomorphisms of the circle with irrational rotation number.

LEMMA 2.5. *A homeomorphism $f \in \mathrm{Homeo}_+(S^1)$ with $\alpha = \mathrm{rot}(f) \in \mathbb{R} \setminus Q$ is conjugate to the rotation R_α if and only if there exists a sequence $\{n_i\}_{i=1}^{\infty}$, $n_i \in \mathbb{N}$, increasing to infinity such that the family $\{f^{n_i}\}_{i=1}^{\infty}$ is uniformly continuous.*

PROOF. NECESSITY. Let $f = h^{-1} \circ R_\alpha \circ h$, where h is a homeomorphism conjugating f and R_α. Then $f^n = h^{-1} \circ R_\alpha^n \circ h$, $n \in \mathbb{N}$. Since h and h^{-1} are uniformly continuous (being continuous mappings of a compact set), and since R_α^n is an isometry, the family $\{f^n\}_{n=1}^{\infty}$ is uniformly continuous.

SUFFICIENCY. Suppose that there is a uniformly continuous family $\{f^{n_i}\}_{i=1}^{\infty}$, $n_i \to +\infty$. According to Lemma 2.4, it suffices to prove that the homeomorphism f^{-1} is transitive. Assume not. Suppose that the semi-orbit $O^-(x_0) = \{f^{-n}(x_0), n \geq 0\}$ is not dense in S^1. Let $\mathcal{J}$ be a component of the set $S^1 \setminus \mathrm{cl}(O^-(x_0))$. Since f is a homeomorphism, $f^{-n}(\mathcal{J})$ is a component of the set $S^1 \setminus \mathrm{cl}[O^-(x_0)]$ for any $n \geq 0$. The rotation number $\mathrm{rot}(f)$ is irrational, so the components $f^{-n}(\mathcal{J})$, $n \geq 0$, are disjoint, and hence the lengths of the intervals $f^{-n_i}(\mathcal{J})$ tend to zero as $n_i \to \infty$. But this contradicts the uniform continuity of the family $\{f^{n_i}\}_{i=1}^{\infty}$. □

2.3. Limit sets. Let $f \in P(S^1)$.

DEFINITION. The ω-limit set of a point $x_0 \in S^1$ with respect to the transformation f is defined to be the set $\omega(x_0) = \{x \in S^1 :$ there exists a sequence $\{n_k\}_{k=1}^{\infty}$, $n_k \in \mathbb{N}$, increasing to infinity such that $f^{n_k}(x_0) \to x$ as $n_k \to \infty\}$.

The union of the ω-limit sets of all the points of the circle is denoted by $\Omega^+(f)$.

If f is a homeomorphism, then the α-limit set $\alpha(x_0)$ of an arbitrary point $x_0 \in S^1$ is defined similarly (with $n_k \to -\infty$ instead of $n_k \to +\infty$). The union of the α-limit sets of all the points of the circle with respect to the homeomorphism f is denoted by $\Omega^-(f)$.

Any semi-orbit of a rotation R_α with irrational α is dense in S^1, so $\Omega^+(R_\alpha) = \Omega^-(R_\alpha) = S^1$.

This implies that if $f \in P(S^1)$ is conjugate to a rotation R_α with $\alpha \in \mathbb{R} \setminus Q$, then $\Omega^+(f) = \Omega^-(f) = S^1$.

LEMMA 2.6. *Suppose that $f \in P(S^1)$ has an irrational rotation number $\alpha = \mathrm{rot}(f)$, and let $h\colon S^1 \to S^1$ be a continuous mapping that is not a homeomorphism and that realizes a semiconjugacy between f and R_α. Then $\Omega^+(f) = \Omega(f) \stackrel{\mathrm{def}}{=} S^1 \setminus \bigcup \mathrm{int}\, h^{-1}(x)$, where the union is over all $x \in \chi(f, h)$. Moreover, if f is a homeomorphism, then $\Omega^+(f) = \Omega^-(f)$.*

PROOF. Take an interval G carried by h into a point, and let $x \in G$. If we assume that $f^n(x) \in G$ for some $n \in \mathbb{N}$, then $R_\alpha^n h(x) = hf^n(x) = h(G) = h(x)$; that is, R_α has a periodic point, which contradicts the irrationality of α. Therefore,

$f^n(x) \notin G$ for all $n \in \mathbb{N}$. This means that any point in G does not belong to the ω-limit set of a point of the circle. Thus, $\Omega^+(f) \subseteq \Omega(f)$.

Take an $x \in \Omega(f)$, and an arbitrary point $x_0 \in S^1$. We show that $x \in \omega(x_0)$. Let $\mathcal{U}$ be a neighborhood of x. Since $x \in \Omega(f)$ the set $h(\mathcal{U})$ contains more than one point. The monotonicity of h implies that there is a closed interval $I \subset h(\mathcal{U})$ covering the point $h(x)$. It follows from the irrationality of α that R_α is minimal, and hence $R_\alpha^n(h(x_0)) \subset I$ for some $n \in \mathbb{N}$. Since $h^{-1}(I) \subset \mathcal{U}$ because h is monotone, we have that $f^n(x_0) \in \mathcal{U}$. From this, $\Omega^+(f) = \Omega(f)$.

We consider the case when f is a homeomorphism. It follows from the relation $h \circ f = R_\alpha \circ h$ that $h \circ f^{-1} = R_\alpha^{-1} \circ h = R_{-\alpha} \circ h$; that is, h realizes a semiconjugacy between f^{-1} and $R_{-\alpha}$. Therefore, $\Omega^-(f) = \Omega(f^{-1}) = \Omega(f) = \Omega^+(f)$. □

COROLLARY 2.3. *If a homeomorphism* $f \in \mathrm{Homeo}_+(S^1)$ *has an irrational rotation number, then either* $\Omega^+(f) = \Omega^-(f) = S^1$, *or* $\Omega^+(f) = \Omega^-(f) = \Omega(f)$ *is a Cantor set that is invariant under* f *and* f^{-1}.

2.4. Classification of transitive homeomorphisms. The following theorem shows that the Poincaré rotation number is a complete topological invariant for a transitive homeomorphism of the circle.

THEOREM 2.3. 1) *Two transitive homeomorphisms of the circle are conjugate if and only if their rotation numbers are the same.*

2) *Any transitive homeomorphism of the circle has an irrational rotation number not exceeding* 1, *and for any irrational* $\alpha \in (0,1)$ *there exists a transitive* $f \in \mathrm{Homeo}_+(S^1)$ *such that* $\mathrm{rot}(f) = \alpha$.

PROOF. 1) The necessity follows from Corollary 1.1. We prove the sufficiency. Let $\alpha = \mathrm{rot}(f) = \mathrm{rot}(g)$, where f and g are transitive homeomorphisms of the circle. By Lemma 2.4, both f and g are conjugate to the rotation R_α, and hence to each other.

2) Suppose that the orbit of a point $x_0 \in S^1$ under a transitive homeomorphism f is dense in S^1. We show that the rotation number $\mathrm{rot}(f)$ is irrational. Assume not. Then by Lemma 1.6, f has a periodic orbit, say of period $k \geq 1$. This implies that the points $x_0, f(x_0), \dots, f^{k-1}(x_0)$ belong to respective intervals $\mathcal{I}_1, \dots, \mathcal{I}_k$ that are invariant under the homeomorphism f^k. Therefore, the sequence $\{f^{nk}(f^i(x_0))\}_{n=-\infty}^{\infty}$ has a limit both as $n \to +\infty$ and as $n \to -\infty$ for each $0 \leq i \leq k-1$. This contradicts the denseness of the orbit of x_0, because any point $f^m(x_0)$ can be represented in the form $f^m(x_0) = f^{nk}[f^i(x_0)]$, $0 \leq i \leq k-1$.

By the definition of the rotation number of a transformation $f \in P(S^1)$, we have that $0 \leq \mathrm{rot}(f) < 1$. Therefore, $\mathrm{rot}(f) \in (0,1)$ for a transitive $f \in \mathrm{Homeo}_+(S^1)$.

Since $\mathrm{rot}(R_\alpha) = \alpha$, we can take $f = R_\alpha$. Since α is irrational, R_α is a transitive (and even minimal) homeomorphism (see Example 1 in §2.3.2 of Chapter 1). □

Two transformations f and g commute if $f \circ g = g \circ f$. The centralizer of a transformation f is defined to be the set of transformations commuting with f. The following lemma shows that for a transitive rotation of the circle the group of all rotations of the circle is the centralizer in the set $P(S^1)$ (obviously, two rotations commute).

LEMMA 2.7. *Suppose that* $R_\alpha \colon S^1 \to S^1$ *is a rotation with* $\alpha \in \mathbb{R} \setminus Q$. *If* $h \circ R_\alpha = R_\alpha \circ h$ *and* $h \in P(S^1)$, *then* $h = R_\beta$ *for some* $\beta \in \mathbb{R}$.

PROOF. Since α is irrational, R_α is a minimal transformation.

Take a point $x_0 \in S^1$ and a number $\beta \in \mathbb{R}$ such that $h(x_0) = R_\beta(x_0)$. Then $h \circ R_\alpha^n(x_0) = R_\alpha^n \circ h(x_0) = R_\alpha^n \circ R_\beta(x_0) = R_\beta \circ R_\alpha^n(x_0)$, $n \in \mathbb{Z}$. The orbit $O(x_0)$ of x_0 with respect to R_α is dense in S^1. It follows from $R_\alpha \circ R_\beta = R_\beta \circ R_\alpha$ that $h[O(x_0)] = R_\beta[O(x_0)]$ is an orbit of R_α, which is also dense in S^1. Since $h[R_\alpha^n(x_0)] = R_\beta[R_\alpha^n(x_0)]$ and since h is monotone, it follows that $h = R_\beta$ on the whole circle. □

2.5. Classification of Denjoy homeomorphisms.

DEFINITION. A homeomorphism f of the circle with irrational rotation number is called a *Denjoy homeomorphism* if its limit set $\Omega(f)$ is nowhere dense.

The set of Denjoy homeomorphisms is denoted by $\mathrm{Den}(S^1)$.

According to Corollary 2.3, the limit set $\Omega(f)$ is a Cantor set that is invariant under f and f^{-1}.

By Lemma 2.3, f is semiconjugate to the rotation R_α, $\alpha \in \mathrm{rot}(f)$.

DEFINITION. Let $f \in \mathrm{Den}(S^1)$ and let $h\colon S^1 \to S^1$ be a continuous mapping realizing a semiconjugacy between f and R_α, $\alpha = \mathrm{rot}(f)$. The set $\chi(f,h) = \{x \in S^1 : h^{-1}(x) \text{ contains more than one point}\}$ is called the *characteristic set* of the homeomorphism f with respect to the semiconjugating mapping h.

LEMMA 2.8. *Let $f \in \mathrm{Den}(S^1)$ and let h be a continuous mapping realizing a semiconjugacy between f and R_α, $\alpha = \mathrm{rot}(f)$. Then $\chi(f,h)$ is made up of an at most countable family of orbits of the rotation R_α.*

PROOF. According to Lemma 2.3, $h^{-1}(x)$ is a closed interval for any point $x \in \chi(f,h)$. Therefore, $h[h^{-1}(x)] = h[\mathrm{int}\, h^{-1}(x)] = x$, and hence $\chi(f,h) = h[\bigcup \mathrm{int}\, h^{-1}(x)] = h[S^1 \setminus \Omega(f)]$, $x \in \chi(f,h)$. Since $\Omega(f)$ is a Cantor set, $S^1 \setminus \Omega(f)$ consists of a countable family of open intervals. Therefore, $\chi(f,h)$ consists of a countable family of points.

In view of Corollary 2.2, $\chi(f,h)$ is invariant under R_α. □

For two sets $\chi_1, \chi_2 \subset S^1$ we write $\chi_1 \equiv \chi_2$ if one of them is the image of the other under a rotation of the circle, that is, $\chi_2 = R_\beta(\chi_1)$, $\beta \in \mathbb{R}$.

We remark that if continuous mappings h_1 and h_2 realize a semiconjugacy between a homeomorphism $f \in \mathrm{Den}(S^1)$ and the rotation R_α, $\alpha = \mathrm{rot}(f)$, then $\chi(f,h_1) \equiv \chi(f,h_2)$ by virtue of Lemma 2.3, 2).

THEOREM 2.4. *Suppose that the continuous mappings h_1 and h_2 realize a semiconjugacy between the Denjoy homeomorphisms f and g and the respective rotations $R_{\mathrm{rot}(f)}$ and $R_{\mathrm{rot}(g)}$. Then f and g are conjugate if and only if* $\mathrm{rot}(f) = \mathrm{rot}(g)$ *and* $\chi(f,h_1) = \chi(g,h_2)$.

PROOF. NECESSITY. Let $h \circ f = g \circ h$, where $h \in \mathrm{Homeo}_+(S^1)$ (Figure 5.4). Then $\mathrm{rot}(f) = \mathrm{rot}(g)$ by Corollary 1.1.

Since h is a homeomorphism realizing a conjugacy between f and g, it follows that $h[\Omega(f)] = \Omega(g)$. We take a point $x \in \chi(f,h_1)$. By Lemma 2.6, the closed interval $h^{-1}(x)$ intersects $\Omega(f)$ only in its endpoints, and thus $h[h_1^{-1}(x)]$ is a closed interval intersecting $\Omega(g)$ only at the endpoints. This implies that h_2 takes the interval $h[h_1^{-1}(x)]$ into a point, that is, $h[h_1^{-1}(x)] \in \chi(g,h_2)$. Since the complete

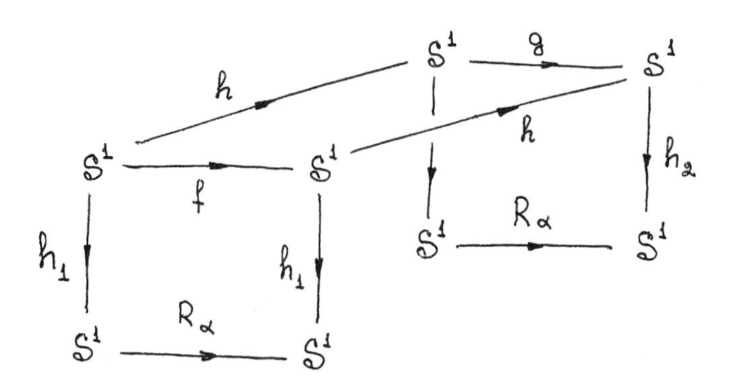

FIGURE 5.4

inverse image $h_1^{-1}(x)$ is a single point for each point $x \in S^1 \setminus \chi(f, h_1)$, the mapping $h_0 = h_2 \circ h \circ h_1^{-1}$ is well defined, and $h_0[\chi(f, h_1)] \subseteq \chi(g, h_2)$. The transformation h_0 is continuous because the mappings h_1, h_2, and h are monotone and continuous. In a completely analogous way it can be shown that the mapping $h_1 \circ h^{-1} \circ h_2^{-1} = h_0^{-1}$ is well defined and continuous, and that $h_0^{-1}[\chi(g, h_2)] \subseteq \chi(f, h_1)$. Consequently, h_0 is a homeomorphism carrying $\chi(f, h_1)$ onto $\chi(g, h_2)$.

We show that h_0 commutes with R_α. Let $x \in S^1 \setminus \chi(f, h_1)$. Then

$$\begin{aligned} R_\alpha \circ h_0(x) &= R_\alpha \circ h_2 \circ h \circ h_1^{-1}(x) = h_2 \circ g \circ h \circ h_1^{-1}(x) \\ &= h_2 \circ h \circ f \circ h_1^{-1}(x) = h_2 \circ h \circ h_1^{-1} \circ R_\alpha(x) = h_0 \circ R_\alpha(x). \end{aligned}$$

By continuity, $R_\alpha \circ h_0(x) = h_0 \circ R_\alpha(x)$ for all $x \in S^1$. According to Lemma 2.7, $h_0 = R_\beta$ for some $\beta \in \mathbb{R}$, and $h_0[\chi(f, h_1)] = R_\beta[\chi(f, h_1)] = \chi(g, h_2)$, that is, $\chi(f, h_1) \equiv \chi(g, h_2)$.

SUFFICIENCY. For a Cantor set Ω on the circle S^1 we set $\overset{\circ}{\Omega} = \Omega \setminus \Gamma(\Omega)$, where $\Gamma(\Omega)$ is the set of endpoints of the adjacent intervals of $S^1 \setminus \Omega$.

By assumption, there exists a $\beta \in \mathbb{R}$ such that $\chi(g, h_2) = R_\beta[\chi(f, h_1)]$. Since $h_1[\overset{\circ}{\Omega}(f)] = S^1 \setminus \chi(f, h_1)$ and $h_2[\overset{\circ}{\Omega}(g)] = S^1 \setminus \chi(g, h_2)$, the mapping $h_2^{-1} \circ R_\beta \circ h_1|_{\overset{\circ}{\Omega}(f)} : \overset{\circ}{\Omega}(f) \to \overset{\circ}{\Omega}(g)$ is defined.

We show that this mapping is uniformly continuous. Let $\varepsilon > 0$. There exists a finite collection of adjacent intervals $\mathcal{J}_1, \ldots, \mathcal{J}_k$ of the Cantor set $\Omega(g)$ such that the length of any component of the set $S^1 \setminus \bigcup_{i=1}^k \mathcal{J}_i$ is less than ε. Since $h_2[\bigcup_{i=1}^k \mathcal{J}_i] \subset \chi(g, h_2)$ and $R_\beta[\chi(f, h_1)] = \chi(g, h_2)$, there are adjacent intervals $G_1, \ldots, G_k$ of the Cantor set $\Omega(f)$ such that $R_\beta[h_1(\bigcup_{i-1}^k G_i)] = h_2(\bigcup_{i=1}^k \mathcal{J}_i)$. Suppose that the number $0 < \delta < \frac{1}{2}$ is less than the length of each G_i, $i = 1, \ldots, k$. In this case if the distance between points $x_1, x_2 \in \overset{\circ}{\Omega}(f)$ is less than δ, then the smaller circular arc with endpoints x_1 and x_2 does not contain the interval G_i $(i = 1, \ldots, k)$. Consequently, one of the arcs with endpoints $h_2^{-1} \circ R_\beta \circ h_1(x_i)$, $i = 1, 2$, does not contain any of the intervals $\mathcal{J}_i$, $i = 1, \ldots, k$. Therefore, $|h_2^{-1} \circ R_\beta \circ h_1(x_1) - h_2^{-1} \circ R_\beta \circ h_1(x_2)| < \varepsilon$, which proves that the mapping $h_2^{-1} \circ R_\beta \circ h_1|_{\overset{\circ}{\Omega}(f)}$ is uniformly continuous.

It is proved similarly that the mapping $h_1^{-1} \circ R_\beta^{-1} \circ h_2|_{\overset{\circ}{\Omega}(g)} : \overset{\circ}{\Omega}(g) \to \overset{\circ}{\Omega}(f)$ is also uniformly continuous.

Since $h_2^{-1} \circ R_\beta \circ h_1|_{\overset{\circ}{\Omega}(f)}$ is uniformly continuous, it can be extended to a continuous mapping $h\colon \mathrm{cl}[\overset{\circ}{\Omega}(f)] = \Omega(f) \to \Omega(g) = \mathrm{cl}[\overset{\circ}{\Omega}(g)]$. We show that $h[\Gamma(\Omega(f))] = \Gamma(\Omega(g))$ (that is, the endpoints of the adjacent intervals of $\Omega(f)$ are carried by h into the endpoints of adjacent intervals of $\Omega(g)$). We take $x \in \Gamma(\Omega(f))$ and assume that $h(x) \in \overset{\circ}{\Omega}(g)$. Then $x_1 = (h_2^{-1} \circ R_\beta \circ h_1)^{-1}(h(x)) \in \overset{\circ}{\Omega}(f)$. Since h is an extension of the monotone mapping $h_2^{-1} \circ R_\beta \circ h_1$, it is also monotone. Consequently, one of the arcs I of S^1 with endpoints x_1 and x has the property that $h[I \cap \Omega(f)] = h(x)$. By the construction of the Cantor set, I contains an infinite family of points in $\overset{\circ}{\Omega}(f)$. Then the last equality contradicts the one-to-oneness of $h_2^{-1} \circ R_\beta \circ h_1|_{\overset{\circ}{\Omega}(f)}$ and the equality $h|_{\overset{\circ}{\Omega}(f)} = h_2^{-1} \circ R_\beta \circ h_1|_{\overset{\circ}{\Omega}(f)}$. Thus, $h[\Gamma(\Omega(f))] \subseteq \Gamma[\Omega(g)]$. The monotonicity of h and the denseness of $\overset{\circ}{\Omega}$ in the Cantor set Ω imply that $h[\Gamma(\Omega(f))] = \Gamma[\Omega(g)]$.

We show that h is one-to-one. To do this it suffices to prove that the restriction $h|_{\Gamma(\Omega(f))}$ is one-to-one. Take two distinct points $x_1, x_2 \in \Gamma(\Omega(f))$. If these points are not endpoints of a single adjacent interval of the Cantor set $\Omega(f)$, then both the arcs in $S^1 \setminus \{x_1, x_2\}$ contain points in $\overset{\circ}{\Omega}(f)$. Since $h|_{\overset{\circ}{\Omega}(f)}$ is monotone and one-to-one, $h(x_1) \neq h(x_2)$. Suppose now that x_1 and x_2 are endpoints of a single adjacent interval, and assume that $h(x_1) = h(x_2) \in \Gamma(\Omega(g))$. Let $\mathfrak{I}$ be the adjacent interval of the Cantor set $\Omega(g)$ with endpoints $h(x_1)$ and y_1. Since $h[\Gamma(\Omega(f))] = \Gamma(\Omega(g))$, it follows that $h^{-1}(y_1) \neq \emptyset$. Let $x \in h^{-1}(y_1)$. By the structure of a Cantor set, each of the arcs in $S^1 \setminus \{x_1, x_2\}$ contains points in $\Omega(f)$. It follows from the monotonicity of h that there are points in $\Omega(f)$ carried by h into $\mathfrak{I}$, which cannot be. This contradiction proves that h is one-to-one, and h preserves the order of points on the circle. Thus, $h\colon \Omega(f) \to \Omega(g)$ is a homeomorphism.

Let us now extend h to a mapping of the circle. We break up the adjacent intervals of the Cantor set $\Omega(f)$ into equivalence classes: intervals G' and G'' are regarded as equivalent if there exists an $n \in \mathbb{Z}$ such that $f^n(G') = G''$. From each equivalence class we choose a representative, obtaining $G_1, G_2, \dots$. We extend h from the endpoints of G_i to an arbitrary homeomorphism $h_i\colon G_i \to S^1 \setminus \Omega(g)$ on the whole interval G_i, $i = 1, 2, \dots$. For the points $x \in f^n(G_i)$, $n \in \mathbb{Z}$, we set $h_{i,n}(x) = g^n \circ h_i \circ f^{-n}(x)$.

We extend h to the whole circle by means of the homeomorphisms $h_{i,n}$, $i = 1, 2, \dots$, $n \in \mathbb{Z}$. Denote the mapping obtained again by h. By construction, $h\colon S^1 \to S^1$ is a homeomorphism.

If $x \in \overset{\circ}{\Omega}(f)$, then $h \circ f(x) = h_2^{-1} \circ R_\beta \circ h_1 \circ f(x) = h_2^{-1} \circ R_\beta \circ R_\alpha \circ h_1(x) = g \circ h_2^{-1} \circ R_\beta \circ h_1(x) = g \circ h(x)$. By continuity, $h \circ f(x) = g \circ h(x)$ for $x \in \Omega(f)$.

If $x \in f^n(G_i)$, then $h \circ f(x) = h_{i,n+1}[f(x)] = g^{n+1} \circ h_i \circ f^{-(n+1)}(f(x)) = g \circ [g^n \circ h_i \circ f^{-n}](x) = g \circ h(x)$. Thus, $h \circ f(x) = g \circ h(x)$ for all $x \in S^1$. □

Theorem 2.4 shows that the characteristic set is a topological invariant of a Denjoy homeomorphism up to a rotation. We remark that the characteristic set itself is determined by the Denjoy homeomorphism also up to a rotation. The next theorem shows that the characteristic set is a complete topological invariant.

THEOREM 2.5. *Let χ be an at most countable family of orbits of the rotation R_α, $\alpha \in \mathbb{R} \setminus Q$. Then there exist an $f \in \operatorname{Den}(S^1)$ and a continuous mapping h semiconjugating f and R_α such that $\chi = \chi(f, h)$.*

PROOF. The proof is by the scheme for constructing a homeomorphism of the circle with Cantor limit set in the description of a Denjoy flow (see §3.1 in Chapter 1). The construction consists in a "blowing up" of points of the orbits in the family χ.

On each orbit in χ we choose one point, obtaining $x_1, \dots, x_l, \dots$. With a point $R_\alpha^n(x_l)$ $(n \in \mathbb{Z})$ we associate a number $a_n^{(l)} > 0$ so that

$$\sum_{n,l} a_n^{(l)} = a < +\infty.$$

(For example, $a_n^{(l)} = (|n| + l + 1)^{-1}(|n| + l + 2)^{-1}$.)

In place of each point $R_\alpha^n(x_l)$ we put on the circle a segment $I_n^{(l)}$ of length $a_n^{(l)}$. This operation can be realized formally as follows. We remove from S^1 a point $R_\alpha^n(x_l)$ (n and l fixed). The closure of the open arc obtained will be homeomorphic to a closed segment whose endpoints we identify with the endpoints of the segment $I_n^{(l)}$ of length $a_n^{(l)}$. As a result we get a circle $S(1+a_n^{(l)})$ of length $1+a_n^{(l)}$. Obviously, $S^1 \setminus \{R_\alpha^n(x_l)\}$ is homeomorphic to $S(1+a_n^{(l)}) \setminus I_n^{(l)}$, and it can be assumed that the remaining points in $\chi \setminus \{R_\alpha^n(x_l)\}$ lie on $S(1+a_n^{(l)})$. Continuing this process, we get a circle $S(1+a)$ of length $1+a$, and by construction the mutual arrangement of the intervals $I_n^{(l)}$, $n \in \mathbb{Z}$, $l = 1, 2, \dots$, on $S(1+a)$ is the same as the mutual arrangement of the points $x_n^{(l)} = R_\alpha^n(x_l)$, $n \in \mathbb{Z}$, $l = 1, 2, \dots$, on S^1. (The construction given is called a blowing up; the circle $S(1+a)$ is obtained by blowing up each point in χ.)

We introduce an equivalence relation on $S(1+a)$: the points $x_1, x_2 \in S(1+a)$ are equivalent if they lie on a single interval $I_n^{(l)}$. Taking the quotient of $S(1+a)$ by this equivalence relation (identifying each interval $I_n^{(l)}$ with a point), we get the circle S^1. Let $\widetilde{h}\colon S(1+a) \to S^1$ be the natural projection (the mapping $\widetilde{h}$ is called a blowing-down). By construction, $\widetilde{h}(I_n^{(l)}) = x_n^{(l)}$ and $\widetilde{h}(\bigcup_{l,n} I_n^{(l)}) = \chi$.

Since χ is dense in S^1 and $\widetilde{h}$ is monotonically nondecreasing, the set $\Omega = S^1 \setminus \bigcup_{n,l} \operatorname{int} I_n^{(l)}$ is a Cantor set.

The mapping $\widetilde{h}$ is one-to-one on $\overset{\circ}{\Omega} = S^1 \setminus \bigcup_{n,l} I_n^{(l)}$, and by monotonicity is a homeomorphism onto its range $S^1 \setminus \chi$. Therefore, $R_\alpha|_{S^1 \setminus \chi}\colon S^1 \setminus \chi \to S^1 \setminus \chi$ induces a homeomorphism $\widetilde{f}|_{\overset{\circ}{\Omega}} = \widetilde{h}^{-1} \circ R_\alpha \circ \widetilde{h}|_{\overset{\circ}{\Omega}}\colon \overset{\circ}{\Omega} \to \overset{\circ}{\Omega}$ by means of $\widetilde{h}$. As in the proof of uniform continuity of the mapping $h_1^{-1} \circ R_\alpha \circ h_1$ for Theorem 2.4, it can be shown that $\widetilde{f}|_{\overset{\circ}{\Omega}}$ is uniformly continuous and can be extended to a homeomorphism $\widetilde{f}|_\Omega\colon \Omega \to \Omega$, which can in turn be extended to a homeomorphism $\widetilde{f}\colon S(1+a) \to S(1+a)$.

The set $\overset{\circ}{\Omega}$ is invariant under $\widetilde{f}$ and $\widetilde{f}^{-1}$, and any orbit $O(x) = \{\widetilde{f}^n(x) : n \in \mathbb{Z}\}$, $x \in \overset{\circ}{\Omega}$, is dense in $\overset{\circ}{\Omega}$. Therefore, $\widetilde{f} \in \operatorname{Den}[S(1+a)]$. Since the lengths of the intervals $\widetilde{f}^n(I_k^{(l)})$ tend to zero as $n \to +\infty$, it follows that $\Omega(\widetilde{f}) = \Omega$. The mapping $\widetilde{h}$ realizes a semiconjugacy between $\widetilde{f}$ and R_α, and $\chi(\widetilde{f}, \widetilde{h}) = \chi$.

A linear diffeomorphism $S(1+a) \to S^1$ carries $\widetilde{f}$ into the required $f \in \mathrm{Den}(S^1)$. $\square$

DEFINITION. Let $\chi(f,h)$ be the characteristic set of a Denjoy homeomorphism f. The number of orbits or the cardinality $|\chi(f,h)|$ of the collection of orbits making up $\chi(f,h)$ is called the *characteristic* of f.

By Lemma 2.3, 2), the characteristic of a homeomorphism f does not depend on the semiconjugating mapping h. It can be defined as the number or cardinality of the set of equivalence classes of adjacent intervals of the Cantor set $\Omega(f)$, where two adjacent intervals G' and G'' are taken to be equivalent if $G'' = f^n(G')$ for some $n \in \mathbb{Z}$. The characteristic is obviously a topological invariant of a Denjoy homeomorphism. But, as the next lemma shows, the characteristic is not a complete topological invariant.

LEMMA 2.9. *For any irrational number $\alpha \in (0,1)$ there is a continuum of nonconjugate Denjoy homeomorphisms with rotation number α and characteristic equal to* 2.

PROOF. Two sets $\chi_1, \chi_2 \subset S^1$ will be said to be equivalent if $\chi_1 \equiv \chi_2$ (that is, one of them can be superimposed on the other by a rotation of the circle). According to Theorems 2.4 and 2.5, it suffices to show that there is a continuum of nonequivalent sets, each of which consists of two orbits of the rotation R_α.

We fix an orbit O_1 of R_α and form the set $\chi_\mu = O_1 \cup O_\mu$, where O_μ is an orbit of R_α different from O_1. Since the set of orbits has the cardinality of a continuum, there is a continuum of distinct sets χ_μ.

Denote by $\{\chi_\mu\}$ the family of all sets of the form $\chi_{\mu'}$ equivalent to χ_μ (that is, consisting of two orbits and with a fixed orbit O_1). Let $\chi' = O_1 \cup O_{\mu_1} \in \{\chi_\mu\}$. Then $R_{\beta_1}(\chi') = \chi_\mu$ for some $\beta_1 \in \mathbb{R}$. Since $R_{\beta_1} \circ R_\alpha = R_\alpha \circ R_{\beta_1}$, the rotation R_{β_1} carries an orbit into an orbit of R_α. Therefore, there are only two possibilities: 1) $R_{\beta_1}(O_1) = O_1$; 2) $R_{\beta_1}(O_1) = O_\mu$. In case 1) $R_{\beta_1}(O_\mu) = O_\mu$ by the minimality of R_α, and hence $O_\mu = O_{\mu_1}$, that is, $\chi' = \chi_\mu$. In case 2) $R_{\beta_1}(O_{\mu_1}) = O_1$.

Let $\chi'' = O_1 \cup O_{\mu_2} \in \{\chi_\mu\}$ be different from χ_μ. It follows from the foregoing that $R_{\beta_2}(O_1) = O_\mu$ and $R_{\beta_2}(O_{\mu_2}) = O_1$ for some $\beta_2 \in \mathbb{R}$. The minimality of R_α and the equalities $O_\mu = R_{\beta_1}(O_1) = R_{\beta_2}(O_1)$ give us that $R_{\beta_2} = R_{\beta_1} \circ R_\alpha^n$ for some $n \in \mathbb{Z}$. Therefore, $O_{\mu_2} = R_{\beta_2}^{-1}(O_1) = R_{\beta_1}^{-1}(O_1) = O_{\mu_1}$, that is, $\chi' = \chi''$.

Consequently, each class $\{\chi_\mu\}$ has only two representatives, and since there is a continuum of distinct sets χ_μ, there is also a continuum of distinct classes $\{\chi_\mu\}$. Sets χ_{μ_1} and χ_{μ_2} in different classes $\{\chi_{\mu_1}\}$ and $\{\chi_{\mu_2}\}$ are not equivalent; therefore, representatives of all the classes $\{\chi_\mu\}$ give the required continuum family of nonequivalent sets, each consisting of two orbits of R_α. $\square$

2.6. Classification of Cherry transformations. In this subsection we consider Cherry transformations of the line and the circle. Cherry transformations of the circle arise in a natural way in the study of the Poincaré mapping on contact-free cycles in Cherry flows. Recall that $P(\mathbb{R})$ denotes the set of monotonically nondecreasing transformations of $\mathbb{R}$ of degree 1.

DEFINITION. A transformation $\overline{f} \in P(\mathbb{R})$ is called a *Cherry transformation of the line* $\mathbb{R}$ if it satisfies the following conditions:

1) on any finite interval, $\overline{f}$ has finitely many intervals of constancy (that is, intervals on which $\overline{f}$ takes a constant value) and finitely many points of discontinuity;

2) $\overline{f}$ is continuous at endpoints of intervals of constancy;

3) if x_0 is a point of discontinuity, then $\overline{f}(x) \uparrow \overline{f}(x_0)$ as $x \uparrow x_0$ (that is, $\overline{f}$ is left-continuous at points of discontinuity; see Figure 5.5);

4) $\overline{f}$ has an irrational rotation number;

5) if $[a, b]$ is an interval of constancy, then for any $n \in \mathbb{N}$ the complete inverse image $\overline{f}^{-n}([a, b])$ is a closed interval with a neighborhood in which $\overline{f}$ is a homeomorphism;

6) if $\overline{f}$ is discontinuous at x_0 (see Figure 5.5) and if $[c, d] = [\overline{f}(x_0), \lim_{x \downarrow x_0} \overline{f}(x)]$, then for any $n \in \mathbb{N} \cup \{0\}$ the image $\overline{f}^n([c, d])$ is a closed interval with a neighborhood in which $\overline{f}$ is a homeomorphism;

7) all the images of the intervals of constancy and all the points of discontinuity of $\overline{f}$ lie in $\Omega(\overline{f})$.

The set of Cherry transformations of the line is denoted by $\mathrm{Ch}(\mathbb{R})$.

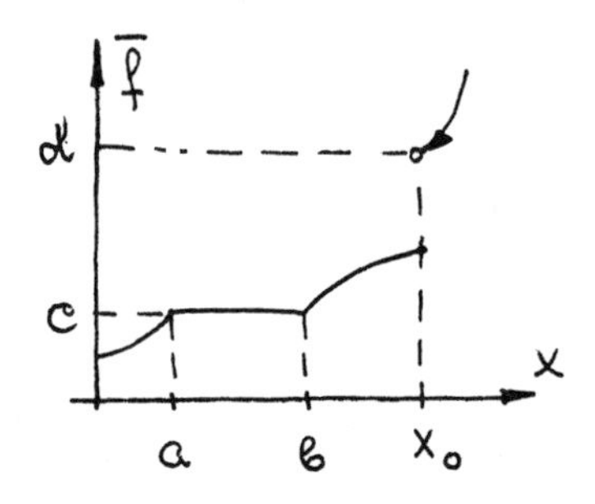

FIGURE 5.5

DEFINITION. A transformation $f \in P(S^1)$ is called a *Cherry transformation of the circle* S^1 if there exists a covering Cherry transformation of the line for f.

The set of Cherry transformations of the circle is denoted by $\mathrm{Ch}(S^1)$.

Unless otherwise stated, we assume below that a Cherry transformation is not a homeomorphism (that is, has at least one point of discontinuity or at least one interval of constancy). Since the rotation number of a Cherry transformation of the circle is irrational, this implies that the positive semi-orbit of any point under this transformation is not dense (for example, if a Cherry transformation has an interval I of constancy, then any positive semi-orbit intersects I in at most one point). Therefore, by Lemma 2.3, a Cherry transformation of the circle is semiconjugate to a rotation by means of a continuous monotonically nondecreasing mapping that is not a homeomorphism.

DEFINITION. Let $f \in \mathrm{Ch}(S^1)$ be semiconjugate to R_α, $\alpha \in \mathrm{rot}(f)$, by means of a continuous mapping $h \in P(S^1)$, and suppose that a closed interval I is mapped by h into a point, with $h^{-1}[h(I)] = I$. If $f^n(I)$ is an interval for all $n \in \mathbb{Z}$, then I is called a *gray interval*.

By Lemma 2.3, the definition of a gray interval does not depend on the semiconjugating mapping h.

Let I be a gray interval. It follows from the equality $R_\alpha^n \circ h = h \circ f^n$, $n \in \mathbb{N}$, that the interval $f^i(I)$ belongs to some gray interval, denoted by $\widehat{f^i}(I)$, for any $i \in \mathbb{Z}$. We remark that the equality $f^i(I) = \widehat{f^i}(I)$ does not always hold (for example, $f(I)$ can be adjacent to an interval of constancy, and then $\widehat{f}(I)$ is equal to the union of the interval $f(I)$ and the interval of constancy).

DEFINITION. Let I be a gray interval of a transformation $f \in \mathrm{Ch}(S^1)$. The union $\mathcal{I}(I) = \bigcup_{n\in\mathbb{Z}} \widehat{f^n}(I)$ is called a *gray cell.*

We now give the definitions of a black interval and a black cell (negative and positive) of a Cherry transformation f of the circle S^1.

Let $[a,b] \subset S^1$ be an interval of constancy of a transformation $f \in \mathrm{Ch}(S^1)$. According to condition 5) in the definition of a Cherry transformation, the complete inverse image $f^{-n}([a,b])$ is an interval for any $n \in \mathbb{Z}^+$ ($\mathbb{Z}^+$ is the set of nonnegative integers), called a *black negative interval.* The union

$$\bigcup_{n\in\mathbb{Z}^+} f^{-n}([a,b]) \stackrel{\text{def}}{=} \mathcal{I}(a,b)$$

is called a *black negative cell.*

If $x_0 \in S^1$ is a point of discontinuity of a transformation $f \in \mathrm{Ch}(S^1)$, then $[f(x_0), \lim_{x\downarrow x_0} f(x)] \stackrel{\text{def}}{=} [c,d]$ is an interval, which we denote by $\widehat{f}(x_0)$. According to condition 6) in the definition of a Cherry transformation, $f^n([c,d])$ is an interval for any $n \in \mathbb{Z}^+$, called a *black positive interval.* The union

$$\bigcup_{n\in\mathbb{Z}^+} f^n([c,d]) \stackrel{\text{def}}{=} \mathcal{I}(x_0)$$

is called a *black positive cell.*

Let $\chi(f,h) = \cup h(I)$, where the union is over all gray intervals I. The set $\chi(f,h)$ is an at most countable family of orbits of the rotation R_α, where $\alpha = \mathrm{rot}(f)$.

Let $\chi^-(f,h) = \cup h(I)$, where the union is over all black negative intervals I. According to the condition 1) in the definition of a Cherry transformation, the set $\chi^-(f,h)$ is made up of finitely many negative semi-orbits of R_α.

Similarly, $\chi^+(f,h) = \cup h(I)$, where the union is over all black positive intervals I, is made up of finitely many positive semi-orbits of R_α.

If a gray interval $[\alpha,\beta]$ contains intervals of constancy, then we assign a code $(\varepsilon_1,\varepsilon_2)^-$ to the point $h([\alpha,\beta])$ as follows. According to 5) and 7), any interval $[a,b] \subset [\alpha,\beta]$ of constancy has a common endpoint with the interval $[\alpha,\beta]$. Therefore, $[\alpha,\beta]$ contains at most two intervals of constancy. If $[\alpha,\beta]$ contains two intervals of constancy, then we set $(\varepsilon_1,\varepsilon_2)^- = (1,1)^-$. If $[\alpha,\beta]$ contains a single interval of constancy $[a,b]$, then either $a = \alpha$ or $b = \beta$. In the first case we set $(\varepsilon_1,\varepsilon_2)^- = (1,0)^-$, and in the second $(\varepsilon_1,\varepsilon_2)^- = (0,1)^-$.

We assign a code $(\varepsilon_1,\varepsilon_2)^+$ to the point $h([\alpha,\beta])$ in an analogous way if the gray interval $[\alpha,\beta]$ contains intervals of the form $\widehat{f}(x_0)$, where x_0 is a point of discontinuity of f.

Note that points equipped with a code are initial points of semi-orbits in $\chi^-(f,h)$ or $\chi^+(f,h)$ which belong to orbits in $\chi(f,h)$. Note also that two codes are assigned to some points in $\chi^-(f,h) \cap \chi^+(f,h)$.

Let $f \in \mathrm{Ch}(S^1)$. The *scheme* $S(f,h)$ of a transformation f with respect to a semiconjugating mapping h is defined to be the collections $\chi(f,h)$ of orbits, $\chi^-(f,h)$ and $\chi^+(f,h)$ of semi-orbits, and $\chi^*(f,h)$ of coded points equipped with the corresponding codes.

The schemes of transformations $f, g \in \mathrm{Ch}(S^1)$ are said to be *isomorphic* if there exists a $\beta \in \mathbb{R}$ such that

$$R_\beta[\chi(f,h_1)] = \chi(g,h_2), \qquad R_\beta[\chi^-(f,h_1)] = \chi^-(g,h_2),$$
$$R_\beta[\chi^+(f,h_1)] = \chi^+(g,h_2), \qquad R_\beta[\chi^*(f,h_1)] = \chi^*(g,h_2),$$

and each point in $\chi^*(f,h_1)$ is carried by R_β into a point with the same code (or codes), where h_1 and h_2 realize semiconjugacies between f, g and the respective rotations $R_{\mathrm{rot}(f)}$, $R_{\mathrm{rot}(g)}$.

By Lemma 2.3, the schemes of a transformation $f \in \mathrm{Ch}(S^1)$ with respect to different mappings h semiconjugating f and $R_{\mathrm{rot}(f)}$ are isomorphic.

THEOREM 2.6. *Suppose that the transformations $f, g \in \mathrm{Ch}(S^1)$ are semiconjugate to rotations by means of mappings h_1 and h_2, and let $S(f,h_1)$ and $S(g,h_2)$ be the schemes of f and g with respect to h_1 and h_2, respectively. Then f and g are conjugate if and only if $\mathrm{rot}(f) = \mathrm{rot}(g)$ and the schemes $S(f,h_1)$ and $S(g,h_2)$ are isomorphic.*

With obvious changes the proof repeats that of Theorem 2.4, and we omit it.

Let us consider a rotation R_α with $\alpha \in \mathbb{R} \setminus Q$. An *admissible scheme* is defined to be a collection made up of an at most countable family χ of orbits of R_α and finite families χ^+ and χ^- of positive and negative semi-orbits, respectively, satisfying the following conditions:

a) a code $(\varepsilon_1, \varepsilon_2)^{+(-)}$ with $\varepsilon_i \in \{0,1\}$ and $\varepsilon_1 + \varepsilon_2 \geq 1$ is assigned to each initial point of a positive (negative) semi-orbit in the intersection $\chi^+ \cap \chi$ ($\chi^- \cap \chi$);

b) all the semi-orbits in $\chi^+ \cup \chi^-$ not belonging to χ are disjoint;

c) for each orbit $O \in \chi$ there are at most four semi-orbits in $\chi^- \cup \chi^+$ lying on O, and at most two in each of χ^- and χ^+, and any point of O belongs to at most two semi-orbits in $\chi^- \cup \chi^+$;

d) if two semi-orbits in $\chi^- \cup \chi^+$ intersect (and thus lie on some orbit in χ), then the codes $(\varepsilon_1, \varepsilon_2)$ and $(\varepsilon_1', \varepsilon_2')$ of their initial points are opposites, that is, $\varepsilon_i + \varepsilon_i' = 1$ for $i = 1, 2$.

A transformation $f \in \mathrm{Ch}(S^1)$ has an admissible scheme with respect to any semiconjugation between f and $R_{\mathrm{rot}(f)}$. Indeed, let $S(f,h)$ be the scheme of f with respect to a semiconjugating mapping h. If two semi-orbits $O_1^{()}$ and $O_2^{()}$ in $\chi^-(f,h) \cup \chi^+(f,h)$ do not belong to $\chi(f,h)$, then the black cells corresponding to them do not intersect the gray intervals, and do not have intersecting black intervals in view of the conditions 5) and 6) in the definition of Cherry transformations. Consequently, the semi-orbits $O_1^{()}$ and $O_2^{()}$ are disjoint.

The conditions c) and d) follow from the condition 7).

THEOREM 2.7. *Let $R_\alpha \colon S^1 \to S^1$ be a rotation with $\alpha \in \mathbb{R} \setminus Q$, and let S be an admissible scheme. Then there exists an $f \in \mathrm{Ch}(S^1)$ that is semiconjugate to R_α (so that $\alpha = \mathrm{rot}(f)$) by means of an h such that $S(f,h) = S$.*

The proof is by the scheme used for Theorem 2.5, and we omit it.

§3. Structurally stable diffeomorphisms

In this section we consider the space of diffeomorphisms of the circle, and in that space we single out the dense open subspace of structurally stable (or weakly structurally stable) diffeomorphisms. The results in this section go back to Poincaré [**68**], Maĭer [**54**], Pliss [**65**], and others, and are reflected in the books [**62**], [**64**], and [**89**]. Our exposition thus bears a schematic character.

3.1. The C^r-topology. For a numerical C^r-smooth function $\overline{f}\colon \mathbb{R} \to \mathbb{R}$ we denote by $\mathcal{D}^r\overline{f}$ the derivative of order $r \in \mathbb{N}$.

Let $f\colon S^1 \to S^1$ be a transformation of degree 1, and let $\overline{f}\colon \mathbb{R} \to \mathbb{R}$ be a covering transformation. Then $\overline{f}(x) = x + h(x)$, where $h(x)$ is a periodic function of period 1. If the derivative $\mathcal{D}^r\overline{f}$ exists, then we set $\mathcal{D}^r f = \mathcal{D}^r\overline{f}$ for $r \in \mathbb{N}$ and call $\mathcal{D}^r f$ the rth-order derivative of f. Since any covering transformation for f has the form $\overline{f}(x) + n$, $n \in \mathbb{Z}$, the definition of $\mathcal{D}^r f$ is independent of the choice of the covering transformation.

For $r = 0$ we take $\mathcal{D}^r f = \mathcal{D}^0 f = f$.

A homeomorphism f of the circle is said to be a *C^r-homeomorphism* if $\mathcal{D}^i f$ exists for $i = 1, \dots, r$.

The set of C^r-homeomorphisms of the circle is denoted by $\mathrm{Homeo}^r(S^1)$.

If $f \in \mathrm{Homeo}^r(S^1)$ for all $r \in \mathbb{N}$, then f is a C^∞-homeomorphism. If a covering $\overline{f}$ for f is an analytic function, then f is said to be an analytic or C^ω-homeomorphism.

A homeomorphism f of the circle is said to be a *C^r-diffeomorphism* if $f, f^{-1} \in \mathrm{Homeo}^r(S^1)$. The set of C^r-diffeomorphisms of the circle is denoted by $\mathrm{Diff}^r(S^1)$. Let $\mathrm{Homeo}(S^1) = \mathrm{Diff}^0(S^1)$.

We introduce a metric ρ_r on the set $\mathrm{Homeo}^r(S^1)$, $0 \le r < \infty$. First let $\rho_0(f,g)$ be defined for $f, g \in \mathrm{Homeo}(S^1)$ by

$$\rho_0(f,g) = \max_{S^1} |f(x) - g(x)| = \max_{0 \le x \le 1} |\overline{f}(x) - \overline{g}(x)| \pmod 1.$$

If $f, g \in \mathrm{Homeo}^r(S^1)$ for $1 \le r < \infty$, then we set

$$\rho_r(f,g) = \rho_0(f,g) + \rho_0(\mathcal{D}f, \mathcal{D}g) + \dots + \rho_0(\mathcal{D}^r f, \mathcal{D}^r g),$$

where $\rho_0(\mathcal{D}^i f, \mathcal{D}^i g) = \max_{0 \le x \le 1} |\mathcal{D}^i\overline{f}(x) - \mathcal{D}^i\overline{g}(x)|$.

The topology induced by ρ_k on the set $\mathrm{Homeo}^r(S^1)$ is called the *C^r-topology*, and this topology turns $\mathrm{Homeo}^r(S^1)$ into a topological space.

The C^∞-topology on the set $\mathrm{Homeo}^\infty(S^1)$ is defined to be the weakest topology induced by the C^r-topologies on $\mathrm{Homeo}^r(S^1)$, $r < \infty$, and by the imbeddings $\mathrm{Homeo}^\infty(S^1) \subset \mathrm{Homeo}^r(S^1)$, $r \in \mathbb{N}$.

The subset $\mathrm{Diff}^r(S^1) \subset \mathrm{Homeo}^r(S^1)$ is open in $\mathrm{Homeo}^r(S^1)$, $r \ge 0$, and is equipped with the induced C^r-topology as a subset of that topological space.

We introduce the C^∞-topology on the set $\mathrm{Homeo}^\omega(S^1)$ of analytic homeomorphisms and the set $\mathrm{Diff}^\omega(S^1)$ of analytic diffeomorphisms.

3.2. Main definitions.

DEFINITION. A C^r-homeomorphism f, $r \ge 1$, is said to be *structurally stable* if for any $\varepsilon > 0$ there is a neighborhood $\mathcal{U}$ of this homeomorphism in $\mathrm{Homeo}^r(S^1)$ such that any $g \in \mathcal{U}$ is conjugate to f by means of a homeomorphism h that is ε-close to the identity in the metric ρ_0.

If the requirement that the conjugating homeomorphism h be ε-close to the identity is dropped in this definition, then we get the definition of a weakly structurally stable C^r-homeomorphism f.

A structurally stable C^r-homeomorphism is obviously weakly structurally stable.

DEFINITION. A fixed point $x_0 \in S^1$ of a C^r-homeomorphism f, $r \geq 1$, is said to be *hyperbolic* if $|\mathcal{D}f(x_0)| \neq 1$.

A point x_0 is called a *sink* or *attracting point* if $|\mathcal{D}f(x_0)| < 1$.

If $|\mathcal{D}f(x_0)| > 1$, then x_0 is called a *source* or *repelling point*.

DEFINITION. For a C^r-homeomorphism f a periodic point of period m is said to be *hyperbolic* (attracting or repelling) if it is a hyperbolic fixed point (attracting or repelling) for f^m.

LEMMA 3.1. *Suppose that $f \in \operatorname{Diff}^r(S^1)$, $r \geq 1$, has fixed points, and each fixed point of f is hyperbolic. Then f is weakly structurally stable.*

PROOF. It follows from the definition of hyperbolicity that a hyperbolic fixed point is isolated in the set of fixed points of the diffeomorphism. Therefore, f has finitely many fixed points. Topological considerations give us that half the fixed points are sinks and half are sources (in particular, the number of fixed points is even), and they alternate on S^1 (Figure 5.6).

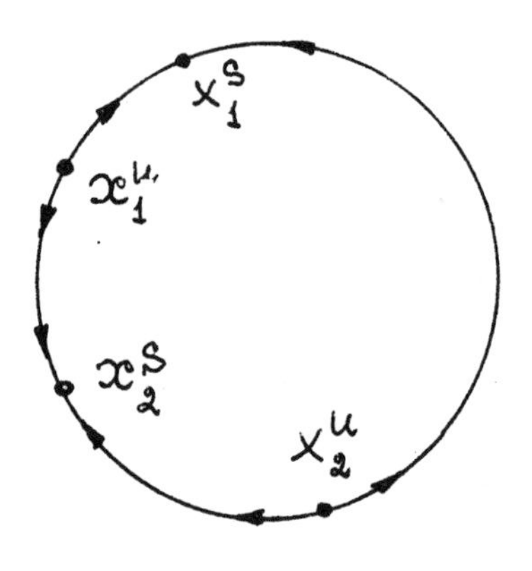

FIGURE 5.6

Let $x_1^s, \ldots, x_l^s$ be the sinks and $x_1^u, \ldots, x_l^u$ the sources of f.

For a subset $\mathcal{U} \subset S^1$ we denote by $m(\mathcal{U})$ its Lebesgue measure.

Let x_i^s be a sink, and $\mathcal{U}_i \ni x_i^s$ a neighborhood of it in which $|\mathcal{D}f| \leq \lambda < 1$. Then $m[f(\mathcal{U}_i)] \leq \lambda m(\mathcal{U}_i) < m(\mathcal{U}_i)$, and hence $\operatorname{cl}[f(\mathcal{U}_i)] \subset \mathcal{U}_i$. Moreover, if μ is the length of a minimal interval in $\mathcal{U}_i \setminus f(\mathcal{U}_i)$, then $g(\mathcal{U}_i) \subset \mathcal{U}_i$ for any $g \in \operatorname{Homeo}^r(S^1)$ that is μ-close to f in the C^0-topology. Therefore, g has a fixed point in $\mathcal{U}_i$. If $\rho_1(f, g) < (1-\lambda)/3$, then $|\mathcal{D}g| \leq \lambda + (1-\lambda)/3 \leq (2\lambda+1)/3 < 1$. Consequently, g has exactly one attracting fixed point in $\mathcal{U}_i$.

An analogous neighborhood can be constructed for each sink and each source.

We take disjoint neighborhoods V_j of the fixed points x_i^s, x_i^u, $i = 1, \ldots, l$, $j = 1, \ldots, 2l$, together with a number $\mu > 0$ such that if $\rho_1(f, g) < \mu$, then g has in each V_j exactly one hyperbolic fixed point of the same character as f. The complement $S^1 \setminus \cup V_j$ consists of disjoint closed segments $I_1, \dots, I_{2l}$. None of these segments contain fixed points of f, so

$$\nu = \min_{x \in \cup I_j} \rho(x, f(x)) > 0,$$

where ρ is the metric on S^1 induced by the covering $\pi \colon \mathbb{R} \to S^1$. Then any $g \in \mathrm{Homeo}^r(S^1)$ that is $(\nu/2)$-close to f in the C^0-topology does not have fixed points on $S^1 \setminus \cup V_j$.

Let $\delta = \min\{\mu, \nu/2\}$. It follows from the foregoing that any $g \in \mathrm{Homeo}^r(S^1)$ with $\rho_r(f, g) < \delta$ has on S^1 only $2l$ fixed points, and they all are hyperbolic: half sinks $\widetilde{x}_i^s$, $i = 1, \dots, l$, and half sources $\widetilde{x}_i^u$, $i = 1, \dots, l$.

For both f and g the attracting and repelling points alternate on S^1. On the arc between a sink and the source next to it all the points move in the direction of the sink under the action of the homeomorphism.

We consider two such (open) arcs: $I_i(f)$ for f and $I_i(g)$ for g. Let $x \in I_i(f)$ and $x' \in I_i(g)$, and let $h \colon [x, f(x)] \to [x', g(x')]$ be an arbitrary orientation-preserving homeomorphism. Since the segments $f^n[x, f(x)]$, $n \in \mathbb{Z}$, (respectively, $g^n[x', g(x')]$, $n \in \mathbb{Z}$) have disjoint interiors, and their union is $I_i(f)$ (respectively, $I_i(g)$), h can be extended to a homeomorphism $I_i(f) \to I_i(g)$ as follows:

$$h|_{f^n[x, f(x)]} = g^n \circ h \circ f^{-n}|_{f^n[x, f(x)]} \colon f^n[x, f(x)] \to g^n[x', g(x')].$$

It is easy to verify that on $I_i(f)$

$$g \circ h = h \circ f.$$

Making this construction on all the arcs $I_i(f)$, $i = 1, \dots, 2l$, and setting $h(x_i^s) = \widetilde{x}_i^s$ and $h(x_i^u) = \widetilde{x}_i^u$, we get a homeomorphism $h \colon S^1 \to S^1$ conjugating f and g. This proves the weak structural stability of f. $\square$

REMARK 1. By suitably choosing the neighborhoods $\mathfrak{U}_i$ and the points $x \in I_i(f)$ and $x' \in I_i(g)$ we can ensure that h moves points by a distance not exceeding a given $\varepsilon > 0$. Therefore, f is also a structurally stable diffeomorphism.

REMARK 2. If $f \in \mathrm{Diff}^r(S^1)$, $r \geq 1$, has only a finite nonzero number of periodic points, then the smallest period is the same for all these points. If all the periodic points are hyperbolic, then their number is even: half sinks and half sources. The proof that f is weakly structurally stable and structurally stable is analogous to that for Lemma 3.1.

3.3. Instability of an irrational rotation number.

THEOREM 3.1. *Suppose that a transformation $f \in P(\mathbb{R})$ is continuous and $\mathrm{rot}(f) \in \mathbb{R} \setminus Q$. Then for any $t > 0$*

$$\mathrm{rot}(\overline{R}_{-t} \circ f) < \mathrm{rot}(f) < \mathrm{rot}(\overline{R}_t \circ f).$$

PROOF. Since $\overline{f}(x+1) = \overline{f}(x) + 1$, $\overline{f}$ is a covering for some transformation $f \in P(S^1)$. We show that there is a point $x_0 \in S^1$ such that for any connected neighborhood $\mathfrak{U}$ of it each of the open intervals in $\mathfrak{U} \setminus \{x_0\}$ contains points in the positive semi-orbit $O^+(x_0)$ (Figure 5.7). Indeed, by Lemmas 2.2 and 2.3, f is conjugate or semiconjugate to the rotation R_α, $\alpha = \mathrm{rot}(f)$. If f is conjugate to R_α, then the existence of the required point x_0 follows from the irrationality of α

(in this case any orbit of f is dense in S^1). Suppose now that f is semiconjugate to R_α by means of a continuous mapping $h \in P(S^1)$. We take an arbitrary point x_0 in the nonempty perfect set $\Omega(f)$ that is not an endpoint of any interval $h^{-1}(x)$, $x \in \chi(f,g)$. There is such a point in view of Lemma 2.3. Let $\mathcal{U}$ be an arbitrary connected neighborhood of x_0, and assume that $O^+(x_0)$ is disjoint from one of the intervals in $\mathcal{U} \setminus \{x_0\}$ (denote this interval by $\mathcal{U}^+$). We show that $h(\mathcal{U}^+)$ is a point. Indeed, if $h(\mathcal{U}^+)$ contains an interval, then the irrationality of α implies that $h(\mathcal{U}^+)$ contains points of the form $R_\alpha^k[h(x_0)]$ for a countable family of numbers $k \in \mathbb{N}$. From this and the monotonicity of h it follows that $\mathcal{U}^+ \cap O^+(x_0) \neq \emptyset$. Therefore, $h(\mathcal{U}^+)$ is a point, and $x_0 \in \Omega(f)$ is an endpoint of the interval $h^{-1}[h(\mathcal{U}^+)]$, which contradicts the choice of x_0. The contradiction proves that $O^+(x_0)$ intersects both the intervals in $\mathcal{U} \setminus \{x_0\}$.

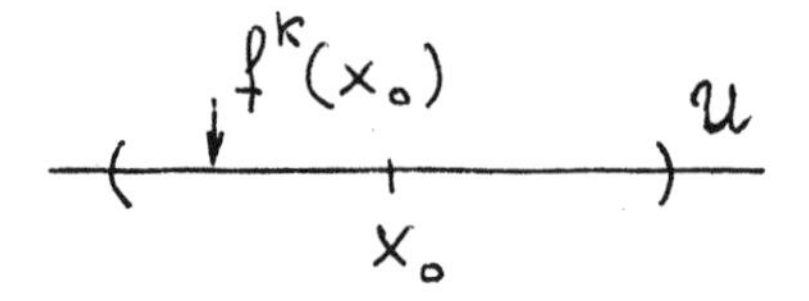

FIGURE 5.7

Assume now that the theorem is false. According to Theorem 1.2, the function $\operatorname{rot}(\overline{R}_t \circ \overline{f}) = \operatorname{rot}(\overline{f} + t)$ is monotonically nondecreasing and continuous. Therefore, there is a $t_0 > 0$ such that $\operatorname{rot}(\overline{R}_t \circ \overline{f}) = \operatorname{rot}(f)$ for all $0 \le t \le t_0$ (or $-t_0 \le t \le 0$, in which case the proof is analogous).

We take a point $\overline{x}_0 \in \pi^{-1}(x_0)$. By the properties of x_0, there exist $k \in \mathbb{N}$ and $r \in \mathbb{Z}$ such that

$$\overline{x}_0 + r - t_0/2 \le \overline{f}^k(\overline{x}_0) < \overline{x}_0 + r.$$

Then $\operatorname{rot}(\overline{f}) \le r/k$ in view of Lemma 1.3.

Since $R_{t_0} \circ \overline{f} \ge \overline{f}$, it follows that $(\overline{R}_{t_0} \circ \overline{f})^n \ge \overline{f}^n$ for all $n \in \mathbb{N}$. Therefore,

$$\begin{aligned}(\overline{R}_{t_0} \circ \overline{f})^k(\overline{x}_0) &= (\overline{R}_{t_0} \circ \overline{f}) \circ (\overline{R}_{t_0} \circ \overline{f})^{k-1}(\overline{x}_0) \\ &\ge (\overline{R}_{t_0} \circ \overline{f}) \circ \overline{f}^{k-1}(\overline{x}_0) = \overline{f}^k(\overline{x}_0) + t_0.\end{aligned}$$

This and the inequality $\overline{f}^k(\overline{x}_0) \ge \overline{x}_0 + r - t_0/2$ imply that $(\overline{R}_{t_0} \circ \overline{f})^k(\overline{x}_0) \ge \overline{x}_0 + r + t_0/2$. Lemma 1.3 again gives us that $\operatorname{rot}(\overline{R}_{t_0} \circ \overline{f}) \ge r/k$. Then $\operatorname{rot}(\overline{f}) = r/k \in Q$, which contradicts the irrationality of $\operatorname{rot}(\overline{f})$. □

It follows from Theorem 3.1 that a structurally stable or weakly structurally stable orientation-preserving diffeomorphism of the circle has a rational rotation number.

LEMMA 3.2. *Suppose that a continuous transformation $f \in P(S^1)$ has a rational rotation number r/k ($k \in \mathbb{N}$). Then f has a periodic point of period k.*

PROOF. We first consider the case $\operatorname{rot}(f) = 0$ and show that f has a fixed point. Take a covering transformation $\overline{f} \in P(\mathbb{R})$ for f with $\operatorname{rot}(\overline{f}) = 0$, and assume that $\overline{f}(x) \neq x$ for all $x \in \mathbb{R}$. For definiteness let $\overline{f}(x) > x$. The monotonicity of $\overline{f}$ gives us that $x < \overline{f}(x) < \cdots < \overline{f}^{n-1}(x) < \overline{f}^n(x) < \cdots$, that is, $\{\overline{f}^n(x)\}_{n=0}^\infty$

is a monotonically increasing sequence. According to the equality $\operatorname{rot}(\overline{f}) = 0$ and Lemma 1.3, $\overline{f}^{n+1}(x) < x + 1$ for $n \in \mathbb{N}$. Therefore, the limit $\lim_{n\to\infty} \overline{f}^n(x) \stackrel{\text{def}}{=} x_0$ exists. Since $\overline{f}$ is continuous, it follows that $\overline{f}(x_0) = \lim_{n\to\infty} \overline{f}^{n+1}(x) = x_0$, and hence $\pi(x_0)$ is a fixed point of f.

Suppose now that $\operatorname{rot}(f) = r/k \neq 0$. We take a covering transformation $\overline{f}$ for f, with $\operatorname{rot}(\overline{f}) = r/k$. Then $\overline{g}(x) \stackrel{\text{def}}{=} \overline{f}^k(x) - r \in P(\mathbb{R})$ and $\operatorname{rot}(\overline{g}) = k\operatorname{rot}(\overline{f}) - r = 0$ (Lemma 1.2). According to what was proved above, the transformation f^k has a fixed point. □

COROLLARY 3.1. *A structurally stable or weakly structurally stable diffeomorphism of the circle has a periodic point.*

3.4. Openness and denseness of the set of weakly structurally stable diffeomorphisms. In the space $\operatorname{Diff}^r_+(S^1)$, $r \geq 1$, of orientation-preserving C^r-diffeomorphisms of the circle we consider the subset $\mathcal{Y}^r_+(S^1)$ of diffeomorphisms having periodic points, each hyperbolic.

LEMMA 3.3. 1) *The set* $\mathcal{Y}^r_+(S^1)$, $r \geq 1$, *is open and dense in* $\operatorname{Diff}^r_+(S^1)$.

2) *The set of weakly structurally stable orientation-preserving diffeomorphisms of the circle coincides with* $\mathcal{Y}^r_+(S^1)$.

There is a proof of this result in the books [**62**], [**64**], and [**65**], so we omit it.

3.5. Classification of weakly structurally stable diffeomorphisms.

THEOREM 3.2. *Two weakly structurally stable diffeomorphisms* $f, g \in \operatorname{Diff}^r_+(S^1)$ *are conjugate if and only if they have the same number (necessarily even) of periodic orbits, with the same minimal period.*

The proof is analogous to that of Lemma 3.1, and we omit it.

Thus, in the space of diffeomorphisms of the circle the weakly structurally stable diffeomorphisms form a dense open set. The (even) number of periodic orbits and the minimal period of any periodic orbit make up a conjugacy invariant of a weakly structurally stable diffeomorphism (the number of periodic points can be taken instead of the minimal period).

Remark. Diffeomorphisms of the first degree of structural instability. Just as relatively structurally stable flows (that is, flows of the first degree of structural instability) are singled out in the space of structurally unstable flows, diffeomorphisms of the first degree of structural instability are singled out in the space of diffeomorphisms of the circle.

A structurally unstable C^r-diffeomorphism $f \in \operatorname{Diff}^r_+(S^1) \setminus \mathcal{Y}^r_+(S^1)$ is called a diffeomorphism of the first degree of structural instability if for any $\varepsilon > 0$ there is a neighborhood $\mathcal{U}$ of this diffeomorphism in $\operatorname{Diff}^r_+(S^1)$ such that any structurally unstable diffeomorphism $g \in \mathcal{U}$ is conjugate to f by means of a homeomorphism that is ε-close to the identity diffeomorphism in the space $\operatorname{Homeo}(S^1)$.

Aranson (*Generic bifurcations of diffeomorphisms of the circle*, Differentsial′nye Uravneniya **23** (1987), 388–394; English transl. in Differential Equations **23** (1987)) obtained necessary and sufficient conditions for a diffeomorphism to belong to the set of diffeomorphisms of the first degree of structural instability and, by using a well-known theorem of Herman [**89**], proved that the set of diffeomorphisms of the first degree of structural instability is open and dense in the set of all structurally unstable diffeomorphisms of the circle.

§4. The connection between smoothness properties and topological properties of transformations of the circle

The first part of the section goes back to Poincaré's conjecture about the existence of analytic nontransitive diffeomorphisms of the circle with irrational rotation number [**68**]. In 1932 Denjoy proved that any C^2-diffeomorphism with irrational rotation number is transitive (thereby disproving Poincaré's conjecture). In 1984 Yoccoz generalized Denjoy's theorem to analytic homeomorphisms.

The second part of the section is devoted to smooth conjugacy of diffeomorphisms of the circle (invariants of smooth conjugacy are singled out for weakly structurally stable diffeomorphisms, and a result of Herman is presented for a transitive diffeomorphism).

4.1. Continued fractions. We present the needed facts about continued fractions (see [**75**] for proofs).

A continued fraction is defined to be an expression of the form

$$a_0 + \cfrac{1}{a_1 + \cfrac{1}{a_2 + \cdots}} \overset{\text{def}}{=} [a_0, a_1, a_2, \dots], \tag{4.1}$$

where the numbers a_i are called the partial quotients or elements of the expansion.

Any real number $\alpha \in (0,1)$ can be uniquely represented as a continued fraction

$$\alpha = [a_1, a_2, \dots] \tag{4.2}$$

with positive integer elements a_i. If α is rational, then the continued fraction (4.2) is finite, and if α is irrational, then (4.2) is infinite.

The elements a_i of the expansion (4.2) can be obtained with the help of the Gauss transformation $G\colon (0,1) \to (0,1)$,

$$G(x) = 1/x - [1/x], \qquad x \in (0,1),$$

where $[z]$ is the integer part of a number z. If for a real number $\alpha \in (0,1)$ the iterates $G(\alpha), \dots, G^{n-1}(\alpha)$ are not equal to zero, then

$$a_1 = \left[\frac{1}{\alpha}\right], a_2 = \left[\frac{1}{G(\alpha)}\right], \dots, a_n = \left[\frac{1}{G^{n-1}(\alpha)}\right].$$

In what follows we assume that $\alpha \in (0,1) \setminus Q$. Then α can be written as an infinite continued fraction.

For $\alpha = [a_1, a_2, \dots]$ the fractions $p_n/q_n = [a_1, \dots, a_n]$ are called the convergents of the continued fraction. The numerators p_n and denominators q_n of the convergents satisfy the recursion relations

$$\begin{cases} p_0 = 0, \quad q_0 = 1, \quad p_1 = 1, \quad q_1 = a_1, \\ p_n = a_n p_{n-1} + p_{n-2}, \qquad n \geq 2, \\ q_n = a_n q_{n-1} + q_{n-2}, \qquad n \geq 2. \end{cases} \tag{4.3}$$

It follows from (4.3) that

$$q_n \geq 2^{\frac{n-1}{2}}, \tag{4.4}$$

$$\frac{1}{q_n(q_n+q_{n+1})} \le \left|\alpha - \frac{p_n}{q_n}\right| \le \frac{1}{q_n q_{n+1}} < \frac{1}{q_n^2}, \tag{4.5}$$

that is, the convergents of the continued fraction are rational approximations of α, and

$$\left|\alpha - \frac{p_n}{q_n}\right| > \left|\alpha - \frac{p_{n+1}}{q_{n+1}}\right|$$

(each convergent is a better approximation than the previous one).

The convergents with even indices form a monotonically increasing sequence, those with odd indices form a monotonically decreasing sequence, and both sequences converge to α (Figure 5.8).

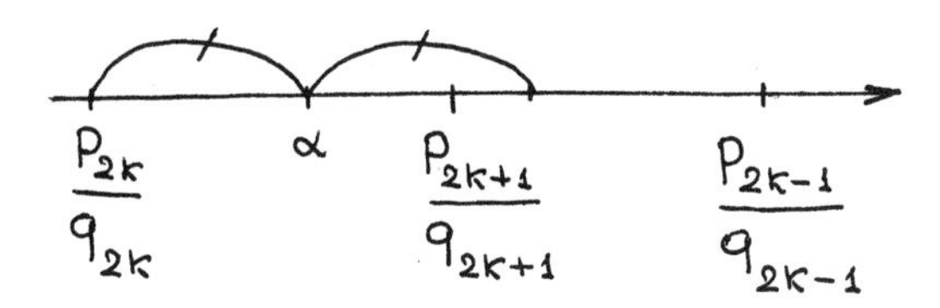

FIGURE 5.8

The convergents of a continued fraction are best approximations of the second kind for α; that is, if $p/q \neq p_n/q_n$ and $0 \le q < q_n$, then

$$|q_n\alpha - p_n| < |q\alpha - p|. \tag{4.6}$$

Conversely, every best approximation of the second kind for α is one of its convergents. This means that the inequality (4.6) is valid for all fractions of the form p/q with denominators $0 \le q < q_{n+1}$, $q \neq q_n$, and the number q_{n+1} is the smallest positive integer among the $q \in \mathbb{N}$ such that $|q_n\alpha - p_n| > |q\alpha - p|$ for some $p \in \mathbb{Z}$:

$$q_{n+1} = \min\{q \in \mathbb{N} : |q_n\alpha - p_n| > |q\alpha - p|\}. \tag{4.7}$$

The fractions $(p_n + ip_{n+1})/(q_n + iq_{n+1})$, $0 < i < a_{n+2}$, are called intermediate quotients. They are situated between the convergents p_n/q_n and p_{n+1}/q_{n+1} on the number line, and they form an increasing sequence for even n (Figure 5.9) and a decreasing sequence for odd n.

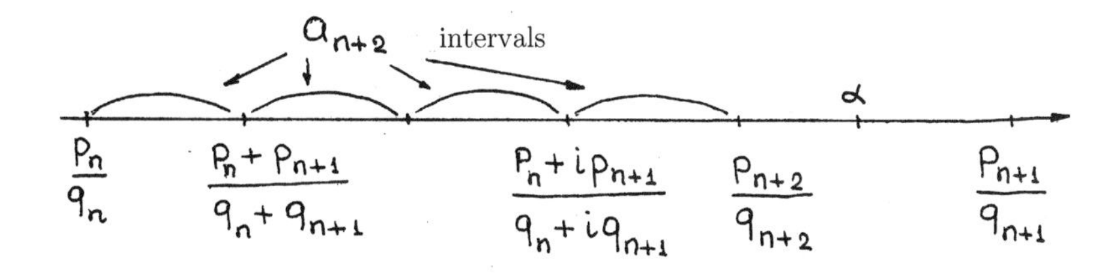

FIGURE 5.9

REMARK. If $a_{n+2} = 1$, then there are no intermediate quotients between the convergents p_n/q_n and p_{n+1}/q_{n+1}.

4.2. The order of the points on the circle. Let $R_\alpha\colon S^1 \to S^1$ be a rotation of the circle by an irrational number $\alpha \in (0,1)$, and let p_n/q_n be the convergents of α, $n \in \mathbb{N}$. The following arrangement of the iterates of an arbitrary point $x_0 \in S^1$ under the action of the rotation follows from the theory of continued fractions. Denote by $[x_0, R_\alpha^{q_n}(x_0)]$ the arc between x_0 and $R_\alpha^{q_n}(x_0)$ containing the point $R_\alpha^{q_{n+2}}(x_0)$. Then on this arc there are no points $R_\alpha^i(x_0)$ with $q_{n+1} \le i < q_{n+1}+q_n$, and, what is more, the first point $R_\alpha^i(x_0)$ with positive minimal $i \ge q_{n+1}$ falling on the arc $[x_0, R_\alpha^{q_n}(x_0)]$ is the point $R_\alpha^{q_{n+1}+q_n}(x_0)$ (Figure 5.10 for even n), and there are no points $R_\alpha^i(x_0)$ with $0 \le i < q_{n+1} + q_n$ on the arc

$$[R_\alpha^{q_{n+1}+q_n}(x_0), R_\alpha^{q_n}(x_0)] \subset [x_0, R_\alpha^{q_n}(x_0)]$$

between $R_\alpha^{q_{n+1}+q_n}(x_0)$ and $R_\alpha^{q_n}(x_0)$. Indeed, if

$$R_\alpha^i(x_0) \in [R_\alpha^{q_{n+1}+q_n}(x_0), R_\alpha^{q_n}(x_0)]$$

for some $0 \le i < q_{n+1}+q_n$, then it follows from (4.6) that $i > q_n$. Thus $0 < i - q_n < q_{n+1}$, and the point $R_\alpha^{i-q_n}(x_0)$ lies on the arc $[R_\alpha^{q_{n+1}}(x_0), x_0]$ between $R_\alpha^{q_{n+1}}(x_0)$ and x_0, closer to x_0 than $R_\alpha^{q_{n+1}}(x_0)$, which contradicts the definition (4.7) of q_{n+1}. If

$$R_\alpha^i(x_0) \in [x_0, R_\alpha^{q_n}(x_0)] \setminus [R_\alpha^{q_{n+1}+q_n}(x_0), R_\alpha^{q_n}(x_0)] = [x_0, R_\alpha^{q_{n+1}+q_n}(x_0)]$$

for some $q_{n+1} \le i < q_{n+1} + q_n$, then $0 \le i - q_{n+1} < q_n$ and $R_\alpha^{i-q_{n+1}}(x_0) \in [x_0, R_\alpha^{q_n}(x_0)]$, which contradicts the definition of q_n.

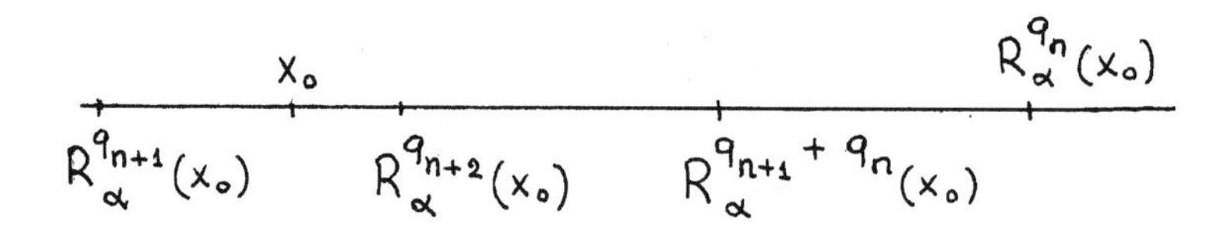

FIGURE 5.10. If $a_{n+2} = 1$, then $q_{n+2} = q_{n+1} + q_n$.

Since R_α is an isometry, the points $R_\alpha^{kq_{n+1}+q_n}(x_0)$, $k = 1, \ldots, a_{n+2}$, lie on the arc $[x_0, R_\alpha^{q_n}(x_0)]$ in the order pictured in Figure 5.11 (n is even, and $a_{n+2} > 1$).

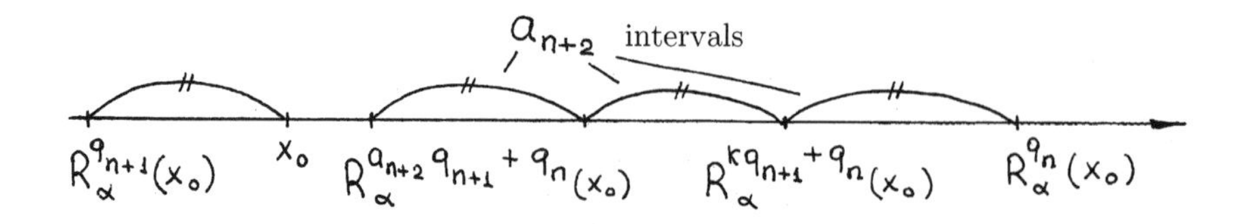

FIGURE 5.11

Denote by $\|q_i\alpha\|$ the length of the circular arc between x_0 and $R_\alpha^{q_i}(x_0)$ not containing $R_\alpha^{q_{i+1}}(x_0)$, $i \ge 1$. Then an analysis of Figure 5.11 leads to the equality

$$\|q_n\alpha\| = a_{n+1}\|q_{n+1}\alpha\| + \|q_{n+2}\alpha\|.$$

LEMMA 4.1. *Let $R_\alpha \colon S^1 \to S^1$ be a rotation with irrational α, and let p_n/q_n be the convergents of the continued fraction expansion of α. Denote by $\mathfrak{I}_n$ the open arc with endpoints $R_\alpha^{-q_n}(x_0)$ and $R_\alpha^{q_n}(x_0)$ that contains x_0, where x_0 is an arbitrary but fixed point, and by $I_n \subset \mathfrak{I}_n$ the open arc with endpoints x_0 and $R_\alpha^{-q_n}(x_0)$ (Figure 5.12). Then:*

1) the arcs $R_\alpha^i(I_n)$, $0 \le i < q_{n+1}$, are disjoint;

2) each point of the circle belongs to at most two arcs in the family $R_\alpha^i(\mathfrak{I}_n)$, $0 \le i < q_{n+1}$;

3) the points $R_\alpha^{-q_n}(x_0)$, $R_\alpha^{-q_{n+1}-q_n}(x_0)$, $R_\alpha^{-q_{n+1}}(x_0)$, and $R_\alpha^{q_n}(x_0)$ are located in cyclical order on the circle.

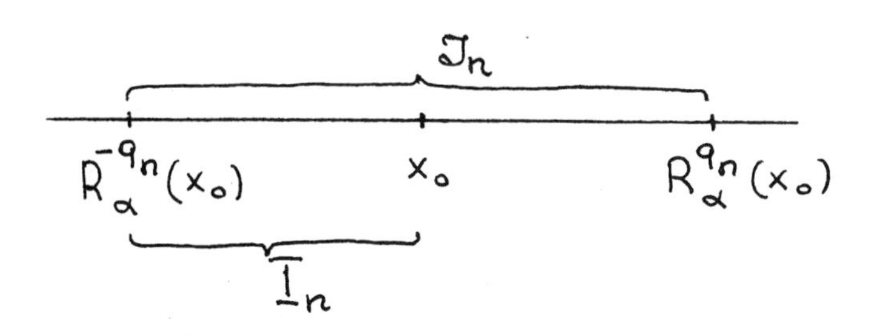

FIGURE 5.12. (n even)

PROOF. Assume that 1) is false, that is, $R_\alpha^k(I_n) \cap R_\alpha^j(I_n) \neq \emptyset$ for some $0 \le k, j < q_{n+1}$. For definiteness assume that $j < k$. Then $R_\alpha^{k-j}(I_n) \cap I_n \neq \emptyset$, and hence $R_\alpha^{j-k}(x_0)$ is closer to x_0 than $R_\alpha^{q_n}(x_0)$. Since $0 \le k - j < q_{n+1}$, we get a contradiction to the definition (4.7) of q_{n+1}.

The assertion 2) follows from 1) and the inclusion $\mathfrak{I}_n \subset I_n \cup R_\alpha^{q_n}(I_n) \cup \{x_0\}$.

The assertion 3) follows from the cyclical arrangement of the points $R_\alpha^{q_n}(x_0)$, $R_\alpha^{q_{n+1}+q_n}(x_0)$, x_0, $R_\alpha^{q_{n+1}}(x_0)$, and $R_\alpha^{-q_n}(x_0)$ on the circle (Figure 5.10) and the fact that R_α is an isometry. □

Since a homeomorphism of the circle with irrational rotation number is conjugate or semiconjugate to a rotation by means of a monotone transformation, Lemma 4.1 gives us

COROLLARY 4.1. *Let f be a homeomorphism of the circle with irrational rotation number α, and let q_n be the denominators of the convergents of the continued fraction expansion of α, $n \in \mathbb{N}$. Denote by $\mathfrak{I}_n$ the open arc with endpoints $f^{-q_n}(x_0)$ and $f^{q_n}(x_0)$ that contains x_0, where x_0 is an arbitrary but fixed point, and denote by $I_n \subset \mathfrak{I}_n$ the open arc with endpoints x_0 and $f^{-q_n}(x_0)$. Then:*

1) the arcs $f^i(I_n)$, $0 \le i < q_{n+1}$, are disjoint;

2) each point of the circle belongs to at most two arcs in the family $f^i(\mathfrak{I}_n)$, $0 \le i < q_{n+1}$;

3) if f is a Denjoy homeomorphism and $\mathfrak{I}$ is an adjacent interval of the Cantor set $\Omega(f)$, then the intervals $f^{-q_n}(\mathfrak{I})$, $f^{-q_n-q_{n+1}}(\mathfrak{I})$, $\mathfrak{I}$, $f^{-q_{n+1}}(\mathfrak{I})$, and $f^{q_n}(\mathfrak{I})$ are arranged in cyclical order on the circle.

4.3. The theorem of Denjoy. In [**68**] Poincaré presupposed that there is an analytic diffeomorphism of the circle without periodic points and with a nowhere dense (hence Cantor) limit set. This conjecture gave birth to an entire direction involving the interrelation of smoothness properties and topological properties in the qualitative theory of dynamical systems (see the survey in [**25**]). A half-century later Poincaré's conjecture was refuted by Denjoy [**82**]; namely, he proved the following theorem.

THEOREM 4.1. *Suppose that $f \in \mathrm{Diff}^r_+(S^1)$, $r \geq 1$, has an irrational rotation number and a derivative $\mathcal{D}f$ of bounded variation. Then $\Omega(f) = S^1$. That is, f is a transitive diffeomorphism.*

PROOF. Assume the contrary. Then $\Omega(f)$ is a Cantor set by Corollary 2.3. Let $G_0 \subset S^1 \setminus \Omega(f)$ be an adjacent interval, and denote by x_0 its left endpoint (the positive direction on S^1 is induced by the positive direction on the line $\mathbb{R}$ and by the covering $\pi \colon \mathbb{R} \to S^1$). Let I_n be an arc satisfying the condition in Corollary 4.1. Since $G_0 \cap \Omega(f) = \emptyset$ and $x_0 \in \Omega(f)$, it follows that $G_0 \subset I_n$ for all odd $n \in \mathbb{N}$ (Figure 5.13). The right-hand endpoint of the arc I_n is equal to $f^{-q_n}(x_0)$ (by the definition of the arc I_n). Therefore, the adjacent interval $f^{-q_n}(G_0) \stackrel{\text{def}}{=} G_{-q_n}$ adjoins I_n.

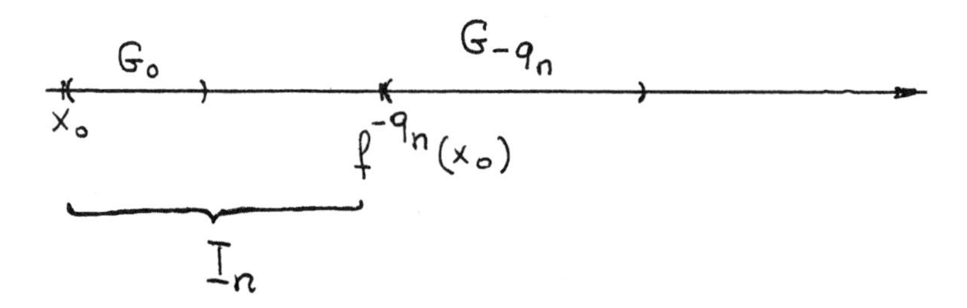

FIGURE 5.13. (n odd)

Let $f^k(G_0) = G_k$, $k \in \mathbb{Z}$. It is a consequence of the mean value theorem that $m(G_{k+1}) = \mathcal{D}f(z_k)m(G_k)$, where $z_k \in G_k$ ($m(\mathcal{U})$ is the Lebesgue measure of a set $\mathcal{U} \subset S^1$).

We form the Denjoy sum

$$\sum_n = \sum_{i=0}^{q_n-1} [\ln \mathcal{D}f(z_{-q_n+i}) - \ln \mathcal{D}f(z_i)].$$

Then

$$\sum_n = \sum_{i=0}^{q_n-1} \ln \frac{\mathcal{D}f(z_{-q_n+i})}{\mathcal{D}f(z_i)} = \ln \prod_{i=0}^{q_n-1} \frac{\mathcal{D}f(z_{-q_n+i})}{\mathcal{D}f(z_i)}$$

$$= \ln \prod_{i=0}^{q_n-1} \frac{m(G_{-q_n+i+1})}{m(G_{-q_n+i})} \cdot \frac{m(G_i)}{m(G_{i+1})} = \ln \frac{[m(G_0)]^2}{m(G_{-q_n})m(G_{q_n})}.$$

Since the lengths $m(G_k)$ of the adjacent intervals G_k tend to zero as $k \to \infty$, it follows that

$$\lim_{n\to\infty} \sum_n = +\infty. \tag{4.8}$$

We estimate the Denjoy sum in another way. By Corollary 4.1, 1), the arcs $f^i(I_n)$ $(0 \le i < q_n)$ are disjoint and have their endpoints in $\Omega(f)$, so the arcs $f^i(I_n) \cup f^i(G_{-q_n}) = f^i(I_n) \cup G_{-q_n+i}$ $(0 \le i < q_n)$ are also disjoint. This implies that $\sum_n \le \mathrm{var}_{S^1} \ln \mathcal{D}f$ for all $n \in \mathbb{Z}$. According to the hypotheses of the theorem, $\mathcal{D}f \ge \mathrm{const} > 0$ and $\mathrm{var}_{S^1} \mathcal{D}f < \infty$, and hence $\mathrm{var}_{S^1} \ln \mathcal{D}f = M < \infty$. This gives us that $\sum_n \le M < \infty$, which contradicts (4.8). $\square$

4.4. The theorem of Yoccoz. In 1981 Hall [**87**] gave a construction of a C^∞-homeomorphism of the circle with a single critical point (that is, a point at which the derivative is zero; this homeomorphism is thus not a C^1-diffeomorphism), an irrational rotation number, and a Cantor limit set. Therefore, the group $\mathrm{Diff}^r_+(S^1)$ in Theorem 4.1 cannot be replaced by $\mathrm{Homeo}^r_+(S^1)$, $r \ge 1$. In 1984 Yoccoz [**112**] showed that, nevertheless, under certain restrictions on the behavior of a smooth homeomorphism in a neighborhood of critical points it cannot have a Cantor limit set. In this subsection we present the result of Yoccoz.

We say that an $f \in \mathrm{Homeo}^r(S^1)$, $r \ge 1$, satisfies the *Yoccoz conditions* if:

1) f has finitely many critical points $x_1, \dots, x_l \in S^1$;

2) $\log \mathcal{D}f$ has bounded variation on any compact interval not containing critical points;

3) for each critical point x_i, $i = 1, \dots, l$, there exist strictly positive constants A_i, B_i, and C_i and an ε_i-neighborhood $\mathcal{U}_i$ of x_i such that

a) $B_i|t|^{C_i} \le \mathcal{D}f(x_i + t) \le A_i|t|^{C_i}$ for $|t| < \varepsilon_i$,

b) the function $(\mathcal{D}f)^{-1/2}$ is convex (downward) on each of the intervals in $\mathcal{U}_i \setminus \{x_i\}$.

REMARKS. I) If $f \in \mathrm{Homeo}^r(S^1)$, $r \ge 2$, then the convexity of $(\mathcal{D}f)^{-1/2}$ means that the function

$$\left(\frac{1}{\sqrt{\mathcal{D}f}}\right)' = -\frac{\mathcal{D}^2 f}{2(\sqrt{\mathcal{D}f})^3}$$

is increasing on each of the intervals in $\mathcal{U}_i \setminus \{x_i\}$.

II) If $f \in \mathrm{Homeo}^r(S^1)$, $r \ge 3$, then the convexity of $(\mathcal{D}f)^{-1/2}$ is equivalent to the condition

$$\left(\frac{1}{\sqrt{\mathcal{D}f}}\right)'' = -\frac{Sf}{2\sqrt{\mathcal{D}f}} \ge 0,$$

where

$$S(f) = \frac{\mathcal{D}^3 f}{\mathcal{D}f} - \frac{3}{2}\left(\frac{\mathcal{D}^2 f}{\mathcal{D}f}\right)^2$$

is the Schwarzian derivative. Consequently, $Sf \le 0$ in the neighborhood $\mathcal{U}_i$, $i = 1, \dots, l$.

III) If $f \in \mathrm{Homeo}^\infty(S^1)$, then the condition 3) holds when at each critical point f is not flat (that is, some finite-order derivative at the critical point is nonzero).

IV) Obviously, the condition 3) holds for an analytic homeomorphism f.

Denote by $I(S^1)$ the set of all possible compact connected segments of S^1.

For brevity we denote the Lebesgue measure $\mathrm{meas}(I)$ of a segment $I \in I(S^1)$ by $m(I)$.

We define the *Yoccoz function* on $\mathrm{Homeo}^r(S^1) \times I(S^1)$, $r \ge 1$, by

$$M(f, I) = \frac{m(f(I))}{m(I)}[\mathcal{D}f(a)\mathcal{D}f(b)]^{-1/2}$$

($f \in \mathrm{Homeo}^r(S^1)$, $I \in I(S^1)$) if the endpoints a and b of I are not critical points of the C^r-homeomorphism f, and by $M(f, I) = +\infty$ otherwise.

It is not hard to see that the Yoccoz function has the *multiplicative property*

$$M(f \circ g, I) = M(f, g(I)) \cdot M(g, I).$$

Indeed,

$$\begin{aligned} M(f \circ g, I) &= \frac{m(f \circ g(I))}{m(I)} [\mathcal{D} f \circ g(a) \mathcal{D} f \circ g(b)]^{-1/2} \\ &= \frac{m[f(g(I))]}{m[g(I)]} [\mathcal{D} f(g(a)) \mathcal{D} f(g(b))]^{-1/2} \cdot \frac{m[g(I)]}{m(I)} [\mathcal{D} f(a) \mathcal{D} f(b)] \\ &= M(f, g(I)) M(g, I). \end{aligned}$$

LEMMA 4.2. *Suppose that $f \in \mathrm{Homeo}^r(S^1)$, $r \geq 1$, satisfies the Yoccoz conditions, and let $I \subset S^1$ be a compact connected interval with endpoints $a, b \in S^1$. Then:*

1) *if I does not contain critical points, then*

$$\exp(-\tfrac{1}{2} \operatorname{var}_I \log \mathcal{D} f) \leq M(f, I) \leq \exp(\tfrac{1}{2} \operatorname{var}_I \log \mathcal{D} f);$$

2) *if I lies in the ε_i-neighborhood $\mathfrak{U}_i$ of a critical point x_i but does not contain x_i, then*

$$M(f, I) \geq 1;$$

3) *if I lies in $\mathfrak{U}_i$ and contains x_i, then*

$$M(f, I) \geq \frac{B_i}{2A_i(C_i + 1)};$$

4) *if I contains one of the intervals $(x_i - \varepsilon_i, x_i - \varepsilon_i/2)$ or $(x_i + \varepsilon_i/2, x_i + \varepsilon_i)$, then there exists a $\delta > 0$ independent of i such that*

$$M(f, I) \geq \delta / \mathcal{D},$$

where $\mathcal{D} = \max_{S^1} \mathcal{D} f$.

PROOF. 1) Let $v = \operatorname{var}_{S^1} \log \mathcal{D} f(x)$. Since $m(f(I)) = \mathcal{D} f(c) m(I)$, $c \in I$, it follows that

$$\begin{aligned} \log M(f, I) &= \log \mathcal{D} f(c) - \tfrac{1}{2} \log \mathcal{D} f(a) - \tfrac{1}{2} \log \mathcal{D} f(b) \\ &= \tfrac{1}{2} [\log \mathcal{D} f(c) - \log \mathcal{D} f(a)] + \tfrac{1}{2} [\log \mathcal{D} f(c) - \log \mathcal{D} f(b)]. \end{aligned}$$

From this, $|\log M(f, I)| \leq \frac{1}{2} v$, and hence

$$e^{-\frac{1}{2} v} \leq M(f, I) \leq e^{\frac{1}{2} v}.$$

2) Let $u(t) = \alpha t + \beta$ be the linear function such that $u(a) = (\mathcal{D} f(a))^{-\frac{1}{2}}$ and $u(b) = (\mathcal{D} f(b))^{-\frac{1}{2}}$ (Figure 5.14). Since $(\mathcal{D} f)^{-\frac{1}{2}}$ is a convex (downward) function, it follows that $u(t) \geq (\mathcal{D} f(t))^{-\frac{1}{2}}$ on I.

It is not hard to verify that

$$\int_a^b \frac{dt}{u^2(t)} = \int_a^b \frac{dt}{(\alpha t + \beta)^2} = \frac{m(I)}{u(a) u(b)}.$$

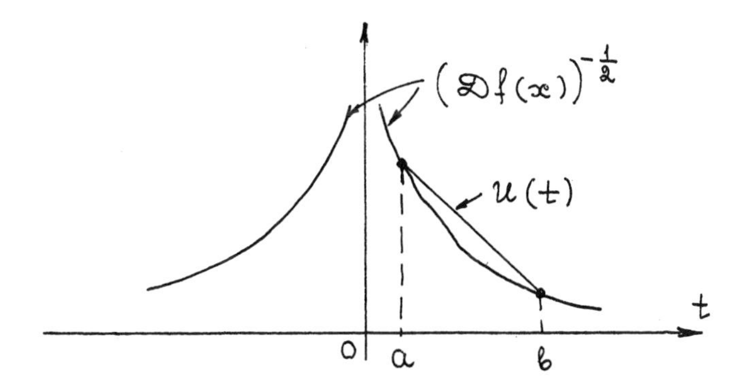

FIGURE 5.14

Then

$$\int_a^b \mathcal{D}f(t)\,dt \geq \int_a^b \frac{dt}{u^2(t)} = \frac{m(I)}{u(a)u(b)}.$$

From this,

$$M(f,I) = \frac{[\mathcal{D}f(a)\mathcal{D}f(b)]^{-\frac{1}{2}}}{m(I)} \cdot \int_a^b \mathcal{D}f(t)\,dt \geq 1.$$

3) The point x_i divides I into two intervals. Suppose that the smallest I_1 of the two has endpoints x_i and b (the proof is analogous for x_i and a). Then

$$\mathcal{D}f(a)\mathcal{D}f(b) \leq A_i^2 t_1^{2C_i},$$

where t_1 is the distance between x_i and b. Further,

$$m[f(I_1)] = \int_{x_i}^b \mathcal{D}f(x_i + t)\,dt \geq B_i \int_0^{t_1} t^{C_i}\,dt = \frac{B_i}{C_i + 1} t_1^{C_i+1},$$

and

$$\begin{aligned} M(f,I) &\geq \frac{m[f(I_1)]}{2m(I_1)} [\mathcal{D}f(a)\mathcal{D}f(b)]^{-\frac{1}{2}} \\ &\geq \frac{B_i}{C_i+1} t_1^{C_i+1} \frac{1}{2t_1} (A_i^2 t_1^{2C_i})^{-\frac{1}{2}} = \frac{B_i}{2A_i(C_i+1)}. \end{aligned}$$

4) Denote by $\mathcal{E}$ the family of intervals $[x_i - \varepsilon_i, x_i - \varepsilon_i/2]$, $[x_i + \varepsilon_i/2, x_i + \varepsilon_i]$, $i = 1, \ldots, l$. Since these intervals do not contain critical points,

$$\delta \stackrel{\text{def}}{=} \min\{m[f(\mathcal{J})] : \mathcal{J} \in \mathcal{E}\} > 0.$$

Obviously, $m(\mathcal{J}) \leq 1$ for $\mathcal{J} \in \mathcal{E}$. Therefore,

$$M(f,I) \geq \frac{\delta}{1} (\mathcal{D} \cdot \mathcal{D})^{-\frac{1}{2}} = \delta/\mathcal{D}. \quad \square$$

THEOREM 4.2. *Suppose that* $f \in \mathrm{Homeo}^r(S^1)$, $r \geq 1$, *has* $\mathrm{rot}(f) \in \mathbb{R} \setminus Q$ *and satisfies the Yoccoz conditions. Then* $\Omega(f) = S^1$.

PROOF. Assume the contrary. Then $\Omega(f)$ is a Cantor set. We take an adjacent interval $\mathcal{J} \subset S^1 \setminus \Omega(f)$ and a point $x_0 \in \mathcal{J}$.

Let q_n be the denominators of the convergents of the continued fraction expansion of $\operatorname{rot}(f)$.

Denote by $\mathcal{J}_n$ the arc between $f^{-q_n}(x_0)$ and $f^{q_n}(x_0)$ containing x_0. According to Corollary 4.1, 3), the intervals $f^{-q_n}(\mathcal{J})$, $f^{-q_{n+1}-q_n}(\mathcal{J})$, $\mathcal{J}$, $f^{-q_{n+1}}(\mathcal{J})$, and $f^{q_n}(\mathcal{J})$ are located in cyclical order on S^1. Therefore, $\mathcal{J}_n$ contains the intervals $f^{-q_{n+1}-q_n}(\mathcal{J})$, $\mathcal{J}$, and $f^{-q_{n+1}}(\mathcal{J})$.

Since $f^{q_{n+1}}[f^{-q_{n+1}-q_n}(\mathcal{J})] = f^{-q_n}(\mathcal{J})$ and $f^{q_{n+1}}[f^{-q_{n+1}}(\mathcal{J})] = \mathcal{J}$, there exist points $a, b \in S^1$ such that

$$m[f^{-q_n}(\mathcal{J})] = \mathcal{D}f(a)m[f^{-q_{n+1}-q_n}(\mathcal{J})],$$
$$m(\mathcal{J}) = \mathcal{D}f(b)m[f^{-q_{n+1}}(\mathcal{J})].$$

Denote by I the interval with endpoints a and b that contains x_0. According to the foregoing, $I \subset \mathcal{J}_n$.

Let us estimate the Yoccoz function $M(f^{q_{n+1}}, I)$. We have that

$$\begin{aligned} M(f^{q_{n+1}}, I) &= \frac{m[f^{q_{n+1}}(I)]}{m(I)}[\mathcal{D}f(a)\mathcal{D}f(b)]^{-\frac{1}{2}} \\ &= \frac{m[f^{q_{n+1}}(I)]}{m(I)}\left(\frac{m[f^{-q_{n+1}-q_n}(\mathcal{J})]}{m[f^{-q_n}(\mathcal{J})]}\cdot\frac{m[f^{-q_{n+1}}(\mathcal{J})]}{m(\mathcal{J})}\right)^{\frac{1}{2}} \\ &\le \frac{1}{[m(\mathcal{J})]^{\frac{3}{2}}}\left(\frac{m[f^{-q_{n+1}}(\mathcal{J})]}{m[f^{-q_n}(\mathcal{J})]}\cdot m[f^{-q_{n+1}-q_n}(\mathcal{J})]\right)^{\frac{1}{2}}. \end{aligned}$$

The adjacent intervals $f^{-q_n}(\mathcal{J})$, $n \in \mathbb{N}$, do not intersect, and therefore the sequence $\{m[f^{-q_n}(\mathcal{J})]\}_{n=1}^{\infty}$ converges to zero. By passing to a subsequence if necessary, we can assume that $m[f^{-q_{n+1}}(\mathcal{J})]/m[f^{-q_n}(\mathcal{J})] \le 1$. Thus, $M(f^{q_{n+1}}, I) \to 0$ as $n \to +\infty$.

We estimate the function $M(f^{q_{n+1}}, I)$ in another way. By the multiplicative property of the Yoccoz function, $M(f^{q_{n+1}}, I) = \prod_{j=0}^{q_{n+1}-1} M(f, I_j)$, where $I_j = f^j(I)$. We partition the family of intervals I_j, $j = 0, \dots, q_{n+1} - 1$, into four subfamilies:

$\mathcal{F}_1 = \{I_j$ disjoint from the $(\varepsilon_i/2)$-neighborhoods of the critical points x_i, $i = 1, \dots, l\}$,

$\mathcal{F}_2 = \{I_j$ lying in the ε_i-neighborhood $\mathcal{U}_i$ of x_i but not covering x_i, $i = 1, \dots, l\}$,

$\mathcal{F}_3 = \{I_j$ lying in $\mathcal{U}_i$ and covering $x_i\}$,

$\mathcal{F}_4 = \{I_j$ intersecting the $(\varepsilon_i/2)$-neighborhoods of the points x_i and not contained in $\mathcal{U}_i\}$

Let $\mathcal{U}$ be the complement of the $(\varepsilon_i/2)$-neighborhoods of the critical points x_i, $i = 1, \dots, l$. In view of the condition $I \subset \mathcal{J}_n$ and Corollary 4.1, 2), each point in $\mathcal{U}$ is covered by at most two intervals I_j in the family $\mathcal{F}_1$. Therefore, $\sum_{I_j \in \mathcal{F}_1} \operatorname{var}_{I_j} \log \mathcal{D}f \le 2 \operatorname{var}_{\mathcal{U}} \log \mathcal{D}f$. Lemma 4.2, 1) gives us that

$$\prod_{I_j \in \mathcal{F}_1} M(f, I_j) \ge \exp\Bigg(-\tfrac{1}{2} \sum_{I_j \in \mathcal{F}_1} \operatorname{var} \log \mathcal{D}f\Bigg) \ge \exp(-\operatorname{var}_{\mathcal{U}} \log \mathcal{D}f) \stackrel{\text{def}}{=} k_1.$$

By Lemma 4.2, 2), $\prod_{I_j \in \mathcal{F}_2} M(f, I_j) \ge 1$.

According to Corollary 4.1, 2), there are at most $2l$ and $4l$ intervals in the respective families $\mathcal{F}_3$ and $\mathcal{F}_4$ (l is the number of critical points). Therefore, by Lemma 4.2, 3), and 4),

$$\prod_{I_j \in \mathcal{F}_3} M(f, I_j) \geq \left[\frac{B}{2A(C+1)}\right]^{2l} \stackrel{\text{def}}{=} k_2,$$

where $A = \max\{A_i,\ i = 1, \ldots, l\}$, $B = \min\{B_i,\ i = 1, \ldots, l\}$, $C = \max\{C_i,\ i = 1, \ldots, l\}$, and

$$\prod_{I_j \in \mathcal{F}_4} M(f, I_j) \geq \left(\frac{\delta}{\mathcal{D}}\right)^{4l} \stackrel{\text{def}}{=} k_3.$$

We note that the constants k_1, k_2, and k_3 are strictly positive and independent of n. Then

$$M(f^{q_{n+1}}, I) \geq k_1 k_2 k_3 > 0$$

for all $n \in \mathbb{N}$, which contradicts the previously proved relation $M(f^{q_{n+1}}, I) \to 0$ as $n \to \infty$. □

4.5. Corollary to the theorem of Yoccoz for Cherry transformations. We formulate the *Yoccoz conditions for a Cherry transformation* $f \in \mathrm{Ch}(S^1)$:

1) f has finitely many points of discontinuity $z_1, \ldots, z_s \in S^1$, and is of smoothness class C^r on the set $S^1 \setminus \bigcup_{i=1}^s z_i$, $r \geq 1$;

2) f has finitely many isolated critical points $x_1, \ldots, x_l \in S^1$ and finitely many intervals of constancy $(a_1, b_1), \ldots, (a_k, b_k) \subset S^1$;

3) $\log \mathcal{D}f$ has bounded variation on any compact interval not containing critical points, intervals of constancy, nor points of discontinuity;

4) if z is a critical point or a point of discontinuity of f, then there exist constants $A = A(z) > 0$, $B = B(z) > 0$, and $C = C(z)$ and an ε-neighborhood $\mathcal{U}$ of z such that

a) $B|t|^C \leq \mathcal{D}f(z+t) \leq A|t|^C$ for $0 < |t| < \varepsilon$,

b) the function $(\mathcal{D}f)^{-\frac{1}{2}}$ is convex (downward) on each of the intervals in $\mathcal{U} \setminus \{z\}$ (Figure 5.15);

5) for each interval of constancy (a_i, b_i), $i = 1, \ldots, k$, of f there exist constants $A_i > 0$, $B_i > 0$, and $C_i > 0$ and ε_i-neighborhoods $\mathcal{U}_i(a_i)$ and $\mathcal{U}_i(b_i)$ of the respective points a_i and b_i such that

a) $B_i|t|^{C_i} \leq \mathcal{D}f(a+t) \leq A_i|t|^{C_i}$ for $-\varepsilon_i < t \leq 0$ and $B_i|t|^{C_i} \leq \mathcal{D}f(b+t) \leq A_i|t|^{C_i}$ for $0 \leq t < \varepsilon_i$,

b) the function $(\mathcal{D}f)^{-\frac{1}{2}}$ is convex (downward) on each of the intervals in $\mathcal{U}_i(a_i) \cup \mathcal{U}_i(b_i) \setminus [a_i, b_i]$.

COROLLARY 4.2. *Suppose that $f \in \mathrm{Ch}(S^1)$ has $\mathrm{rot}(f) \in \mathbb{R} \setminus Q$, satisfies the Yoccoz conditions, and does not have intervals of constancy. Then f does not have gray cells.*

The proof is completely analogous to that of Theorem 4.2 (Yoccoz), and we omit it.

4.6. The Herman index of smooth conjugacy to a rotation. In this subsection we consider the problem of C^r-conjugacy of a smooth diffeomorphism of the circle to a rotation. Significant results toward solving this problem were

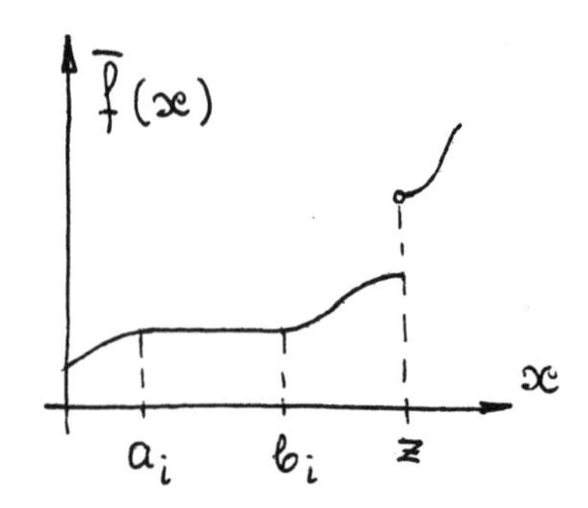

FIGURE 5.15

obtained by Herman [**89**], who proved Arnol′d's conjecture about the existence of a set $A \subset [0,1]$ of measure 1 such that if an analytic diffeomorphism of the circle has rotation number in A, then it is analytically conjugate to a rotation. Herman also introduced the index of C^r-conjugacy of a diffeomorphism to a rotation. We present Herman's result only in connection with C^1-conjugacy of an orientation-preserving C^1-diffeomorphism of the circle (with irrational rotation number) to a rotation.

DEFINITION. Two diffeomorphisms $f, g \in \mathrm{Diff}^k(S^1)$, $k \geq 1$, are said to be C^r-conjugate ($r \geq 1$) if there exists an orientation-preserving C^r-diffeomorphism $h\colon S^1 \to S^1$ such that $h \circ f = g \circ h$.

For a C^r-diffeomorphism $f \in \mathrm{Diff}^k(S^1)$ we define the *Hermann index* $H_r(f) \in \mathbb{R} \cup \{+\infty\}$ to be ($r \geq 1$)

$$H_r(f) = \sup_{n \in \mathbb{Z}} |\mathcal{D} f^n|_{C^{r-1}} = \sup_{n \in \mathbb{Z}} (|\mathcal{D} f^n|_0 + \cdots + |\mathcal{D}^r f^n|_0).$$

LEMMA 4.3. *If $f \in \mathrm{Diff}^r(S^1)$, $r \geq 1$, is C^r-conjugate to a rotation, then $H_r(f) < +\infty$.*

PROOF. By assumption, $f = h^{-1} \circ R_\alpha \circ h$, where $h\colon S^1 \to S^1$ is an orientation-preserving C^r-diffeomorphism. Then $f^n = h^{-1} \circ R_\alpha^n \circ h$, $n \in \mathbb{Z}$. Since $\mathcal{D} R_\alpha^n \equiv 1$ and $\mathcal{D}^i R_\alpha^n \equiv 0$ ($i = 2, \ldots, r$) for $n \in \mathbb{Z}$, the required follows from the formula for differentiating a composition function and the fact that $\max\{|h|_{C^r}, |h^{-1}|_{C^r}\} < +\infty$. $\square$

Herman proved that the inequality $H_r(f) < +\infty$ is also a sufficient condition for a C^r-diffeomorphism f to be C^r-conjugate to a rotation. We give a proof for the case $r = 1$. Let us first find an equivalent formulation for C^r-conjugacy of a diffeomorphism f to a rotation. If $h \circ f = R_\alpha \circ f$ for some $h \in \mathrm{Diff}^1(S^1)$, then $(\mathcal{D} f \circ f)\mathcal{D} f = (\mathcal{D} R_\alpha \circ h)\mathcal{D} h = \mathcal{D} h$. From this, $\log \mathcal{D} h \circ f + \log \mathcal{D} f = \log \mathcal{D} h$, or $\log \mathcal{D} f = \log \mathcal{D} h - \log \mathcal{D} h \circ f$.

We now proceed to the proof of a lemma due to Gottschalk and Hedlund [**84**].

LEMMA 4.4. *Let f be a minimal homeomorphism of a compact manifold $\mathcal{M}$, and let $\chi\colon \mathcal{M} \to \mathbb{R}$ be a continuous function. Then the following statements are equivalent:*

1) *there exists a continuous function $\varphi\colon \mathcal{M} \to \mathbb{R}$ such that $\chi = \varphi - \varphi \circ f$;*

2) *there exists a point* $x_0 \in \mathcal{M}$ *such that*

$$\sup_{n\in\mathbb{N}} \left| \sum_{i=0}^{n} \chi \circ f^i(x_0) \right| < +\infty.$$

PROOF. 1) $\Longrightarrow$ 2) because

$$\left|\sum_{i=0}^{n} \chi \circ f^i(x_0)\right| = \left|\sum_{i=0}^{n} (\varphi - \varphi \circ f) \circ f^i(x_0)\right| = \left|\sum_{i=0}^{n} \varphi \circ f^i(x_0) - \sum_{i=0}^{n} \varphi \circ f^{i+1}(x_0)\right|$$
$$= |\varphi(x_0) - \varphi \circ f^{n+1}(x_0)| < 2 \max_{x\in\mathcal{M}} |\varphi(x)| < +\infty.$$

We show that 2) implies 1). Let $F\colon \mathcal{M} \times \mathbb{R} \to \mathcal{M} \times \mathbb{R}$ be defined by $F(x,t) = (f(x), -\chi(x) + t)$, and $R_\lambda\colon \mathcal{M} \times \mathbb{R} \to \mathcal{M} \times \mathbb{R}$ by $R_\lambda(x,t) = (x, t+\lambda)$. Since f is a homeomorphism, so is F. It can be checked directly that

$$F \circ R_l = R_\lambda \circ F.$$

We get from the definition of F that $F^n(x_0, 0) = (f^n(x_0), -\sum_{i=0}^{n-1} \chi \circ f^i(x_0))$. According to the condition 2), the closure $\widetilde{N}$ of the set $\{F^n(x_0,0),\ n \in \mathbb{N}\}$ is a nonempty compact subset of the space $\mathcal{M} \times \mathbb{R}$. Since F is a homeomorphism, the set $\widetilde{N}$ is invariant. Therefore, $\widetilde{N}$ contains a minimal subset N with respect to F.

Since the homeomorphism f is minimal, the projection $(x,\cdot) \mapsto x$ of N on the first factor coincides with the set $\mathcal{M}$. We show that the subset $N \subset \mathcal{M} \times \mathbb{R}$ is the graph of a function; that is, for any $x \in \mathcal{M}$ there is a unique $y \in \mathbb{R}$ such that $(x,y) \in N$. Assume that $(x,y), (x, y+\lambda) \in N$. Since $R_\lambda(N) = R_\lambda \circ F(N) = F \circ R_\lambda(N)$, it follows that $R_\lambda(N)$ is an invariant compact set with respect to F. By assumption, $R_\lambda(N) \cap N \neq \emptyset$, so $R_\lambda(N) = N$ by the minimality of N. From this, $R_{n\lambda}(N) = N$ for all $n \in \mathbb{N}$. If $\lambda \neq 0$, then the equality $R_{n\lambda}(N) = N$ contradicts the compactness of N.

Thus, N is the graph of a function $\varphi\colon \mathcal{M} \to \mathbb{R}$. Since N is compact, φ is continuous. It can be verified directly that the equality $\chi = \varphi - \varphi \circ f$ holds on the set $\{f^n(x_0),\ n \in \mathbb{N}\}$. Indeed, $\varphi[f^n(x_0)] = -\sum_{i=0}^{n-1} \chi \circ f^i(x_0)$ and $\varphi \circ f[f^n(x_0)] = \varphi[f^{n+1}(x_0)] = -\sum_{i=0}^{n} \chi \circ f^i(x_0)$, and therefore $\varphi[f^n(x_0)] - \varphi \circ f[f^n(x_0)] = -\sum_{i=0}^{n-1} \chi \circ f^i(x_0) + \sum_{i=0}^{n} \chi \circ f^i(x_0) = \chi[f^n(x_0)]$. The set $\{f^n(x_0),\ n \in \mathbb{N}\}$ is dense in $\mathcal{M}$ because f is a minimal homeomorphism, so the equality $\chi = \varphi - \varphi \circ f$ holds on the whole manifold $\mathcal{M}$. □

THEOREM 4.3. *An orientation-preserving C^1-diffeomorphism of the circle with irrational rotation number is C^1-conjugate to a rotation if and only if $H_1(f) < +\infty$.*

PROOF. NECESSITY. If f is C^1-conjugate to a rotation, then $H_1(f) < +\infty$ by Lemma 4.3.

SUFFICIENCY. Let $H_1(f) < +\infty$. Then $|f^n(x) - f^n(y)| \leq |\mathcal{D}f^n|_0 |x-y| \leq H_1(f)|x-y|$ for any $x, y \in S^1$ and any positive integer n. Therefore, the family $\{f^n\}_{n=1}^{\infty}$ of mappings f^n is uniformly continuous. In view of Lemma 2.5 the homeomorphism f is conjugate to the rotation R_α with $\alpha = \operatorname{rot}(f) \in \mathbb{R}\backslash Q$. Consequently, f is minimal.

Since $f \in \operatorname{Diff}^1(S^1)$, $\chi \overset{\text{def}}{=} \log \mathcal{D}f$ is continuous. Moreover, $1/H_1(f) \leq \mathcal{D}f^n \leq H_1(f)$ for any $n \in \mathbb{N}$ because $\mathcal{D}f^{-n}[f^n(x)] = 1/\mathcal{D}f^n(x) \leq \sup_{k\in\mathbb{Z}} |\mathcal{D}f^k|_0 = H_1(f)$

for $n \in \mathbb{N}$, and hence $\sup_{n\in\mathbb{N}} |\log \mathcal{D}f^n|_0 < +\infty$. But $\log \mathcal{D}f^n = \sum_{i=0}^{n-1} \log \mathcal{D}f \circ f^i = \sum_{i=0}^{n-1} \chi \circ f^i$. According to Lemma 4.4, there exists a continuous function $\varphi \colon S^1 \to \mathbb{R}$ such that $\log \mathcal{D}f = \chi = \varphi - \varphi \circ f$. Obviously, the last equality is satisfied by any function $\varphi_1 = \varphi + c$ for any constant c. Take c such that $\int_0^1 e^{\varphi(x)+c}\,dx = e^c \int_0^1 e^{\varphi(x)}\,dx = 1$. Then the transformation $\overline{h}(x) \stackrel{\text{def}}{=} \int_0^x e^{\varphi(x)+c}\,dx$ is covering for some C^1-diffeomorphism $h\colon S^1 \to S^1$, and $\log \mathcal{D}h = \varphi(x)+c \stackrel{\text{def}}{=} \varphi_1(x)$. Substituting the function $\varphi_1 = \log \mathcal{D}h$ in place of φ in the equality $\log \mathcal{D}f = \varphi - \varphi \circ f$, we get that $\log \mathcal{D}f = \log \mathcal{D}h - \log \mathcal{D}h \circ f$, which gives us that $\mathcal{D}h = (\mathcal{D}h \circ f)\mathcal{D}f$. Consequently, the diffeomorphisms h and $h \circ f$ have the same derivative. Therefore, $R_\beta \circ h = h \circ f$ for some $\beta \in \mathbb{R}$. By Lemma 1.5, $\beta = \mathrm{rot}(f)$. □

We note that Theorem 4.3 is valid also for a diffeomorphism f with rational rotation number.

We give without proof another result of Herman.

THEOREM 4.4. *An orientation-preserving C^1-diffeomorphism f of the circle is C^1-conjugate to the rotation R_α, $\alpha = \mathrm{rot}(f)$, if and only if $|\mathcal{D}f^n - 1|_0 \to 0$ as $n\alpha \pmod 1 \to 0$.*

§5. Smooth classification of structurally stable diffeomorphisms

Recall that structurally stable diffeomorphisms of the circle are characterized by rational rotation numbers and hyperbolicity of all cycles. If the rotation number of a diffeomorphism F is equal to $\tau = p/q$, then F^q has only fixed points. Therefore, the problem of classification with respect to conjugacy reduces to classification of just such diffeomorphisms.

The number of fixed points is a unique topological invariant of orientation-preserving structurally stable diffeomorphisms. The *multipliers* (derivatives) at fixed points are obvious supplementary invariants in the smooth classification. However, it turns out that in addition to these invariants there is a further "massive" functional invariant ("modulus") of the smooth classification. Its construction uses the existence of an invariant covering on whose elements the diffeomorphism is smoothly conjugate to a standard model—a linear mapping of the line.

Here we consider classification with respect to the group of orientation-preserving diffeomorphisms.

5.1. Pasting cocycles. Let us fix a collection

$$\overline{P} = \{p_1, \dots, p_{2k}\}$$

of points on the circle. We assume that in correspondence to the orientation

$$p_1 < \cdots < p_{2k}.$$

Further, let

$$0 < \lambda_{2i-1} < 1 < \lambda_{2i} \qquad (1 \le i \le k)$$

be real numbers. Denote by $D^r(\overline{p}, \overline{\lambda})$ the set of all C^r-diffeomorphisms $F\colon S^1 \to S^1$ with fixed points $p_1, \dots, p_{2k}$ and multipliers

$$\lambda_i = F'(p_i).$$

Thus, the diffeomorphisms F are orientation-preserving, the fixed points p_{2i-1} are sinks for them, and the points p_{2i} are sources.

We remark first of all that a diffeomorphism $F \in D^r(\overline{p}, \overline{\lambda})$ is "pasted together" from linear diffeomorphisms of the line.

Namely, we set

$$U_i = (p_{i-1}, p_{i+1}), \qquad p_0 \equiv p_{2k}, \quad p_{2k+1} \equiv p_1.$$

The arc U_i is clearly invariant under F. In the case $k = 1$ (two fixed points)

$$U_1 = S^1 \setminus \{p_2\}, \quad U_2 = S^1 \setminus \{p_1\}.$$

LEMMA 5.1. *For any $\alpha < 1$ there exist orientation-preserving homeomorphisms*

$$\Phi_i \colon (U_i, p_i) \to (\mathbb{R}^1, 0), \qquad \Phi_i \in C^{0,\alpha},$$

such that

$$(\Phi_i F \Phi_i^{-1})(x) = \lambda_i x \qquad (x \in \mathbb{R}^1).$$

If $r \geq 2$, then the Φ_i can be chosen to be C^r-diffeomorphisms.

PROOF. Let Φ_i be local transformations of the corresponding class (see Theorem 1.1 in Chapter 4). Then

$$\Phi_i(Fz) = \lambda_i \Phi_i(z)$$

for points $z \in S^1$ sufficiently close to p_i. By this equality the transformation Φ_i can be extended successively to the whole arc.

The rest of our arguments are somewhat different for the case of two fixed points and the case of more than two.

First let $k = 1$. We consider the diffeomorphism

$$M_F \colon \mathbb{R}^1 \setminus \{0\} \to \mathbb{R}^1 \setminus \{0\}$$

of the punctured line defined by[1]

$$M_F(x) = \Phi_2 \Phi_1^{-1}(x).$$

We call this a *pasting cocycle for F*. This diffeomorphism belongs to the same class as Φ_i and satisfies the functional equation

$$M_F(\lambda_1 x) = \lambda_2 M_F(x) \qquad (x \neq 0).$$

Consequently,

$$M_F(x) = |x|^\theta \cdot \begin{cases} \gamma_+\left(\frac{\ln x}{\ln \lambda_1}\right), & x > 0, \\ \gamma_-\left(\frac{\ln |x|}{\ln \lambda_1}\right), & x < 0, \end{cases}$$

where the $\gamma_\pm(t)$ are 1-periodic functions of the same smoothness class as M_F, and

$$\theta = \frac{\ln \lambda_2}{\ln \lambda_1}.$$

Moreover, the functions $\gamma_\pm$ satisfy the conditions

$$\gamma_+(t) < 0 < \gamma_-(t) \qquad (t \in \mathbb{R}^1) \tag{5.1}$$

[1] That is, (U_1, Φ_1) and (U_2, Φ_2) are interpreted as a C^r-atlas on S^1 with transition function M_F.

and

$$\gamma'_{\pm}(t) + \gamma_{\pm}(t) \ln \lambda_2 \neq 0 \qquad (t \in \mathbb{R}^1). \tag{5.2}$$

Suppose now that $k \geq 2$. We define a pasting cocycle for F to be the collection $M_F = (M_i)$ of diffeomorphisms

$$M_i \colon \mathbb{R}_+ \to \mathbb{R}_- \qquad (1 \leq i \leq 2k)$$

defined by[2]

$$M_i = \Phi_{i+1}\Phi_i^{-1} \quad (1 \leq i \leq 2k), \qquad \Phi_{2k+1} \equiv \Phi_1.$$

The diffeomorphisms M_i are of the same class as Φ_i and satisfy the equations

$$M_i(\lambda_i x) = \lambda_{i+1} M_i(x).$$

Consequently,

$$M_i(x) = x^{\theta_i}\gamma_i\left(\frac{\ln x}{\ln \lambda_i}\right),$$

where

$$\theta_i = \frac{\ln \lambda_{i+1}}{\ln \lambda_i},$$

and γ_i is a negative periodic function satisfying the condition

$$\gamma'_i(t) + \gamma_i(t) \ln \lambda_{i+1} \neq 0. \quad \square \tag{5.3}$$

5.2. $C^{0,\alpha}$-conjugacy. As already mentioned, all the diffeomorphisms $F \in D^1(\overline{p}, \overline{\lambda})$ are topologically conjugate to each other. This assertion can be refined:

THEOREM 5.1. *All the diffeomorphisms $F \in D^1(\overline{p}, \overline{\lambda})$ are mutually conjugate in the class $C^{0,\alpha}$ for any $\alpha < 1$.*

PROOF. We first consider the case of two fixed points. Let $F, \widetilde{F} \in D^r(\overline{p}, \overline{\lambda})$ and let M_F and $M_{\widetilde{F}}$ be pasting cocycles belonging to the class $C^{0,\alpha}$. We set

$$G(x) = M_{\widetilde{F}}^{-1} M_F(x) \qquad (x \in \mathbb{R}^1).$$

Then $G(\lambda_1 x) = \lambda_1 G(x)$. Therefore,

$$G(x) = x \cdot \begin{cases} \delta_+\left(\frac{\ln x}{\ln \lambda_1}\right), & x > 0, \\ \delta_-\left(\frac{\ln |x|}{\ln \lambda_1}\right), & x < 0, \end{cases}$$

where the $\delta_\pm$ are periodic functions of class $C^{0,\alpha}$. Consequently, extending the definition of G by

$$G(0) = 0,$$

we get a $C^{0,\alpha}$-homeomorphism of the line. Let

$$\Psi|_{U_i} = \widetilde{\Phi}_i^{-1} G \Phi_i|_{U_i} \qquad (i = 1, 2),$$

where $\widetilde{\Phi}_i$ and Φ_i are linearizing diffeomorphisms for $\widetilde{F}$ and F, respectively. Since

$$\widetilde{\Phi}_1^{-1} G \Phi_1|_{U_1 \cap U_2} = \widetilde{\Phi}_2^{-1} G \Phi_2|_{U_1 \cap U_2}$$

[2] As in the case $k = 1$, the collection of (U_i, Φ_i) forms a C^r-atlas on S^1, and the M_i are the transition functions.

in view of the definition of G, it follows that $\Psi\colon S^1 \to S^1$ is a well-defined homeomorphism of class $C^{0,\alpha}$. Further,

$$\begin{aligned}\Psi F|_{U_1} &= \widetilde{\Phi}_1^{-1} G\Phi_1\Phi_1^{-1}(\lambda_1\Phi_1)|_{U_1} = \widetilde{\Phi}_1^{-1}(\lambda_1 G\Phi_1)|_{U_1} \\ &= \widetilde{\Phi}_1^{-1}(\lambda_1\widetilde{\Phi}_1\widetilde{\Phi}_1^{-1}G\Phi_1)|_{U_1} = \widetilde{F}\Psi|_{U_1}.\end{aligned}$$

Consequently, Ψ conjugates F and $\widetilde{F}$.

Suppose now that there are at least four fixed points, and let

$$M_F = \{M_i\}, \qquad M_{\widetilde{F}} = \{\widetilde{M}_i\}$$

be pasting cocycles for F and $\widetilde{F}$, respectively.

Consider the homeomorphism

$$H_2(x) = \widetilde{M}_1 \circ M_1^{-1}(x) \qquad (x \in \mathbb{R}_-).$$

Then H_2 maps $\mathbb{R}_-$ onto itself, and

$$H_2(\lambda_2 x) = \lambda_2 H_2(x). \tag{5.4}$$

Consequently,

$$H_2(x) = x\tau\left(\frac{\ln|x|}{\ln\lambda_2}\right) \qquad (x \in \mathbb{R}^-),$$

where τ is a 1-periodic function of class $C^{0,\alpha}$. Setting

$$H_2(x) = x\tau\left(\frac{\ln x}{\ln\lambda_2}\right) \quad (x \in \mathbb{R}^+), \qquad H(0) = 0,$$

we get a $C^{0,\alpha}$-homeomorphism of the axis that commutes with a linear homeomorphism (see (5.4)). If we now set

$$H_3(x) = \widetilde{M}_2 \circ H_2 \circ M_2^{-1}(x),$$

we get a $C^{0,\alpha}$-homeomorphism of the semi-axis $\mathbb{R}_-$ which, when extended in the same way, gives a homeomorphism of the axis commuting with multiplication by λ_3.

Continuing this process, we get a sequence of $C^{0,\alpha}$-homeomorphisms

$$H_i\colon \mathbb{R}^1 \to \mathbb{R}^1, \qquad H_i(\lambda_i x) = \lambda_i H_i(x),$$

such that

$$H_{i+1}M_i|_{\mathbb{R}_+} = \widetilde{M}_i H_i|_{\mathbb{R}_+}.$$

We now set

$$\Psi|_{U_i} = \widetilde{\Phi}_i^{-1} H_i \Phi_i \qquad (1 \le i \le 2k),$$

where $\widetilde{\Phi}_i$ and Φ_i are linearizing homeomorphisms for $\widetilde{F}$ and F, respectively. By the construction of the homeomorphisms H_i, Ψ is a well-defined $C^{0,\alpha}$-homeomorphism of the circle. Further, we have that

$$\begin{aligned}\Psi F|_{U_i} &= \widetilde{\Phi}_i^{-1} H_i\Phi_i\Phi_i^{-1}(\lambda_i\Phi_i)|_{U_i} = \widetilde{\Phi}_i^{-1}H_i(\lambda_i\Phi_i)|_{U_i} \\ &= \widetilde{\Phi}_i^{-1}(\lambda_i H_i\Phi_i)|_{U_i} = \widetilde{\Phi}_i^{-1}(\lambda_i\widetilde{\Phi}_i\widetilde{\Phi}_i^{-1}H_i\Phi_i)|_{U_i} = \widetilde{F}\Psi|_{U_i}.\end{aligned}$$

Consequently, Ψ conjugates F and $\widetilde{F}$, and the theorem is proved. $\square$

5.3. Smooth classification. Here we construct an invariant smooth classification for C^r-diffeomorphisms. As before, we separate the case of two fixed points from the case of more than two.

First assume that $k = 1$. We introduce the following equivalence relations in the set $\{\gamma_\pm\}$ of pairs of periodic C^r- functions satisfying (5.1) and (5.2):

$$\{\gamma_\pm\} \sim \{\widetilde{\gamma}_\pm\}$$

if

$$\widetilde{\gamma}_\pm(t) = a\gamma_\pm(t + b)$$

for some $a > 0$ and $b \in \mathbb{R}^1$. The set of equivalence classes is denoted by $L^r(\lambda_1, \lambda_2)$.

Similarly, for $k \geq 2$ and for collections $\{\gamma_i\}$ of periodic functions of class C^r satisfying (5.3) we introduce the equivalence relation

$$\{\widetilde{\gamma}_i\} \sim \{\gamma_i\} \Longleftrightarrow \widetilde{\gamma}_i(t) = \lambda_{i+1}^{\beta_{i+1}+\beta_i}\gamma_i(t - \beta_i)$$

for some $\beta_i \in \mathbb{R}$ with $\beta_{2k+1} \equiv \beta_1$. As above, the set of equivalence classes is denoted by $L^r(\overline{\lambda})$. The equivalence of the collections $\{\gamma_\pm\}$, $\{\widetilde{\gamma}_\pm\}$ and $\{\gamma_i\}$, $\{\widetilde{\gamma}_i\}$ means that the cocycles M and $\widetilde{M}$ constructed from them are connected by the relations

$$\widetilde{M}_i(\alpha_i x) = \alpha_{i+1} M_i(x), \qquad \alpha_i > 0$$

(let $\alpha_i = \lambda_i^{\beta_i}$ for $k \geq 2$).

For a diffeomorphism $F \in D^r(\overline{p}, \overline{\lambda})$ we denote by $I(F)$ the equivalence class of the collection $\{\gamma_i\}$ (of the pair $\{\gamma_\pm\}$) determined by the pasting cocycle M_F.

THEOREM 5.2. *Let $r \geq 2$. Then:*

a) *for conjugacy of diffeomorphisms $F, \widetilde{F} \in D^r(\overline{p}, \overline{\lambda})$ in the class C^1 it is necessary that*

$$I(F) = I(\widetilde{F}),$$

and, conversely, if this equality holds, then F and $\widetilde{F}$ are conjugate in the class C^r;

b) *for each element $I \in L^r(\overline{\lambda})$ there is a diffeomorphism $F \in D^r(\overline{p}, \overline{\lambda})$ such that $I(F) = I$.*

Thus, a) shows that $I(F)$ is a complete invariant of the smooth classification. The asssertion b) is usually formulated as "a realization of the invariant".

Denote by $\widetilde{D}^r(\overline{p}, \overline{\lambda})$ the set of classes of C^1-conjugacy of diffeomorphisms in $D^r(\overline{p}, \overline{\lambda})$. Then Theorem 5.2 asserts that the mapping

$$I \colon \widetilde{D}^r(\overline{p}, \overline{\lambda}) \to L^r(\overline{\lambda})$$

is bijective. In this sense the set $L^r(\overline{\lambda})$ is a "moduli space" (of invariants) of the smooth classification.

PROOF OF THEOREM 5.2. Suppose that the diffeomorphisms F and $\widetilde{F}$ are conjugate, and

$$\Psi \colon S^1 \to S^1$$

is a conjugating C^1-diffeomorphism. It can be assumed that the arcs U_i are invariant under Ψ. Then we get from the equality

$$\Psi F|_{U_i} = \widetilde{F}\Psi|_{U_i}$$

by using the linearizations of F and $\widetilde{F}$ on the arcs U_i that

$$\Psi\Phi_i^{-1}\lambda_i\Phi_i = \widetilde{\Phi}_i^{-1}\lambda_i\widetilde{\Phi}_i\Psi|_{U_i}.$$

We set

$$H_i = \widetilde{\Phi}_i\Psi\Phi_i^{-1}.$$

Then H_i is a C^1-diffeomorphism of the line commuting with multiplication by λ_i. In view of the smoothness of this diffeomorphism, it is linear:

$$H_i(x) = \alpha_i x \qquad (x \in \mathbb{R}^1),$$

where $\alpha_i \in \mathbb{R}_+$, because it preserves orientation. We now get from the equalities

$$\Psi|_{U_i} = \widetilde{\Phi}_i^{-1}H_i\Phi_i|_{U_i}$$

that $\widetilde{\Phi}_i^{-1}H_i\Phi_i|_{U_i\cap U_{i+1}} = \widetilde{\Phi}_{i+1}^{-1}H_{i+1}\Phi_{i+1}|_{U_i\cap U_{i+1}}$. Consequently,

$$\widetilde{M_i}(\alpha_i x) = \alpha_{i+1}M_i(x). \tag{5.5}$$

This gives us the direct assertion in a).

Conversely, suppose that (5.5) holds. Let

$$\Psi|_{U_i} = \widetilde{\Phi}_i^{-1}H_i\Phi_i, \quad H_i(x) = \alpha_i x \qquad (i = 1, \dots, k).$$

Then $\Psi\colon S^1 \to S^1$ is a well-defined C^r-diffeomorphism conjugating F and $\widetilde{F}$.

We prove the assertion b). Let $\{\gamma_\pm\}$ (or $\{\gamma_i\}$) be a representative of the given class $I \in L^r(\overline{\lambda})$. From this representative we construct a cocycle M:

$$M(x) = |x|^\theta \begin{cases} \gamma_+\left(\frac{\ln x}{\ln \lambda}\right), & x > 0, \\ \gamma_-\left(\frac{\ln |x|}{\ln \lambda}\right), & x < 0 \end{cases}$$

for $k = 1$, and

$$M_i(x) = x^{\theta_i}\gamma_i\left(\frac{\ln x}{\ln \lambda_i}\right), \qquad x > 0$$

for $k \geq 2$. We "factor" the cocycle M_i, setting

$$M_i(x) = \Phi_{i+1}\Phi_i^{-1}(x),$$

where $\Phi_i\colon U_i \to \mathbb{R}^1$ is a C^r-diffeomorphism. Now let

$$F|_{U_i} = \Phi_i^{-1}(\lambda_i\Phi_i) \qquad (1 \leq i \leq k).$$

Then F is a well-defined diffeomorphism of the circle in $D^r(\overline{p}, \overline{\lambda})$. Obviously, $M_F = I$. The theorem is proved. □

5.4. Corollaries. Theorem 5.2 implies the existence for any $r \in [2, \infty]$ of diffeomorphisms in $D^r(\overline{p}, \overline{\lambda})$ that are not conjugate in the class C^1. What is more, we have the following because $L^r(\overline{\lambda})$ has the cardinality of a continuum.

COROLLARY 5.1. *The set $\widetilde{D}^r(\overline{p}, \overline{\lambda})$ of C^1-conjugacy classes has the cardinality of a continuum for any r.*

COROLLARY 5.2. *If diffeomorphisms* $F, \widetilde{F} \in C^r(\overline{p}, \overline{\lambda})$ *are conjugate in the class* C^1, *then they are also conjugate in the class* C^r.

In other words, there are no "intermediate" conjugacy invariants between C^1 and C^r.

Since the equivalence of collections $\{\gamma_i\}$ preserves smoothness class, we get a criterion for "smoothability" of a diffeomorphism F.

COROLLARY 5.3. *A diffeomorphism* $F \in D^r(\overline{p}, \overline{\lambda})$ *is smoothly* (C^1-) *conjugate to a diffeomorphism of class* C^q *if and only if the pasting cocycle* M_F (*that is, the functions* $\gamma_\pm$ *or* γ_i) *belongs to this class.*

COROLLARY 5.4. *The set* $D^r(\overline{p}, \overline{\lambda})$ *contains diffeomorphisms not conjugate in the class* C^1 *to any* C^{r+1}*-diffeomorphism.*

5.5. Conjugacy of flows. We recall that structurally stable flows on the circle are flows either without any equilibrium states at all, or with hyperbolic rest points.

As in the local situation the orbital classification of one-dimensional flows is obvious. In the problem of conjugacy of flows, numerical invariants arise instead of functional invariants.

Let F^t be a flow without equilibrium states. We denote by $\varkappa(F^t)$ the "time of a complete circuit" of the circle by the trajectories $F^t(z)$; in other words, the smallest number $t > 0$ such that

$$F^t(z) = z, \quad z \in S^1.$$

This is well defined: $\varkappa(F^t)$ does not depend on z and can be computed by the formula

$$\varkappa(F^t) = \int_{S^1} \frac{dz}{a(z)}, \tag{5.6}$$

where

$$a(z) = \left.\frac{dF^t(z)}{dt}\right|_{t=0}.$$

The number $\varkappa(F^t)$ is the only invariant of topological and smooth conjugacy of flows without equilibrium states. Namely, if C^r- flows F^t and $\widetilde{F}^t$ are topologically conjugate, then

$$\varkappa(F^t) = \varkappa(\widetilde{F}^t).$$

Conversely, if this equality holds, then the flows F^t and $\widetilde{F}^t$ are conjugate in the class C^r. A conjugating diffeomorphism can be given by the formula

$$\Phi(z) = \widetilde{F}^{t(z)} F^{-t(z)}(z).$$

Here $t(z)$ is the smallest $t > 0$ such that

$$F^t(z_0) = z,$$

and the "initial" point z_0 of the construction is an arbitrary but fixed point.

Suppose now that F^t is a flow of class C^r with fixed points $p_1, \dots, p_{2k}$ and multipliers $\nu_1, \dots, \nu_{2k}$. As in the case of diffeomorphisms, there are linearizing transformations

$$\Phi_i\colon (U_i, p_i) \to (\mathbb{R}^1, 0)$$

such that

$$\Phi_i F^t \Phi_i^{-1}(x) = e^{t\nu_i} x \qquad (x \in \mathbb{R}^1).$$

These transformations belong to the class $C^{0,\alpha}$ for $r = 1$, and to the class C^r for $r \geq 2$.

Setting

$$M_i(x) = \Phi_{i+1}\Phi_i^{-1}(x)$$

($M(x) = \Phi_2\Phi_1^{-1}(x)$ in the case $k = 1$), we get the diffeomorphisms

$$M_i\colon \mathbb{R}_+ \to \mathbb{R}_-$$

(diffeomorphisms of the punctured line in the case $k = 1$). Since

$$M_i(e^{\nu_i t}x) = e^{\nu_{i+1}t} M_i(x) \qquad (t \in \mathbb{R}^1),$$

it follows that

$$M_i(x) = x^{\theta_i} \cdot c_i \qquad (x \in \mathbb{R}_+,\ c_i < 0),$$

where $\theta_i = \nu_{i+1}/\nu_i$. In the case $k = 1$

$$M(x) = |x|^\theta \cdot \begin{cases} c_+, & c > 0, \\ c_-, & c < 0, \end{cases}$$

where $c_+ < 0 < c_-$.

Thus, the pasting cocycle of a flow has a very special form. For this reason there is a simplification of the invariant. Namely, we set

$$\varkappa(F^t) = -c_+/c_-$$

in the case $k = 1$, and

$$\varkappa(F^t) = \prod_{i=1}^{2k} (-c_i)^{\nu_{i+1}} \tag{5.7}$$

for $k \geq 2$.

THEOREM 5.3. *Two structurally stable C^1-flows with the same equilibrium states and the same multipliers are conjugate in the class $C^{0,\alpha}$ for any $\alpha < 1$.*

Taking two C^r-flows F^t and $\widetilde{F}^t$ with the same rest points $p_1, \dots, p_{2k}$ and the same multipliers $\nu_1, \dots, \nu_{2k}$, we get the next result.

THEOREM 5.4. *Let $r \geq 2$. Then:*

a) *for the flows F^t and $\widetilde{F}^t$ to be conjugate in the class C^1 it is necessary that*

$$\varkappa(F^t) = \varkappa(\widetilde{F}^t),$$

and if this equality holds, then F^t and $\widetilde{F}^t$ are conjugate in the class C^r;

b) *for each $\varkappa \in \mathbb{R}_+$ there is a C^r-flow F^t ($r \geq 2$) such that $\varkappa(F^t) = \varkappa$.*

Theorems 5.3 and 5.4 can be proved by the same scheme as for diffeomorphisms.

5.6. Inclusion of a diffeomorphism in a flow. We recall that a diffeomorphism F is said to be *includable in a flow* F^t of class C^r if $F = F^{t_0}$ for some t_0.

The property of being includable in a flow of class C^r is an invariant with respect to conjugacy. A diffeomorphism F with irrational rotation number is includable in a C^r-flow if and only if it is conjugate in C^r to a rotation. Indeed, if Φ is a conjugating diffeomorphism, then

$$F^t = \Phi\tau^t\Phi^{-1}$$

is a flow containing F. Here τ^t is a flow of linear diffeomorphisms (rotations). Conversely, each C^r-flow without equilibrium states is conjugate in C^r to a flow of rotations (see §5.5).

Further, since all diffeomorphisms $F \in D^r(\overline{p}, \overline{\lambda})$ are conjugate in the class $C^{0,\alpha}$, they can be included in a flow of this class. It suffices to observe that $D^r(\overline{p}, \overline{\lambda})$ contains at least one diffeomorphism includable in a C^1-flow of diffeomorphisms.

However, if F is includable in a flow of class C^1, then its pasting cocycle is such that the functions γ_i ($\gamma_\pm$ for $k = 1$) are constants. This is a necessary and sufficient condition for includability in a C^1-flow. If it holds, then F is includable in a C^r-flow.

Includability in a smooth flow thus turns out to be a fairly rare property (see [**37**]).

5.7. Comments. 1. The problem of a C^1-classification of C^1-diffeomorphisms in $D^1(\overline{p}, \overline{\lambda})$ has been left out of our considerations, as has classification of analytic structurally stable diffeomorphisms.

The main obstacle to the construction of a C^1-classification of diffeomorphisms in $D^1(\overline{p}, \overline{\lambda})$ is the absence of "local models". It is unclear how to describe the "normal forms" of such diffeomorphisms with respect to local C^1-conjugacy. For a particular multiplier there is probably a continuum of C^1-diffeomorphisms that are pairwise not smoothly conjugate in a neighborhood of a fixed point.

As for analytic conjugacy, the scheme we have presented is perfectly suitable also here when the local (one-dimensional) analytic theorem of Poincaré is taken into account (see §8 of Chapter 4).

2. It is no accident that the same notation $\varkappa(F^t)$ appears in (5.6) (for a field without singularities) and in (5.7): it is shown in [**38**] that the invariant $\varkappa(F^t)$ is equal to the integral in (5.6), understood in the principal value sense. This makes it possible to produce sufficiently simple normal forms of flows in $D^r(\overline{p}, \overline{\lambda})$, $r \geq 2$. Some corresponding fields are given by trigonometric polynomials.

3. In contrast to the case of flows, a serious deficiency of the description given here of the moduli space $L^r(\overline{\lambda})$ is its nonconstructivity: the problem of "computing" the invariant $I(F)$ from a given diffeomorphism $F \in D^r(\overline{p}, \overline{\lambda})$ and thereby making it possible, in particular, to explicitly determine diffeomorphisms that are not smoothly conjugate. One roundabout possibility is to specify a diffeomorphism by its cocycle.

CHAPTER 6

Classification of Flows on Surfaces

This chapter is devoted to the questions of topological equivalence and classification of flows on closed orientable surfaces. We confine ourselves mainly to flows with finitely many singular trajectories. It was shown in the third chapter that a flow (with finitely many equilibrium states and separatrices) on a closed orientable surface can be decomposed into flows without nontrivial recurrent semitrajectories, and irreducible flows (that is, flows with a single quasiminimal set: the closure of a nontrivial recurrent semitrajectory). Therefore, it is natural to investigate flows without nontrivial recurrent semitrajectories and irreducible flows separately.

Irreducible flows exist only on surfaces of genus at least 1 (among closed orientable surfaces). However, the study of irreducible flows on the torus (genus 1) is essentially different from that on surfaces of genus greater than 1, due to the asymptotic behavior of trajectories of covering flows on the universal coverings.

In the first section we consider irreducible flows on the torus. The second, third, and fourth sections are devoted to irreducible flows on closed orientable surfaces of genus greater than 1. Flows without nontrivial recurrent semitrajectories are treated in the fifth section.

§1. Topological classification of irreducible flows on the torus

Any flow with a nontrivial recurrent semitrajectory on the torus is irreducible (Lemma 3.1 in Chapter 3). The asymptotic behavior of a nontrivial recurrent semitrajectory is described by the Poincaré rotation number: a topological invariant that is the same for all nontrivial recurrent positive (respectively, negative) semitrajectories of a fixed irreducible flow.

In the set of irreducible flows on the torus we single out two classes—the class of minimal flows and the class of Denjoy flows—for which the problem of topological classification has been solved.

1.1. Preliminary facts. Let $\pi\colon \mathbb{R}^2 \to T^2$ be the universal covering of the torus T^2 by the Euclidean plane $\mathbb{R}^2$ (with Cartesian coordinates x, y). The integer translations of the Euclidean plane form the group Γ of covering transformations of π, which is isomorphic to $\mathbb{Z}^2$ and to the fundamental group $\pi_1(T^2)$ of the torus.

DEFINITION. Two sets $A, B \subset \mathbb{R}^2$ (possibly single points) are said to be congruent (with respect to the group Γ) if $\gamma(A) = B$ for some $\gamma \in \Gamma$.

Together with the universal covering $\mathbb{R}^2$ we use the unit disk $\mathcal{D}^2 = \{(x,y) : x^2+y^2<1\}$ with coordinates (ξ,η) that are connected with the coordinates (x,y) by the relations

$$\xi = \frac{x}{\sqrt{1+x^2+y^2}}, \qquad \eta = \frac{y}{\sqrt{1+x^2+y^2}}, \tag{1.1}$$

$$x = \frac{\xi}{\sqrt{1-\xi^2-\eta^2}}, \qquad y = \frac{\eta}{\sqrt{1-\xi^2-\eta^2}}. \tag{1.2}$$

The homeomorphism $\mathcal{D}^2 \to \mathbb{R}^2$ given by the formulas (1.2) is denoted by τ.

The unit circle $S_\infty = \partial\mathcal{D}^2 = \{(x,y) : x^2+y^2=1\}$ is called the *absolute*.

The following result is a consequence of the properties of a covering.

LEMMA 1.1. *Let l be a simple closed curve nonhomotopic to zero on T^2. Then the inverse image $\pi^{-1}(l)$ of this curve breaks up into a countable set $\{\bar{l}_i\}_1^\infty$ of congruent (with respect to Γ) disjoint nonclosed curves such that:*

1) *for each curve $\bar{l}_i \in \pi^{-1}(l)$ there exists an infinite cyclic subgroup $G(\bar{l}_i) \subset \Gamma$ of elements carrying $\bar{l}_i$ into itself;*

2) *an element of Γ carrying some point on $\bar{l}_i \in \pi^{-1}(l)$ into a point on the same curve leaves invariant the whole curve $\bar{l}_i$ and belongs to $G(\bar{l}_i)$;*

3) *$G(\bar{l}_i) = G(\bar{l}_j)$ for any $\bar{l}_i, \bar{l}_j \in \pi^{-1}(l)$;*

4) *there exists a line $\bar{l}$ with rational slope that is invariant under any transformation in $G(\bar{l}_i)$ and such that the deviation of any curve $\bar{l}_i \in \pi^{-1}(l)$ from $\bar{l}$ is bounded;*

5) *each curve $\tau^{-1}(\bar{l}_i)$, $\bar{l}_i \in \pi^{-1}(l)$, has two boundary points, which lie on the absolute S_∞ and are diametrically opposite;*

6) *all the curves $\{\tau^{-1}(\bar{l}_i)\}_1^\infty$ have common boundary points coinciding with the boundary points of the curve $\tau^{-1}(\bar{l})$ (Figure 6.1).*

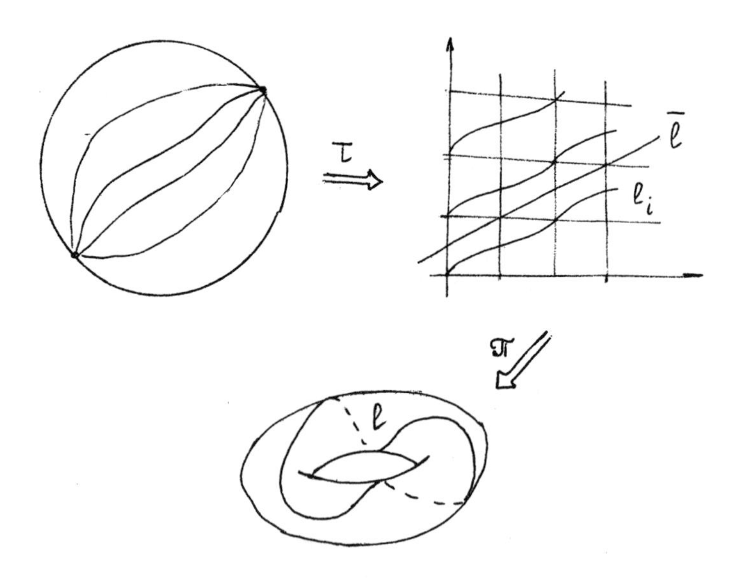

FIGURE 6.1

PROOF. We take a point $m_0 \in l$ and one of its inverse images $\overline{m}_0 \in \mathbb{R}^2$, $\pi(\overline{m}_0) = m_0$. According to the covering homotopy theorem [**43**], there exists a curve $\bar{l}_{01}$ with endpoints $\overline{m}_0, \overline{m}_1 \in \pi^{-1}(m_0)$ such that $\pi(\bar{l}_{01}) = l$ and no proper part of the curve $\bar{l}_{01} \setminus \{\overline{m}_1\}$ is mapped by π onto l.

Since l is a closed curve nonhomotopic to zero, it follows that $\overline{m}_0 \neq \overline{m}_1$ and there exists an element $\gamma \in \Gamma$ with $\gamma \neq \text{id}$ such that $\gamma(\overline{m}_0) = \overline{m}_1$. Since no proper part of the arc $\bar{l}_{01} \setminus \{\overline{m}_1\}$ is mapped by π onto the whole curve l, and l does not have self-intersections, γ is not a power of any other element of Γ.

Denote by $\bar{l}$ the line passing through $\overline{m}_0$ and $\overline{m}_1$. Since $\bar{l}_{01}$ is compact, the deviation of it from the line $\bar{l}$ is bounded. Then the deviation of the curve $\bar{l}_0 = \cup\gamma^n(\bar{l}_{01})$ $(n \in \mathbb{Z})$ from $\bar{l}$ is also bounded (Figure 6.2).

It follows from the definitions of the covering π and the group Γ that $\pi(\bar{l}_0) = l$. Since l is a simple curve, so is $\bar{l}_0$.

Denote by $G(\bar{l}_0) \stackrel{\text{def}}{=} G$ the cyclic subgroup generated by the element γ. It follows from $\gamma(\overline{m}_0) = \overline{m}_1$ that $\bar{l}_0$ and $\bar{l}$ are invariant under the action of G. Since γ is not a power of some element of Γ, any element of Γ carrying some point on $\bar{l}_0$ into a point on the same curve $\bar{l}_0$ leaves the whole of $\bar{l}_0$ invariant and belongs to G.

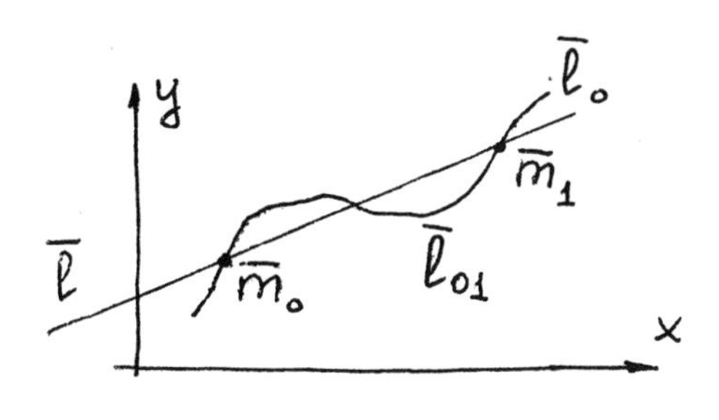

FIGURE 6.2

The facts that the deviation of $\bar{l}_0$ from $\bar{l}$ is bounded and l is simple imply that for any $\widetilde{\gamma} \in \Gamma \setminus G$ the curves $\widetilde{\gamma}(\bar{l}_0)$ and $\bar{l}_0$ are different and disjoint. This and the countability of the factor group Γ/G give us that the complete inverse image $\pi^{-1}(l) = \{\widetilde{\gamma}(\bar{l}_0) : \widetilde{\gamma} \in \Gamma\}$ breaks up into a countable set of congruent (with respect to Γ) disjoint nonclosed curves satisfying the conditions 1)–3) because Γ is commutative.

The rationality of the slope of $\bar{l}$ follows from the equality $\gamma(\overline{m}_0) = \overline{m}_1$ $(\overline{m}_0, \overline{m}_1 \in \bar{l})$, and the fact that γ is a translation by an integer vector.

The properties 5) and 6) follow from (1.1), (1.2), and the property 4).

REMARK. By the covering homotopy theorem, if $l \subset T^2$ is a simple closed curve homotopic to zero, then the complete inverse image $\pi^{-1}(l)$ breaks up into a countable set of congruent (with respect to Γ) disjoint closed (hence homotopic to zero on $\mathbb{R}^2$) curves.

We recall that a transformation $\overline{h}\colon \mathbb{R}^2 \to \mathbb{R}^2$ is called a covering (or lifting) for a transformation $h\colon T^2 \to T^2$ if $\pi \circ \overline{h} = h \circ \pi$.

LEMMA 1.2. *Suppose that $h\colon T^2 \to T^2$ is a continuous mapping. Then:*

1) *there exists a continuous covering $\overline{h}\colon \mathbb{R}^2 \to \mathbb{R}^2$ for h, and if h is a homeomorphism, then so is $\overline{h}$;*

2) *if $\overline{h}$ is a covering for h, then for any element $\gamma \in \Gamma$ the transformation $\gamma \circ \overline{h}$ is also a covering for h;*

3) *if $\overline{h}_1$ and $\overline{h}_2$ are two continuous liftings of h, then there exists a $\gamma \in \Gamma$ such that $\overline{h}_2 = \gamma \circ \overline{h}_1$.*

PROOF. Fix a point $m_0 \in T^2$ and an inverse image $\overline{m}_0 \in \mathbb{R}^2$ of it, $\pi(\overline{m}_0) = m_0$. Also, take a point $\widehat{m}_0$ in $\pi^{-1}[h(m_0)]$. We now define the mapping $\overline{h}$ at an arbitrary point $\overline{m} \in \mathbb{R}^2$. To do this we join $\overline{m}_0$ and $\overline{m}$ by an arc d (Figure 6.3). Since π is a covering, there exists a unique lifting $\widehat{d}$ of the arc $h[\pi(d)]$ with endpoint $\widehat{m}_0$ [**43**]. Denote by $\widehat{m}$ the second endpoint of the arc $\widehat{d}$ and let $\overline{h}(\overline{m}) = \widehat{m}$.

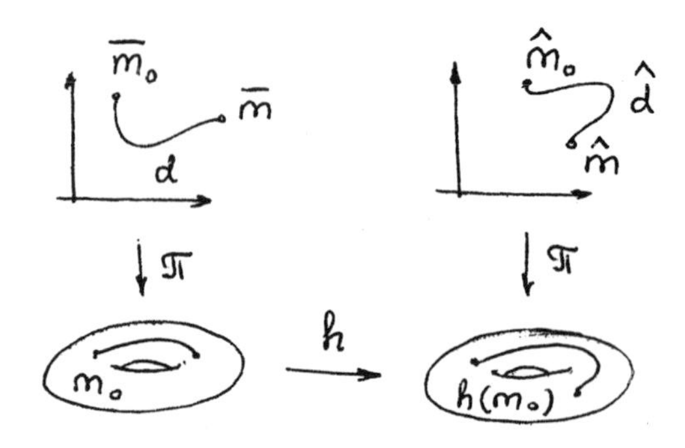

FIGURE 6.3

A closed loop on $\mathbb{R}^2$ is mapped under the action of π into a curve on T^2 homotopic to zero. This and the remark after Lemma 1.1 imply that $\overline{h}$ is well defined, that is, that the point $\widehat{m}$ is independent of the form of the arc d.

It follows immediately from the construction and the continuity of h that $\overline{h}$ is continuous.

Any arc on $\mathbb{R}^2$ with endpoints $\overline{m}$ and $\gamma(\overline{m})$, where $\gamma \in \Gamma$, projects into a closed curve on T^2. Therefore, by Lemma 1.1 and the regularity of the covering π, the transformation $\overline{h}$ is a covering for h (regularity of the covering means that all the liftings of any loop are either closed or open). From this and the fact that π is a local homeomorphism it follows that if h is a homeomorphism, then so is $\overline{h}$.

The assertion 2) is an immediate consequence of the definitions of the covering π and the group Γ.

Since Γ is a discontinuous group, any two liftings of h coinciding at some point coincide on the whole plane. This implies 3). $\square$

Any homomorphism of the two-dimensional integer lattice $\mathbb{Z}^2 = \mathbb{Z} \times \mathbb{Z}$ is given by a 2×2 integer matrix $\begin{pmatrix} a & b \\ c & d \end{pmatrix}$, $a, b, c, d \in \mathbb{Z}$, and the homomorphism has the form $(m, n) \mapsto (am + bn, cm + dn)$, $(m, n) \in \mathbb{Z}^2$. Conversely, any 2×2 integer matrix determines a homomorphism of the group $\mathbb{Z}$ (of the indicated form).

The group Γ of integer translations of the plane $\mathbb{R}^2$ is naturally isomorphic to the group $\mathbb{Z}^2$. Therefore, each homomorphism of Γ is given by a 2×2 integer matrix.

We remark that the fundamental group $\pi_1(T^2)$ of the torus can be identified with $\mathbb{Z}^2 \cong \Gamma$ for a fixed basis in $\pi_1(T^2)$. As a basis in $\pi_1(T^2)$ we take elements containing the zero parallel $\pi(y = 0, 0 \le x \le 1)$ and the zero meridian $\pi(0 \le y \le 1, x = 0)$. In the identification $\mathbb{Z}^2 \cong \pi_1(T^2)$ we identify the zero parallel and the zero meridian with $(1, 0), (0, 1) \in \mathbb{Z}^2$, respectively.

It is well known that a continuous mapping of any manifold into itself induces a homomorphism of its fundamental group. Consequently, any continuous mapping

$h\colon T^2 \to T^2$ induces a homomorphism $h_*\colon \pi_1(T^2) \to \pi_1(T^2)$, which is given by a integer 2×2 matrix.

LEMMA 1.3. *Let $h\colon T^2 \to T^2$ be a continuous transformation, and $\overline{h}\colon \mathbb{R}^2 \to \mathbb{R}^2$ a covering mapping for h. Then:*

1) *there exists a homomorphism $\overline{h}_*\colon \Gamma \to \Gamma$ such that $\overline{h}_*(\gamma) \circ \overline{h} = \overline{h} \circ \gamma$ for all $\gamma \in \Gamma$;*

2) *$\overline{h}_* = h_*$, where h_* is the homomorphism of the fundamental group $\pi_1(T^2) \cong \Gamma$ induced by h (consequently, $(\overline{h}_1)_* = (\overline{h}_2)_*$ for any covering transformations $\overline{h}_1$ and $\overline{h}_2$ of h);*

3) *if $\overline{h}_*$ is written as a 2×2 integer matrix $\begin{pmatrix} a & b \\ c & d \end{pmatrix}$, then the covering mapping $\overline{h} = (\overline{h}_1, \overline{h}_2)$ has the form*

$$\overline{h}_1(x, y) = ax + by + \psi_1(x, y),$$
$$\overline{h}_2(x, y) = cx + dy + \psi_2(x, y),$$

where $\psi_1(x, y)$ and $\psi_2(x, y)$ are continuous periodic functions with period 1 in each argument.

PROOF. Since a covering transformation carries congruent points into congruent points, for any $m \in \mathbb{R}^2$ and any element $\gamma \in \Gamma$ there is an element $\gamma' \in \Gamma$ such that $\gamma' \circ \overline{h}(m) = \overline{h} \circ \gamma(m)$. It follows from the continuity of $\overline{h}$ and the discontinuity of the group Γ that the last equality is valid for all points $m \in \mathbb{R}^2$, and hence γ' does not depend on $m \in \mathbb{R}^2$. We leave it to the reader as an exercise to prove that the correspondence $\gamma \mapsto \gamma' = \overline{h}_*(\gamma)$ is a homomorphism.

The assertion 2) follows from the equality $\overline{h}_*(\gamma) \circ \overline{h} = \overline{h} \circ \gamma$ and the commutativity of Γ.

We prove 3). The square $0 \le x \le 1$, $0 \le y \le 1$ is a fundamental domain F for the transformation group Γ. For any point $(x, y) \in F$ we set $\psi_1(x, y) = \overline{h}_1(x, y) - ax - by$, $\psi_2(x, y) = \overline{h}_2(x, y) - cx - dy$.

Denote by $\gamma_{1,0} \in \Gamma$ an element carrying a point (x, y) into $(x + 1, y)$. By assumption, $\overline{h}_*$ maps $\gamma_{1,0}$ into the element $\gamma_{a,b}\colon (x, y) \mapsto (x + a, y + b)$. Then the equality $\gamma_{a,b} \circ \overline{h}(0, y) = \overline{h} \circ \gamma_{1,0}(0, y)$ takes the form $\overline{h}_1(1, y) = a + \overline{h}_1(0, y)$, $\overline{h}_2(1, y) = b + \overline{h}_2(0, y)$. Therefore, $\psi_1(1, y) = \psi_1(0, y)$ and $\psi_2(1, y) = \psi_2(0, y)$. The equalities $\psi_i(x, 1) = \psi_i(x, 0)$, $i = 1, 2$, are proved similarly. We extend the functions ψ_1 and ψ_2 from F to $\mathbb{R}^2$ by periodicity with period 1 in both arguments.

Let $m(x, y) \in \mathbb{R}^2$ be arbitrary. Since F is a fundamental domain, there exists a $\gamma \in \Gamma$ such that $\gamma^{-1}(m) \in F$ (Figure 6.4). Let $\gamma\colon (x, y) \mapsto (x + n, y + k)$. Then $\overline{h}_*(\gamma)\colon (x, y) \mapsto (x + an + bk, y + cn + dk)$. We have that $\overline{h}_1(\gamma^{-1}(m)) = \overline{h}_1(x-n, y-k) = a(x-n)+b(y-k)+\psi_1(x, y)$. In view of the equality $\overline{h}_*(\gamma)\circ\overline{h} = \overline{h}\circ\gamma$ (Figure 6.4), we get that $\overline{h}_1(x, y) = \overline{h}_1[\gamma^{-1}(m)] + an + bk = ax + by + \psi_1(x, y)$. Similarly, $\overline{h}_2(x, y) = cx + dy + \psi_2(x, y)$. □

COROLLARY 1.1. *Assume the conditions of Lemma 1.3, and let h be a homotopy equivalence (in particular, a homeomorphism). Then the matrix $\overline{h}_* = \begin{pmatrix} a & b \\ c & d \end{pmatrix}$ is unimodular.*

PROOF. Since h is a homotopy equivalence, the homomorphism h_* is an automorphism. Therefore, the matrix $\overline{h}_*$ is invertible. □

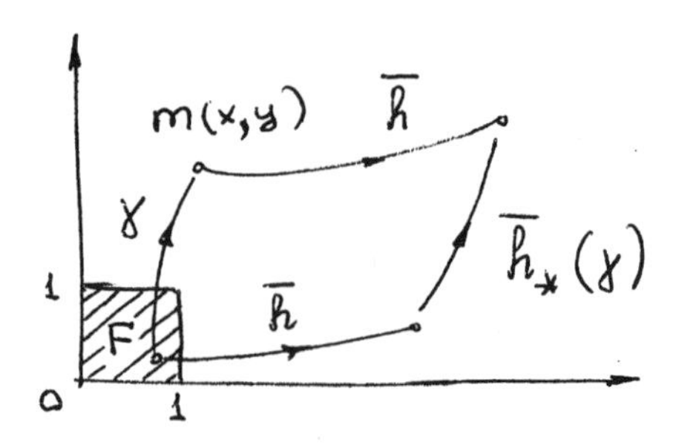

FIGURE 6.4

LEMMA 1.4. *Let $h\colon T^2 \to T^2$ be a continuous mapping homotopic to the identity, and let $\overline{h}\colon \mathbb{R}^2 \to \mathbb{R}^2$ be a covering mapping. Then there exists a constant $\alpha > 0$ such that the distance between the points $\overline{m}$ and $\overline{h}(\overline{m})$ does not exceed α for all $\overline{m} \in \mathbb{R}^2$.*

PROOF. Since h is continuous, and the set $F = \{(x, y) \in \mathbb{R}^2 : 0 \leq x \leq 1, 0 \leq y \leq 1\}$ is compact, it follows that $\alpha = \max |\overline{m} - \overline{h}(\overline{m})| < +\infty$ for $\overline{m} \in F$. The fact that h is homotopic to the identity implies that $\overline{h}_* = \mathrm{id}$, and hence $\overline{h} \circ \gamma = \gamma \circ \overline{h}$ for $\gamma \in \Gamma$.

For any $\overline{m} \in \mathbb{R}^2$ there is a $\gamma \in \Gamma$ such that $\gamma(\overline{m}) \in F$. Therefore, $|\overline{m} - \overline{h}(\overline{m})| = |\gamma(\overline{m}) - \gamma \circ \overline{h}(\overline{m})| = |\gamma(\overline{m}) - \overline{h} \circ \gamma(\overline{m})| \leq \alpha$. $\square$

1.2. Curvilinear rays. Denote by $\mathbb{R}_+$ the set of nonnegative real numbers, and let $\varphi\colon \mathbb{R}_+ \to \mathcal{M}$ be a homeomorphism onto its image, where $\mathcal{M}$ is a manifold. The set $\varphi(\mathbb{R}_+)$ is called a half-closed (or half-open) infinite curve. The mapping φ determines on the curve $\varphi(\mathbb{R}_+) \overset{\text{def}}{=} l^+$ a parameter $t \mapsto \varphi(t)$, $t \in \mathbb{R}_+$.

Let $\pi\colon \overline{\mathcal{M}} \to \mathcal{M}$ be a covering with universal covering space $\overline{\mathcal{M}}$. Since φ is one-to-one, the complete inverse image $\pi^{-1}(l^+)$ is a family of disjoint half-closed infinite curves $\overline{l}_1^+, \dots$, with a parametrization on each of them determined by φ.

DEFINITION. A half-closed infinite curve $l^+ \subset \mathcal{M}$ is called a *curvilinear ray* if there exists a lifting $\overline{l}^+ \subset \overline{\mathcal{M}}$ of it (which is also a half-closed infinite curve) onto the universal covering $\overline{\mathcal{M}}$ such that $\overline{l}^+$ leaves any compact set as the parameter increases unboundedly, and does not return to it, beginning with some value of the parameter.

We mention the possibility that $\mathcal{M} = \overline{\mathcal{M}}$ and $l^+ = \overline{l}^+$. It follows from the definition that a lifting $\overline{l}^+$ of a curvilinear ray is also a curvilinear ray.

It is sometimes more convenient to parametrize a half-closed curve by the set of nonpositive numbers. The concept of a curvilinear ray is defined similarly for such curves.

Each semitrajectory of a flow has a natural parametrization, which we have in mind when we refer to a semitrajectory that is a curvilinear ray.

LEMMA 1.5. *Let f^t be a flow on the torus, and let $\overline{f}^t$ be a flow on $\mathbb{R}^2$ covering it. Then a semitrajectory $l^{()}$ of f^t is a curvilinear ray if and only if each semitrajectory of $\overline{f}^t$ in the inverse image $\pi^{-1}[l^{()}]$ is a curvilinear ray.*

PROOF. This follows from the fact that any two semitrajectories in $\pi^{-1}[l^{()}]$ can be carried one into the other by an integer translation. □

For brevity we call a trajectory a curvilinear ray if both its semitrajectories are curvilinear rays.

LEMMA 1.6. *Suppose that f^t is a flow on the torus T^2, and $\overline{f}^t$ is a flow on $\mathbb{R}^2$ covering it. Then the following semitrajectories of f^t and $\overline{f}^t$ are curvilinear rays:*

1) *a semitrajectory of $\overline{f}^t$ that projects into a closed trajectory of f^t nonhomotopic to zero;*

2) *a nonclosed semitrajectory of f^t containing in its limit set a closed trajectory nonhomotopic to zero, or a one-sided contour nonhomotopic to zero, or a nontrivial recurrent semitrajectory;*

3) *a nontrivial recurrent semitrajectory of f^t.*

PROOF. 1) and 2) follow from 3) and Lemma 1.1. We prove 3). Let l^+ be a positive nontrivial recurrent semitrajectory of f^t (the proof is analogous for a negative semitrajectory). Assume that a semitrajectory $\overline{l}^+ \in \pi^{-1}(l^+)$ does not leave the compact set $K \subset \mathbb{R}^2$ or that it returns to K infinitely many times as the time increases. Then $\omega(\overline{l}^+) \neq \emptyset$. Since l^+ is a nontrivial recurrent semitrajectory, $\omega(l^+)$ contains at least one regular point $\overline{m}$. Since it is nonclosed, the semitrajectory $\overline{l}^+$ intersects more than once a contact-free segment passing through $\overline{m}$. Therefore, there exists a contact-free cycle C intersecting $\overline{l}^+$ (Corollary 2.1 in Chapter 2).

Denote by $\mathcal{D}$ the disk bounded by the curve C. We consider two cases: 1) $\overline{m}$ belongs to a disk congruent to $\mathcal{D}$ (possibly $\mathcal{D}$ itself); 2) there is no $\gamma \in \Gamma$ such that $\overline{m}$ belongs to $\gamma(\mathcal{D})$. In case 1) there is a semitrajectory $\overline{l}_1^+$ congruent to $\overline{l}^+$ whose ω-limit set belongs to $\mathcal{D}$. Therefore, $\omega(\overline{l}_1^+)$ is a single equilibrium state, or a closed trajectory, or a one-sided contour. Then the ω-limit set of the semitrajectory $l^+ = \pi(\overline{l}_1^+)$ also is a single equilibrium state, or a closed trajectory, or a one-sided contour, and this contradicts the fact that l^+ is a nontrivial recurrent semitrajectory. In case 2) we get a set $\pi(\mathcal{D})$ that is bounded by a contact-free cycle $\pi(C)$ homotopic to zero, and $l^+ \cap \pi(C) \neq \emptyset$, contradicting Lemma 2.3 in Chapter 2. □

1.3. Asymptotic directions. Let f^t be a flow on the torus. Along with a covering flow $\overline{f}^t$ on $\mathbb{R}^2$ we consider a flow $\widetilde{f}^t$ on the unit disk $\mathcal{D}^2$ that is a covering flow for f^t with respect to the covering $\pi \circ \tau$ (see §1.1). The flow $\widetilde{f}^t$ is isomorphic to $\overline{f}^t$ by means of the diffeomorphism τ^{-1} defined by the formulas (1.1) in §1.1.

Corresponding objects of the flow $\widetilde{f}^t$ on $\mathcal{D}^2$ will be equipped with a tilde above them.

DEFINITION. Suppose that $\overline{l}^+$ is a positive semitrajectory of a flow $\overline{f}^t$ on $\mathbb{R}^2$ and is a curvilinear ray, and let $\widetilde{l}^+ = \tau^{-1}(\overline{l}^+)$ be the corresponding semitrajectory of the flow $\widetilde{f}^t$ on $\mathcal{D}^2$. The semitrajectories $\overline{l}^+$ and $\widetilde{l}^+$ have an *asymptotic direction* if the ω-limit set of $\widetilde{l}^+$ consists of a single point $\sigma(\widetilde{l}^+)$ belonging to the absolute $S_\infty = \partial\mathcal{D}^2$.

An analogous definition for the existence of an asymptotic direction applies to the negative semitrajectories $\widetilde{l}^-$ and $\overline{l}^-$.

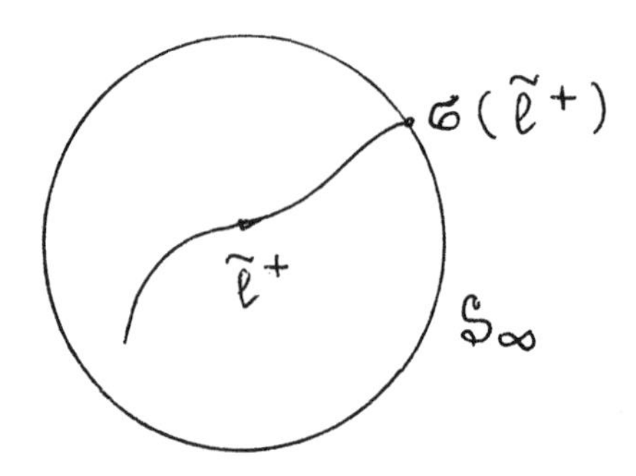

FIGURE 6.5

The point $\sigma(\widetilde{l}^{()})$ is said to be *accessible by the semitrajectory* $\widetilde{l}^{()}$.

We remark that if a semitrajectory $\overline{l}^{()}$ is a curvilinear ray, then the ω- (α-) limit set of the semitrajectory $\widetilde{l}^{()} = \tau^{-1}(\overline{l}^{()})$ is nonempty and belongs to the absolute. The existence of an asymptotic direction indicates that the ω- (α-) limit set consists of a single point.

In Theorem 1.1 we prove that if a semitrajectory of a covering flow on $\mathbb{R}^2$ is a curvilinear ray, then it has an asymptotic direction.

EXAMPLE 1. A covering flow $\overline{f}^t_\mu$ for a rational or irrational winding of the torus is given by the system $\dot{x} = 1$, $\dot{y} = \mu$. Each trajectory $\overline{l}$ $(x(t) = x_0 + t$, $y(t) = y_0 + \mu t)$ is a curvilinear ray. We leave it to the reader as an exercise to prove that all the semitrajectories of $\overline{f}^t_\mu$ have an asymptotic direction, and the semitrajectories of the corresponding flow $\widetilde{f}^t_\mu$ on $\mathcal{D}^2$ have accessible points on the absolute as shown in Figure 6.6 (all the positive semitrajectories have the accessible point $(1/\sqrt{1+\mu^2}, \mu/\sqrt{1+\mu^2})$, and all the negative ones have the diametrically opposite point).

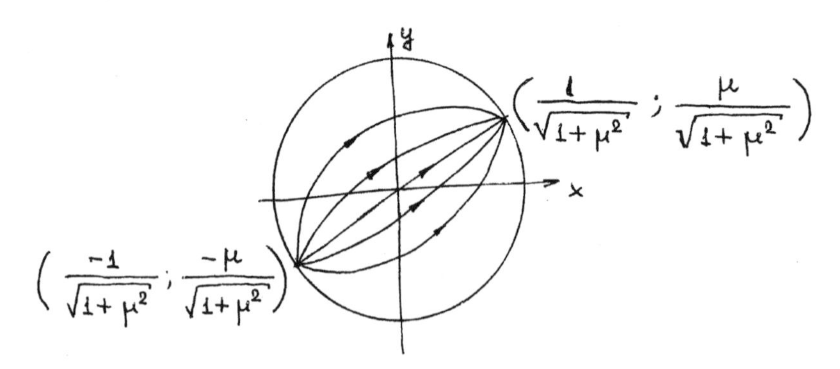

FIGURE 6.6

EXAMPLE 2. We consider a covering flow $\overline{f}^t$ whose phase portrait is pictured in Figure 6.7, a). All the semitrajectories of $\overline{f}^t$ are curvilinear rays and have an asymptotic direction. The phase portrait of the corresponding flow $\widetilde{f}^t$ is pictured in Figure 6.7, b). The points $(-1, 0)$ and $(1, 0)$ are accessible both for the positive and for the negative semitrajectories of $\widetilde{f}^t$.

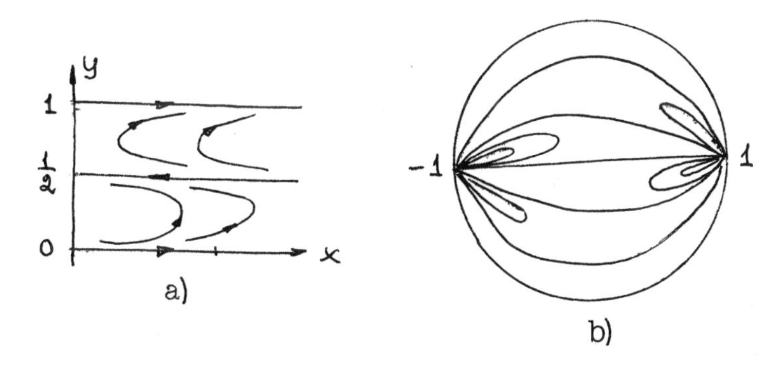

FIGURE 6.7

LEMMA 1.7. *Let $\overline{l}^+(x = x(t), y = y(t))$ be a positive semitrajectory of some covering flow on $\mathbb{R}^2$, and assume that it is a curvilinear ray and has an asymptotic direction. Then $\sigma(\widetilde{l}^+) = (\mu_1, \mu_2) \in S_\infty$ is accessible by the semitrajectory $\widetilde{l}^+ = \tau^{-1}(\overline{l}^+)$ if and only if one of the following conditions holds:* a) *$lim_{t\to\infty} y(t)/x(t) = 0$, in which case $\sigma(\widetilde{l}^+) = (\pm 1, 0)$;* b) *$\lim_{t\to\infty} x(t)/y(t) = 0$, in which case $\sigma(\widetilde{l}^+) = (0, \pm 1)$;* c) *both the limits $\lim_{t\to\infty} y(t)/x(t) = \mu_2/\mu_1 \overset{\text{def}}{=} \mu$ and $\lim_{t\to\infty} x(t)/y(t) = \mu_1/\mu_2$ exist and are finite and nonzero, and*

$$\mu_1 = \frac{1}{\sqrt{1+\mu^2}}, \qquad \mu_2 = \frac{\mu}{\sqrt{1+\mu^2}}. \tag{1.3}$$

PROOF. Since $\widetilde{l}^+$ is a curvilinear ray, $x^2(t) + y^2(t) \to +\infty$ as $t \to +\infty$.

NECESSITY. Let $\sigma = (\mu_1, \mu_2)$, where

$$\mu_1 = \lim_{t\to\infty} \frac{x(t)}{\sqrt{1 + x^2(t) + y^2(t)}}, \qquad \mu_2 = \lim_{t\to\infty} \frac{y(t)}{\sqrt{1 + x^2(t) + y^2(t)}},$$

and $\mu_1^2 + \mu_2^2 = 1$. If $|\mu_1| = 1$, then $\mu_2 = 0$; therefore, $\lim_{t\to\infty} |x(t)| = \infty$. The relation

$$|\mu_1| = \lim_{t\to\infty} \frac{1}{\sqrt{1/x^2(t) + 1 + y^2(t)/x^2(t)}} = 1$$

implies that $\lim_{t\to\infty} y(t)/x(t) = 0$. Similarly, $\lim_{t\to\infty} x(t)/y(t) = 0$ if $\mu_1 = 0$ and $|\mu_2| = 1$.

In the case $0 < |\mu_i| < 1$, $i = 1, 2$, we get from (1.1) that $\lim_{t\to\infty} |x(t)| = \lim_{t\to\infty} |y(t)| = \infty$. It follows from the relation $\mu_1^2 + \mu_2^2 = 1$ that (1.3) holds, and the case c) is realized.

SUFFICIENCY. Let $\xi = \xi(t)$, $\eta = \eta(t)$ be the equations of the semitrajectory $\widetilde{l}^+$. Since $\overline{l}^+$ is a curvilinear ray, it follows that $\xi^2(t) + \eta^2(t) \to 1$ as $t \to \infty$.

If a) holds, then $\lim_{t\to\infty} \eta(t)/\xi(t) = \lim_{t\to\infty} y(t)/x(t) = 0$, from which $|\mu_1| = \lim_{t\to\infty} |\xi(t)| = 1$ and $\mu_2 = \lim_{t\to\infty} \eta(t) = 0$. Similarly, it follows from the condition b) that $\mu_1 = 0$ and $|\mu_2| = 1$.

In the case c) we get from $\lim_{t\to\infty}[x^2(t) + y^2(t)] = \infty$ and the finiteness of the limit $\lim_{t\to\infty} y(t)/x(t) \overset{\text{def}}{=} \mu$, $\mu \neq 0$, that $\lim_{t\to\infty} |y(t)| = \lim_{t\to\infty} |x(t)| = \infty$. This and the formulas (1.1) imply (1.3). □

COROLLARY 1.2. *Let $\overline{l}^+$ be a positive semitrajectory of some covering flow on $\mathbb{R}^2$, and assume that it is a curvilinear ray and has an asymptotic direction. Then any semitrajectory $\overline{l}_1^+$ congruent to it is a curvilinear ray and has an asymptotic direction, and $\sigma(\widetilde{l}^+) = \sigma(\widetilde{l}_1^+)$.*

PROOF. Assume that the case a) of Lemma 1.7 is realized for the semitrajectory $\overline{l}^+(x = x(t), y = y(t))$, that is, $\lim_{t\to\infty} y(t)/x(t) = 0$. Since $\overline{l}_1^+$ is a semitrajectory congruent to $\overline{l}^+$, $\overline{l}_1^+$ is given by the equations $x = x(t) + k$, $y = y(t) + n$ for some $k, n \in \mathbb{Z}$. It follows from

$$\lim_{t\to\infty} [x^2(t) + y^2(t)] = \infty$$

that

$$\lim_{t\to\infty} |x(t)| = \infty,$$

and therefore,

$$\lim_{t\to\infty} [y(t) + n]/[x(t) + k] = 0.$$

By Lemma 1.7, $\overline{l}_1^+$ has an asymptotic direction, and $\sigma(\widetilde{l}^+) = \sigma(\widetilde{l}_1^+) = (\pm 1, 0)$.

The remaining cases b) and c) are treated similarly. □

To prove Theorem 1.1 we need the following result.

LEMMA 1.8. *A minimal flow on the torus is topologically equivalent (and even topologically orbitally equivalent) to an irrational winding by means of a homeomorphism homotopic to the identity.*

PROOF. Let f^t be a minimal flow on the torus. Since each trajectory of f^t is dense, there exists a contact-free cycle C that intersects any semitrajectory of f^t (that is, C is a global section). Therefore, f^t induces a Poincaré mapping $P\colon C \to C$ that is a homeomorphism. Since f^t is minimal, each semi-orbit of P is dense. Then by Corollary 1.2 and Theorem 2.3 in Chapter 5, P is conjugate to a rotation $R_\alpha\colon S^1 \to S^1$ with irrational α.

We consider an irrational winding f_α^t on the torus with covering system $\dot{x} = 1$, $\dot{y} = \alpha$. The zero median $C_0 = \pi(0 \le y \le 1,\ x = 0)$ is a contact-free cycle of f_α^t on which f_α^t induces the Poincaré mapping R_α. Then a homeomorphism realizing a conjugacy between P and R_α can be extended to a homeomorphism $h\colon T^2 \to T^2$ realizing a topological equivalence (and even a topological orbital equivalence) of the flows f^t and f_α^t.

If the homeomorphism h is nonhomotopic to zero, then we take a linear diffeomorphism h_0 of the torus with covering of the form $\overline{x} = ax + by$, $\overline{y} = cx + dy$, where $h_* = \begin{pmatrix} a & b \\ c & d \end{pmatrix}$ is the isomorphism of $\pi_1(T^2) = \Gamma$ induced by h. It follows from Lemma 1.3 that $(h_0)_* = h_*$.

Under the action of the linear diffeomorphism h_0^{-1} the irrational winding f_α^t clearly passes into an irrational winding f_0^t. Then f^t is topologically equivalent to f_0^t by means of the homeomorphism $h_0^{-1} \circ h$, which is homotopic to the identity because $(h_0^{-1} \circ h)_* = (h_0^{-1})_* \circ h_* = \mathrm{id}$. □

THEOREM 1.1. *Let f^t be a flow on the torus, and let $\overline{f}^t$ and $\widetilde{f}^t$ be flows covering it on $\mathbb{R}^2$ and $\mathbb{D}^2$, respectively. Then:*

1) *each semitrajectory of $\overline{f}^t$ that is a curvilinear ray has an asymptotic direction;*

2) *for any two semitrajectories of* $\widetilde{f}^t$ *having an asymptotic direction the accessible points on the absolute either coincide or are diametrically opposite;*

3) *if* f^t *has a nontrivial recurrent semitrajectory, then all the positive (respectively, negative) semitrajectories of* $\widetilde{f}^t$ *that are curvilinear rays have a common accessible point* $\sigma^+ \in S_\infty$ $(\sigma^- \in S_\infty)$, *and the points* σ^+ *and* σ^- *are diametrically opposite.*

PROOF. Suppose that a positive semitrajectory $\overline{l}^+$ of $\overline{f}^t$ is a curvilinear ray. We show that it has an asymptotic direction (the proof is analogous for a negative semitrajectory).

Consider the semitrajectory $l^+ = \pi(\overline{l}^+)$. It follows from the properties of a covering and the fact that $\overline{l}^+$ is a curvilinear ray that $\omega(l^+)$ does not consist of a single equilibrium state, and is neither a closed trajectory homotopic to zero nor a one-sided contour homotopic to zero. Therefore, by Theorem 3.6 in Chapter 2, $\omega(l^+)$ is:

a) a closed trajectory l_0 nonhomotopic to zero;

b) a one-sided contour nonhomotopic to zero;

c) the closure of a nontrivial recurrent trajectory.

In the case a) we take a trajectory $\overline{l}_0$ of $\overline{f}^t$ such that $\pi(\overline{l}_0) = l_0$. It follows from Lemma 1.1 that both semitrajectories $\overline{l}_0^+$ and $\overline{l}_0^-$ of $\overline{l}_0$ have an asymptotic direction, and the points $\sigma(\widetilde{l}_0^+)$ and $\sigma(\widetilde{l}_0^-)$ are diametrically opposite. This implies that $\overline{l}^+$ also has an asymptotic direction, and $\sigma(\widetilde{l}^+) \in \sigma(\widetilde{l}_0^+) \cup \sigma(\widetilde{l}_0^-)$.

We prove 2) in the case a). If $\overline{l}_1^{()}$ is a semitrajectory of f^t that is a curvilinear ray, then the limit set of the semitrajectory $\pi[\overline{l}_1^{()}]$ is a closed trajectory nonhomotopic to zero or a one-sided contour k_0 nonhomotopic to zero, because in this case f^t does not have nontrivial recurrent trajectories. The curves l_0 and k_0 are homotopic, and hence one of the points $\sigma(\widetilde{l}_0^+)$ and $\sigma(\widetilde{l}_0^-)$ is accessible for the semitrajectory $\widetilde{l}_1^{()}$.

The case b) is handled similarly.

We proceed to the case c). According to Lemma 3.1 in Chapter 3, the flow f^t is irreducible. Therefore, by Theorem 3.1 in Chapter 3, there exists a continuous mapping $h\colon T^2 \to T^2$ homotopic to the identity (a blowing-down operation) carrying the trajectories of f^t into trajectories or unions of trajectories of some transitive flow f_0^t. According to Lemma 3.4 in Chapter 3, a transitive flow is obtained from a minimal flow by adjoining a certain number of impassable grains (possibly none). Therefore, in view of Lemma 1.8 the flow f_0^t can be assumed to be an irrational winding with a certain number of impassable grains (possibly none). By Lemma 1.4, a covering transformation homotopic to the identity moves the points of the plane a distance bounded by a fixed constant. This and Example 1 imply the assertions 1)–3) in the case c). □

1.4. The Poincaré rotation number. Let f^t be a flow on T^2, and let $\overline{f}^t$ and $\widetilde{f}^t$ be flows covering it on $\mathbb{R}^2$ and $\mathfrak{D}^2$, respectively. Suppose that a semitrajectory $\overline{l}^{()}$ of $\overline{f}^t$ is a curvilinear ray and hence has an asymptotic direction. This means that the corresponding semitrajectory $\widetilde{l}^{()} = \tau^{-1}(\overline{l}^{()})$ tends to a single point $\sigma(\widetilde{l}^{()})$ on the absolute S_∞.

DEFINITION. Let the point $\sigma(\widetilde{l}^{()}) \in S_\infty = \partial\mathfrak{D}^2$ have coordinates (μ_1, μ_2). If $\mu_1 \neq 0$, then the number $\mu = \mu_2/\mu_1$ is called the *Poincaré rotation number* of the flow f^t. If $\mu_1 = 0$, then the Poincaré rotation number μ is set equal to ∞.

By Theorem 1.1, this is well defined: it does not depend on the choice of the semitrajectory $\overline{l}^{()}$ of $\overline{f}^t$ having an asymptotic direction.

The point $\sigma(\widetilde{l}^{()})$ and the diametrically opposite point on the absolute are called the *rotation points* of f^t.

The Poincaré rotation number of a flow f^t is denoted by $\mathrm{rot}(f^t)$.

Not every flow on the torus has a Poincaré rotation number. For example, a flow with covering flow pictured in Figure 6.8 does not have a rotation number because no semitrajectory of the covering flow is a curvilinear ray.

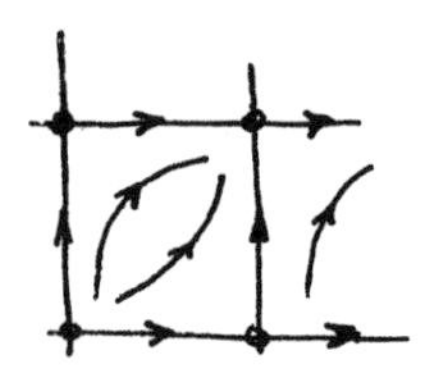

FIGURE 6.8

LEMMA 1.9. *Let f^t be a flow on the torus, and let $\overline{f}^t$ be a flow on $\mathbb{R}^2$ covering it. In this case:*

1) *if f^t has a rotation number $\mu = \mathrm{rot}(f^t)$, then for any positive semitrajectory $\overline{l}^+(x = x(t), y = y(t))$ of f^t that is a curvilinear ray*

$$\mu = \lim_{t\to\infty} y(t)/x(t), \tag{1.4}$$

and the points (μ_1, μ_2) and $(-\mu_1, -\mu_2)$ with $\mu_1 = 1/\sqrt{1+\mu^2}$ and $\mu_2 = \mu/\sqrt{1+\mu^2}$ are the rotation points of f^t;

2) *if there exists a positive semitrajectory $\overline{l}^+(x = x(t), y = y(t))$ of $\overline{f}^t$ that is a curvilinear ray, then the limit (1.4) exists and is equal to the rotation number of f^t, and the points (μ_1, μ_2) and $(-\mu_1, -\mu_2)$ with μ_1 and μ_2 given by (1.3) are the rotation points of f^t.*

PROOF. This follows from Lemma 1.7, Theorem 1.1, and the definition of the rotation number of a flow. □

Lemma 1.9 yields the following connection between the rotation number of a flow and the rotation number of the Poincaré mapping the flow induces on the zero meridian.

COROLLARY 1.3. *Suppose that a flow f^t on the torus has a global section C_0 (that is, a contact-free cycle intersecting each regular trajectory) coinciding with the zero meridian of the torus, and assume that f^t induces a Poincaré mapping $P\colon C_0 \to C_0$. Then* $\mathrm{rot}(P) = \mathrm{rot}(f^t) \pmod 1$.

PROOF. For a covering flow $\overline{f}^t$ the lines $x = n$, $n \in \mathbb{Z}$, in the inverse image $\pi^{-1}(C_0)$ are contact-free lines. By the hypotheses, there exists a semitrajectory l^+ of f^t intersecting C_0 countably many times (perhaps at a single point if l^+ is a closed trajectory). Then $\overline{f}^t$ has a semitrajectory $\overline{l}^+(y_0)$ intersecting all the lines $x = n$, $n \in \mathbb{N}$, with y_0 on the line $x = 0$.

Denote by $\overline{P}_n$ the Poincaré mapping induced by $\overline{f}^t$ from the line $x = 0$ to the line $x = n$. Since $\overline{f}^t$ is a covering flow, the relations $\pi \circ \overline{P}_n = P^n \circ \pi$, $n \in \mathbb{N}$, hold on the line $x = 0$. From Lemma 1.9 and the definition of the rotation number of a transformation of the circle (§1 in Chapter 5) we get that

$$\operatorname{rot}(f^t) = \lim_{t\to\infty} y(t)/x(t) = \lim_{n\to+\infty} y(n)/n = \lim_{n\to+\infty} \overline{P}_n(y_0)/n = \operatorname{rot}(\overline{P}_1),$$

where $x = x(t)$, $y = y(t)$ are the equations of the semitrajectory $\overline{l}^+(y_0)$. □

REMARK. A connection between the rotation number of a flow without compact trajectories on the torus and the rotation number of the Poincaré mapping on an arbitrary global section was found by I. Strelcyn in the paper, *Flots sur le tore et nombres de rotation* (Bull. Soc. Math. France **100** (1972), 195–208).

The following theorem is analogous to Lemma 1.6 in Chapter 5, and indicates a connection between the dynamical properties of a flow and the arithmetic properties of the rotation number.

THEOREM 1.2. *Suppose that a flow f^t on the torus has a rotation number μ. Then*:

1) *μ is rational or $\mu = \infty$ if and only if f^t has a closed trajectory nonhomotopic to zero or a one-sided contour nonhomotopic to zero*;

2) *μ is irrational if and only if f^t has a nontrivial recurrent semitrajectory.*

PROOF. Since f^t has a rotation number, f^t has a) a closed trajectory nonhomotopic to zero, or b) a one-sided contour nonhomotopic to zero, or c) a nontrivial recurrent semitrajectory, and the case c) cannot be realized simultaneously with one of the cases a) or b).

It follows from Lemma 1.1 that in the cases a) and b) the rotation number of f^t is rational or $\mu = \infty$.

Suppose that the case c) is realized. Then there exists a continuous mapping homotopic to the identity (a blowing-down operation) that carries the trajectories of f^t into those of an irrational winding with some number of impassable grains (possibly none). Therefore, $\operatorname{rot}(f^t)$ is irrational. □

1.5. The rotation orbit. Suppose that a flow f^t on the torus has a rotation number $\mu = \operatorname{rot}(f^t)$, and let $(\mu_1, \mu_2), (-\mu_1, -\mu_2) \in S_\infty$ be the rotation points of f^t, with their coordinates related to μ by the formulas (1.3):

$$\mu_1 = 1/\sqrt{1+\mu^2}, \qquad \mu_2 = \mu/\sqrt{1+\mu^2}$$

(we note that if $\mu = \infty$, then the rotation points of the flow are $(0,1)$ and $(0,-1)$, and vice versa).

DEFINITION. The *rotation orbit* of a flow f^t is defined to be the set of points (μ_1, μ_2), $(-\mu_1, -\mu_2)$ of the absolute S_∞ with coordinates determined from the relations (1.3)

$$\mu_1 = 1/\sqrt{1+\mu^2}, \qquad \mu_2 = \mu/\sqrt{1+\mu^2},$$

where μ is computed according to the formula

$$\mu = \frac{c + d\operatorname{rot}(f^t)}{a + b\operatorname{rot}(f^t)} \tag{1.5}$$

for all possible unimodular integer matrices $\begin{pmatrix} a & b \\ c & d \end{pmatrix}$. If $\operatorname{rot}(f^t) = \infty$ or if $\operatorname{rot}(f^t)$ is rational, then μ runs through the set of rational numbers, and the points $(0, 1)$ and $(0, -1)$ are adjoined to the rotation orbit.

In the next theorem it is shown that the rotation numbers of two topologically equivalent flows can be calculated one from the other by means of a unimodular integer matrix, that is, according to the formula (1.5), and hence the rotation orbit is an invariant of topological equivalence.

THEOREM 1.3. *Suppose that flows f^t and g^t on T^2 have Poincaré rotation numbers $\operatorname{rot}(f^t) = \mu$ and $\operatorname{rot}(g^t) = \lambda$, and that f^t is topologically equivalent to g^t by means of a homeomorphism h. In this case if $\mu \neq \infty$, then*

$$\lambda = \frac{c + d\mu}{a + b\mu}, \tag{1.6}$$

where $h_* = \begin{pmatrix} a & b \\ c & d \end{pmatrix}$ is the homomorphism induced by h on the fundamental group $\pi(T^2) \cong \mathbb{Z}$. If $\mu = \infty$, then $\lambda = d/b$.

PROOF. Let $\overline{f}^t$ and $\overline{g}^t$ be covering flows for f^t and g^t, respectively, and let $\overline{h}$ be a covering homeomorphism for h. Then $\overline{h}$ carries trajectories of $\overline{f}^t$ into trajectories of $\overline{g}^t$.

Since f^t has a rotation number, there exists a semitrajectory $\overline{l}^{()}$ of this flow having an asymptotic direction. Let $\overline{l}^{()}$ be a positive semitrajectory. If $x = x(t)$, $y = y(t)$ are the parametric equations of $\overline{l}^+$, then $\mu = \lim_{t\to\infty} y(t)/x(t)$ by Lemma 1.9.

Let $\overline{l}' = \overline{h}(\overline{l}^+)$ be the semitrajectory of $\overline{g}^t$ into which $\overline{l}^+$ passes under the action of $\overline{h}$. We show that $\overline{l}'$ is a curvilinear ray.

According to Lemma 1.3, the parametric equations $x = x'(t)$, $y = y'(t)$ of the semitrajectory $\overline{l}'$ have the form

$$\begin{aligned} x'(t) &= ax(t) + by(t) + \psi_1(x(t), y(t)), \\ y'(t) &= cx(t) + dy(t) + \psi_2(x(t), y(t)), \end{aligned} \tag{1.7}$$

where the $\psi_i(x, y)$ are continuous 1-periodic functions in each argument, and $h_* = \overline{h}_* = \begin{pmatrix} a & b \\ c & d \end{pmatrix}$ is a unimodular integer matrix (Corollary 1.1).

We remark that the motion of the point $(x'(t), y'(t)) \in \mathbb{R}^2$ along the semitrajectory $\overline{l}'$ as $t \to \infty$ by virtue of the formulas (1.7) does not necessarily correspond to the motion along $\overline{l}'$ as a phase trajectory of $\overline{g}^t$ as time increases. This motion can correspond to decreasing time, which means that $\overline{h}$ reverses the direction of motion along trajectories. For definiteness and for simplicity we assume that $\overline{h}$ preserves the positive direction of motion along trajectories.

To prove that $\overline{l}'$ is a curvilinear ray it is necessary to show that $[x'(t)]^2 + [y'(t)]^2 \to \infty$ as $t \to \infty$. Three cases are possible:

a) $\mu = 0$;

b) $\mu = \infty$;

c) $\mu \neq 0, \infty$.

In case a) we have from the equalities $\lim_{t\to\infty} y(t)/x(t) = 0$ and $\lim_{t\to\infty}[x^2(t) + y^2(t)] = \infty$ that $\lim_{t\to\infty} |x(t)| = \infty$. Then we get from (1.7) that

$$\lim_{t\to\infty} \frac{x'(t)}{x(t)} = \lim_{t\to\infty} \left[a + b \cdot \frac{y(t)}{x(t)} + \frac{\psi_1}{x(t)}\right] = a + b\mu = a,$$

$$\lim_{t\to\infty} \frac{y'(t)}{x(t)} = \lim_{t\to\infty} \left[c + d \cdot \frac{y(t)}{x(t)} + \frac{\psi_2}{x(t)}\right] = c + d\mu = c.$$

Since the matrix $\overline{h}_*$ is unimodular, $a^2 + c^2 \neq 0$. Therefore, $\lim_{t\to\infty} |x'(t)| = \infty$ or $\lim_{t\to\infty} |y'(t)| = \infty$.

In case b) the argument is analogous to the preceding, with use of the equality $\lim_{t\to\infty} x(t)/y(t) = 0$ (each of the equalities in (1.7) is divided by $y(t)$).

In case c) it suffices to show that one of the limits

$$\lim_{t\to\infty} \frac{x'(t)}{x(t)} = a + b\mu, \qquad \lim_{t\to\infty} \frac{y'(t)}{x(t)} = c + d\mu$$

is nonzero. But if $a + b\mu = 0$ and $c + d\mu = 0$, then $\begin{vmatrix} a & b \\ c & d \end{vmatrix} = \begin{vmatrix} -b\mu & b \\ -d\mu & d \end{vmatrix} = 0$, which contradicts the fact that $\overline{h}_*$ is a unimodular matrix.

Thus, $\overline{l}'$ is a curvilinear ray.

If $\mu \neq \infty$, then $\lim_{t\to\infty} |x(t)| = \infty$. In this case

$$\lim_{t\to\infty} \frac{y'(t)}{x'(t)} = \lim_{t\to\infty} \frac{c + d\frac{y(t)}{x(t)} + \frac{\psi_2}{x(t)}}{a + b\frac{y(t)}{x(t)} + \frac{\psi_1}{x(t)}} = \frac{c + d\mu}{a + b\mu} = \lambda.$$

If $\mu = \infty$, then $\lim_{t\to\infty} |y(t)| = \infty$, and

$$\lim_{t\to\infty} \frac{y'(t)}{x'(t)} = \lim_{t\to\infty} \frac{c\frac{x(t)}{y(t)} + d + \frac{\psi_2}{y(t)}}{a\frac{x(t)}{y(t)} + b + \frac{\psi_1}{y(t)}} = \frac{d}{b}. \quad \square$$

1.6. Classification of minimal flows. We recall that a flow is said to be minimal if each trajectory of it is dense in the manifold. According to Theorem 1.2, a minimal flow on the torus has an irrational (and hence not infinite) Poincaré rotation number.

THEOREM 1.4. *Two minimal flows f^t and g^t on the torus with rotation numbers $\mu = \operatorname{rot}(f^t)$ and $\lambda = \operatorname{rot}(g^t)$ are topologically equivalent if and only if there exists a unimodular integer matrix $\begin{pmatrix} a & b \\ c & d \end{pmatrix}$ such that*

$$\lambda = \frac{c + d\mu}{a + b\mu}. \tag{1.8}$$

PROOF. NECESSITY. It follows from Theorem 1.3 and Corollary 1.1.

SUFFICIENCY. Assume (1.8). We show that f^t and g^t are topologically equivalent.

According to Lemma 1.8, a minimal flow on the torus is topologically equivalent to an irrational winding by means of a homeomorphism homotopic to the identity. Since a homeomorphism homotopic to the identity induces the identity mapping of the fundamental group of the torus, it follows from Theorem 1.3 that f^t and g^t can be assumed to be irrational windings with respective rotation numbers μ and λ and with covering flows

$$\overline{f}^t : \begin{cases} \dot{x} = 1 \\ \dot{y} = \mu \end{cases}, \qquad \overline{g}^t : \begin{cases} \dot{x} = 1 \\ \dot{y} = \lambda \end{cases}.$$

By (1.8), the linear transformation $h \colon \mathbb{R}^2 \to \mathbb{R}^2$ given by

$$\overline{x} = ax + by, \qquad \overline{y} = cx + dy$$

realizes a topological equivalence of the flows $\overline{f}^t$ and $\overline{g}^t$. By assumption, $\left(\begin{smallmatrix} a & b \\ c & d \end{smallmatrix}\right)$ is a unimodular integer matrix. Therefore, $\overline{h}$ is a covering for some homeomorphism $h \colon T^2 \to T^2$ that realizes a topological equivalence of the flows f^t and g^t. □

COROLLARY 1.4. *Two minimal flows on the torus are topologically equivalent if and only if their rotation orbits coincide.*

1.7. Classification of Denjoy flows. We recall that a Denjoy flow is defined to be a flow on the torus without equilibrium states and closed trajectories and with a limit set of Cantor type (see §3.1 in Chapter 1).

The set of Denjoy flows is denoted by $\mathrm{Den}^t(T^2)$.

Since the limit set $\Omega(f^t)$ of a flow $f^t \in \mathrm{Den}^t(T^2)$ is nowhere dense and since f^t does not have equilibrium states nor closed orbits by definition, $\Omega(f^t)$ consists of nontrivial recurrent trajectories which are nowhere dense on the torus. Obviously, each nontrivial recurrent trajectory of f^t lies in $\Omega(f^t)$. Since the torus has genus 1, any flow $f^t \in \mathrm{Den}^t(T^2)$ is irreducible, and hence any nontrivial recurrent trajectory of f^t is dense in $\Omega(f^t)$. Therefore, $\Omega(f^t)$ is a minimal set of the flow f^t.

According to Lemma 3.3 in Chapter 3, any Denjoy flow can be obtained from some minimal flow on the torus by a blowing-up operation. This result enables us to classify Denjoy flows.

The next result follows from Lemma 3.3 in Chapter 3 and Theorem 1.3 because a minimal flow on a torus is topologically orbitally equivalent to an irrational winding by means of a homeomorphism homotopic to the identity (Lemma 1.8).

LEMMA 1.10. *Let f^t be a Denjoy flow, and let f_0^t be an irrational winding with rotation number* $\mathrm{rot}(f^t)$. *Then there exists a continuous mapping $h \colon T^2 \to T^2$ (a blowing-down operation) that is homotopic to the identity and has the following properties*:

1) *h carries each trajectory of f^t into a trajectory of f_0^t with preservation of direction in time*;

2) *any nontrivial recurrent trajectory of f^t is mapped by h homeomorphically onto its image*;

3) $h[\Omega(f^t)] = T^2$;

4) *for any component w of the set $T^2 \setminus \Omega(f^t)$*

a) *w is simply connected,*

b) *the accessible (from within) boundary of w consists of exactly two trajectories* $l_1, l_2 \in \Omega(f^t)$,

c) $h(w \cup l_1 \cup l_2)$ *is a trajectory of the flow* f_0^t;

5) *if* $\overset{\circ}{\Omega}(f^t) \subset \Omega(f^t)$ *stands for the trajectories not lying on the accessible (from within) boundary of any component of* $T^2 \setminus \Omega(f^t)$, *then the restriction* $h|_{\overset{\circ}{\Omega}(f^t)}$ *is a homeomorphism of* $\overset{\circ}{\Omega}(f^t)$ *onto its image.*

NOTATION. Let f^t be a Denjoy flow, and suppose that the mapping h satisfies Lemma 1.10. Let

$$\chi(f^t, h) = \cup h(w),$$

where the union is over all the components w of the set $T^2 \setminus \Omega(f^t)$.

The set $\chi(f^t, h)$ is an at most countable family of trajectories of an irrational winding.

DEFINITION. Two sets $A, B \subset T^2$ are said to be *equivalent* if there exists a homeomorphism $F\colon T^2 \to T^2$ with a covering of the form

$$\overline{x} = x, \qquad \overline{y} = y + \xi, \quad \xi = \text{const} \in \mathbb{R}, \tag{1.9}$$

and such that $F(A) = B$.

It follows from the lemma that for a fixed flow $f^t \in \mathrm{Den}^t(T^2)$ the sets $\chi(f^t, h)$ are equivalent for different blowing-down operations h.

LEMMA 1.11. *Suppose that flows* $f_1^t, f_2^t \in \mathrm{Den}^t(T^2)$ *are topologically equivalent by means of a homeomorphism* $\varphi\colon T^2 \to T^2$ *homotopic to the identity, and let* h_i *be a blowing-down operation constructed by virtue of Lemma* 1.10 *for the flow* f_i^t, $i = 1, 2$. *Then the sets* $\chi(f_1^t, h_1)$ *and* $\chi(f_2^t, h_2)$ *are equivalent.*

PROOF. Suppose for definiteness that φ carries trajectories of f_1^t into trajectories of f_2^t. Since φ is homotopic to the identity, Theorem 1.3 gives us that $\mathrm{rot}(f_1^t) = \mathrm{rot}(f_2^t) \overset{\text{def}}{=} \mu$.

Denote by f_0^t an irrational winding with covering flow $\overline{f}_0^t$ given by the system $\dot{x} = 1$, $\dot{y} = \mu$. Then h_i is a blowing-down operation of the flow f_i^t to the flow f_0^t, $i = 1, 2$.

Obviously, a mapping $\overline{F}\colon \mathbb{R}^2 \to \mathbb{R}^2$ of the form (1.9) carries trajectories of $\overline{f}_0^t$ into trajectories of the same flow $\overline{f}_0^t$. We show that there exists a $\xi \in \mathbb{R}$ such that the $\overline{F}$ in (1.9) maps $\pi^{-1}(\chi(f_1^t, h_1))$ onto $\pi^{-1}[\chi(f_2^t, h_2)]$.

Denote by $\overline{h}_i$ a mapping covering h_i, $i = 1, 2$, and take some component $\overline{w} \subset \mathbb{R}^2 \setminus \pi^{-1}[\Omega(f_1^t)]$. Then $\overline{\varphi}(\overline{w})$ is a component of the set $\mathbb{R}^2 \setminus \pi^{-1}[\Omega(f_2^t)]$, where $\overline{\varphi}$ is a homeomorphism covering φ. According to Lemma 1.10, 4), the sets $\overline{h}_1(\overline{w})$ and $\overline{h}_2 \circ \overline{\varphi}(\overline{w})$ are trajectories of the flow $\overline{f}_0^t$. By the form of the trajectories of $\overline{f}_0^t$, there exists a number ξ such that the mapping $F\colon \mathbb{R}^2 \to \mathbb{R}^2$ defined by (1.9) carries $\overline{h}_1(\overline{w})$ into $\overline{h}_2 \circ \overline{\varphi}(\overline{w})$.

Suppose that $\overline{F}$ projects to the mapping $F\colon T^2 \to T^2$. It follows from (1.9) that F is homotopic to the identity. Since h_1, h_2, and φ are also homotopic to the identity, Lemma 1.3 gives us that

$$\overline{F} \circ \overline{h}_1 \circ \gamma(\overline{w}) = \gamma \circ \overline{F} \circ \overline{h}_1(\overline{w}) = \gamma \circ \overline{h}_2 \circ \overline{\varphi}(\overline{w}) = \overline{h}_2 \circ \overline{\varphi} \circ \gamma(\overline{w})$$

for any element $\gamma \in \Gamma$ (recall that Γ is the group of integer translations). From this and the denseness of $\{\gamma(\overline{w}) : \gamma \in \Gamma\}$ in the family of components of the set $\mathbb{R}^2 \setminus \pi^{-1}[\Omega(f_1^t)]$ it follows that $\overline{F} \circ \overline{h}_1(\overline{v}) = \overline{h}_2 \circ \overline{\varphi}(\overline{v})$ for any component $v \subset \mathbb{R}^2 \setminus \pi^{-1}[\Omega(f_1^t)]$. Then $\overline{F}(\pi^{-1}[\chi(f_1^t, h_1)]) = \pi^{-1}[\chi(f_2^t, h_2)]$, and hence $F[\chi(f_1^t, h_1)] = \chi(f_2^t, h_2)$. □

DEFINITION. Let f^t be a Denjoy flow, and h the blowing-down operation constructed by virtue of Lemma 1.10. The *characteristic class* $\chi(f^t)$ of f^t is defined to be the family of sets equivalent to the set $\chi(f^t, h)$.

According to Lemma 1.11, the characteristic class of a flow $f^t \in \mathrm{Den}^t(T^2)$ is the collection of sets $\chi(f^t, h)$ constructed for all possible blowing-down operations h.

DEFINITION. Two characteristic classes $\chi(f_1^t)$ and $\chi(f_2^t)$ of respective Denjoy flows f_1^t and f_2^t are said to be *commensurable* if for any representatives $\chi_1 \in \chi(f_1^t)$ and $\chi_2 \in \chi(f_2^t)$ of these classes there exists a homeomorphism $F\colon T^2 \to T^2$, $F(\chi_1) = \chi_2$, with a covering of the form

$$\overline{x} = ax + by, \qquad \overline{y} = cx + dy + \xi, \tag{1.10}$$

where $\begin{pmatrix} a & b \\ c & d \end{pmatrix}$ is a unimodular integer matrix, and $\xi \in \mathbb{R}$.

The next lemma shows that up to commensurability the characteristic class is a topological invariant of a Denjoy flow.

LEMMA 1.12. *If flows $f_1^t, f_2^t \in \mathrm{Den}^t(T^2)$ are topologically equivalent, then their characteristic classes $\chi(f_1^t)$ and $\chi(f_2^t)$ are commensurable.*

PROOF. Suppose that the homeomorphism $\varphi\colon T^2 \to T^2$ carries trajectories of f_1^t into trajectories of f_2^t. According to Lemma 1.3 and Corollary 1.1, φ induces an isomorphism $\varphi_*\colon \pi_1(T^2) \to \pi_1(T^2)$ given by a unimodular integer matrix $\begin{pmatrix} a & b \\ c & d \end{pmatrix}$. Denote by F the diffeomorphism with covering of the form

$$\overline{x} = ax + by, \qquad \overline{y} = cx + dy.$$

The flow $f_3^t = F \circ f_1^t \circ F^{-1}$ is topologically equivalent to the flow f_1^t by means of the homeomorphism F. If h_1 is a blowing- down operation of f_1^t to an irrational winding f_0^t, then $h_3 = F \circ h_1 \circ F^{-1}$ is a blowing-down operation of f_3^t to the irrational winding $F \circ f_0^t \circ F^{-1}$, and $\chi(f_3^t, h_3) = h_3[T^2 \setminus \Omega(f_3^t)] = F \circ h_1 \circ F^{-1}[T^2 \setminus \Omega(f_3^t)] = F \circ h_1[T^2 \setminus \Omega(f_1^t)] = F[\chi(f_1^t, h_1)]$.

The flows f_3^t and f_2^t are topologically equivalent by means of the homeomorphism $\varphi \circ F^{-1}$, which is homotopic to the identity, and therefore, by Lemma 1.11, $\chi(f_3^t, h_3)$ is equivalent to $\chi(f_2^t, h_2)$, where h_2 is a blowing-down operation of f_2^t to an irrational winding. Consequently, there exists a homeomorphism carrying $\chi(f_1^t, h_1)$ into $\chi(f_2^t, h_2)$, and it has a covering of the form (1.9). □

In what follows it will be proved that commensurability of characteristic classes implies topological equivalence of Denjoy flows. For this we first prove two lemmas.

LEMMA 1.13. *Any Denjoy flow is topologically equivalent to the suspension over some Denjoy homeomorphism.*

PROOF. Since a Denjoy flow f^t has nontrivial recurrent trajectories, there exists a contact-free cycle C of f^t that is nonhomotopic to zero (Lemma 2.3 in Chapter 2). The set $T^2 \setminus C$ is homeomorphic to an annulus, so in view of Lemma 2.4 in Chapter 2 each nontrivial recurrent semitrajectory intersects C, and hence C is a global section of f^t.

Denote by $P\colon C \to C$ the Poincaré mapping induced by f^t. Since f^t does not have equilibrium states, P is a homeomorphism. It follows from the nowhere denseness of the limit set of f^t that P is conjugate to a Denjoy homeomorphism. Consequently, f^t is topologically equivalent to the suspension over a Denjoy homeomorphism. $\square$

LEMMA 1.14. *Suppose that C is a simple closed curve on the torus that is nonhomotopic to zero, and let f^t be a Denjoy flow. Then there exists a contact-free cycle of f^t that is homotopic to C.*

PROOF. By Lemma 1.13, it can be assumed that f^t is the suspension $\operatorname{sus}(f)$ over a Denjoy homeomorphism f. We recall that $\operatorname{sus}(f)$ is defined on the torus obtained from the cylinder $S^1 \times [0,1]$ after identifying the points $(x,1)$ and $(f(x),0)$ for $x \in S^1$, and the segments (x,t), $0 \le t \le 1$, project into trajectories of the flow $\operatorname{sus}(f)$.

Since f is a Denjoy homeomorphism, $f^n(x) \neq x$ for all $x \in S^1$, $n \in \mathbb{Z}$. This implies that any closed geodesic on the torus that is homotopic to C is a transversal of the flow $\operatorname{sus}(f)$. $\square$

COROLLARY 1.5. *Any Denjoy flow is topologically equivalent to the suspension over some Denjoy homeomorphism by means of a homeomorphism homotopic to the identity.*

PROOF. It suffices to prove that there is a contact-free cycle C homotopic to the zero meridian n_0 of the torus. Let the homeomorphism $\varphi\colon T^2 \to T^2$ carry trajectories of the Denjoy flow into trajectories of a suspension $\operatorname{sus}(f)$ (Lemma 1.13). According to Lemma 1.14, the flow $\operatorname{sus}(f)$ has a contact-free cycle $\widehat{C}$ that is homotopic to the simple closed curve $\varphi(n_0)$. Then $C = \varphi^{-1}(\widehat{C})$ is the desired contact-free cycle of f^t. $\square$

THEOREM 1.5. *Denjoy flows f_1^t and f_2^t are topologically equivalent if and only if their characteristic classes $\chi(f_1^t)$ and $\chi(f_2^t)$ are commensurable.*

PROOF. By Lemma 1.12, topological equivalence of flows implies commensurability of the characteristic classes.

Suppose that the classes $\chi(f_1^t)$ and $\chi(f_2^t)$ are commensurable; that is, there exists a transformation $F\colon T^2 \to T^2$, with a covering of the form (1.10), that carries $\chi(f_1^t, h_1)$ into $\chi(f_2^t, h_2)$, where h_i is a blowing-down operation of f_i^t to an irrational winding, $i = 1, 2$. Then for the flow $f_3^t = F \circ f_1^t \circ F^{-1}$ we have that $\chi(f_3^t, h_3) = \chi(f_2^t, h_2)$, where $h_3 = F \circ h_1 \circ F^{-1}$.

We show that the flows f_2^t and f_3^t are topologically equivalent. By Corollary 1.5, f_2^t and f_3^t are topologically equivalent to the suspensions $\operatorname{sus}(f_2)$ and $\operatorname{sus}(f_3)$

over some Denjoy homeomorphisms f_2 and f_3, respectively, by means of homeomorphisms homotopic to the identity. It suffices to prove that f_2 and f_3 are conjugate.

It follows from $\chi(f_2^t, h_2) = \chi(f_3^t, h_3)$ that $\mathrm{rot}(f_2^t) = \mathrm{rot}(f_3^t)$, so $\mathrm{rot}[\mathrm{sus}(f_2)] = \mathrm{rot}[\mathrm{sus}(f_3)] \stackrel{\mathrm{def}}{=} \mu$. From this, $\mathrm{rot}(f_2) = \mathrm{rot}(f_3) = \mu \pmod 1$.

Denote by R_μ the rotation of the circle with rotation number $\mu \pmod 1$. According to Lemma 2.3 in Chapter 5, the homeomorphism f_i is topologically semiconjugate to R_μ by means of a continuous transformation $\widehat{h}_i \colon S^1 \to S^1$, $i = 2$, 3. This enables us to construct a special blowing-down operation from $\mathrm{sus}(f_i)$ to $\mathrm{sus}(R_\mu)$. For a point $m \in T^2$ we denote by $S_i^t(m)$ (respectively, $S_0^t(m)$) the shift of m along the corresponding trajectory of $\mathrm{sus}(f_i)$ (respectively, $\mathrm{sus}(R_\mu)$) by the time t. Let $t(m)$ be the smallest nonnegative number such that $S_i^{-t(m)}(m) \in S^1 = \pi(\text{the line } x = 0)$ (respectively, $S_0^{-t(m)}(m) \in S^1 = \pi(\text{the line } x = 0)$). Let $\widetilde{h}_i(m) = S_0^{t(m)} \circ \widehat{h}_i \circ S_i^{-t(m)}(m)$, $m \in T^2$. By Lemma 2.3 in Chapter 5, $\widetilde{h}_i$ is a blowing-down operation from the flow $\mathrm{sus}(f_i)$ to the flow $\mathrm{sus}(R_\mu)$, $i = 2$, 3.

It follows from $\chi(f_2^t, h_2) = \chi(f_3^t, h_3)$ that the sets $\chi[\mathrm{sus}(f_2), \widetilde{h}_2]$ and $\chi[\mathrm{sus}(f_3), \widetilde{h}_3]$ are equivalent, and therefore the characteristic sets $\chi(f_2, \widehat{h}_2)$ and $\chi(f_3, \widehat{h}_3)$ of the Denjoy homeomorphisms f_2 and f_3 are carried one into the other by a rotation of the circle S^1, that is, $\chi(f_2, \widehat{h}_2) \equiv \chi(f_3, \widehat{h}_3)$ in the notation of §2.5 in Chapter 5. According to Theorem 2.4 in Chapter 5, f_2 and f_3 are conjugate. This implies that the suspensions $\mathrm{sus}(f_2)$ and $\mathrm{sus}(f_3)$ are topologically equivalent. □

Thus, the characteristic class of a Denjoy flow f^t is a complete topological invariant. It includes information about the Poincaré rotation number of f^t, the cardinality of the set of components of $T^2 \setminus \Omega(f^t)$, and their mutual arrangement.

We recall that a characteristic class is a family of pairwise commensurable sets, each of which is an at most countable collection of trajectories of an irrational winding. The next theorem solves the problem of constructing from a given characteristic class a Denjoy flow with that characteristic class.

THEOREM 1.6. *Let χ_0 be an at most countable collection of trajectories of an irrational winding f_0^t, and let χ be the family of sets commensurable with χ_0. Then there exists a Denjoy flow f^t such that $\chi(f^t) = \chi$.*

PROOF. The irrational winding f_0^t induces on the zero meridian n_0 a Poincaré mapping that is a rotation R_α with irrational α. Obviously, $\chi_0' = \chi_0 \cap n_0$ is an at most countable family of orbits of R_α. According to Theorem 2.5 in Chapter 5, there exist a Denjoy homeomorphism $f \colon S^1 \to S^1$ and a continuous mapping $\widehat{h} \colon S^1 \to S^1 = n_0$ semiconjugating f and R_α such that $\chi_0' = \chi(f, \widehat{h})$. Therefore, there is a blowing-down operation h from the Denjoy flow $f^t \stackrel{\mathrm{def}}{=} \mathrm{sus}(f)$ to the irrational winding $f_0^t = \mathrm{sus}(R_\alpha)$ such that $\chi_0 = \chi(f^t, h)$. By Theorem 1.5, $\chi(f^t) = \chi$. □

DEFINITION. Let f^t be a Denjoy flow. The number of components of the set $T^2 \setminus \Omega(f^t)$ is called the *characteristic* of the flow f^t. Obviously, the characteristic is a topological invariant of the flow, but not a complete invariant, as the next result shows.

LEMMA 1.15. *There exists a continuum of topologically nonequivalent Denjoy flows with the same rotation number and with characteristic equal to* 2.

PROOF. According to Lemma 2.9 in Chapter 5, there exists a continuum $\{f_\nu\}$ of nonconjugate Denjoy homeomorphisms with the same rotation number μ and with characteristic equal to 2. We show that the suspensions $\operatorname{sus}(f_\nu)$ are pairwise not topologically equivalent if μ is a transcendental number, that is, is not a root of an algebraic equation with integer coefficients. Indeed, if $\operatorname{sus}(f_{\nu_1})$ and $\operatorname{sus}(f_{\nu_2})$ are topologically equivalent, then by Theorem 1.4

$$\mu = \frac{c + d\mu}{a + b\mu},$$

where $\begin{pmatrix} a & b \\ c & d \end{pmatrix}$ is a unimodular integer matrix. But the last equality contradicts the fact that μ is transcendental. Thus, $\{\operatorname{sus}(f_\nu)\}$ is the desired family of flows. □

REMARK. For Denjoy flows f^t with characteristic 1 (that is, the set $T^2 \backslash \Omega(f^t)$ is connected), just as for transitive flows, the Poincaré rotation number is a complete invariant of topological equivalence up to recomputation by the formula (1.8).

Appendix. Polynomial Cherry flows. Following [**79**], we show that there are Cherry flows given by polynomial vector fields of degree 1.

To simplify the notation we define a universal covering $\widehat{\pi}\colon \mathbb{R}^2 \to T^2$ by the formula $\widehat{\pi}(x, y) = (e^{ix}, e^{iy})$. Then the dynamical system

$$(\mathrm{A}_\alpha) \qquad \begin{cases} \dot{x} = 1 + \cos x + \sin y, \\ \dot{y} = \alpha(1 + \sin x) + \cos y \end{cases}$$

determines on the torus a polynomial vector field $\vec{X}_\alpha$ of degree 1, which in turn induces a flow f_α^t on the torus. We show that there is a continuum of values α in $[0, 1]$ for which the flows f_α^t are Cherry flows.

It suffices to investigate the system (A_α) on the square $K : 0 \le x, y \le 2\pi$, since $\widehat{\pi}$ maps this square onto the torus, and K is a fundamental domain of the group of covering transformations of $\widehat{\pi}$.

We shall find and investigate the equilibrium states of the system (A_α). To do this we form the system

$$\begin{cases} P_\alpha(x, y) = 1 + \cos x + \sin y = 0, \\ Q_\alpha(x, y) = \alpha(1 + \sin x) + \cos y = 0. \end{cases}$$

Both equalities are possible when $\cos x \le 0$, $\sin y \le 0$, and $\cos y \le 0$; that is, the equilibrium states of the system (A_α) (in K) lie in the rectangle $Q : \pi/2 \le x \le 3\pi/2$, $\pi \le y \le 3\pi/2$. The curves $P_\alpha(x, y) = 0$, $Q_\alpha(x, y) = 0$ in Q are determined by the respective relations $y = 2\pi - \arcsin(1 + \cos x)$, $y = \pi + \arccos[\alpha(1 + \sin x)]$ and intersect at the two points $S_\alpha(x_\alpha, y_\alpha)$, $F(3\pi/2, 3\pi/2)$ (Figure 6.9).

We have that

$$\Delta = \begin{vmatrix} \frac{\partial P_\alpha}{\partial x} & \frac{\partial P_\alpha}{\partial y} \\ \frac{\partial Q_\alpha}{\partial x} & \frac{\partial Q_\alpha}{\partial y} \end{vmatrix} = \begin{vmatrix} -\sin x & \cos y \\ \alpha \cos x & -\sin y \end{vmatrix} = \sin x \sin y - \alpha \cos x \cos y,$$

$$\sigma = \frac{\partial P_\alpha}{\partial x} + \frac{\partial Q_\alpha}{\partial y} = -\sin x - \sin y,$$

$$\Delta[F(3\pi/2, 3\pi/2)] = 1 > 0, \quad \sigma[F(3\pi/2, 3\pi/2)] = 2 > 0,$$

and hence F is a simple unstable node.

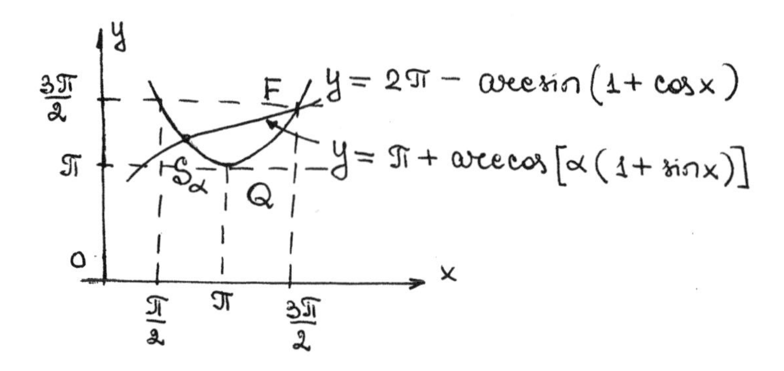

FIGURE 6.9

Since $\Delta(S_\alpha) = [(\alpha^2 - 1)\cos x - 1]\sin x + \alpha^2 \cos x < 0$ and $\sigma(S_\alpha) = 1 + \cos x - \sin x = 1 - \sqrt{2}\sin(x - \pi/4) \leq 0$ for $\pi/2 \leq x \leq 3\pi/2$, it follows that the roots of the equation $\lambda^2 - \sigma\lambda + \Delta = 0$ are real and have different signs. Therefore, S_α is a saddle for all $\alpha \in [0,1]$ [**3**].

Further, $P_\alpha(0,y) = P_\alpha(2\pi, y) = 2 + \sin y > 0$ and $Q_\alpha(x,0) = Q_\alpha(x, 2\pi) = 1 + \alpha(1 + \sin x) > 0$, and hence the zero meridian n_0 and the zero parallel n_1 of the torus (that is, the curves into which the respective lines $x = 0$ and $y = 0$ project) are contact-free cycles of the flows f^t_α, $\alpha \in [0,1]$. This yields a qualitative picture of the phase portrait of the dynamical system (A_α) (Figure 6.10).

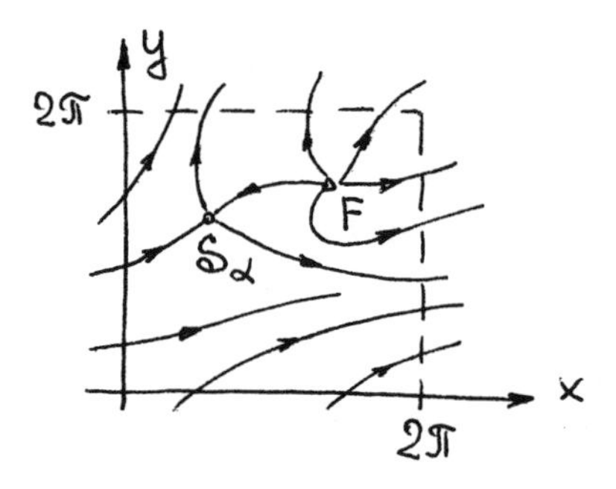

FIGURE 6.10

The ω-separatrix of S_α intersecting the line $x = 0$ intersects the family of lines $x = k$, $k \in \mathbb{Z}$, at most at countably many points. Therefore, there exists a positive semitrajectory of the dynamical system (A_α) that intersects the lines $x = 2\pi k$, $k \in \mathbb{N}$, that is, is a curvilinear ray (moreover, there is a continuum of such semitrajectories). Consequently, the flow f^t_α has rotation number μ_α for each value of $\alpha \in [0,1]$.

We show that μ_α depends continuously on α. Indeed, fix $\alpha_0 \in [0,1]$ and let l^+_0 be a positive semitrajectory of (A_{α_0}) intersecting the lines $x = 2\pi k$, $k \in \mathbb{N}$. Denote by $\overset{\frown}{x_0 x_k}$ the arc of l^+_0 with endpoints x_0 and x_k lying on the respective lines $x = 0$ and $x = 2\pi k$ (note that l^+_0 intersects each line $x = 2\pi k$, $k \in \mathbb{N}$ exactly once). On the line $x = 2\pi k$ take the point x^+_0 (respectively, x^-_0) that lies above (respectively, below) x_k, is congruent to x_0, and is closest to x_k ($x^\pm_0 \neq x_k$ if x_k is congruent to x_0) (Figure 6.11). We approximate the union of the arc $\overset{\frown}{x_0 x_k}$ and the segment $[x_k, x^+_0]$

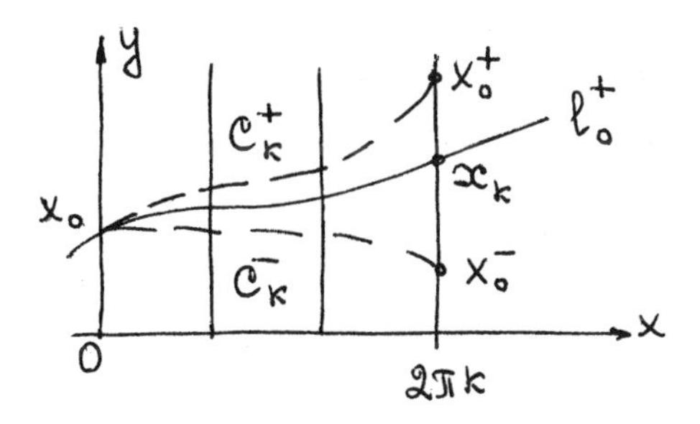

FIGURE 6.11

(respectively, $[x_0^-, x_k]$) by a curve C_k^+ (respectively, C_k^-) that is transversal to the trajectories of the system (A_{α_0}).

Since the point x_k^+ (x_k^-) is congruent to x_0, there exists a transformation γ_k^+ (γ_k^-) in the covering transformation group that carries x_0 into the point x_k^+ (x_k^-). Let $x = x^+(t)$, $y = y^+(t)$ be parametric equations of the curve

$$C^+ = \cup(\gamma_k^+)^n(C_k^+),$$

where the union is over all $n \in \mathbb{Z}$. Then C^+ is transversal to the trajectories of the system (A_{α_0}) and projects into a closed curve on the torus. It follows from the congruence of x_0 and x_0^+ that the limit $\lim_{t\to\infty} y^+(t)/x^+(t) \stackrel{\text{def}}{=} \mu_k^+$ is a rational number.

The number $\mu_k^- \in Q$ is introduced similarly for the curve $C^- = \cup(\gamma_k^-)^n(C_k^-)$, $n \in \mathbb{Z}$.

It follows from the construction that $\mu_k^- < \mu_{\alpha_0} < \mu_k^+$. For values of α sufficiently close to α_0 the curves C^+ and C^- are transversal to the trajectories of the system (A_α), so $\mu_k^- \leq \mu_\alpha \leq \mu_k^+$. On the other hand, μ_k^+ and μ_k^- tend to μ_{α_0} as $k \to +\infty$. This implies that the rotation number μ_α depends continuously on the parameter α.

Moreover, we have the following theorem.

THEOREM. *Suppose that in the space $\mathfrak{X}^r(T^2)$ of flows, $r \geq 1$, there is a family of flows f_α^t, $\alpha \in [0,1]$, depending continuously on a parameter α (that is, the mapping $\alpha \mapsto f_\alpha^t$ defines a continuous curve in the space $\mathfrak{X}^r(T^2)$), and suppose that the flows f_α^t, $\alpha \in [0,1]$, have a common contact-free cycle C. Assume also that for each $\alpha \in [0,1]$ the flow f_α^t has a semitrajectory intersecting C countably many times. Then there is a rotation number* $\operatorname{rot}(f_\alpha^t)$ *for all $\alpha \in [0,1]$, and it depends continuously on α.*

The proof of the theorem is left to the reader as an exercise.

A direct qualitative investigation of the dynamical systems $(\mathrm{A}_0) : \dot{x} = 1 + \cos x + \sin y$, $\dot{y} = \cos y$ and $(\mathrm{A}_1) : \dot{x} = 1 + \cos x + \sin y$, $\dot{y} = 1 + \sin x + \cos y$ shows that their phase portraits have the form pictured in Figure 6.12, a), b), respectively. Consequently, $\mu_0 = \operatorname{rot}(f_0^t) = 0$ and $\mu_1 = \operatorname{rot}(f_1^t) = 1$. Since the rotation number depends continuously on the parameter, there is a continuum of values α for which the rotation number $\mu_\alpha = \operatorname{rot}(f_\alpha^t)$ is irrational. According to Theorem 1.2, the corresponding flows f_α^t have nontrivial recurrent semitrajectories. These flows are Cherry flows because of the presence of a hyperbolic (or structurally stable) saddle.

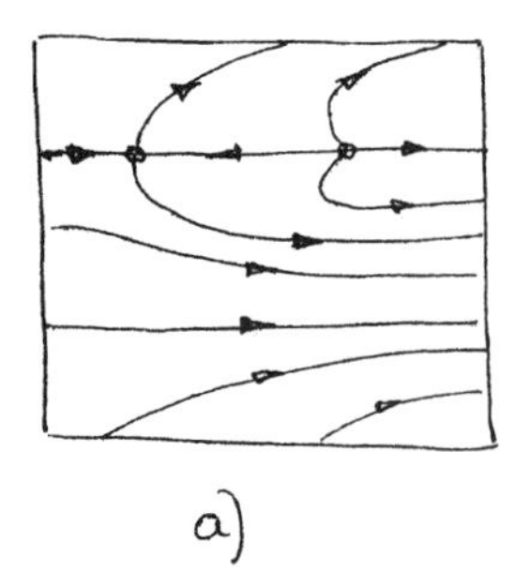

a)

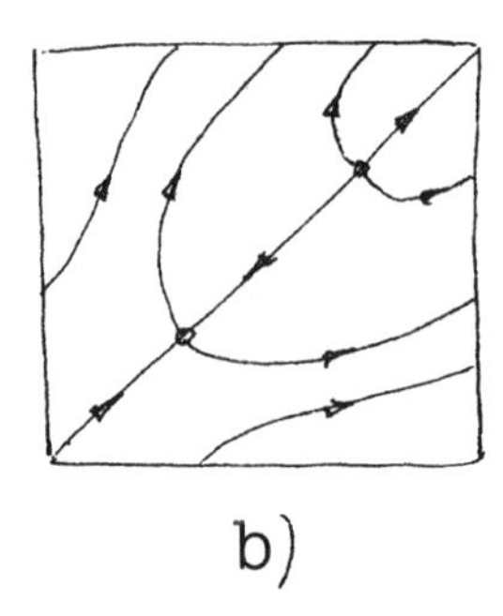

b)

FIGURE 6.12

§2. The homotopy rotation class

The asymptotic behavior of the trajectories of a flow covering an irreducible flow on a closed orientable surface of genus greater than 1 differs essentially from the asymptotic behavior of the trajectories of a flow covering an irreducible flow on the torus (genus 1). While the trajectories of a flow covering an irreducible flow on the torus have two diametrically opposite asymptotic directions (the corresponding flow has two accessible points on the absolute, and this enables us to introduce a topological invariant like the Poincaré rotation number), the trajectories of a flow covering an irreducible flow on a closed orientable surface of genus greater than 1 have a continuum of asymptotic directions, and this leads to a new topological invariant—the homotopy rotation class. This section is devoted to the definition of the homotopy rotation class and to an investigation of its properties.

We first describe the Lobachevsky plane, which is a universal covering of a closed orientable surface of genus > 1.

2.1. Lobachevsky geometry and uniformization. The non-Euclidean geometry of Lobachevsky is realized as the following model, proposed by Poincaré. Take the open unit disk $\Delta = \{z : |z| < 1\}$ in the complex z-plane and introduce on Δ the metric

$$d^2 s = \frac{4(dx^2 + dy^2)}{(1 - x^2 - y^2)^2},$$

where $z = x + iy$, (x, y) being the Cartesian coordinates in $\mathbb{R}^2$.

The disk Δ with this metric is called the *Lobachevsky plane* (the Poincaré model).

Arcs of Euclidean circles orthogonal to the boundary $\partial\Delta$ of Δ (in particular, the arcs of Euclidean lines passing through the origin of coordinates and orthogonal to $\partial\Delta$) are the geodesics of the Lobachevsky plane.

The boundary $S_\infty \stackrel{\text{def}}{=} \partial\Delta = \{z : |z| = 1\}$ of Δ is called the *absolute.*

The set $I(\Delta)$ of orientation-preserving isometries of the Lobachevsky plane coincides with the group of fractional linear transformations

$$z \mapsto \frac{pz + q}{\overline{q}z + \overline{p}},$$

where $p, q \in \mathbb{C}$ and $|p|^2 - |q|^2 = 1$. Each such nonidentity transformation has one or two fixed points in the closed disk $\Delta \cup S_\infty$. This leads to the following classification of the isometries of the Lobachevsky plane.

An isometry $\gamma \in I(\Delta)$ is said to be *elliptic* if it has a fixed point interior to Δ.

An isometry $\gamma \in I(\Delta)$ is said to be *hyperbolic* if it has two fixed points on the absolute.

An isometry $\gamma \in I(\Delta)$ is said to be *parabolic* if it has exactly one fixed point, which lies on the absolute.

Any orientation-preserving isometry of the Lobachevsky plane is one of these transformations.

An important class of subgroups of $I(\Delta)$ is made up of the Schottky groups.

A *Schottky group* Γ_p of genus $p \geq 1$ is defined to be a finitely generated subgroup of $I(\Delta)$ consisting solely of hyperbolic isometries and having $2p$ generators $\{a_1, b_1, \dots, a_p, b_p\} \subset I(\Delta)$ which satisfy the relation

$$a_1 b_1 a_1^{-1} b_1^{-1} \cdots a_p b_p a_p^{-1} b_p^{-1} = \mathrm{id}\,.$$

According to the Klein–Poincaré–Koebe uniformization theorem [**43**], for any closed orientable surface $\mathcal{M}_p$ of genus $p \geq 2$ there exists a Schottky group Γ_p of genus p acting freely on Δ and with Δ as a domain of discontinuity (see Chapter 1) such that $\mathcal{M}_p \cong \Delta/\Gamma_p$, and the natural projection $\pi\colon \Delta \to \Delta/\Gamma_p \cong \mathcal{M}_p$ is a universal covering of the surface $\mathcal{M}_p$. Moreover, the limit set of the group Γ_p (the set of accumulation points of the orbits $\Gamma_p(z_0) = \{\gamma(z_0) : \gamma \in \Gamma_p\}$) coincides with the absolute S_∞, and the set of fixed points of the isometries in Γ_p is countable and dense in S_∞.

The Lobachevsky plane Δ is a Riemannian manifold of constant negative curvature -1. The covering $\pi\colon \Delta \to \Delta/\Gamma_p$ induces on $\mathcal{M}_p$ a metric of constant negative curvature -1.

Points $a, b \in \Delta$ are said to be congruent with respect to the group Γ_p if there exists an element $\gamma \in \Gamma_p$ such that $\gamma(a) = b$. Obviously, $\pi(a) = \pi(b)$ for congruent points. Conversely, if points $a, b \in \Delta$ belong to the complete inverse image $\pi^{-1}(m_0)$ of a point $m_0 \in \mathcal{M}_p$, then a and b are congruent.

Unless stated otherwise, it is assumed below that $\mathcal{M}_p = \Delta/\Gamma_p$ is a closed orientable surface of genus $p \geq 2$, and Γ_p is a Schottky group of genus $p \geq 2$ acting freely on Δ and with domain of discontinuity Δ.

2.2. The axes of hyperbolic isometries. Consider a hyperbolic isometry $\gamma \in \Gamma_p$ in a Schottky group Γ_p. One fixed point γ^+ of γ is attracting, and one γ^- is repelling. The iterates $\gamma^n(z_0)$ of any point z_0 different from γ^+ and γ^- (z_0 can also lie on S_∞) tend to γ^+ as $n \to +\infty$ and to γ^- as $n \to -\infty$.

We draw the geodesic $O(\gamma)$ through γ^+ and γ^-. Since there is a unique (up to orientation) geodesic joining the points γ^- and γ^+ in Δ, and since the isometry γ carries geodesics into geodesics, it follows that $\gamma[O(\gamma)] = O(\gamma)$.

The geodesic $O(\gamma)$ is called the *axis* of the element $\gamma \in \Gamma_p$.

The axis $O(\gamma)$ is the unique geodesic on Δ that is invariant under γ.

The elements of Γ_p with respect to which the geodesic $O(\gamma)$ is invariant form an infinite cyclic subgroup $\{\gamma_1^n\}$, $n \in \mathbb{Z}$, generated by some $\gamma_1 \in \Gamma_p$.

The element γ_1 is called a *minimal* element of the axis $O(\gamma)$. This element (and also γ_1^{-1}) moves the axis along itself by a minimal distance.

LEMMA 2.1. *On a closed orientable surface $\mathcal{M}_p$ of genus $p \geq 2$ let L be a closed geodesic that is nonhomotopic to zero. In this case:*

1) *the inverse image $\pi^{-1}(L)$ is made up of countably many geodesics $\{\overline{L}_i\}_{i=1}^\infty$ of the Lobachevsky plane;*

2) *each geodesic* $\overline{L}_i \in \pi^{-1}(L)$ *is the axis of some element* $\gamma_i \in \Gamma_p$;

3) *any geodesics* $\overline{L}_i, \overline{L}_j \in \pi^{-1}(L)$, $i \neq j$, *do not have common endpoints*;

4) *if* $\{\overline{L}_{i_n}\}_{n=1}^{\infty}$ *is a subsequence of distinct geodesics* $\overline{L}_{i_n}$ *in* $\pi^{-1}(L)$ *with endpoints* $\sigma_1(\overline{L}_{i_n})$ *and* $\sigma_2(\overline{L}_{i_n})$, *and the sequence* $\{\sigma_1(\overline{L}_{i_n})\}$ *converges to a point* $\sigma \in S_\infty$, *then the sequence* $\{\sigma_2(\overline{L}_{i_n})\}$ *also converges to* σ, *and, moreover, the topological limit*[1] *of the geodesics* $\{\overline{L}_{i_n}\}$ *is equal to* σ.

PROOF. By the covering homotopy theorem and the fact that π is a local isometry, there exists a geodesic $\overline{L} \subset \Delta$ such that $\pi(\overline{L}) = L$. Since L is closed, there are two points z_1 and z_2 on $\overline{L}$ such that $\pi(z_1) = \pi(z_2) = z$, and the distance between z_1 and z_2 is equal to the length of the geodesic L. Obviously, L has only finitely many self-intersections, and it can be assumed that there is not a self-intersection at z. By the equality $\pi(z_1) = \pi(z_2)$, there exists a $\gamma \in \Gamma_p$ such that $\gamma(z_1) = z_2$. Since there is not a self-intersection at $z \in L$, it follows that $\gamma(\overline{L}) = \overline{L}$; that is, $\overline{L}$ is the axis of the element γ, $\overline{L} = O(\gamma)$. The assertion 2) is proved.

By construction, the distance between z_1 and z_2 is equal to the length of L. Therefore, γ generates an infinite cyclic subgroup $\Gamma(\overline{L})$ of elements with respect to which $\overline{L}$ is invariant.

The group Γ_p is a group with $2p \geq 4$ generators and a single defining relation. Therefore, the factor group $\Gamma_p/\Gamma(\overline{L})$ is countable [**53**], and the complete inverse image $\pi^{-1}(L) = \{\gamma(\overline{L}) : \gamma \in \Gamma_p\}$ consists of countably many geodesics. The assertion 1) is proved.

We prove 3). Assume the contrary. Suppose that geodesics $\overline{L}_i, \overline{L}_j \in \pi^{-1}(L)$, $i \neq j$, have a common endpoint $\sigma_1 \in S_\infty$. The endpoint σ_{2i} of $\overline{L}_i$ is different from the endpoint σ_{2j} of $\overline{L}_j$ because otherwise $\overline{L}_i = \overline{L}_j$ (Figure 6.13).

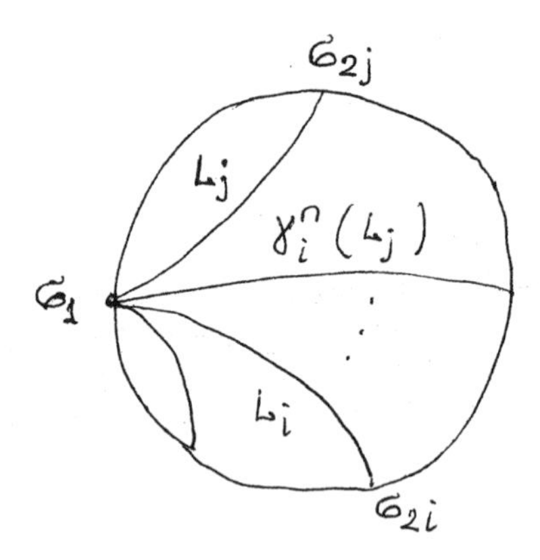

FIGURE 6.13

According to 2), $\overline{L}_i$ is the axis of some element $\gamma_i \in \Gamma_p$. Suppose for definiteness that $\sigma_1 = \gamma_i^-$ and $\sigma_{2i} = \gamma_i^+$. Then $\gamma_i^n(\sigma_{2j}) \to \sigma_{2i}$, and $\gamma_i^n(\overline{L}_j) \to \overline{L}_i$ as $n \to \infty$. The geodesics $\gamma_i^n(\overline{L}_j)$ project onto the geodesic L on $\mathcal{M}_p$. Consequently, a neighborhood of any point $m \in L$ intersects L in a countable set of arcs, which contradicts the closedness of L. This proves 3).

[1] The topological limit of a sequence of sets $\{A_i\}_{i=1}^{\infty}$ is defined to be the union of all possible limits of sequences $\{a_i\}_{i=1}^{\infty}$ with $a_i \in A_i$.

It remains to prove 4). If the sequence $\{\sigma_2(\overline{L}_{i_n})\}$ has a subsequence converging to $\sigma_* \neq \sigma$, then the subsequence of corresponding geodesics would converge to the geodesic joining σ to σ_*. Then some neighborhood of a point on $\mathcal{M}_p$ would intersect L in countably many arcs, which is impossible. Therefore, $\sigma_2(\overline{L}_{i_n}) \to \sigma$.

From this and the fact that the geodesics in Δ are Euclidean circles perpendicular to S_∞ it follows that the topological limit of the sequence $\{\overline{L}_{i_n}\}$ is equal to σ. □

The next result is a direct consequence of the proof of 3).

COROLLARY 2.1. *If two elements $\gamma_1, \gamma_2 \in \Gamma_p$ have a common fixed point, then their axes coincide (that is, they have two common fixed points). Two axes either coincide or do not have common endpoints.*

LEMMA 2.2. *Let C be a simple closed curve that is nonhomotopic to zero on a closed orientable surface $\mathcal{M}_p$ of genus $p \geq 2$. In this case:*

1) *the inverse image $\pi^{-1}(C)$ consists of a countable set of disjoint curves, each with two endpoints lying on the absolute;*

2) *for each curve $\overline{C} \in \pi^{-1}(C)$ there exists an element $\gamma \in \Gamma_p$, $\gamma \neq \mathrm{id}$, with respect to which $\overline{C}$ is invariant, and $\overline{C}$ and the axis $O(\gamma)$ have common endpoints;*

3) *an element $\gamma \in \Gamma_p$ with $\gamma \neq \mathrm{id}$ that carries some point of a curve $\overline{C} \in \pi^{-1}(C)$ into a point on $\overline{C}$ leaves the whole curve $\overline{C}$ invariant;*

4) *distinct curves $\overline{C}_1, \overline{C}_2 \in \pi^{-1}(C)$ do not have common endpoints;*

5) *if $\{\overline{C}_i\}_1^\infty$ is a sequence of distinct curves $\overline{C}_i$ in $\pi^{-1}(C)$ with endpoints $\sigma_1(\overline{C}_i)$ and $\sigma_2(\overline{C}_i)$ and if the sequence $\{\sigma_1(\overline{C}_i)\}_1^\infty$ of points converges to a point $\sigma \in S_\infty$, then the sequence $\{\sigma_2(\overline{C}_i)\}_1^\infty$ also converges to σ, and, moreover, the topological limit of the curves $\{\overline{C}_i\}_1^\infty$ is σ.*

PROOF. As in the proof of Lemma 1.1, we construct a curve $\overline{C} \subset \Delta$ with endpoints lying on the absolute and such that $\pi(\overline{C}) = C$ and there exists an element $\gamma \in \Gamma_p$ with respect to which $\overline{C}$ is invariant. There is a unique geodesic $\overline{L} \subset \Delta$ with endpoints coinciding with those of $\overline{C}$. Since $\gamma(\overline{C}) = \overline{C}$, it follows that $\gamma(\overline{L}) = \overline{L}$. This implies that $\pi(\overline{L}) = L$ is a closed geodesic on $\mathcal{M}_p$ that is homotopic to C. Since C is simple, so is L. We get from the properties of a covering that for each curve $\overline{C}_\nu \in \pi^{-1}(C)$ with endpoints on the absolute there exists a geodesic $\overline{L}_\nu \in \pi^{-1}(L)$ such that $\overline{C}_\nu$ and $\overline{L}_\nu$ have common endpoints. The required assertions follow from this and Lemma 2.1. □

2.3. Asymptotic directions. The concept of a curvilinear ray is defined as in §1.2, to be a half-closed infinite curve l (on the surface $\mathcal{M}_p$ or on the universal covering Δ) that has a lifting $\bar{l} \subset \Delta$ (if $l \subset \Delta$, then $\bar{l} = l$) such that $\bar{l}$ leaves any compact set of the Lobachevsky plane Δ).

It follows from the properties of the group Γ_p (in particular, the fact that Γ_p consists of isometries) that all the curves in Δ congruent to a curvilinear ray $\bar{l} \subset \Delta$ are also curvilinear rays. Consequently, if $l \subset \mathcal{M}_p$ is a curvilinear ray, then all the curves in the complete inverse image $\pi^{-1}(l)$ are curvilinear rays.

As in §1.2, when we call a semitrajectory of some flow on $\mathcal{M}_p$ or Δ a curvilinear ray, we shall have in mind the natural parametrization of this semitrajectory with respect to time.

If a semitrajectory $\overline{l}^{\pm}$ of the flow $\overline{f}^t$ on Δ is a curvilinear ray, then its ω- (α-) limit set lies on the absolute.

DEFINITION. Let $\overline{l}^{()}$ be a positive (negative) semitrajectory of a flow $\overline{f}^t$ on Δ, and let it be a curvilinear ray. We say that $\overline{l}^{()}$ has an *asymptotic direction* if its ω- (α-) limit set consists of a single point $\sigma(\overline{l}^{()})$ belonging to the absolute.

The point $\sigma(\overline{l}^{()})$ is said to be *accessible* by the semitrajectory $\overline{l}^{()}$.

Let $l^{()}$ be a semitrajectory of a flow f^t on $\mathcal{M}_p$, and let it be a curvilinear ray. If at least one semitrajectory $\overline{l}^{()} \in \pi^{-1}(l^{()})$ of a covering flow $\overline{f}^t$ on Δ has an asymptotic direction, then any semitrajectory in $\pi^{-1}(l^{()})$ also has an asymptotic direction. In this case we say that $l^{()}$ has an asymptotic direction.

Everywhere below we consider flows on $\mathcal{M}_p = \Delta/\Gamma_p$ with finitely many equilibrium states (unless otherwise stated).

THEOREM 2.1. *Suppose that a semitrajectory $l^{()}$ of a flow f^t on $\mathcal{M}_p$ $(p \geq 2)$ intersects a contact-free cycle C infinitely many times. Then $l^{()}$ is a curvilinear ray having an asymptotic direction.*

PROOF. For definiteness we assume that $l^{()}$ is a positive semitrajectory l^+.

By Lemma 2.2 in Chapter 2, C is nonhomotopic to zero (even nonhomologous to zero). Therefore, any curve $\overline{C} \in \pi^{-1}(C)$ separates Δ in view of Lemma 2.2.

Let $\overline{f}^t$ be a covering flow for f^t. Since $\overline{C}$ is a contact-free arc of $\overline{f}^t$, any covering semitrajectory $\overline{l}^+ \in \pi^{-1}(l^+)$ intersects any curve $\overline{C} \in \pi^{-1}(C)$ at no more than one point. Consequently, $\overline{l}^+$ intersects a countable family $\{\overline{C}_i\}$ of disjoint curves in $\pi^{-1}(C)$ (Figure 6.14). Denote by $\sigma_1(\overline{C}_i)$ and $\sigma_2(\overline{C}_i)$ the endpoints of $\overline{C}_i$ as pictured in Figure 6.14. Simple geometric considerations show that the points in $\{\sigma_1(\overline{C}_i)\}_1^\infty$ form a bounded monotone sequence, and hence this sequence has a limit $\sigma \in S_\infty$. We get from Lemma 2.2, 5) that $\overline{C}_i \to \sigma$, and therefore $\omega(\overline{l}^+) = \sigma$. $\square$

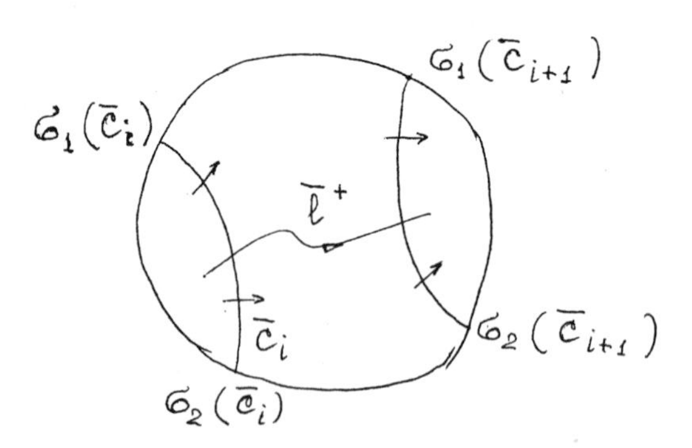

FIGURE 6.14

DEFINITION. We say that a pair of points $x_1, x_2 \in S_\infty$ is separated on the absolute by a pair $x_3, x_4 \in S_\infty$ if one arc in the set $S \setminus \{x_3, x_4\}$ contains x_1 and the other contains x_2.

The next result follows immediately from the proof of Theorem 2.1.

COROLLARY 2.2. *Suppose that the conditions of Theorem* 2.1 *hold and a covering semitrajectory* $\overline{l}^{()}$ *for* $l^{()}$ *intersects the family* $\{\overline{C}_i\}$ *of curves in the inverse image* $\pi^{-1}(C)$. *Then* $\sigma^{()}(\overline{l}^{()}) = \omega(\overline{l}^{()})$ $(\alpha(\overline{l}^{()}))$ *coincides with the topological limit of the family* $\{\overline{C}_i\}$, *and hence for any point* $\overline{m} \in S_\infty$ *different from* $\sigma^{()}(\overline{l}^{()})$ *there is an index* i_0 *such that the pair of points* $\overline{m}$, $\sigma^{()}(\overline{l}^{()})$ *is separated by the pair of points* $\sigma_1(\overline{C}_i)$, $\sigma_2(\overline{C}_i)$ *for all* $i \geq i_0$, *where* $\sigma_1(\overline{C}_i)$ *and* $\sigma_2(\overline{C}_i)$ *are the endpoints of* $\overline{C}_i$.

LEMMA 2.3. *Let* f^t *be a flow on a closed orientable surface* $\mathcal{M}_p$ *of genus* $p \geq 2$, *and let* $\overline{f}^t$ *be a flow on* Δ *covering it. Then the semitrajectories of* f^t *and* $\overline{f}^t$ *listed below are curvilinear rays and have an asymptotic direction*:

1) *a nontrivial recurrent semitrajectory of* f^t;

2) *a semitrajectory of* $\overline{f}^t$ *projecting into a closed trajectory of* f^t *that is nonhomotopic to zero*;

3) *a nonclosed semitrajectory of* f^t *whose limit set contains a nontrivial recurrent semitrajectory, or a closed trajectory that is nonhomotopic to zero, or a one-sided contour that is nonhomotopic to zero.*

PROOF. Let $l^{()}$ be a nontrivial recurrent semitrajectory of f^t. According to Lemma 2.3 in Chapter 2, there exists a contact-free cycle intersecting $l^{()}$. This and Theorem 2.1 yield 1). The assertion 2) follows from Lemma 2.2, and 3) follows from 1) and 2). □

COROLLARY 2.3. *Assume the conditions of Lemma* 2.3. *Then any semitrajectory of* f^t *or* $\overline{f}^t$ *that is a curvilinear ray has an asymptotic direction. Moreover, only those semitrajectories in Lemma* 2.3 *are curvilinear rays.*

PROOF. This follows from Lemma 2.3 and Theorem 3.6 in Chapter 2 (the catalogue of limit sets). See also the proof of Theorem 1.1. □

REMARK 1. We remark that Corollary 2.3 fails in general for flows f^t on $\mathcal{M}_p$ with an infinite set of equilibrium states.

REMARK 2. The question of the existence of an asymptotic direction of a curvilinear ray has been answered most thoroughly in the following papers.

1. D. V. Anosov, *On the behavior of trajectories in the Euclidean plane that cover trajectories of flows on closed surfaces*, I, II, Izv. Akad. Nauk SSSR Ser. Mat. **51** (1987), 16–43, **52** (1988), 451–478; English transl. in Math. USSR Izv. **30** (1988), **32** (1989).

2. D. V. Anosov, *On infinite curves on the torus and closed surfaces of negative Euler characteristic*, Trudy Mat. Inst. Steklov **185** (1988), 30–53; English transl. in Proc. Steklov Inst. Math. **1990**, issue 2.

In particular, Anosov proved the following result.

THEOREM. *If the set of equilibrium states of a flow* f^t *on a closed surface* $\mathcal{M}$ *with nonpositive Euler characteristic is contractible to a point on* $\mathcal{M}$, *then each semitrajectory of a covering flow* $\overline{f}^t$ *either is bounded or tends to some point of the absolute.*

2.4. Arithmetic properties of the homotopy rotation class. We fix a group Γ_p, $p \geq 2$, of hyperbolic isometries of the Lobachevsky plane Δ. With respect to this group the points of the absolute S_∞ break up into two classes.

DEFINITION. A point $\sigma \in S_\infty$ is said to be *rational* if σ is a fixed point of some element $\gamma \in G_p$ with $\gamma \neq \mathrm{id}$. Points of the absolute that are not rational are said to be *irrational.*

The set of rational points (and hence the set of irrational points) is invariant under the action of Γ_p. Indeed, if $\sigma \in S_\infty$ is a fixed point of an element $\gamma \in \Gamma_p$ and $\delta \in \Gamma_p$, then $\delta(\sigma)$ is a fixed point of the element $\delta \circ \gamma \circ \delta^{-1} \in \Gamma_p$.

Suppose that a flow f^t on $\mathcal{M}_p = \Delta/\Gamma_p$ has a semitrajectory $l^{()}$ that is a curvilinear ray and hence has an asymptotic direction (Corollary 2.3). We consider an arbitrary semitrajectory $\bar{l}^{()}$ of a flow $\bar{f}^t$ on Δ such that $\pi(\bar{l}^{()}) = l^{()}$, and let $\sigma(\bar{l}^{()})$ be a limit point of $\bar{l}^{()}$ belonging to the absolute.

DEFINITION. The *homotopy rotation class* of the semitrajectory $l^{()}$ of the flow f^t on $\mathcal{M}_p$ is defined to be the set

$$\mu(l^{()}) = \bigcup_{\gamma \in \Gamma_p} \gamma[\sigma(\bar{l}^{()})].$$

In other words, the homotopy rotation class (HRC) of the semitrajectory $l^{()}$ is obtained by taking the union of the limit points of all semitrajectories covering $l^{()}$ (Figure 6.15), because $\gamma[\sigma(\bar{l}^{()})] = \sigma[\gamma(\bar{l}^{()})]$.

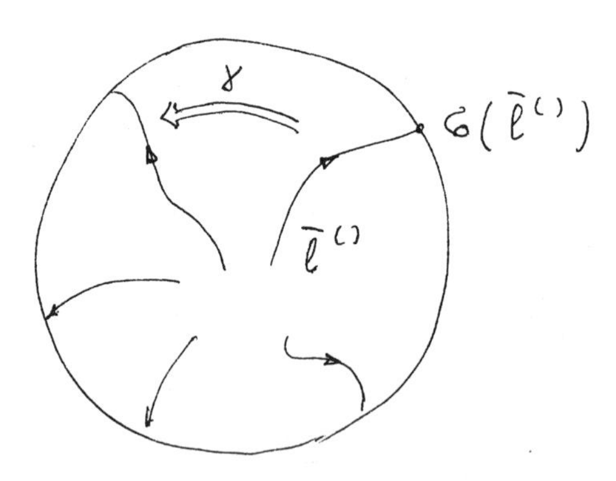

FIGURE 6.15

It follows from the invariance of the set of rational (and irrational) points with respect to the group Γ_p and the definition of the HRC that the HRC of a particular semitrajectory $l^{()}$ lies either in the set of rational points, or in the set of irrational points. Therefore, the following definition is unambiguous.

DEFINITION. The homotopy rotation class of a semitrajectory $l^{()}$ of a flow on $\mathcal{M}_p$ is said to be *rational* (*irrational*) if it consists of rational (irrational) points of the absolute.

THEOREM 2.2. 1) *The homotopy rotation class of a closed trajectory that is nonhomotopic to zero is rational, as is the HRC of a nonclosed semitrajectory whose limit set contains a closed trajectory nonhomotopic to zero or a closed one-sided contour nonhomotopic to zero.*

2) *The HRC of a nontrivial recurrent semitrajectory is irrational.*

PROOF. If l is a closed trajectory nonhomotopic to zero, then by Lemma 2.2, 2), each curve $\bar{l} \in \pi^{-1}(l)$ is invariant with respect to some element of the group Γ_p. Therefore, $\mu(l)$ is rational.

If $l^{()}$ is a nonclosed semitrajectory whose limit set contains a closed trajectory l_0 nonhomotopic to zero, then for any ε-neighborhood $\mathcal{U}_\varepsilon(l_0)$ of l_0 there is a moment of time beginning with which $l^{()}$ enters $\mathcal{U}_\varepsilon(l_0)$ and does not leave it again. This implies that the HRC $\mu(l^{()}) = \mu(l_0)$ is rational. Analogous arguments prove that the HRC is rational in the case when the limit set of $l^{()}$ contains a one-sided contour nonhomotopic to zero.

We prove 2). Let l^+ be a P^+ nontrivial recurrent semitrajectory, and let $\bar{l}^+$ be a semitrajectory on Δ covering it. Assume the contrary. Then $\sigma^+(\bar{l}^+) = \omega(\bar{l}^+) \in S_\infty$ is a fixed point of some element $\gamma \in \Gamma_p$. It can be assumed that $\sigma^+(\bar{l}^+) = \gamma^+$ is an attracting point of γ (otherwise γ^{-1} could be taken instead of γ), and γ a minimal element.

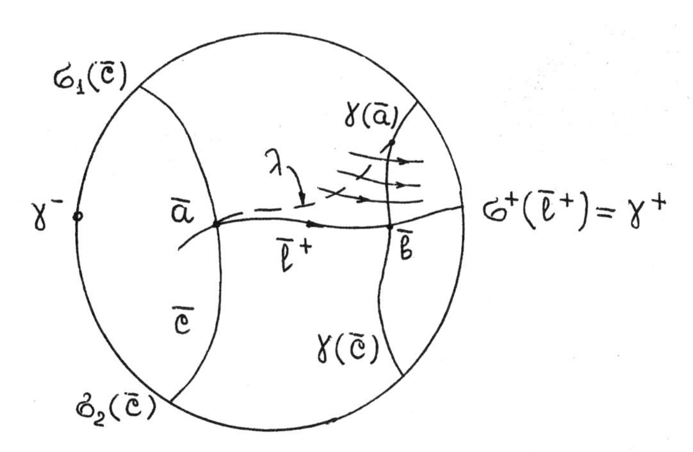

FIGURE 6.16

According to Lemma 2.3 in Chapter 2, there exists a contact-free cycle C intersecting l^+. Then from Theorem 2.1 and Corollary 2.2 it follows that there is a curve $\overline{C} \in \pi^{-1}(C)$ with endpoints $\sigma_1(\overline{C}), \sigma_2(\overline{C}) \in S_\infty$ such that $\bar{l}^+$ intersects $\overline{C}$, and the pair of points γ^+, γ^- is separated by the pair of points $\sigma_1(\overline{C})$, $\sigma_2(\overline{C})$ (Figure 6.16). Therefore, $\bar{l}^+$ intersects the curve $\gamma(\overline{C})$.

The curves $\overline{C}$ and $\gamma(\overline{C})$ separate Δ and are contact-free arcs of the covering flow $\bar{f}^t$. Consequently, $\bar{l}^+$ intersects $\overline{C}$ and $\gamma(\overline{C})$ in only a single point. Let $\bar{a} = \bar{l}^+ \cap \overline{C}$ and $\bar{b} = \bar{l}^+ \cap \gamma(\overline{C})$. We approximate the curve made up of the segments $\bar{a}\bar{b} \subset \bar{l}^+$ and $\gamma(\bar{a})\bar{b} \subset \gamma(\overline{C})$ by an arc λ joining $\bar{a}$ and $\gamma(\bar{a})$ and transversal to the flow $\bar{f}^t$ (see Figure 6.16). Then the curve $\overline{S} = \bigcup_{n\in\mathbb{Z}} \gamma^n(\lambda)$ is transversal to $\bar{f}^t$ and invariant with respect to γ, and it has endpoints γ^+ and γ^-.

Since $\gamma(\sigma^+) = \sigma^+$, it follows that $\sigma^+[\gamma^n(\overline{l}^+)] = \sigma^+(\overline{l}^+)$ for any $n \in \mathbb{Z}$. Therefore, the semitrajectories $\overline{l}^+(\gamma(\overline{a})) = \gamma(\overline{l}^+(\overline{a}))$ and $\overline{l}^+(\gamma^{-1}(\overline{a})) = \gamma^{-1}(\overline{l}^+(\overline{a}))$ and the arc $\overline{d}$ of $\overline{S}$ between $\overline{a}_1 = \gamma(\overline{a})$ and $\overline{a}_{-1} = \gamma^{-1}(\overline{a})$ form a curvilinear triangle T (Figure 6.17). A positive semitrajectory of $\overline{f}^t$ upon entering T across the arc $\overline{d}$ cannot leave T with increasing time.

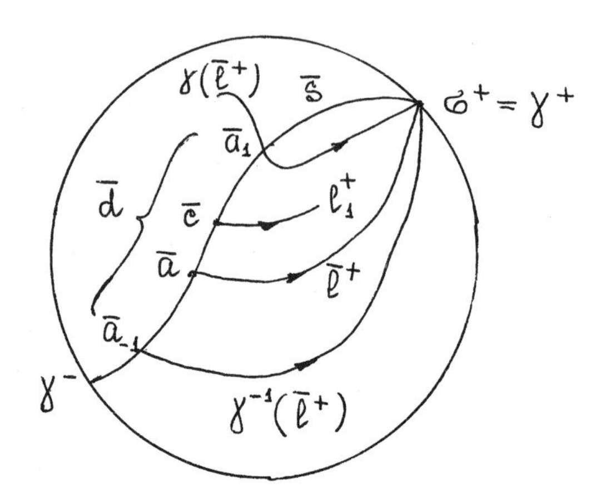

FIGURE 6.17

Since $\pi(\overline{l}^+)$ is a nontrivial recurrent semitrajectory, a semitrajectory $\overline{l}_1^+$ congruent to $\overline{l}^+$ (that is, there is an element $\gamma_1 \in \Gamma_p$ such that $\gamma_1(\overline{l}^+) = \overline{l}_1^+$) intersects $\overline{d}$ at a point $\overline{c}$ arbitrarily close to $\overline{a}$. It follows from the properties of the triangle T that $\omega(\overline{l}_1^+) = \sigma^+ = \gamma^+$; therefore, $\gamma_1(\sigma^+) = \sigma^+$, and hence either $\gamma_1^+ = \gamma^+$ or $\gamma_1^- = \gamma^+$. According to Corollary 2.1, the axes of γ and γ_1 coincide. Since γ is a minimal element, $\gamma_1 = \gamma^k$ for some $k \in \mathbb{Z} \setminus \{0\}$.

By construction, $\gamma(\overline{S}) = \overline{S}$. Therefore, $\gamma_1(\overline{S}) = \overline{S}$. But then $\gamma_1(\overline{a}) = \gamma_1(\overline{S} \cap \overline{l}^+) = \gamma_1(\overline{S}) \cap \gamma_1(\overline{l}^+) = \overline{S} \cap \overline{l}_1^+ = \overline{c}$; that is, the points $\overline{a}$ and $\overline{c}$ are congruent. Thus, any neighborhood of $\overline{a}$ contains points congruent to $\overline{a}$, and this contradicts the fact that π is a covering. □

COROLLARY 2.4. *Let $l^{()}$ be a nontrivial recurrent semitrajectory, and let $\overline{l}_1^{()}$ and $\overline{l}_2^{()}$ be distinct coverings of it. Then $\sigma^{()}(\overline{l}_1^{()}) \neq \sigma^{()}(\overline{l}_2^{()})$.*

PROOF. Since $\overline{l}_1^{()}$ and $\overline{l}_2^{()}$ are congruent, it follows that $\gamma[\overline{l}_1^{()}] = \overline{l}_2^{()}$ for some $\gamma \in \Gamma_p$. If $\sigma^{()}(\overline{l}_1^{()}) = \sigma^{()}(\overline{l}_2^{()}) \stackrel{\text{def}}{=} \sigma$, then $\gamma(\sigma) = \sigma$, and hence σ is a rational point, which contradicts Theorem 2.2. □

2.5. The homotopy rotation class of a nontrivial recurrent semitrajectory. All nontrivial recurrent positive (negative) semitrajectories of a flow on the torus have the same asymptotic direction; that is, on the disk $\mathcal{D}^2$ any covering semitrajectories for P^+ (P^-) nontrivial recurrent semitrajectories tend to the same point of the absolute. On closed orientable surfaces of genus $p \geq 2$ the situation is essentially different from that on the torus.

DEFINITION. A *boundary* nontrivial recurrent semitrajectory or trajectory l of a flow f^t is one for which any point $m \in l$ has a neighborhood $\mathcal{U}(m) \ni m$ such that the arc d in $\mathcal{U}(m) \cap l$ containing m separates $\mathcal{U}(m)$ into two domains, one of which does not contain points lying on nontrivial recurrent semitrajectories of f^t.

REMARK. Boundary trajectories are called singular trajectories in the paper of Aranson and Grines, *Topological classification of flows on closed two-dimensional manifolds* (Uspekhi Mat. Nauk **41** (1986), no. 1, 149–169; English transl. in Russian Math. Surveys **41** (1986), no. 1).

A nontrivial recurrent semitrajectory or trajectory is a boundary semitrajectory or trajectory if it intersects any contact-free cycle or segment at the endpoints of intervals that do not intersect any nontrivial recurrent semitrajectories.

In a Denjoy flow f^t the accessible (from within) boundary of each component of $T^2 \setminus \Omega(f^t)$, where $\Omega(f^t)$ is the minimal set of f^t, consists of boundary nontrivial recurrent trajectories. In the general case if N is a quasiminimal set of a flow f^t on a surface $\mathcal{M}$, then the boundary nontrivial recurrent semitrajectories in N are in the accessible (from within) boundary of the components of $\mathcal{M} \setminus N$.

We remark that a boundary nontrivial recurrent trajectory is exceptional (that is, its closure is nowhere dense in the manifold).

An *interior* nontrivial recurrent semitrajectory or trajectory is one that is not a boundary semitrajectory or trajectory (in other words, such a trajectory 1‘approaches” itself from both sides).

In a transitive flow all the nontrivial recurrent semitrajectories and trajectories are interior.

THEOREM 2.3. *Suppose that a flow f^t on a closed orientable surface $\mathcal{M}_p$ with $p \geq 2$ has nontrivial recurrent semitrajectories, and let $\overline{f}^t$ be a covering flow on the Lobachevsky plane Δ. Then the following hold.*

1) *Any point of the absolute is a limit point of at most two semitrajectories of $\overline{f}^t$ that cover nontrivial recurrent semitrajectories of f^t.*

2) *If $\sigma \in S_\infty$ is a limit point for a lifting of an interior nontrivial recurrent semitrajectory of f^t, then σ is the limit of exactly one semitrajectory of $\overline{f}^t$. In particular, if f^t is a transitive flow, then at most one semitrajectory of $\overline{f}^t$ tends to any point of the absolute.*

3) *If $\sigma \in S_\infty$ is a limit point for two semitrajectories of $\overline{f}^t$ covering nontrivial recurrent semitrajectories $l_1^{()}$ and $l_2^{()}$ of f^t, then $l_1^{()}$ and $l_2^{()}$ are boundary nontrivial recurrent semitrajectories (both positive or both negative) that belong to a single quasiminimal set N of f^t and are in the accessible boundary of the same component of the set $\mathcal{M}_p \setminus N$. Conversely, if $\sigma \in S_\infty$ is a limit point for a semitrajectory $\overline{l}_1^{()}$ of $\overline{f}^t$ covering a boundary nontrivial recurrent semitrajectory $l_1^{()}$ in some quasiminimal set N of f^t, then there exists a unique boundary nontrivial recurrent semitrajectory $l_2^{()} \neq l_1^{()}$ that lies in N and has a covering semitrajectory $\overline{l}_2^{()}$ such that $\sigma = \sigma(\overline{l}_2^{()})$.*

PROOF. 1) Assume the contrary; that is, let $\sigma = \sigma(\overline{l}_i^{()}) \in S_\infty$, $i = 1, 2, 3$, where $\pi(\overline{l}_i^{()})$ is a nontrivial recurrent semitrajectory. By Theorem 2.1 and Corollary 2.2, there is a contact-free arc $\overline{C}$ of $\overline{f}^t$ with endpoints on the absolute which is intersected by all three semitrajectories $\overline{l}_1^{()}$, $\overline{l}_2^{()}$, and $\overline{l}_3^{()}$ (Figure 6.18). This implies that the semitrajectories $\overline{l}_i^{()}$, $i = 1, 2, 3$, are either all positive or all negative. For definiteness we assume they are positive. Let $\overline{a}_i = \overline{l}_i^{()} \cap \overline{C}$, $i = 1, 2, 3$, and suppose that the point $\overline{a}_2$ lies on the arc $\overset{\frown}{\overline{a}_1\overline{a}_3}$ of $\overline{C}$ between the points $\overline{a}_1$ and $\overline{a}_3$. Since $\pi(\overline{l}_2^{()})$ is a nontrivial recurrent semitrajectory, the arc $\overset{\frown}{\overline{a}_1\overline{a}_3}$ intersects a positive semitrajectory $\overline{l}^+$ congruent to $\overline{l}_2^{()}$. But then $\omega(\overline{l}^+) = \sigma(\overline{l}^+) = \sigma$, which contradicts Corollary 2.4.

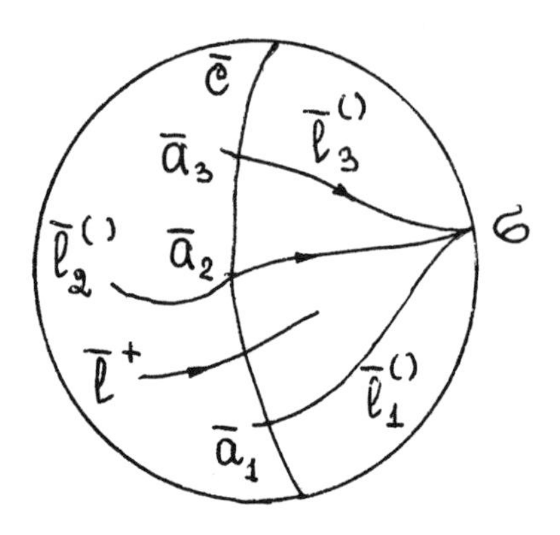

FIGURE 6.18

2) Assume the contrary, that is, let σ be the limit for the semitrajectories $\overline{l}^{()}$ and $\overline{l}_1^{()}$, where $\pi(\overline{l}^{()})$ is an interior nontrivial recurrent semitrajectory. As in the proof of 1), we construct the Bendixson "bag" bounded by $\overline{l}^{()}$, $\overline{l}_1^{()}$, and the arc $\overset{\frown}{\overline{a}\,\overline{a}_1}$ of the contact-free segment $\overline{C}$, where $\overline{a} = \overline{l}^{()} \cap \overline{C}$ and $\overline{a}_1 = \overline{l}_1^{()} \cap \overline{C}$. Since $\pi(\overline{l}^{()})$ is an interior nontrivial recurrent semitrajectory, the arc $\overset{\frown}{\overline{a}\,\overline{a}_1}$ intersects a semitrajectory $\overline{l}_2^{()}$ congruent to $\overline{l}^{()}$. Then $\sigma(\overline{l}_2^{()}) = \sigma$, and this contradicts Corollary 2.4.

We prove 3). Suppose that $\sigma(\overline{l}_1^{()}) = \sigma(\overline{l}_2^{()})$, where $l_i^{()} = \pi(\overline{l}_i^{()})$, $i = 1, 2$, are nontrivial recurrent semitrajectories. By 2), $l_1^{()}$ and $l_2^{()}$ are boundary semitrajectories.

Denote by N the quasiminimal set containing $\overline{l}_1^{()}$, that is, $N = \mathrm{cl}(l_1^{()})$. According to Lemma 4.1 in Chapter 2, there exists a contact-free cycle C that intersects N and is disjoint from quasiminimal sets different from N. By Theorem 2.1 and Corollary 2.2, there is a contact-free arc $\overline{C} \in \pi^{-1}(C)$ of $\overline{f}^t$ intersecting $\overline{l}_1^{()}$ and $\overline{l}_2^{()}$. Therefore, $l_2^{()} \cap C \neq \emptyset$, and hence $l_2^{()} \subset N$.

It follows from 1) that the arc $\overline{d}$ of $\overline{C}$ between the points $\overline{l}_1^{()} \cap \overline{C}$ and $\overline{l}_2^{()} \cap \overline{C}$ is disjoint from all inverse images of nontrivial recurrent semitrajectories. Therefore, $\pi(\overline{d})$ is disjoint from quasiminimal sets, and thus $l_1^{()}$ and $l_2^{()}$ are in the accessible boundary of the same component of the set $\mathcal{M}_p \setminus N$.

Conversely, suppose that $\sigma = \sigma(\overline{l}_1^{()}) \in S_\infty$, where $\pi(\overline{l}_1^{()}) = l_1^{()}$ is a boundary nontrivial recurrent semitrajectory lying in a quasiminimal set N of f^t. We take a contact-free cycle C with $C \cap N \neq \emptyset$ that does not intersect quasiminimal sets different from N (Lemma 4.1 in Chapter 2), and we denote by $P\colon C \to C$ the

Poincaré mapping induced by f^t. Since f^t has boundary nontrivial recurrent semitrajectories by assumption, the intersection $C \cap N \stackrel{\text{def}}{=} \Omega$ is a Cantor set, and $l_1^{()}$ intersects C at the endpoints of adjacent intervals of Ω.

For definiteness we assume that $l_1^{()}$ is a positive semitrajectory. Denote by $\{I_n\}_{n=1}^{\infty}$ the sequence of adjacent intervals of Ω whose endpoints intersect l_1^+ as time increases. Let $I_n = (a_n, b_n) \subset C$, where $a_n \in l_1^+ \cap C$ and $P(a_n) = a_{n+1}$, $n \in \mathbb{N}$. Trajectories in the quasiminimal set N pass through the points b_n. According to Theorem 3.4 in Chapter 2, the regular trajectories in a quasiminimal set are separatrices joining equilibrium states, or separatrices that are nontrivial recurrent semitrajectories, or nontrivial recurrent trajectories. Therefore, in view of the assumption that there are finitely many equilibrium states, the points b_n lie on a nontrivial recurrent semitrajectory l_2^+ for a sufficiently large index ($n \geq n_0$). Obviously, l_2^+ is a boundary semitrajectory.

We show that there is a semitrajectory $\overline{l}_2^+$ covering l_2^+ such that $\sigma(\overline{l}_2^+) = \sigma$.

By Corollary 2.2, the semitrajectory $\overline{l}_1^+$ intersects a countable family $\{\overline{C}_i\}_{i=1}^{\infty}$ of contact-free arcs in $\pi^{-1}(C)$ with σ as their topological limit. Let $\overline{a}_i = \overline{l}_1^+ \cap \overline{C}_i$. Then $\pi(\overline{a}_i) = a_i$, and $\overline{a}_i$ is an endpoint of an interval $\overline{I}_i \subset \overline{C}_i$ such that $\pi(\overline{I}_i) = I_i \subset C$. Denote by $\overline{b}_i$ the endpoint of $\overline{I}_i$ different from $\overline{a}_i$. Then $\pi(\overline{b}_i) = b_i$.

By the assumption that the number of separatrices is finite, there is an index n_1 such that the Poincaré mapping P is defined on I_n for $n \geq n_1$. Therefore, the domain bounded by $I_n, I_{n+1} \subset C$ and the arcs $\overset{\frown}{a_n a_{n+1}}$ and $\overset{\frown}{b_n b_{n+1}}$ of the semitrajectories l_1^+ and l_2^+ is simply connected. It follows from the properties of the covering π that the points $\overline{b}_n$ with $n \geq \max\{n_0, n_1\}$ belong to a single semitrajectory $\overline{l}_2^+$ covering l_2^+. Since σ is the topological limit of $\{\overline{C}_i\}$, $\sigma = \sigma(\overline{l}_2^+)$.
□

2.6. The connection between quasiminimal sets and geodesic laminations. We fix a Riemannian metric of constant curvature -1 on a closed orientable surface $\mathcal{M}_p$ of genus $p \geq 2$.

DEFINITION. A *geodesic lamination* $\mathcal{F}$ on $\mathcal{M}_p$ is defined to be a closed set consisting of disjoint simple[2] geodesics.

A simple closed geodesic provides a trivial example of a geodesic lamination. The topological closure of a simple nonclosed geodesic is a more complicated example.

We recall the definition of the geodesic flow. Denote by $T_1\mathcal{M}_p$ the space of unit vectors tangent to the surface $\mathcal{M}_p$. The geodesic flow ξ^t is defined as follows to be a one-parameter group of diffeomorphisms of the space $T_1\mathcal{M}_p$: in the time t each vector $e \in T_1\mathcal{M}_p$ is shifted along the geodesic tangent to it by a distance t while remaining tangent to this geodesic (Figure 6.19).

To each orientable geodesic on $\mathcal{M}_p$ there corresponds a unique trajectory of the geodesic flow, and conversely.

We recall that a trajectory of a flow on a compact manifold is B-recurrent if its closure is a compact minimal set.

DEFINITION. A geodesic on $\mathcal{M}_p$ is called a *nontrivial B-recurrent* geodesic if a nonclosed B-recurrent trajectory corresponds to it in the geodesic flow.

[2]A geodesic (closed or nonclosed) is said to be simple if it does not have self-intersections.

FIGURE 6.19

A geodesic lamination is said to be *nontrivial* if it coincides with the closure of a simple nontrivial B-recurrent geodesic.

DEFINITION. A nontrivial B-recurrent geodesic $L \subset \mathcal{M}_p$ is called a *boundary* geodesic if any point $m \in L$ has a neighborhood $\mathcal{U}(m) \ni m$ such that the arc d in $\mathcal{U}(m) \cap L$ containing m divides $\mathcal{U}(m)$ into two domains, one of which does not have points lying on L.

A nontrivial B-recurrent geodesic L is said to be *interior* if, for any neighborhood $\mathcal{U}(m)$ of any point $m \in L$, there are points on L in both the domains into which the arc d in $\mathcal{U}(m) \cap L$ containing m divides $\mathcal{U}(m)$.

In a nontrivial geodesic lamination $\mathcal{F} \subset \mathcal{M}_p$ (which is locally homeomorphic to the direct product of the line and the Cantor set) the boundary geodesics form the accessible (from within) boundaries of the components of $\mathcal{M}_p \setminus \mathcal{F}$.

Two boundary geodesics in the accessible (from within) boundary of a single component of $\mathcal{M}_p \setminus \mathcal{F}$ are said to be *associated*.

The definition of a boundary (interior) geodesic of a nontrivial geodesic lamination is completely analogous to the concept of a boundary (interior) trajectory in a quasiminimal set. It is clear that a simple boundary (interior) nontrivial B-recurrent geodesic L is a boundary (interior) geodesic of the nontrivial geodesic lamination $\mathcal{F} = \mathrm{cl}(L)$.

In this subsection we indicate a way of constructing a nontrivial geodesic lamination from a quasiminimal set of a flow.

We consider a flow f^t on $\mathcal{M}_p$, $p \geq 2$, with a quasiminimal set N, and let $\overline{f}^t$ be a covering flow on Δ for f^t. Take a trajectory $\bar{l}$ of $\overline{f}^t$ covering a nontrivial recurrent trajectory $l \subset N$. According to Theorem 2.3, the points $\sigma^-(\bar{l}) = \alpha(\bar{l}), \sigma^+(\bar{l}) = \omega(\bar{l}) \in S_\infty$ are distinct. We join them by an orientable geodesic $\overline{L}(\bar{l})$ from σ^- to σ^+ (Figure 6.20).

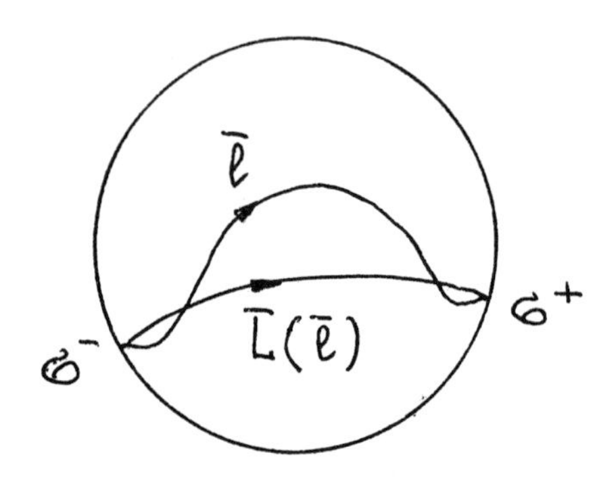

FIGURE 6.20

By the properties of the group Γ_p, the geodesic $\pi[\overline{L}(\bar{l})] \stackrel{\text{def}}{=} L(l)$ on the surface $\mathcal{M}_p$ is independent of the choice of the covering trajectory $\bar{l} \in \pi^{-1}(l)$. In this notation we have the following result.

LEMMA 2.4. *Let a flow with quasiminimal set N be given on $\mathcal{M}_p$, $p \geq 2$, and let $l \subset N$ be a nontrivial recurrent trajectory. In this case:*

1) *$L(l)$ is a simple nontrivial B-recurrent geodesic;*

2) *if l is an interior trajectory in N, then $L(l)$ is an interior geodesic;*

3) *if l_1 is a boundary trajectory in N belonging to the accessible (from within) boundary of a component w of $\mathcal{M}_p \setminus N$, and if a nontrivial recurrent trajectory l_2 is in the accessible (from within) boundary of w and has a covering trajectory $\bar{l}_2$ for which $\omega(\bar{l}_2) = \omega(\bar{l}_1)$, where $\bar{l}_1$ is a covering trajectory for l_1, then*

a) *$L(l_1)$ and $L(l_2)$ are associated boundary geodesics when $\alpha(\bar{l}_1) = \sigma^-(\bar{l}_1) \neq \sigma^-(\bar{l}_2) = \alpha(\bar{l}_2)$,*

b) *$L(l_1)$ is an interior geodesic when $\alpha(\bar{l}_1) = \alpha(\bar{l}_2)$.*

PROOF. 1) Suppose that $\sigma^+ = \omega(\bar{l})$ and $\sigma^- = \alpha(\bar{l})$, where $\bar{l}$ is a covering trajectory for l. If $L(l)$ is not a simple geodesic, then $\overline{L}(\bar{l}) \cap \gamma[\overline{L}(\bar{l})] \neq \emptyset$ for some nonidentity element $\gamma \in \Gamma_p$. Then the pairs of points (σ^-, σ^+) and $(\sigma^-(\gamma(\overline{L})), \sigma^+(\gamma(\overline{L})))$ are separated on the absolute, and hence $\bar{l} \cap \gamma(\bar{l}) \neq \emptyset$, which is impossible (Figure 6.21).

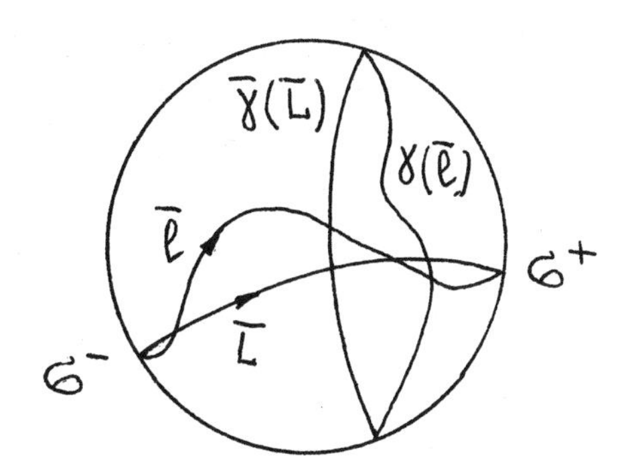

FIGURE 6.21

According to Theorem 2.2, α^- and σ^+ are irrational; therefore, the geodesic $L(l)$ is nonclosed.

It follows from Corollary 2.2 that there exists a sequence $\{\overline{C}_i\}_{-\infty}^{\infty}$ of contact-free arcs of $\overline{f}^t$ such that $\overline{C}_i \cap \bar{l} \neq \emptyset$, $i \in \mathbb{Z}$, and the point σ^+ (σ^-) is the topological limit of the arcs $\{\overline{C}_i\}_{-\infty}^{\infty}$ as $i \to +\infty$ (respectively, $i \to -\infty$). The continuous dependence of trajectories of the initial conditions and the self-limit property of the trajectory l imply the existence of a sequence of elements $\gamma_k \in \Gamma_p$, $k \in \mathbb{N}$, such that $\sigma^{\pm}[\gamma_k(\bar{l})] \to \sigma^{\pm}$ as $k \to \infty$ (Figure 6.22). Therefore, the sequence of geodesics $L_k = \overline{L}(\gamma_k(\bar{l}))$ converges to the geodesic $\overline{L}(\bar{l})$ as $k \to \infty$. This means that in the geodesic flow a nontrivial recurrent trajectory corresponds to the geodesic $L(l)$. Since there are no equilibrium states in the geodesic flow, the recurrent trajectories are B-recurrent. Consequently, $L(l)$ is a B-recurrent geodesic. The assertion 1) is proved.

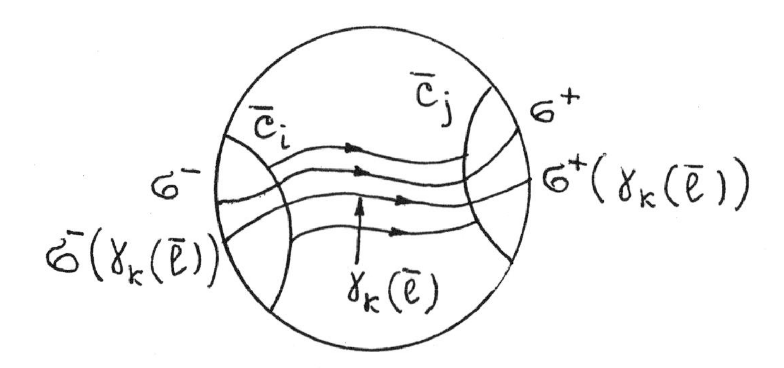

FIGURE 6.22

2) Since l is an interior trajectory, there exists a sequence of geodesics $\overline{L}_k$ congruent to $\overline{L}$ that converge to $\overline{L}$ from both sides. Therefore, $L(l)$ is an interior geodesic.

We prove 3), a). The geodesics $\pi[\overline{L}(\bar{l}_1)]$ and $\pi[\overline{L}(\bar{l}_2)]$ are disjoint because a nonempty intersection would imply (the argument is analogous to the proof of 1)) that trajectories of $\overline{f}^t$ can intersect. Therefore, in the domain $R \subset \Delta$ bounded by $\overline{L}(\bar{l}_1)$ and $\overline{L}(\bar{l}_2)$ there are no geodesics congruent to $\overline{L}(\bar{l}_1)$ or $\overline{L}(\bar{l}_2)$ and having endpoint σ^+. From this, $\pi[\overline{L}(\bar{l}_1)]$ and $\pi[\overline{L}(\bar{l}_2)]$ are associated boundary geodesics.

It remains to prove 3), b). Denote by $\overline{w}$ the domain bounded by trajectories $\bar{l}_1$ and $\bar{l}_2$. Since $\pi(\bar{l}_1)$ and $\pi(\bar{l}_2)$ are nontrivial recurrent trajectories, there are domains congruent to $\overline{w}$ arbitrarily close to $\overline{w}$ from both sides. Consequently, there are geodesics congruent to $\overline{L} = \overline{L}(\bar{l}_1) = \overline{L}(\bar{l}_2)$ arbitrarily close to $\overline{L}$ from both sides, and hence $\pi(\overline{L})$ is an interior geodesic. □

DEFINITION. A family consisting of separatrices $l_1, \dots, l_s$ and equilibrium states $m_1, \dots, m_{s-1}$ ($s \geq 2$) is called a *right-sided* (*left-sided*) *Poisson pencil* if the following conditions hold:

1) $\omega(l_i) = m_i$, $i = 1, \dots, s-1$;

2) $\alpha(l_i) = m_{i-1}$, $i = 2, \dots, s$;

3) l_i is a Bendixson extension of the separatrix l_{i-1} to the right (left), $i = 2, \dots, s$;

4) the separatrices l_1 and l_s are P^- and P^+ nontrivial recurrent trajectories, respectively (Figure 6.23).

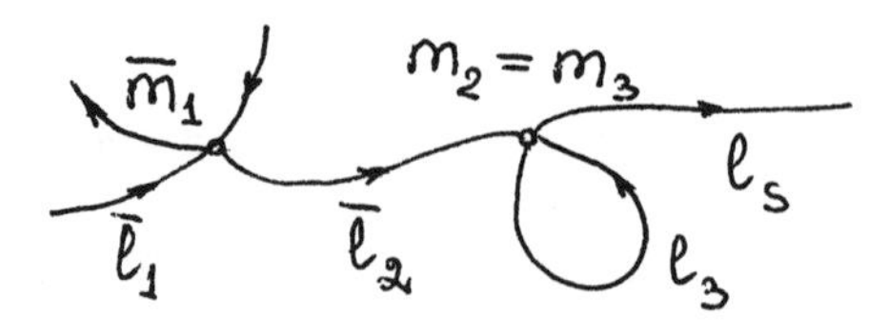

FIGURE 6.23

A right-sided or left-sided Poisson pencil is said to be *one-sided.*

Denote by $\mathcal{P}(l_1, \dots, l_s)$ the one-sided Poisson pencil containing the separatrices $l_1, \dots, l_s$.

The complete inverse image $\pi^{-1}[\mathcal{P}(l_1, \dots, l_s)]$ breaks up into a countable family of disjoint curves $\overline{\mathcal{P}}_i(\bar{l}_1, \dots, \bar{l}_s)$, $i \in \mathbb{N}$. Each such curve $\overline{\mathcal{P}}_i(\bar{l}_1, \dots, \bar{l}_s) \stackrel{\text{def}}{=} \overline{\mathcal{P}}_i$ consists of equilibrium states $\overline{m}_1, \dots, \overline{m}_{s-1}$ and separatrices $\bar{l}_1, \dots, \bar{l}_s$ of the covering flow such that:

1) $\omega(\bar{l}_i) = \overline{m}_i$, $i = 1, \dots, s-1$;

2) $\alpha(\bar{l}_i) = \overline{m}_{i-1}$, $i = 2, \dots, s$;

3) $\bar{l}_i$ is a Bendixson extension of the separatrix $\bar{l}_{i-1}$ to the right or to the left, $i = 2, \dots, s$;

4) $\pi(\overline{m}_i) = m_i$ and $\pi(\bar{l}_j) = l_j$, $i = 1, \dots, s-1$, $j = 1, \dots, s$.

Since the trajectories $\bar{l}_1$ and $\bar{l}_s$ are coverings for the nontrivial recurrent semi-trajectories l_1 and l_2, the points $\alpha(\bar{l}_1) \stackrel{\text{def}}{=} \sigma^-(\overline{\mathcal{P}}_i)$ and $\omega(\bar{l}_s) \stackrel{\text{def}}{=} \sigma^+(\overline{\mathcal{P}}_i)$ lie on the absolute and are distinct.

We join these points by a geodesic, which we denote by $\overline{L}(\overline{\mathcal{P}}_i)$ (Figure 6.24).

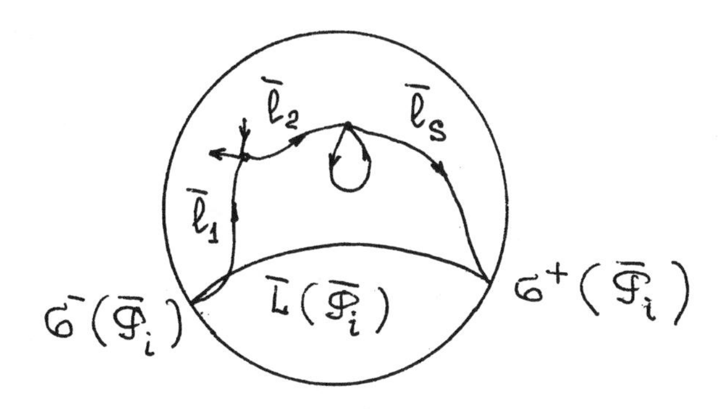

FIGURE 6.24

The geodesic $\pi[\overline{L}(\overline{\mathcal{P}}_i)] \stackrel{\text{def}}{=} L(\mathcal{P})$ does not depend on the choice of the curve $\overline{\mathcal{P}}_i \in \pi^{-1}[\mathcal{P}(l_1, \dots, l_s)]$. In this notation we have the following result.

LEMMA 2.5. *Let f^t be a flow on $\mathcal{M}_p$, $p \geq 2$, for which there is a one-sided Poisson pencil $\mathcal{P} = \mathcal{P}(l_1, \dots, l_s)$. Then:*

1) *$L(\mathcal{P})$ is a simple nontrivial B-recurrent geodesic;*

2) *if both trajectories l_1 and l_2 in $\mathcal{P}(l_1, \dots, l_s)$ are interior, then $L(\mathcal{P})$ is an interior geodesic.*

The proof is analogous to that of 1) and 2) in Lemma 2.4, and we omit it.

We remark that determining the cases for which $L(\mathcal{P})$ is an interior or boundary geodesic is more difficult than in Lemma 2.4 for a nontrivial recurrent trajectory. For example, it is possible that the trajectories l_1 and l_2 in the one-sided Poisson pencil $\mathcal{P}(l_1, \dots, l_s)$ are both boundary trajectories, while the geodesic $L(\mathcal{P})$ is interior (Figure 6.25, a)). Or one of l_1 or l_2 can be interior, while $L(\mathcal{P})$ is a boundary geodesic (Figure 6.25, b)).

LEMMA 2.6. *Suppose that a flow f^t on $\mathcal{M}_p$, $p \geq 2$, has a quasiminimal set N. Then the collection $\bigcup_l L(l) \bigcup_{\mathcal{P}} L(\mathcal{P}) \stackrel{\text{def}}{=} \mathcal{F}(N)$ forms a nontrivial geodesic lamination, where the union is over all nontrivial recurrent trajectories $l \subset N$ and all one-sided Poisson pencils $\mathcal{P} \subset N$.*

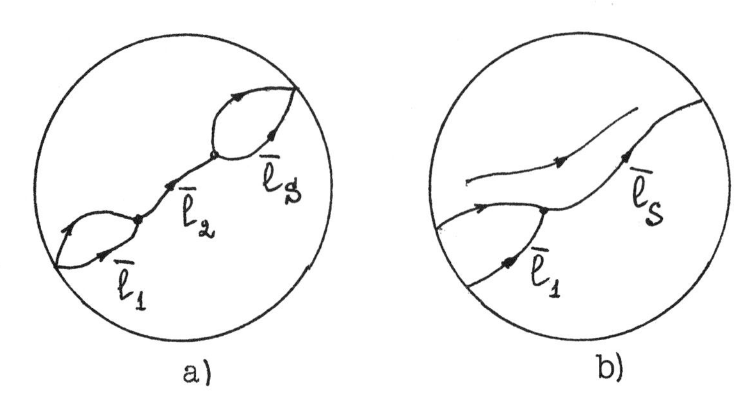

FIGURE 6.25

PROOF. It follows from Lemmas 2.4 and 2.5 that each geodesic in $\mathcal{F}(N)$ is simple. The geodesics in $\mathcal{F}(N)$ are disjoint, because otherwise either the trajectories of the covering flow $\overline{f}^t$ would intersect, or some geodesic would correspond to a Poisson pencil that is not one-sided.

The fact that $\mathcal{F}(N)$ is closed follows from the fact that N is closed and the fact that any nontrivial recurrent semitrajectory $l^{()} \subset N$ is dense in N.

According to Lemmas 2.4 and 2.5, any geodesic in $\mathcal{F}(N)$ is a nontrivial B-recurrent geodesic. □

Suppose that a flow f^t on $\mathcal{M}_p$, $p \geq 2$, has nontrivial recurrent semitrajectories. Geodesics $L(l)$ or $L(\mathcal{P})$ constructed from nontrivial recurrent trajectories l or one-sided Poisson pencils $\mathcal{P}$ of f^t are called *trajectory* geodesics. The geodesics in the complete inverse images $\pi^{-1}[L(l)]$ and $\pi^{-1}[L(\mathcal{P})]$ in the Lobachevsky plane will also be called trajectory geodesics.

It will be assumed that all the geodesics are orientable, that is, equipped with a positive direction. Trajectory geodesics are oriented as follows. If $\bar{l}$ is a covering for a nontrivial recurrent trajectory l of some flow on $\mathcal{M}_p$, $p \geq 2$, then the positive direction from the point $\alpha(\bar{l}) \in S_\infty$ to the point $\omega(\bar{l}) \in S_\infty$ on the geodesic $\overline{L}(\bar{l}) \subset \Delta$ induces a positive direction on the geodesic $L(l)$, and this introduction of an orientation does not depend on the choice of covering trajectory $\bar{l} \in \pi^{-1}(l)$. The geodesics $L(\mathcal{P})$ are oriented similarly for one-sided Poisson pencils.

Suppose that the geodesics $L_1, L_2 \subset \mathcal{M}_p$ intersect transversally at a point $m \in \mathcal{M}_p$, and denote by $e(L_i, m)$ the unit tangent vector to L_i at m, $i = 1, 2$. If the vectors $e(L_1, m)$ and $e(L_2, m)$ form a right-hand (left-hand) frame in the tangent space $T_m\mathcal{M}_p$ to the orientable surface $\mathcal{M}_p$ at m, then the intersection index $\#(L_1, L_2, m)$ of the geodesics L_1 and L_2 at m is equal to $+1$ (-1) [**76**]. We say that L_1 intersects L_2 *orientably* if the intersection index is the same ($+1$ or -1) at all points of $L_1 \cap L_2$. Orientability of an intersection of C^∞-smooth curves is defined analogously.

We remark that two distinct geodesics cannot be tangent (otherwise they coincide), and intersect transversally wherever they intersect.

LEMMA 2.7. *Let $\mathcal{E}$ be a nontrivial recurrent trajectory or a one-sided Poisson pencil of a flow f^t on $\mathcal{M}_p$, $p \geq 2$, and let C be a contact-free cycle intersecting* *a nontrivial recurrent semitrajectory in $\mathcal{E}$. Then the geodesic $L(\mathcal{E})$ intersects any closed geodesic L_0 homotopic to C orientably.*

PROOF. By Lemma 2.3 of Chapter 2, the contact-free cycle C is nonhomotopic to zero. Therefore, the closed geodesic L_0 is also nonhomotopic to zero.

According to Lemmas 2.4 and 2.5, $L(\mathcal{E})$ is a nonclosed geodesic. Consequently, $L(\mathcal{E}) \neq L_0$, and $L(\mathcal{E})$ and L_0 intersect transversally.

Assume that $L(\mathcal{E})$ intersects L_0 nonorientably. Then there exist geodesics $\overline{L} \in \pi^{-1}[L(\mathcal{E})]$ and $\overline{L}_1, \overline{L}_2 \in \pi^{-1}(L_0)$ such that the intersection indices of $\overline{L}$ with $\overline{L}_1$ and $\overline{L}_2$ are opposite in sign (Figure 6.26). We introduce an orientation on C. By assumption, there exist curves $\overline{C}_1, \overline{C}_2 \in \pi^{-1}(C)$ and a trajectory $\overline{\mathcal{E}}$ in $\pi^{-1}(\mathcal{E})$ that intersect with opposite indices (see Figure 6.26), and this is impossible. □

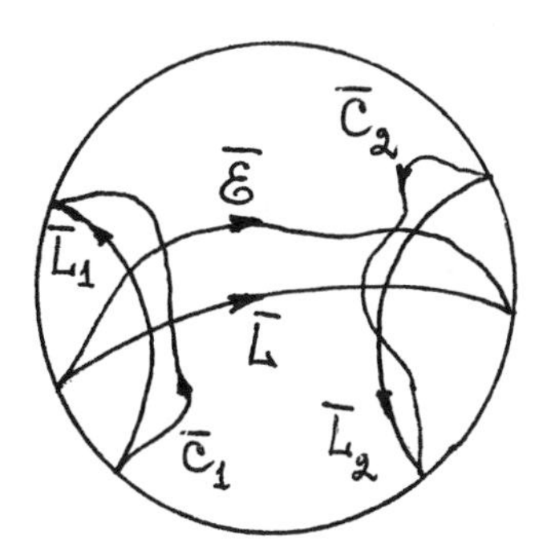

FIGURE 6.26

2.7. Accessible points of the absolute. We fix a covering $\pi\colon \Delta \to \mathcal{M}_p \cong \Delta/\Gamma_p$ of a closed orientable surface $\mathcal{M}_p$ of genus $p \geq 2$ by the Lobachevsky plane Δ.

DEFINITION. A point $\sigma \in S_\infty$ is said to be *accessible* if there exists a semitrajectory $\overline{l}^{()}$ of some covering flow on Δ such that $\sigma = \omega(\overline{l}^{()})$ or $\sigma = \alpha(\overline{l}^{()})$.

The set of accessible points (with respect to the covering $\pi\colon \Delta \to \mathcal{M}_p$) is denoted by $\mathcal{D}(S_\infty)$. This set is invariant under Γ_p and is nonempty. In this subsection we show that $\mathcal{D}(S_\infty) \neq S_\infty$ (that is, there are points on the absolute that are not accessible by semitrajectories of covering flows), and, moreover, the set $S_\infty \setminus \mathcal{D}(S_\infty)$ contains a subset with Lebesgue measure 2π (that is, the Lebesgue measure on the absolute).

The initial and terminal points of an orientable geodesic on Δ are defined in the natural way.

DEFINITION. A geodesic $\overline{L} \subset \Delta$ is said to be *transitive* if for any open intervals $I_1, I_2 \subset S_\infty$ there is a geodesic congruent to $\overline{L}$ with initial point in I_1 and terminal point in I_2.

Denote by $T(S_\infty)$ the set of terminal points of all transitive geodesics.

In 1936 Hedlund [**88**] proved that the set $T(S_\infty)$ has the measure of the absolute. In the next lemma we show that any point in $T(S_\infty)$ is not accessible by a trajectory of a covering flow. This will prove that the set $S_\infty \setminus \mathcal{D}(S_\infty)$ of nonaccessible points contains a subset of Lebesgue measure 2π.

LEMMA 2.8. $T(S_\infty) \subseteq S_\infty \setminus \mathcal{D}(S_\infty)$.

PROOF. Assume the contrary. Suppose that a point $\sigma \in T(S_\infty)$ is a limit point of a semitrajectory $\bar{l}^{()}$ of some covering flow on Δ. The following cases are possible: 1) σ is a rational point; 2) σ is an irrational point.

In case 1) the limit set of the semitrajectory $\pi(\bar{l}^{()})$ consists either of a closed trajectory l_0 nonhomotopic to zero, or a one-sided contour K nonhomotopic to zero (for some flow on $\mathcal{M}_p$). If the limit set of the semitrajectory $\pi(\bar{l}^{()})$ consists of a one-sided contour K, then there exists a contact-free cycle C homotopic to K because $\mathcal{M}_p$ is orientable. Since l_0 and C are simple curves, there is a geodesic $\overline{L}$ on Δ with endpoint σ such that $\pi(\overline{L})$ is a simple closed geodesic on $\mathcal{M}_p$. It follows from $\sigma \in T(S_\infty)$ that there exists a transitive geodesic $\overline{L}_0$ with endpoint σ. It can be assumed without loss of generality that σ is the terminal point of $\overline{L}_0$. Geometric considerations give us that, beginning with some parameter t_0 on $\overline{L}_0$, any geodesic intersecting $\overline{L}_0$ at a point with parameter $t \geq t_0$ intersects also the geodesic $\overline{L}$. But by transitivity, the geodesic $\pi(\overline{L}_0)$ intersects the geodesic $\pi(\overline{L})$ for arbitrarily large values of the parameter, and hence there exists a geodesic in $\pi^{-1}(\pi(\overline{L}))$ that intersects $\overline{L}$. This contradicts the fact that $\pi(\overline{L})$ is simple.

In the case 2) the semitrajectory $\pi(\bar{l}^{()})$ is a nontrivial recurrent semitrajectory in view of Theorem 2.2. Lemma 3.4 in Chapter 2 gives us that $\pi(\bar{l}^{()})$ belongs either to some one-sided Poisson pencil or to some nontrivial recurrent trajectory. According to Lemmas 2.4 and 2.5, σ is an endpoint of a geodesic $\overline{L}$ projecting into a simple geodesic on $\mathcal{M}_p$ in both situations. We arrive at a contradiction to the condition $\sigma \in T(S_\infty)$ in a way completely analogous to that in the case 1). □

REMARK. It can be shown that the inclusion $\mathcal{D}(S_\infty) \subset S_\infty \setminus T(S_\infty)$ is proper. That is, there are both rational and irrational points on the absolute that do not lie in $T(S_\infty)$ and are not accessible by semitrajectories of covering flows (that cover flows with finitely many equilibrium states and separatrices).

2.8. Classification of accessible irrational points. As before, we fix a covering $\pi\colon \Delta \to \mathcal{M}_p \cong \Delta/\Gamma_p$ of a closed orientable surface $\mathcal{M}_p$ of genus $p \geq 2$ by the Lobachevsky plane Δ.

For simplicity, a geodesic on Δ covering an interior geodesic will also be said to be interior.

DEFINITION. An irrational point $\sigma \in \mathcal{D}(S_\infty)$ is said to be a point *of the first kind* if it is an endpoint of an interior trajectory geodesic. The remaining accessible irrational points are called points *of the second kind.*

LEMMA 2.9. *Suppose that $\sigma \in \mathcal{D}(S_\infty)$ is an irrational accessible point of the first kind, and $\overline{L}$ is an interior trajectory geodesic with endpoint σ. Then any geodesic different from $\overline{L}$ and with endpoint σ projects into a geodesic with a self-intersection on $\mathcal{M}_p$ (and consequently is not a trajectory geodesic).*

PROOF. Let $\overline{L}_1$ be a geodesic with terminal point σ, $\overline{L}_1 \neq \overline{L}$.

By Lemma 2.7, there is a family of geodesics $\overline{C}_i \in \pi^{-1}(C)$, $i \in \mathbb{N}$, (C a closed geodesic on $\mathcal{M}_p$) with topological limit σ such that $\overline{L}$ intersects each $\overline{C}_i$ at some point $\overline{x}_i$. Then there is an index i_0 such that $\overline{L}_1$ also intersects $\overline{C}_i$ for $i \geq i_0$. Let $\overline{y}_i = \overline{L}_1 \cap \overline{C}_i$, $i \geq i_0$. Since $\overline{L}$ is an interior geodesic, there exists a subsequence

$\{i_k\}_{i=1}^{\infty}$ of indices such that the points $\overline{x}_{i_k}$ are congruent to points $\overline{x}'_{i_k}$ on the arc $(\overline{x}_{i_0}, \overline{y}_{i_0}) \subset \overline{C}_{i_0}$ tending to $\overline{x}_{i_0}$ as $i_k \to \infty$ (Figure 6.27). Since $\overline{L}$ and $\overline{L}_1$ have a common endpoint, the distance between $\overline{x}_i$ and $\overline{y}_i$ in the non-Euclidean metric tends to zero as $i \to +\infty$ [**77**]. Therefore, the distance between $\overline{y}'_{i_k} = \gamma_{i_k}(\overline{y}_{i_k})$ and $\overline{x}'_{i_k}$ also tends to zero as $i_k \to +\infty$, where $\gamma_{i_k} \in \Gamma_p$ is an element carrying $\overline{x}_{i_k}$ into $\overline{x}'_{i_k}$. From Lemma 2.7 (orientability of the intersection of a trajectory geodesic and a closed geodesic) it follows that $\overline{y}'_{i_k} \in (\overline{x}_{i_0}, \overline{y}_{i_0}) \subset C_{i_0}$ for a sufficiently large index i_k.

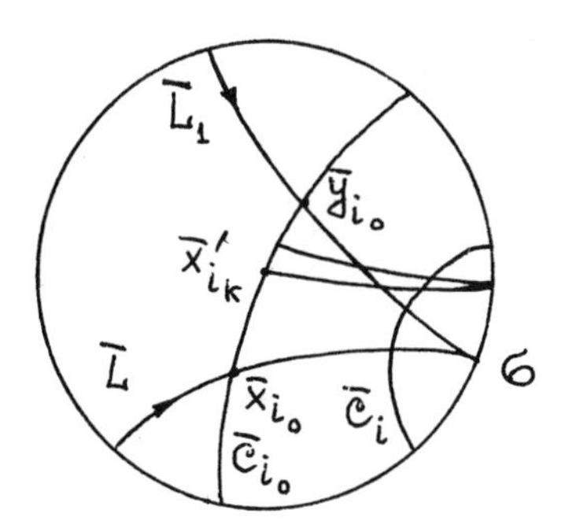

FIGURE 6.27

The geodesic $\overline{L}_{i_k} \stackrel{\text{def}}{=} \gamma_{i_k}(\overline{L})$ is a trajectory geodesic. Therefore, $\sigma^+(\overline{L}_{i_k}) \neq \sigma$ by virtue of Corollary 2.4. Consequently, $\overline{L}_{i_k}$ intersects $\overline{L}_1$ (see Figure 6.27). Since $\overline{y}'_{i_k} \in (\overline{x}_{i_0}, \overline{y}_{i_0})$, this gives us that the geodesic $\gamma_{i_k}(\overline{L}_1)$ intersects $\overline{L}_1$. □

For a trajectory geodesic $\overline{L} \subset \Delta$ denote by $\sigma^-(\overline{L})$ and $\sigma^+(\overline{L})$ its initial and terminal points, respectively.

COROLLARY 2.5. *Let $\sigma \in S_\infty$ be an accessible irrational point of the first kind, and let $\overline{L}$ be an interior trajectory geodesic with endpoint $\sigma = \sigma^+(\overline{L})$. Then the following conditions hold for any flow $\overline{f}^t$ on Δ covering a transitive flow f^t on $\mathcal{M}_p$ without impassable grains and having a semitrajectory $\overline{l}^+$ for which σ is accessible*:

1) *the semitrajectory $l^+ = \pi(\overline{l}^+)$ belongs to a nontrivial recurrent trajectory l of f^t*;

2) *if $\overline{l} \in \pi^{-1}(l)$ contains $\overline{l}^+$, then $\alpha(\overline{l}) = \sigma^-(\overline{L})$.*

PROOF. 1) Assume the contrary. Then l^+ is an σ-separatrix of some saddle. Since f^t does not have impassable grains, l^+ belongs to two different one-sided Poisson pencils $\mathcal{P}_1$ and $\mathcal{P}_2$. We take curves $\overline{\mathcal{P}}_i \in \pi^{-1}(\mathcal{P}_i)$, $i = 1, 2$, containing $\overline{l}^+$. It follows from Lemmas 2.5 and 2.9 and the fact that σ is a point of the first kind that $\sigma^-(\overline{\mathcal{P}}_1) = \sigma^-(\overline{\mathcal{P}}_2) \stackrel{\text{def}}{=} \sigma^-$. Then the curves $\overline{\mathcal{P}}_1$ and $\overline{\mathcal{P}}_2$ and the point σ^- bound a simply connected domain $\mathcal{U}$ on Δ. The transitivity of f^t gives us that there is a trajectory $\overline{l}_1$ in $\mathcal{U}$ covering a nontrivial recurrent trajectory of f^t. Then $\sigma^- = \alpha(\overline{l}_1)$, which contradicts Theorem 2.3.

The assertion 2) follows from Lemma 2.9. □

We proceed to the consideration of accessible points of the second kind. It follows from Lemmas 2.3, 2.4, and 2.5 and Theorem 2.2 that if $\sigma \in S_\infty$ is an accessible irrational point of the second kind, then there is at least one pair of associated boundary geodesics with common endpoint σ.

LEMMA 2.10. *Let $\sigma \in S_\infty$ be an irrational accessible point of the second kind, and let $\overline{L}_1$ and $\overline{L}_2$ be associated boundary trajectory geodesics with endpoint $\sigma = \sigma^+(\overline{L}_1) = \sigma^+(\overline{L}_2)$. Then any trajectory geodesic having endpoint σ and projecting into a B-recurrent geodesic on $\mathcal{M}_p$ coincides with either $\overline{L}_1$ or $\overline{L}_2$.*

PROOF. Denote by $R(\overline{L}_1, \overline{L}_2) \stackrel{\text{def}}{=} R$ the domain in Δ bounded by $\overline{L}_1$ and $\overline{L}_2$ and the corresponding arc of the absolute. Denote by R^+ (respectively, R^-) the component of $\Delta \setminus R$ adjacent to $\overline{L}_1$ (respectively, $\overline{L}_2$) (Figure 6.28).

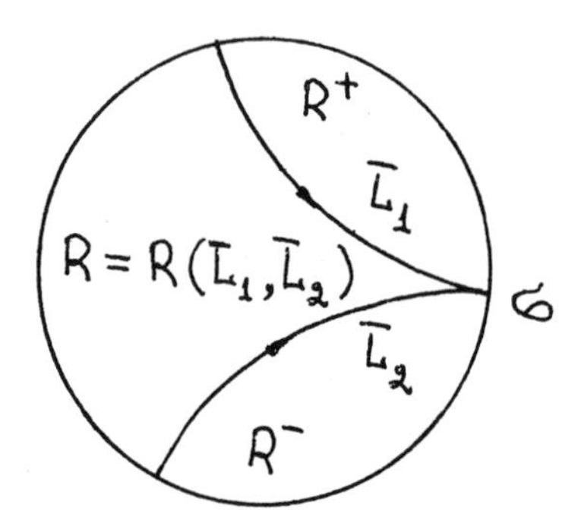

FIGURE 6.28

Under the action of Γ_p the domain R is transformed into domains arbitrarily close to R and lying in both domains R^+ and R^-. Therefore, it can be shown as in the proof of Lemma 2.9 that any geodesic with endpoint σ lying in $R^+ \cup R^-$ projects into a geodesic with a self-intersection on $\mathcal{M}_p$, and hence cannot be a trajectory geodesic.

Assume that there is a trajectory geodesic $\overline{L}_3$ with endpoint σ that is distinct from $\overline{L}_1$ and from $\overline{L}_2$ and projects into a B-recurrent geodesic on $\mathcal{M}_p$. Then $\overline{L}_3 \subset R$. According to Lemma 2.9, $\overline{L}_3$ is not interior. Therefore, it is a boundary geodesic, and there exists a boundary trajectory geodesic $\overline{L}_4$ that is associated with $\overline{L}_3$ and has endpoint σ.

Denote by $R(\overline{L}_3, \overline{L}_4)$ the domain bounded by $\overline{L}_3$ and $\overline{L}_4$ and the corresponding arc of the absolute. Since $\overline{L}_3 \subset R$, either $\overline{L}_1$ or $\overline{L}_2$ lies outside $R(\overline{L}_3, \overline{L}_4)$. By the foregoing, either $\overline{L}_1$ or $\overline{L}_2$ projects into a geodesic with a self-intersection on $\mathcal{M}_p$, and this is impossible. $\square$

COROLLARY 2.6. *Suppose that $\sigma \in S_\infty$ is an accessible irrational point of the second kind, and let $\overline{L}_1$ and $\overline{L}_2$ be associated boundary trajectory geodesics with endpoint $\sigma = \sigma^+(\overline{L}_1) = \sigma^+(\overline{L}_2)$. Then the following conditions hold for any flow $\overline{f}^t$ covering a transitive flow f^t on $\mathcal{M}_p$, $p \geq 2$, and having a semitrajectory $\overline{l}^+$ for which σ is accessible*:

1) *$l^+ = \pi(\overline{l}^+)$ is an α-separatrix and belongs to two one-sided Poisson pencils $\mathcal{E}_1$ and $\mathcal{E}_2$ of f^t;*

2) *if $\overline{\mathcal{E}}_i \in \pi^{-1}(\mathcal{E})$ contains $\overline{l}^+$, $i = 1, 2$, then either $\sigma^-(\overline{\mathcal{E}}_i) = \sigma^-(\overline{L}_i)$, $i = 1, 2$, or $\sigma^-(\overline{\mathcal{E}}_2) = \sigma^-(\overline{L}_1)$ and $\sigma^-(\overline{\mathcal{E}}_1) = \sigma^-(\overline{L}_2)$.*

PROOF. 1) If l^+ belongs to some trajectory that is nontrivial recurrent in both directions, then, since f^t is transitive, l is dense in $\mathcal{M}_p$ and is interior. This contradicts the condition that σ be a point of the second kind. Consequently, l^+ is an α-separatrix.

The semitrajectory l^+ belongs to exactly one one-sided Poisson pencil $\mathcal{E}$ only in the case when all the equilibrium states on $\mathcal{E}$ are impassable grains. Then it follows from the transitivity of f^t that l^+ is an interior semitrajectory. This again contradicts the condition that σ be a point of the second kind. Thus, l^+ belongs to two one-sided Poisson pencils $\mathcal{E}_1$ and $\mathcal{E}_2$.

The assertion 2) follows from Lemma 2.10. □

COROLLARY 2.7. *Let $\sigma \in S_\infty$ be an accessible irrational point of the second kind, and let $\overline{L}_1$ and $\overline{L}_2$ be associated boundary trajectory geodesics with terminal point $\sigma = \sigma^+(\overline{L}_1) = \sigma^+(\overline{L}_2)$. Suppose that a flow $\overline{f}^t$ covering a flow f^t on $\mathcal{M}_p$, $p \geq 2$, has trajectories $\overline{l}_1$ and $\overline{l}_2$ such that $\sigma^+(\overline{l}_1) = \sigma^+(\overline{l}_2)$, and $\pi(\overline{l}_1)$ and $\pi(\overline{l}_2)$ are associated boundary trajectories in a minimal set of f^t. Then either $\sigma^-(\overline{l}_1) = \sigma^-(\overline{L}_1)$ and $\sigma^-(\overline{l}_2) = \sigma^-(\overline{L}_2)$, or $\sigma^-(\overline{l}_1) = \sigma^-(\overline{L}_2)$ and $\sigma^-(\overline{l}_2) = \sigma^-(\overline{L}_1)$ (Figure 6.29).*

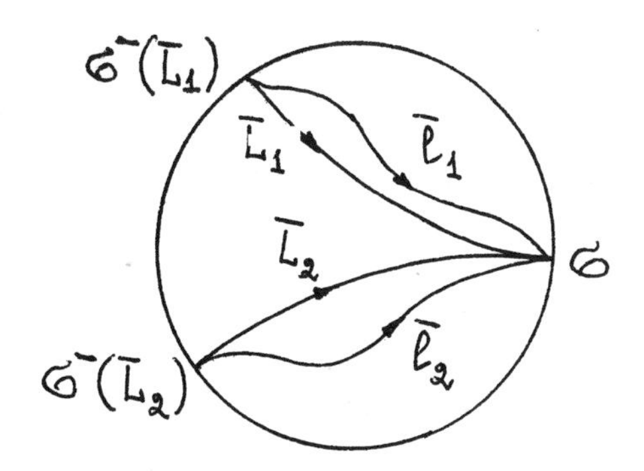

FIGURE 6.29

2.9. The orbit of a homotopy rotation class. Let $\varphi\colon \mathcal{M}_p \to \mathcal{M}_p$ be a homeomorphism of a closed orientable surface $\mathcal{M}_p \cong \Delta/\Gamma_p$, $p \geq 2$, and let $\overline{\varphi}\colon \Delta \to \Delta$ be a homeomorphism covering it. For any point $\overline{m} \in \Delta$ and any element $\gamma \in \Gamma_p$ the points $\overline{\varphi}(\overline{m})$ and $\overline{\varphi}(\gamma(\overline{m}))$ are congruent, and therefore there is an element $\gamma' \in \Gamma_p$ such that $\gamma' \circ \overline{\varphi}(\overline{m}) = \overline{\varphi} \circ \gamma(\overline{m})$ (Figure 6.30). Since the group Γ_p is discontinuous, the last equation is valid for all points $\overline{m} \in \Delta$.

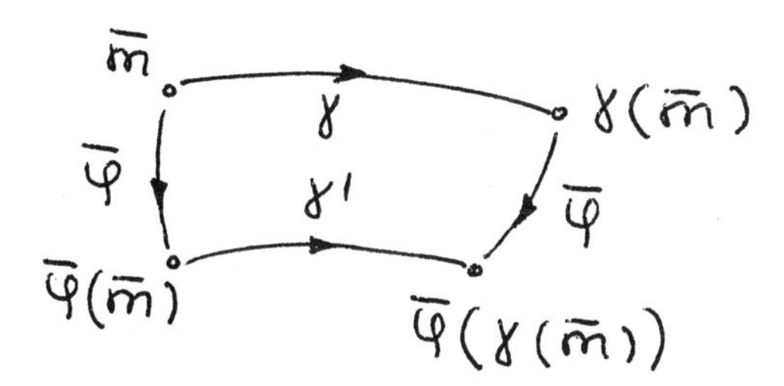

FIGURE 6.30

The correspondence $\gamma \mapsto \gamma' = \overline{\varphi} \circ \gamma \circ \overline{\varphi}^{-1}$ is an automorphism of the group Γ_p (we leave the proof of this to the reader as an exercise) and is denoted by $\overline{\varphi}_*$. Thus, $\overline{\varphi}_*(\gamma) \circ \overline{\varphi} = \overline{\varphi} \circ \gamma$.

The next theorem is due to Nielsen ([**101**], [**102**]).

THEOREM. 1) *An arbitrary automorphism $\tau\colon \Gamma_p \to \Gamma_p$ has the form $\overline{\varphi}_*$; that is, there exists a homeomorphism $\overline{\varphi}\colon \Delta \to \Delta$ covering some homeomorphism of the surface $\mathcal{M}_p \cong \Delta/\Gamma_p$, $p \geq 2$, such that $\tau = \overline{\varphi}_*$.*

2) *Any homeomorphism $\overline{\varphi}\colon \Delta \to \Delta$ covering some homeomorphism of $\mathcal{M}_p \cong \Delta/\Gamma_p$, $p \geq 2$, can be extended to a homeomorphism $\overline{\varphi}^*\colon \Delta \cup S_\infty \to \Delta \cup S_\infty$ (that is, can be extended to the absolute).*

3) *Let $\overline{\varphi}_1$ and $\overline{\varphi}_2$ be covering homeomorphisms for some homeomorphism of $\mathcal{M}_p = \Delta/\Gamma_p$, $p \geq 2$, and let $\overline{\varphi}_1^*$ and $\overline{\varphi}_2^*$ be extensions of them to the absolute. If $\overline{\varphi}_{1*} = \overline{\varphi}_{2*}$, then $\overline{\varphi}_1^*|_{S_\infty} = \overline{\varphi}_2^*|_{S_\infty}$.*

It follows from the theorem that an arbitrary automorphism $\tau\colon \Gamma_p \to \Gamma_p$ induces a homeomorphism of the absolute, which we denote by τ^*.

Let H^* be the set of homeomorphisms of the absolute induced by all possible automorphisms of the group Γ_p.

DEFINITION. Suppose that a semitrajectory $l^{()}$ of a flow f^t on $\mathcal{M}_p$, $p \geq 2$, has a homotopy rotation class $\mu(l^{()}) \subset S_\infty$. The *orbit $O(l^{()})$ of the homotopy rotation class* of $l^{()}$ is defined to be the set

$$H^*[\mu(l^{()})] = \cup \tau^*[\mu(l^{()})],$$

where the union is over all $\tau^* \in H^*$.

The next theorem shows that the orbit of the homotopy rotation class (HRC) is a topological invariant.

THEOREM 2.4. *Suppose that a flow f_1^t on $\mathcal{M}_p$, $p \geq 2$, is topologically equivalent to a flow f_2^t by means of a homeomorphism $\varphi\colon \mathcal{M}_p \to \mathcal{M}_p$, and let $\overline{\varphi}$ be a covering homeomorphism for φ that extends to a homeomorphism $\overline{\varphi}^*\colon \Delta \cup S_\infty \to \Delta \cup S_\infty$. In this case if a semitrajectory $l^{()}$ of f_1^t has homotopy rotation class $\mu(l^{()})$, then the semitrajectory $\varphi(l^{()})$ of f_2^t also has a homotopy rotation class, equal to $\overline{\varphi}^*[\mu(l^{()})]$, and $L(l^{()}) = O[\varphi(l^{()})]$.*

PROOF. Since $\overline{\varphi}$ extends to a homeomorphism of the absolute, the semitrajectory $\overline{\varphi}(\overline{l}^{()})$, where $\overline{l}^{()} \in \pi^{-1}(l^{()})$, has a limit point on the absolute. Consequently, the semitrajectory $\varphi(l^{()}) = \pi[\overline{\varphi}(\overline{l}^{()})]$ has an HRC, equal to $\overline{\varphi}^*[\mu(l^{()})]$. This and the inclusion $\overline{\varphi}^*|_{S_\infty} \in H^*$ give us that $O(l^{()}) = O[\varphi(l^{()})]$. □

COROLLARY 2.8. *Suppose that the flows f_1^t and f_2^t on $\mathcal{M}_p$, $p \geq 2$, are topologically equivalent by means of a homeomorphism homotopic to the identity. Then for any semitrajectory $l_1^{()}$ of f_1^t having HRC $\mu(l_1^{()})$ there is a semitrajectory $l_2^{()}$ of f_2^t with $\mu(l_1^{()}) = \mu(l_2^{()})$.*

PROOF. We use the notation of Theorem 2.4. Since φ is homotopic to the identity, the automorphism $\overline{\varphi}_*$ is interior, and $\overline{\varphi}^*|_{S_\infty} = \gamma|_{S_\infty}$ for some element $\varphi \in G_p$ [**68**]. This and Theorem 2.4 gives us that $\mu[\varphi(l_1^{()})] = \gamma[\mu(l_1^{()})] = \mu(l_1^{()})$. □

§3. Topological equivalence of transitive flows

The problem of topological equivalence of transitive flows without impassable grains and without separatrices joining equilibrium states can be solved with the help of the concept of homotopy rotation class and the concept of the orbit of a homotopy rotation class, introduced for a closed orientable surface of genus $p \geq 2$ in the preceding section.

3.1. Homotopic contact-free cycles. Everywhere in this section $\mathcal{M}_p \cong \Delta/\Gamma_p$ is a closed orientable surface of genus $p \geq 2$.

LEMMA 3.1. *Suppose that flows f_1^t and f_2^t on $\mathcal{M}_p$, $p \geq 2$, have nontrivial recurrent semitrajectories l_1^+ and l_2^+ with the same homotopy rotation class. Then f_1^t and f_2^t have mutually homotopic contact-free cycles C_1 and C_2 such that $C_i \cap l_i^+ \neq \emptyset$, $i = 1, 2$.*

PROOF. According to Lemma 2.3 in Chapter 2, f_2^t has a contact-free cycle C that is nonhomotopic to zero and such that $C \cap l_2^+ \neq \emptyset$. It follows from Corollary 2.2 and the equality $\mu(l_1^+) = \mu(l_2^+)$ that there exist semitrajectories $\overline{l}_1^+ \in \pi^{-1}(l_1^+)$ and $\overline{l}_2^+ \in \pi^{-1}(l_2^+)$ with $\omega(\overline{l}_1^+) = \omega(\overline{l}_2^+) \overset{\text{def}}{=} \sigma$ and a family $\{\overline{C}_i\}_1^\infty$ of contact-free arcs $\overline{C}_i \in \pi^{-1}(C)$ of a flow $\overline{f}^t$ that have σ as their topological limit. There is an index i_0 such that $\overline{l}_1^+$ and $\overline{l}_2^+$ intersect all the arcs $\overline{C}_i$ with $i \geq i_0$. For simplicity we assume that $i_0 = 1$.

Denote by $\overline{m}_1 \in \overline{l}_1^+ \cap \overline{C}_1$ a point after which $\overline{l}_1^+$ does not intersect $\overline{C}_1$ with increasing time (Figure 6.31). By a small perturbation of C in a neighborhood Σ of the point $m_1 = \pi(\overline{m}_1)$ we ensure that the trajectories of f_1^t intersect Σ transversally, while keeping the curve C a contact-free cycle of f_2^t. Since l_1^+ is a nontrivial recurrent semitrajectory, it intersects Σ infinitely many times. Denote by s_1 the first point where $l_1^+(m_1)$ intersects Σ. Then the arc $\overset{\frown}{m_1 s_1} \subset l_1^+$ and the segment $\overline{m_1 s_1} \subset \Sigma$ form a simple closed curve C_1' from which a closed transversal C_1 of f_1^t intersecting l_1^+ can be obtained by the standard method (Lemma 1.2 in Chapter 2). By Lemma 2.3 in Chapter 2, C_1 is homotopic to zero.

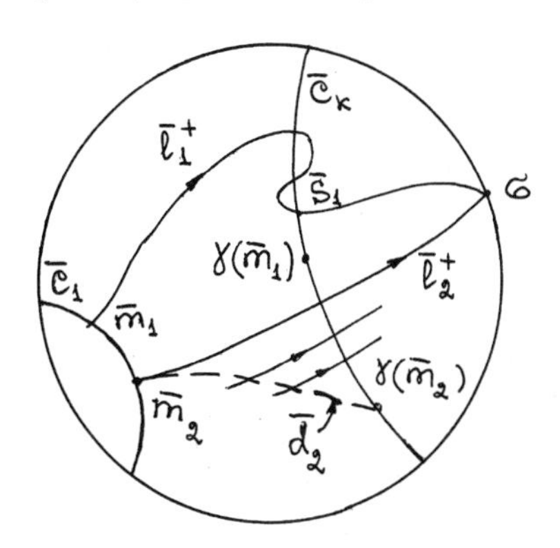

FIGURE 6.31

Denote by $\overline{s}_1$ a point in $\overline{l}_1^+ \cap \overline{C}_k$ such that $\pi(\overline{s}_1) = s_1$ (Figure 6.31), and let $\overline{\Sigma} \in \pi^{-1}(\Sigma)$ be a segment containing $\overline{s}_1$. The segment $\overline{\Sigma}$ contains a point congruent to $\overline{m}_1$. Then $\gamma(\overline{m}_1) \in \overline{\Sigma}$ for some element $\gamma \in \Gamma_p$, and therefore $\overline{C}_k = \gamma(\overline{C}_1)$.

Let $\overline{s}_2 = \overline{l}_2^+ \cap \overline{C}_k$. We use an arc $\overline{d}_2$ transversal to the flow $\overline{f}_2^t$ to approximate the union of the arc $\overset{\frown}{\overline{m}_2\overline{s}_2} \subset \overline{l}_2^+$ and the segment $\overline{s}_2\gamma(\overline{m}_2) \subset \overline{C}_k$, where $\gamma(\overline{m}_2) \in \overline{C}_k$. Let $\overline{C}_2' = \bigcup_{n\in\mathbb{Z}} \gamma^n(\overline{d}_2)$. Then $C_2' = \pi(\overline{C}_2')$ is a closed curve transversal to the flow f_2^t and homotopic to C_1.

The curve C_2' is not simple in general, but since it is homotopic to the simple curve C_1, we can use an isotopy to carry it into a simple curve C_2 transversal to f_2^t. Indeed, by a small perturbation we first ensure that the self-intersections of C_2' are transversal, and we consider two curves $\overline{C}_{2i}', \overline{C}_{2j}' \in \pi^{-1}(C_2')$.

Let $\overline{a}, \overline{b} \in \overline{C}_{2i}' \cap \overline{C}_{2j}'$ be two adjacent points of intersection of the curves $\overline{C}_{2i}'$ and $\overline{C}_{2j}'$ such that the arcs of them with endpoints $\overline{a}$ and $\overline{b}$ bound a disk on Δ that does not contain points in $\pi^{-1}(C_2')$ (Figure 6.32). Such points exist because C_2' is homotopic to the simple curve C_1, and hence the endpoints of $\overline{C}_{2i}'$ and $\overline{C}_{2j}'$ are not separated on the absolute. In Figure 6.32 it is shown how $\overline{C}_{2i}'$ and $\overline{C}_{2j}'$ can be transformed into curves $\overline{C}_{2i}''$ and $\overline{C}_{2j}''$ that have two fewer points of intersection and are transversal to $\overline{f}_2^t$.

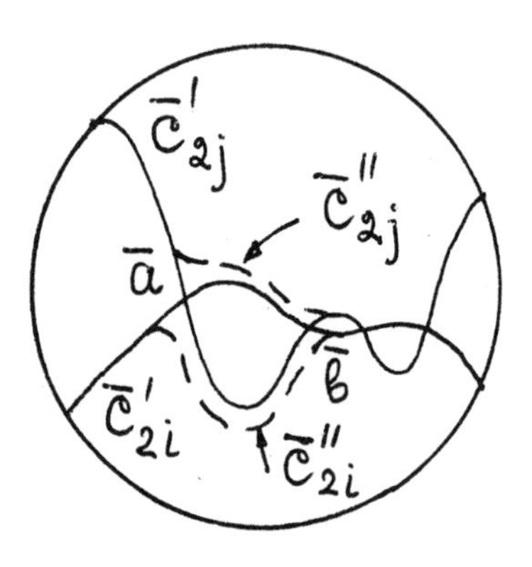

FIGURE 6.32

We now consider the intersections of $\overline{C}_{2i}'$ with the remaining curves in $\pi^{-1}(C_{2i}')$. By successively carrying out the process described above, we obtain a curve $\overline{C}_2$ transversal to $\overline{f}^t$ and such that $\pi(\overline{C}_2) = C_2$ is simple. □

3.2. Auxiliary results.

LEMMA 3.2. *Suppose that flows f_1^t and f_2^t on $\mathcal{M}_p$, $p \geq 2$, have nontrivial recurrent semitrajectories l_1^+ and l_2^+ with the same homotopy rotation class. Then for any nontrivial recurrent positive semitrajectory L_1^+ of f_1^t lying in $\omega(l_1^+)$ there is a nontrivial recurrent positive semitrajectory L_2^+ of f_2^t in $\omega(l_2^+)$ such that $\mu(L_1^+) = \mu(L_2^+)$.*

PROOF. According to Lemma 3.1, there exist mutually homotopic contact-free cycles C_1 and C_2 of f_1^t and f_2^t that are nonhomotopic to zero and intersect l_1^+ and l_2^+, respectively. It follows from $L_1^+ \subset \omega(l_1^+)$ and Theorem 2.3 in Chapter 2 that $l_1^+ \subset \omega(L_1^+)$. Therefore, L_1^+ intersects C_1.

We consider a curve $\overline{C}_1 \in \pi^{-1}(C_1)$ and a semitrajectory $\overline{L}_1^+ \in \pi^{-1}(L_1^+)$ intersecting $\overline{C}_1$. Since C_1 and C_2 are homotopic, there is a curve $\overline{C}_2 \in \pi^{-1}(C_2)$ having common endpoints with $\overline{C}_1$.

We show that there exist sequences of semitrajectories $\overline{l}_{n,1}^+ \in \pi^{-1}(l_1^+)$ and $\overline{l}_{n,2}^+ \in \pi^{-1}(l_2^+)$ with the following properties: 1) $\omega(\overline{l}_{n,1}^+) = \omega(\overline{l}_{n,2}^+) \overset{\text{def}}{=} \sigma_n \in S_\infty$; 2) $\overline{l}_{n,j}^+$ intersects $\overline{C}_j$ for $j = 1, 2$; 3) the points $\overline{x}_n = \overline{l}_{n,1}^+ \cap \overline{C}_1$ converge to the point $\overline{m}_1 = \overline{L}_1^+ \cap \overline{C}_1$ as $n \to \infty$ (Figure 6.33).

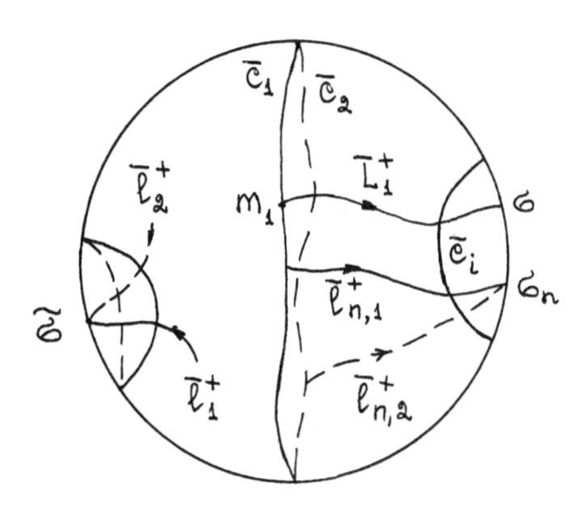

FIGURE 6.33

Suppose that the semitrajectories $\overline{l}_1^+ \in \pi^{-1}(l_1^+)$ and $\overline{l}_2^+ \in \pi^{-1}(l_2^+)$ have a common ω-limit point $\widetilde{\sigma} \in S_\infty$. Since C_1 and C_2 are homotopic, Corollary 2.2 implies the existence of families $\{\widetilde{C}_{i1}\}_1^\infty \in \pi^{-1}(C_1)$ and $\{\widetilde{C}_{i2}\}_1^\infty \in \pi^{-1}(C_2)$ such that $\widetilde{C}_{i1}$ and $\widetilde{C}_{i2}$ have common endpoints on the absolute, and $\widetilde{\sigma}$ is the topological limit of these families. There is an index i_0 such that $\overline{l}_j^+$ intersects $\widetilde{C}_{ij}$ for $i \geq i_0$, $j = 1$, 2. Since $L_1^+ \subset \omega(l_1^+)$, there is a sequence of points $\overline{z}_n \in \overline{l}_1^+ \cap \widetilde{C}_{i_n 1}$ $(i_n \geq i_0)$ such that $\pi(\overline{z}_n) \to \pi(\overline{m}_1)$ as $n \to \infty$. Therefore, there exist elements $\gamma_n \in \Gamma_p$ such that $\gamma_n(\overline{z}_n) \overset{\text{def}}{=} \overline{x}_n \in \overline{C}_1$ and $\overline{x}_n \to \overline{m}_1$ as $n \to \infty$. Then the sequences of semitrajectories $\overline{l}_{n,1}^+ = \gamma_n(\overline{l}_1^+)$ and $\overline{l}_{n,2}^+ = \gamma_n(\overline{l}_2^+)$ satisfy the conditions 1)–3), where $\sigma_n = \gamma_n(\widetilde{\sigma})$.

Let $\overline{y}_n = \overline{l}_{n,2}^+ \cap \overline{C}_2$. We show that the sequence $\{\overline{y}_n\}$ is bounded on $\overline{C}_2$. Since $\overline{x}_n \to \overline{m}_1$, the sequence $\{\overline{x}_n\}$ is bounded on $\overline{C}_1$. Take an element γ leaving invariant the curve $\overline{C}_1$ (and hence also $\overline{C}_2$). There exists a $k \in \mathbb{Z}$ such that the sequence $\{\overline{x}_n\}$ lies on the arc of $\overline{C}_1$ bounded by the points $\gamma^k(\overline{x}_1)$ and $\gamma^{-k}(\overline{x}_1)$. Then by the property 2), the sequence $\{\overline{y}_n\}$ lies on the arc of $\overline{C}_2$ bounded by $\gamma^k(\overline{y}_1)$ and $\gamma^{-k}(\overline{y}_1)$ (Figure 6.34).

It follows from the continuous dependence of trajectories on the initial conditions that there is an index n_i for which the semitrajectories $\overline{l}_{n,1}^+$ all intersect the contact-free arc $\overline{C}_i$ for $n \geq n_i$ (Figure 6.33). Since σ is the topological limit of the curves $\overline{C}_i$, $i \in \mathbb{N}$, we have $\sigma_n \to \sigma$ as $n \to \infty$.

The bounded sequence $\{\overline{y}_n\}$ of points has a subsequence $\{\overline{y}_{n_k}\}_{k=1}^\infty$ converging to some point $\overline{m}_2 \in \overline{C}_2$. It can be assumed without loss of generality that all the points $\overline{y}_{n_k}$ lie in one of the intervals in $\overline{C}_2 \setminus \{\overline{m}_2\}$.

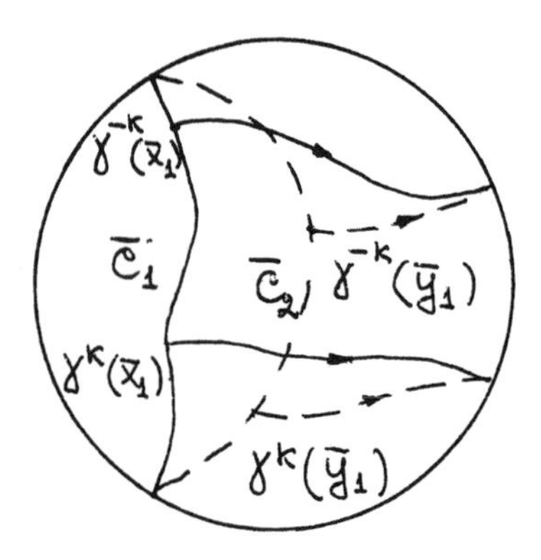

FIGURE 6.34

We consider a semitrajectory $\overline{L}_2^+(\overline{m}_2)$ of the flow f_2^t. Since $\overline{y}_{n_k} \to \overline{m}_2$, the semitrajectory $\pi[\overline{L}_2^+(\overline{m}_2)]$ lies in the quasiminimal set $\mathrm{cl}(l_2^+)$. Therefore, by Theorem 3.4 in Chapter 2, $\pi[\overline{L}_2^+(\overline{m}_2)]$ is either a nontrivial recurrent semitrajectory or an ω-separatrix. In the first case $\omega[\overline{L}_2^+(\overline{m}_2)]$ consists of a single point $\widetilde{\sigma}_2 \in S_\infty$ by Lemma 2.3. From $\overline{y}_{n_k} \to \overline{m}_2$ it follows that $\omega(\overline{l}_{n_k,2}^+) = \sigma_{n_k} \to \widetilde{\sigma}_2$ as $k \to \infty$. Therefore, $\widetilde{\sigma}_2 = \sigma$, and $L_2^+ = \pi[\overline{L}_2^+(\overline{m}_2)]$ is the desired semitrajectory.

If $\pi[\overline{L}_2^+(\overline{m}_2)]$ is an ω-separatrix, then $\overline{L}_2^+(\overline{m}_2) = \overline{L}_2^+$ is also an ω-separatrix. This case can be reduced to the preceding one with the help of a Bendixson extension of the separatrix $\overline{L}_2^+$ to the same side (to the left or to the right) on which the points $\overline{y}_{n_k}$ are located with respect to $\overline{m}_2$. By the assumption that there are finitely many equilibrium states, a finite number of Bendixson extensions lead to a semitrajectory $\widetilde{L}_2^+$ projecting into a nontrivial recurrent semitrajectory on $\mathcal{M}_p$. Then $\sigma^+(\widetilde{L}_2^+) = \sigma$, and $\pi(\widetilde{L}_2^+) = L_2^+$ is the desired semitrajectory. □

LEMMA 3.3. *Suppose that the transitive flows f_1^t and f_2^t on $\mathcal{M}_p$, $p \geq 2$, do not have impassable grains nor separatrices joining equilibrium states, and let the semitrajectories $\overline{l}_1^+$ and $\overline{l}_2^+$ of covering flows $\overline{f}_1^t$ and $\overline{f}_2^t$ have a common ω-limit point $\sigma \in S_\infty$. In this case:* 1) *if $\overline{l}_1^+$ is not an α-separatrix, then neither is $\overline{l}_2^+$, and $\alpha(\overline{l}_1) = \alpha(\overline{l}_2) = \sigma^- \in S_\infty$, where $\overline{l}_i$ is the trajectory containing $\overline{l}_i^+$, $i = 1, 2$;* 2) *if $\overline{l}_1^+$ is an α-separatrix of a saddle $\overline{O}_1$, then $\overline{l}_2^+$ is also an α-separatrix of a saddle $\overline{O}_2$. Moreover, for each ω- (α-) separatrix of $\overline{O}_1$ there is an ω- (α-) separatrix of $\overline{O}_2$ having with it a common α- (ω-) limit point on the absolute, and conversely.*

PROOF. 1) If $\overline{l}_1^+$ is not an α-separatrix, then the semitrajectory $\pi(\overline{l}_1^+)$ belongs to a nontrivial recurrent trajectory of f_1^t. This and the transitivity of the flow f_1^t give us that σ is an accessible irrational point of the first kind. The required assertion now follows from Corollary 2.5.

The assertion 2) is obtained by successively using Corollary 2.6. □

3.3. Construction of a fundamental domain. Everywhere in this subsection f^t is a transitive flow without impassable grains and without separatrices joining equilibrium states, given on a closed orientable surface $\mathcal{M}_p \cong \Delta/\Gamma_p$ of genus $p \geq 2$. Denote by $\overline{f}^t$ a flow on the Lobachevsky plane covering f^t.

For a fixed closed transversal of f^t and one of its liftings to Δ we construct a fundamental domain $\overline{\mathcal{M}}$ of Γ_p whose boundary $\partial\overline{\mathcal{M}}$ consists of contact-free arcs, saddles of the flow $\overline{f}^t$, and separatrices of these saddles.

Let C be a contact-free cycle of f^t, and let $\overline{C}_1 \in \pi^{-1}(C)$ be one of its inverse images. Since f^t is transitive and does not have separatrices joining equilibrium states, each ω-separatrix of any saddle intersects C for unboundedly small times. Therefore, there exists on $\overline{C}_1$ a family of successively located points $\overline{m}_1, \dots, \overline{m}_{k+1}$ satisfying the following conditions: 1) $\overline{l}^+(\overline{m}_i) \overset{\text{def}}{=} \overline{l}_i$ is an ω-separatrix of some saddle $\overline{O}_i$ and does not intersect any curves in $\pi^{-1}(C)$ after the point $\overline{m}_i$, $i = 1, \dots, k+1$; 2) the points $\overline{m}_1$ and $\overline{m}_{k+1}$ are congruent, and there are no other pairs of congruent points on the arc $\overline{m}_1\overline{m}_{k+1} \subset \overline{C}_1$ (between $\overline{m}_1$ and $\overline{m}_{k+1}$); 3) there are no points other than $\overline{m}_1, \dots, \overline{m}_{k+1}$ on the arc $\overline{m}_i\overline{m}_{k+1} \subset \overline{C}_1$ that satisfy the condition 1).

Let $\overline{l}_{il}$ and $\overline{l}_{ir}$ be separatrices of $\overline{O}_i$ that are Bendixson extensions of the separatrix $\overline{l}_i$ to the left and to the right, $i = 1, \dots, k+1$. It follows from the condition 3) and from Lemma 3.6 in Chapter 2 that with increasing time $\overline{l}_{ir}$ and $\overline{l}_{i+1,l}$ first intersect one and the same curve (denote it by $\overline{C}_{i+1}$) in the complete inverse image $\pi^{-1}(C)$, $i = 1, \dots, k$ (Figure 6.35). Let $\overline{m}_{ir} = \overline{l}_{ir} \cap \overline{C}_{i+1}$ and $\overline{m}_{i+1,l} = \overline{l}_{i+1,l} \cap \overline{C}_{i+1}$, and let $\overline{B}_i$ be the domain bounded by the ω-separatrices $\overline{l}_i^+(\overline{m}_i)$ and $\overline{l}_{i+1}^+(\overline{m}_{i+1})$, the α-separatrices $\overline{l}_{ir}^-(\overline{m}_{ir})$ and $\overline{l}_{i+1,l}^-(\overline{m}_{i+1,l})$, and the arcs $\overline{m}_i\overline{m}_{i+1} \subset \overline{C}_1$ and $\overline{d}_{i+1} = \overline{m}_{ir}\overline{m}_{i+1,l} \subset \overline{C}_{i+1}$ (Figure 6.35).

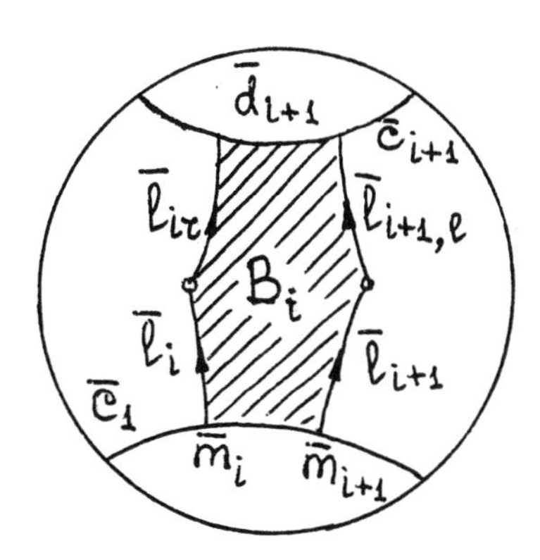

FIGURE 6.35

LEMMA 3.4. *The set* $\overline{\mathcal{M}} = \bigcup_{i=1}^k \operatorname{cl}(\overline{B}_i)$ *is a fundamental polygon of the group* Γ_p (*Figure* 6.36).

PROOF. We first show that there are no pairs of congruent points in $\operatorname{int}\overline{\mathcal{M}}$. Assume the contrary; that is, let $\overline{x}_1, \overline{x}_2 \in \operatorname{int}\overline{\mathcal{M}}$ be congruent points. Then the semitrajectories $\overline{l}^-(\overline{x}_1)$ and $\overline{l}^-(\overline{x}_2)$ are also congruent. By the construction of $\overline{\mathcal{M}}$, $\overline{l}^-(\overline{x}_1)$ and $\overline{l}^-(\overline{x}_2)$ intersect the arc $\overline{m}_1\overline{m}_{k+1} \subset \overline{C}_1$ at some points $\overline{y}_1$ and $\overline{y}_2$, and the arcs $\overset{\frown}{\overline{x}_i\overline{y}_i} \subset \overline{l}^-(\overline{x}_i)$ do not intersect the curves in $\pi^{-1}(C)$ at interior points, $i = 1$, 2. Consequently, $\overline{y}_1$ and $\overline{y}_2$ are congruent, which contradicts the condition 2).

It remains to show that for any point $m \in \mathcal{M}_p$ the complete inverse image $\pi^{-1}(m)$ intersects $\overline{\mathcal{M}}$. Let m be a regular point. The following cases are possible:

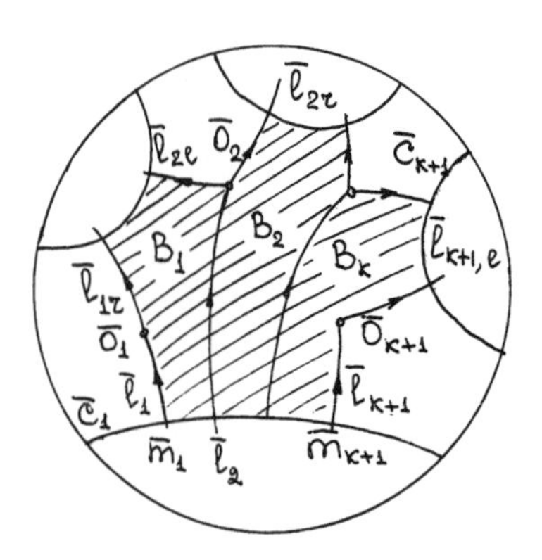

FIGURE 6.36

a) $l(m)$ is not an α-separatrix; b) $l(m)$ is an α-separatrix. In case a) we denote by $z \in C$ the first point where $l^-(m)$ intersects C with decreasing time. Since $\pi[\overline{m}_1, \overline{m}_{k+1}] = C$, the arc $\overline{m}_1\overline{m}_{k+1} \subset \overline{C}$ contains a point $\overline{z}$ such that $\pi(\overline{z}) = z$. The open arc $\overset{\frown}{zm} \subset l(m)$ is disjoint from C, so there is an open arc $\overline{z}\,\overline{m} \subset \overline{l}(\overline{m})$ that is disjoint from $\pi^{-1}(C)$ and such that $\pi(\overset{\frown}{\overline{z}\,\overline{m}}) = \overset{\frown}{zm}$. From this, $\overline{m} \in \overline{\mathcal{M}}$ and $\overline{m} \in \pi^{-1}(m)$.

In case b) if $l^-(m)$ intersects C, then the argument is completely analogous to case a). Therefore, we assume that $l^-(m)$ is disjoint from C, and we denote by l_r an ω-separatrix for which $l(m)$ is a Bendixson extension to the right. Since f^t does not have separatrices joining equilibrium states, l_r intersects C with decreasing time. Denote by $z \in C$ the first such intersection point. Then by the conditions 1) and 2), one of the points $\overline{m}_1, \dots, \overline{m}_{k+1}$, say $\overline{m}_i$, covers z. Consequently, $l_r = \pi(\overline{l}_i)$ and $\pi(\overline{l}_{ir}) = l(m)$. Since $l^-(m)$ does not intersect C, it follows that $\pi^{-1}(m) \cap \overline{l}_{ir} \in \partial\overline{\mathcal{M}} \subset \overline{\mathcal{M}}$.

If m is a saddle O, then in view of the conditions 1)–3) any ω-separatrix of O is covered by one of the separatrices $\overline{l}_1, \dots, \overline{l}_{k+1}$. Therefore, $\pi^{-1}(m) \cap \overline{\mathcal{M}} \neq \emptyset$. □

REMARK. In the treatment of case b) it is possible to take an ω-separatrix l_l for which $l(m)$ is a Bendixson extension to the left instead of the ω-separatrix l_r. Then $l(m) = \pi(\overline{l}_{jl})$ for some $j = 2, \dots, k+1$. This implies that any α-separatrix $\overline{l}_{ir}$ (i is one of the numbers $1, \dots, k$) is congruent to some (unique) α-separatrix $\overline{l}_{jl}$ (j is one of the numbers $2, \dots, k+1$). Therefore, the arc $\overline{d}_{i+1} \subset \partial\overline{\mathcal{M}}$ is congruent to some arc $\overline{d}_j \subset \partial\overline{\mathcal{M}}$.

3.4. Necessary and sufficient conditions for topological equivalence of transitive flows.

THEOREM 3.1. *Let f_1^t and f_2^t be transitive flows without impassable grains and without separatrices joining equilibrium states on a closed orientable surface $\mathcal{M}_p \cong \Delta/\Gamma_p$, $p \geq 2$. Then f_1^t and f_2^t are topologically equivalent by means of a homeomorphism homotopic to the identity if and only if there exist semitrajectories $l_1^{()}$ and $l_2^{()}$ of f_1^t and f_2^t with the same homotopy rotation class.*

PROOF. NECESSITY. If the flows are topologically equivalent by means of a homeomorphism homotopic to the identity, then the existence of $l_1^{()}$ and $l_2^{()}$ with $\mu(l_1^{()}) = \mu(l_2^{()})$ follows from Lemma 2.3 and Corollary 2.8.

SUFFICIENCY. Let $\mu(l_1^{()}) = \mu(l_2^{()})$, where $l_i^{()}$ is a semitrajectory of the flow f_i^t, $i = 1, 2$. Since replacing t by $-t$ gives a flow topologically equivalent to the original one, it can be assumed that $l_1^{()}$ and $l_2^{()}$ are positive semitrajectories.

According to Lemma 3.1, f_1^t and f_2^t have mutually homotopic simple contact-free cycles C_1 and C_2 that are nonhomotopic to zero. Let $\overline{C}^{(1)} \in \pi^{-1}(C_1)$ and $\overline{C}^{(2)} \in \pi^{-1}(C_2)$ be curves with common endpoints c^- and c^+ on the absolute. We construct a homeomorphism $\theta(\overline{C}^{(1)}, \overline{C}^{(2)}): \overline{C}^{(1)} \to \overline{C}^{(2)}$. Take a point $\overline{m} \in \overline{C}^{(1)}$. Two cases are possible: a) $\overline{l}^+(\overline{m})$ is not an ω-separatrix; b) $\overline{l}^+(\overline{m})$ is an ω-separatrix.

In case a) the semitrajectory $\pi(\overline{l}^+(\overline{m}))$ is a nontrivial recurrent semitrajectory. Therefore, $\omega(\overline{l}^+(\overline{m})) \overset{\text{def}}{=} \sigma \in S_\infty$. It follows from the transitivity of f_1^t and f_2^t and Lemma 3.2 that $\overline{f}_2^t$ has a semitrajectory $\overline{l}'^+$ with $\omega(\overline{l}'^+) = \sigma^+$.

We show that the trajectory $\overline{l}'$ containing $\overline{l}'^+$ intersects the arc $\overline{C}^{(2)}$. If the trajectory $\overline{l}(\overline{m})$ is not an α-separatrix, then in view of Lemma 3.3 neither is $\overline{l}'$, and $\alpha(\overline{l}(\overline{m})) = \alpha(\overline{l}') \overset{\text{def}}{=} \sigma^-$. Since $\overline{l}(\overline{m})$ intersects $\overline{C}^{(1)}$ at exactly one point, the pairs (σ^+, σ^-) and (c^-, c^+) of points are separated on the absolute. Therefore, $\overline{l}'$ intersects $\overline{C}^{(2)}$. If $\overline{l}(\overline{m})$ is an α-separatrix of some saddle $\overline{O}$, then $\overline{l}'$ is also an α-separatrix of some saddle $\overline{O}'$ in view of Lemma 3.3. Denote by $\mathfrak{D}$ (respectively, $\mathfrak{D}'$) the domain in the Lobachevsky plane bounded by the arc $\overline{C}^{(1)}$ (respectively, $\overline{C}^{(2)}$) and the arc $S^- = c^-c^+$ of the absolute not containing the point σ^+ (Figure 6.37). Since $\overline{l}(\overline{m})$ intersects $\overline{C}^{(1)}$, it follows that $\overline{O} \in \mathfrak{D}$. This and the fact that f_1^t does not have separatrices joining equilibrium states imply that the ω-limit set of an arbitrary α-separatrix of $\overline{O}$ different from $\overline{l}(\overline{m})$ lies on S^-. By Lemma 3.3, the ω-limit set of an arbitrary α-separatrix of $\overline{O}'$ different from $\overline{l}'$ also lies on S^-. Therefore, $\overline{O}' \in \mathfrak{D}'$, and hence $\overline{l}'$ intersects $\overline{C}^{(2)}$ (Figure 6.37).

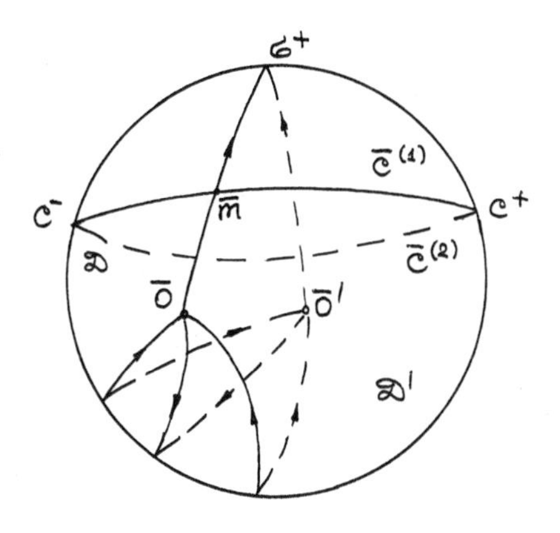

FIGURE 6.37

Thus, in case a) the trajectory $\bar{l}'$ intersects the curve $\overline{C}^{(2)}$ at some (unique) point $\overline{m}'$, and we set $\theta(\overline{C}^{(1)}, \overline{C}^{(2)})(\overline{m}) = \overline{m}'$.

In case b) $\bar{l}(\overline{m})$ is not an α-separatrix in view of the absence of separatrices joining equilibrium states. Therefore, $\alpha[\bar{l}(\overline{m})] \overset{\text{def}}{=} \sigma^- \in S_\infty$. According to Lemma 3.3, there exists a trajectory $\bar{l}'$ of $\overline{f}_2^t$ such that $\alpha(\bar{l}') = \sigma^-$. Then $\bar{l}'$ intersects $\overline{C}^{(2)}$ at some point $\overline{m}'$, and we set $\theta(\overline{C}^{(1)}, \overline{C}^{(2)})(\overline{m}) = \overline{m}'$.

It follows from the continuous dependence of trajectories on the initial conditions that $\theta(\overline{C}^{(1)}, \overline{C}^{(2)})\colon \overline{C}^{(1)} \to \overline{C}^{(2)}$ is a homeomorphism.

The construction of a homeomorphism $\theta(\overline{C}_i^{(1)}, \overline{C}_i^{(2)})\colon \overline{C}_i^{(1)} \to \overline{C}_i^{(2)}$ for any curves $\overline{C}_i^{(1)} \in \pi^{-1}(C_1)$ and $\overline{C}_i^{(2)} \in \pi^{-1}(C_2)$ with common endpoints on the absolute is completely analogous.

If the trajectories $\bar{l}$ and $\bar{l}'$ of the respective flows f_1^t and f_2^t have a common ω- (α-) limit point on the absolute, then for any element $\gamma \in \Gamma_p$ the trajectories $\gamma(\bar{l})$ and $\gamma(\bar{l}')$ also have a common ω- (α-) limit point on the absolute. Therefore,

$$\gamma \circ \theta(\overline{C}^{(1)}, \overline{C}^{(2)}) = \theta[\gamma(\overline{C}^{(1)}), \gamma(\overline{C}^{(2)})] \circ \gamma. \tag{3.1}$$

According to §3.3 we construct the fundamental polygon $\overline{\mathcal{M}}_1$ bounded by arcs of the separatrices $\bar{l}_1, \bar{l}_{1r}, \bar{l}_{2l}, \bar{l}_{2r}, \dots, \bar{l}_{k+1,l}, \bar{l}_{k+1}$ of $\overline{f}_1^t$ and by the contact-free segments $\overline{m}_1\overline{m}_{k+1} \subset \overline{C}_1^{(1)}, \overline{d}_2^{(1)} \subset \overline{C}_2^{(1)}, \dots, \overline{d}_{k+1}^{(1)} \subset \overline{C}_{k+1}^{(1)}$ lying on the curves $\overline{C}_i^{(1)} \in \pi^{-1}(C_1)$. Since C_1 and C_2 are homotopic, there is an arc $\overline{C}_i^{(2)} \in \pi^{-1}(C_2)$ with the same endpoints on the absolute for each $\overline{C}_i^{(1)}$, $i = 1, \dots, k+1$. Therefore, the homeomorphisms $\theta(\overline{C}_i^{(1)}, \overline{C}_i^{(2)})$ are defined, $i = 1, \dots, k+1$. From Lemma 3.3 and the definition of the homeomorphisms $\theta(\cdot,\cdot)$ it follows that an ω-separatrix $\bar{l}_j'$ of the flow $\overline{f}_2^t$ passes through the point $\overline{m}_j' = \theta(\overline{C}_1^{(1)}, \overline{C}_1^{(2)})(\overline{m}_j)$, $j = 1, \dots, k+1$. By the congruence of the points $\overline{m}_1$ and $\overline{m}_{k+1}$ and the relation (3.1), the points $\overline{m}_1'$ and $\overline{m}_{k+1}'$ are congruent, and there are no other pairs of congruent points on the segment $\overline{m}_1'\overline{m}_{k+1}' \subset \overline{C}_1^{(2)}$. Therefore, according to §3.3, there exists a fundamental polygon $\overline{\mathcal{M}}'$ bounded by the separatrices $\bar{l}_1' = \bar{l}'(\overline{m}_1'), \bar{l}_{1r}', \bar{l}_{2l}', \bar{l}_{2r}', \dots, \bar{l}_{k+1,l}', \bar{l}_{k+1}' = \bar{l}'(\overline{m}_{k+1}')$ of $\overline{f}_2^t$ and by the contact-free segments $\overline{m}_1'\overline{m}_{k+1}' \subset \overline{C}_1^{(2)}, \overline{d}_2^{(2)}, \dots, \overline{d}_{k+1}^{(2)}$ lying on the curves in $\pi^{-1}(C_2)$. It follows from the equalities $\overline{m}_j' = \theta(\overline{C}_1^{(1)}, \overline{C}_1^{(2)})(\overline{m}_j)$, $j = 1, \dots, k+1$, and Lemma 3.3 that $\omega(\bar{l}_{jr}') = \omega(\bar{l}_{jr})$, $j = 1, \dots, k$, and $\omega(\bar{l}_{il}') = \omega(\bar{l}_{il})$, $i = 2, \dots, k+1$. From this we get that $\theta(\overline{C}_i^{(1)}, \overline{C}_i^{(2)})(\overline{d}_i^{(1)}) = \overline{d}_i^{(2)}$, $j = 2, \dots, k+1$.

By (3.1) and the definition of the homeomorphisms $\theta(\cdot,\cdot)$, there exists a homeomorphism $\overline{\psi}\colon \partial\overline{\mathcal{M}} \to \partial\overline{\mathcal{M}}'$ with the following properties: 1) $\overline{\psi}$ coincides with $\theta(\overline{C}_1^{(1)}, \overline{C}_1^{(2)})$ on $\overline{m}_1\overline{m}_{k+1} \subset \overline{C}_1^{(1)}$ and with $\theta(\overline{C}_i^{(1)}, \overline{C}_i^{(2)})$ on the segments $\overline{d}_i^{(1)} \subset \overline{C}_i^{(1)}$, $i = 2, \dots, k+1$; 2) $\overline{\psi}(\overline{O}_i) = \overline{O}_i'$, $i = 1, \dots, k+1$; 3) $\overline{\psi}(\bar{l}_{jr} \cap \partial\overline{\mathcal{M}}) = \bar{l}_{jr}' \cap \partial\overline{\mathcal{M}}'$, $j = 1, \dots, k$; 4) $\overline{\psi}(\bar{l}_{jl} \cap \partial\overline{\mathcal{M}}) = \bar{l}_{jl}' \cap \partial\overline{\mathcal{M}}'$, $j = 2, \dots, k+1$; 5) if the points $\overline{x}, \overline{y} \in \partial\overline{\mathcal{M}}$ are congruent, then $\overline{\psi}(\overline{x})$ and $\overline{\psi}(\overline{y})$ are also congruent.

It follows from 1)–4) that $\overline{\psi}$ extends to a homeomorphism $\overline{\mathcal{M}} \to \overline{\mathcal{M}}'$ (we again denote it by $\overline{\psi}$) which carries arcs of trajectories of the flow $\overline{f}^t$ into arcs of trajectories of the flow $\overline{f}_2^t$. According to the property 5), $\overline{\psi}$ projects into a homeomorphism

$\psi\colon \mathcal{M}_p \to \mathcal{M}_p$ carrying trajectories of f_1^t into trajectories of f_2^t; that is, f_1^t and f_2^t are topologically equivalent by means of the homeomorphism ψ.

Denote by $\widetilde{\psi}$ a covering homeomorphism for ψ such that $\widetilde{\psi}|_{\overline{\mathcal{M}}} = \overline{\psi}|_{\overline{\mathcal{M}}}$. By the property 1), $\widetilde{\psi}$ extends to a homeomorphism of the absolute that is the identity on a dense set of points accessible by trajectories of the flows $\overline{f}_1^t$ and $\overline{f}_2^t$. That is, $\widetilde{\psi}^*|_{S_\infty} = \mathrm{id}$. Consequently, ψ is homotopic to the identity. □

THEOREM 3.2. *Let f_1^t and f_2^t be transitive flows without impassable grains and without separatrices joining equilibrium states on a closed orientable surface $\mathcal{M}_p$, $p \geq 2$. Then f_1^t and f_2^t are topologically equivalent if and only if they have respective semitrajectories $l_1^{()}$ and $l_2^{()}$ with the same rotation orbit.*

PROOF. NECESSITY. It follows from Theorem 2.4.

SUFFICIENCY. It follows from Theorem 3.1 and the fact that any automorphism of Γ_p is induced by a homeomorphism of the Lobachevsky plane that covers some homeomorphism of $\mathcal{M}_p$ (see Nielsen's theorem in §2.9). □

COROLLARY 3.1. *Let f_1^t and f_2^t be transitive flows without impassable grains and without separatrices joining equilibrium states on a closed orientable surface $\mathcal{M}_p$, $p \geq 2$. Then f_1^t and f_2^t are topologically equivalent if and only if they have respective separatrices l_1 and l_2 with the same rotation orbit.*

Remark. Levitt's counterexample to a conjecture of Katok. For a transitive flow f^t on an orientable surface $\mathcal{M}$, Katok defined in [**46**] the cone $K(f^t) \subset H^1(\mathcal{M}, F, \mathbb{R})$ generated by the nontrivial invariant measures of f^t (where F is the set of equilibrium states of f^t), and announced the following theorem.

THEOREM. *Let f_1^t and f_2^t be transitive flows on a closed orientable surface $\mathcal{M}_p$ of genus $p \geq 1$, and suppose that the sets of equilibrium states coincide for these flows, with each equilibrium state a nondegenerate saddle (consequently, there are $2p-2$ equilibrium states for each flow). If f_1^t and f_2^t are sufficiently close and their cones $K(f_1^t)$ and $K(f_2^t)$ intersect, then the flows are topologically equivalent.*

Katok conjectured that this theorem is true not only for sufficiently close flows. In 1983 Levitt refuted Katok's conjecture, proving the following theorem.

THEOREM [**94**]. *Let F be a set of $2p-2$ distinguished points on a closed orientable surface $\mathcal{M}_p$ of genus $p \geq 2$. Then for almost every (in the sense of Lebesgue measure) element $[\mu] \in H^1(\mathcal{M}_p, F, \mathbb{R})$ there exists a family of transitive flows f_i^t, $i \in \mathbb{N}$, on $\mathcal{M}_p$ such that:*

1) each equilibrium state of f_i^t, $i \in \mathbb{N}$, lies at one of the distinguished points and is a nondegenerate saddle;

2) the cone $K(f_i^t)$ of the flow f_i^t, $i \in \mathbb{N}$, is equal to the ray $\{t \cdot [\mu] : t > 0\}$;

3) the flows f_i^t and f_j^t are not topologically equivalent for $i \neq j$.

An element $[\mu] \in H^1(\mathcal{M}_p, F, \mathbb{R})$ can be represented as a class of closed differentiable cohomologous 1-forms differing by a differential of a function, with the differential zero on F. Denote by $\mathcal{F}(\omega)$ the foliation determined by a closed 1-form $\omega \in [\mu]$. Then the preceding theorem is equivalent to the following theorem, proved in [**94**].

THEOREM. *Let F be a set of $2p-2$ distinguished points on a closed orientable surface $\mathcal{M}_p$ of genus $p \geq 2$. Then for almost every (in the sense of Lebesgue measure in the space $H^1(\mathcal{M}_p, F, \mathbb{R})$) element $[\mu] \in H^1(\mathcal{M}_p, F, \mathbb{R})$ there exists an infinite family of closed differentiable 1-forms ω_i, $i \in \mathbb{N}$, such that:*

1) *each singularity of the foliation $\mathcal{F}(\omega_i)$, $i \in \mathbb{N}$, is nondegenerate and lies in F;*

2) *$\mathcal{F}(\omega_i)$ is transitive, $i \in \mathbb{N}$;*

3) *$\mathcal{F}(\omega_i)$ is strictly ergodic;*

4) *the foliations $\mathcal{F}(\omega_i)$ and $\mathcal{F}(\omega_j)$ are not topologically equivalent for $i \neq j$.*

§4. Classification of nontrivial minimal sets

The nontrivial minimal sets not containing special pairs of trajectories can be classified on closed orientable surfaces of genus $p \geq 2$ with the help of the homotopy rotation class and the orbit of the homotopy rotation class. In solving the realization problem we construct flows for which the trajectories in a nontrivial minimal set are geodesic curves.

4.1. Special and basic trajectories. We recall that a minimal set of a flow is defined to be a nonempty closed invariant set not containing proper closed invariant subsets.

DEFINITION. A *nontrivial minimal set* is defined to be a nowhere dense set that is not a closed trajectory nor an equilibrium state.

According to the catalogue of minimal sets (§3.9 in Chapter 2) a nontrivial minimal set consists of nontrivial recurrent trajectories, with each of its trajectories dense in the minimal set.

The nontrivial minimal sets on the torus are imbedded in Denjoy flows. That is, if a flow f^t on the torus has a nontrivial minimal set Ω, then there exists a Denjoy flow $\widetilde{f^t}$ with nontrivial minimal set Ω. The Denjoy flows were classified in §1.7.

In this section we consider nontrivial minimal sets of flows on closed orientable surfaces $\mathcal{M}_p$ of genus $p \geq 2$.

DEFINITION. Let f_1^t and f_2^t be flows (perhaps the same) with minimal sets Ω_1 and Ω_2 on a surface $\mathcal{M}$. The minimal sets Ω_1 and Ω_2 are said to be *topologically equivalent* if there exists a homeomorphism of $\mathcal{M}$ carrying the trajectories in Ω_1 into trajectories in Ω_2 and mapping Ω_1 onto Ω_2.

Any nontrivial minimal set of a flow on a closed orientable surface of genus ≥ 2 is locally homeomorphic to the direct product of a closed bounded interval and the Cantor set. Therefore, a nontrivial minimal set in $\mathcal{M}_p$, $p \geq 2$, contains boundary and interior nontrivial recurrent trajectories (see §1.5 for the definition of boundary and interior trajectories).

DEFINITION. A pair of boundary trajectories l_1, l_2 in a nontrivial minimal set Ω is called a *special pair*, and the trajectories themselves are said to be *special*, if there exists a simply connected component of $\mathcal{M}_p \setminus \Omega$ whose accessible (from within) boundary consists of l_1 and l_2.

A boundary trajectory in a nontrivial minimal set is said to be *basic* if it is not special.

LEMMA 4.1. *Suppose that a flow f^t on a closed orientable surface $\mathcal{M}_p$, $p \geq 2$, has a nontrivial minimal set Ω. Two trajectories l_1 and l_2 in Ω form a special pair if and only if there exist covering trajectories $\bar{l}_i \in \pi^{-1}(l_i)$, $i = 1, 2$, such that $\omega(\bar{l}_1) = \omega(\bar{l}_2)$ and $\alpha(\bar{l}_1) = \alpha(\bar{l}_2)$.*

PROOF. NECESSITY. It follows from the properties of a covering.

SUFFICIENCY. Suppose that there are trajectories $\bar{l}_i \in \pi^{-1}(l_i)$ $i = 1, 2$, with $\omega(\bar{l}_1) = \omega(\bar{l}_2)$ and $\alpha(\bar{l}_1) = \alpha(\bar{l}_2)$. Then $\bar{l}_1$ and $\bar{l}_2$ bound a domain $\overline{w}$ on Δ (Figure 6.38). By Corollary 2.4, $\overline{w}$ does not contain inverse images of nontrivial recurrent semitrajectories of f^t. Therefore, l_1 and l_2 are boundary trajectories. Since the points $\omega(\bar{l}_i)$ and $\alpha(\bar{l}_i)$ are irrational for $i = 1, 2$, the domain $\pi(\overline{w})$ is simply connected. □

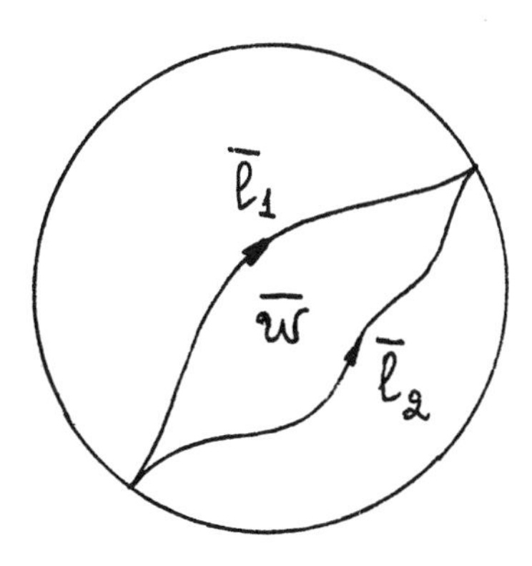

FIGURE 6.38

LEMMA 4.2. *Any nontrivial minimal set of a flow on a closed orientable surface $\mathcal{M}_p$, $p \geq 2$, contains finitely many basic trajectories.*

PROOF. Let Ω be a nontrivial minimal set. According to Lemma 2.3, there exists a contact-free cycle C that intersects Ω. Following §4.3 in Chapter 2, we introduce a partition ξ_Ω of C into closed disjoint subsets (elements). The elements of ξ_Ω are closed intervals in $C \setminus \Omega$ and points not belonging to these closed intervals.

It follows from the definition of basic trajectories that any basic trajectory passes through an endpoint of at least one closed interval forming an element of ξ_Ω of type two. Then Lemma 4.2 in Chapter 2 gives us that there are finitely many basic trajectories. □

REMARK. It is shown in [**17**] that the number of basic trajectories of a nontrivial minimal set of a flow on a closed orientable surface of genus $p \geq 2$ does not exceed $8(p-1)$ and is at least 2.

4.2. The canonical set. In this subsection we construct a closed set containing a nontrivial minimal set Ω which is said to be canonical. The results here overlap in part those in §4 of Chapter 2.

LEMMA 4.3. *Suppose that a flow f^t on $\mathcal{M}_p$, $p \geq 2$, has a nontrivial minimal set Ω, and let C be a contact-free cycle of f^t intersecting Ω. Then there exists a closed set $\mathcal{D}(\Omega, C)$ satisfying the following conditions:*

1) $\Omega \subset \mathcal{D}(\Omega, C)$;

2) *each component of the set* $\mathcal{D}(\Omega, C) \setminus \Omega$ *is simply connected;*

3) *the boundary of* $\mathcal{D}(\Omega, C)$ *consists of finitely many simple closed curves, each a union of an even number of arcs of basic trajectories in* Ω *and the same number of contact-free segments lying on* C (*Figure* 6.39).

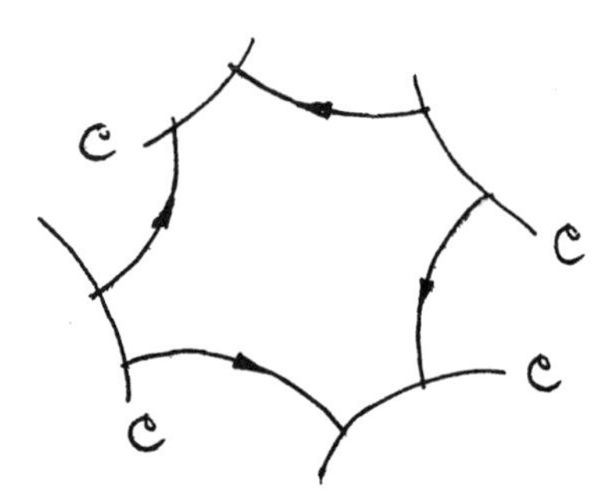

FIGURE 6.39

PROOF. We take a curve $\overline{C} \in \pi^{-1}(C)$ and congruent points $\overline{a}, \overline{b} \in \overline{C} \cap \overline{\Omega}$, where $\overline{\Omega} = \pi^{-1}(\Omega)$, such that there are no other pairs of congruent points on the arc $\overline{a}\overline{b} \subset \overline{C}$. Denote by $\overline{C}_1$ (respectively, $\overline{C}_2$) the first curve in $\pi^{-1}(C)$ that is intersected by the semitrajectory $\overline{l}^+(\overline{a})$ (respectively, $\overline{l}^+(\overline{b})$) after the point $\overline{a}$ ($\overline{b}$). We show that $\overline{C}_1 \neq \overline{C}_2$. Assume not. Then the points $\overline{a}_1 = \overline{C}_1 \cap \overline{l}^+(\overline{a})$ and $\overline{b}_1 = \overline{C}_1 \cap \overline{l}^+(\overline{b})$ are congruent, and there are no other pairs of congruent points on the arc $\overline{a}_1\overline{b}_1 \subset \overline{C}_1$. The trajectory arcs $\overset{\frown}{\overline{a}\,\overline{a}_1} \subset \overline{l}^+(\overline{a})$ and $\overset{\frown}{\overline{b}\,\overline{b}_1} \subset \overline{l}^+(\overline{b})$ are also congruent, and therefore the curvilinear quadrangle bounded by the arcs $\overline{a}\overline{b}$, $\overline{a}_1\overline{b}_1$, $\overset{\frown}{\overline{a}\,\overline{a}_1}$, $\overset{\frown}{\overline{b}\,\overline{b}_1}$ projects into a torus—a surface of genus 1. This contradicts the fact that $\mathcal{M}_p$ is a surface of genus $p \geq 2$.

The set $\overline{\Omega} \cap \overline{C}$ is a Cantor set. Let $(\overline{\alpha}, \overline{\beta})$ be an adjacent interval of $\overline{\Omega} \cap \overline{C}$ such that the trajectory $\pi[\overline{l}(\overline{\beta})]$ is not basic, and let $\overline{\beta}_1 \in \overline{C}$ be a point congruent to $\overline{\beta}$ such that there are no other pairs of congruent points on the segment $\overline{B} \overset{\text{def}}{=} \overline{\beta\beta_1} \subset \overline{C}$.

Let $\overline{\Omega}_0 = \overline{B} \cap \Omega$, and denote by $\overline{C}(\overline{m})$ the first curve in $\pi^{-1}(C)$ intersected by the semitrajectory $\overline{l}^+(\overline{m})$, $\overline{m} \in \overline{\Omega}_0$, with increasing time after the point $\overline{m}$. By the compactness of $\overline{\Omega}_0$ and the continuous dependence of trajectories on the initial conditions, there are only finitely many open intervals $\mathcal{U}_1, \dots, \mathcal{U}_k$ that cover $\overline{\Omega}_0$ and are such that for any $m_i', m_i'' \in \overline{\Omega}_0 \cap \mathcal{U}_i$ the curves $\overline{C}(m_i')$ and $\overline{C}(m_i'')$ coincide ($i = 1, \dots, k$), while for any $m_i, \in \mathcal{U}_i \cap \overline{\Omega}_0$ and $m_j \in \mathcal{U}_j \cap \overline{\Omega}_0$ with $i \neq j$ the curves $\overline{C}(m_i)$ and $\overline{C}(m_j)$ are different. It follows from the first paragraph of the proof that $k \geq 2$.

Denote by $\overline{C}_i$ the curve $\overline{C}(\overline{m})$, $\overline{m} \in \mathcal{U}_i \cap \overline{\Omega}_0$, $i = 1, \dots, k$.

By the Cantor property of the set $\overline{\Omega}_0$, there exists for each $i \in \{1, \dots, k\}$ a closed interval $[\overline{a}_i, \overline{b}_i] \subset \overline{C}_i$ with endpoints $\overline{a}_i, \overline{b}_i \in \mathcal{U}_i \cap \overline{\Omega}_0$ that contains all the points in $\mathcal{U}_i \cap \overline{\Omega}_0$. Since $\overline{C}_i \neq \overline{C}_j$ for $i \neq j$, the trajectories $\pi[\overline{l}(\overline{a}_i)]$ and $\pi[\overline{l}(\overline{b}_i)]$ are basic for $i = 1, \dots, k$.

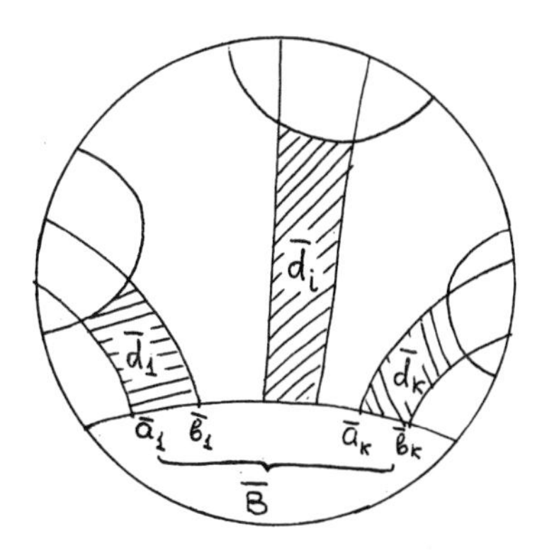

FIGURE 6.40

Let $\overline{A}_i = \overline{C}_i \cap \overline{l}(\overline{a}_i)$ and $\overline{B}_i = \overline{C}_i \cap \overline{l}(\overline{b}_i)$. Denote by $\overline{d}_i$ the closure of the domain bounded by the segments $[\overline{a}_i, \overline{b}_i] \subset \overline{C}$ and $\overline{A}_i\overline{B}_i \subset \overline{C}_i$ and by the arcs $\overset{\curvearrowright}{\overline{a}_i\overline{A}_i} \subset \overline{l}(\overline{a}_i)$ and $\overset{\curvearrowright}{\overline{b}_i\overline{B}_i} \subset \overline{l}(\overline{b}_i)$ for $i = 1, \dots, k$. Let $\overline{\mathcal{D}}(\overline{\Omega}, \overline{C}) = \bigcup_{i=1}^{k} \overline{d}_i$ (Figure 6.40).

We show that $\mathcal{D}(\Omega, C) = \pi[\overline{\mathcal{D}}(\overline{\Omega}, \overline{C})]$ is the desired set.

Since $\pi(\overline{B}) = C$, it follows that $\Omega \subset \mathcal{D}(\Omega, C)$.

Let w be a component of $\mathcal{D}(\Omega, C) \setminus \Omega$. It follows from the construction of the set $\overline{\mathcal{D}}(\overline{\Omega}, \overline{C})$ that the inverse image $\pi^{-1}(w)$ breaks up into curvilinear quadrangles bounded by segments of the curves $\overline{C}$ and $\overline{C}_i$ and arcs of trajectories. This implies that w is simply connected.

We take a point $\overline{b}_{i_1} \in \overline{B}$, $i_1 \neq k$, with a basic trajectory $\overline{l}(\overline{b}_{i_1})$ passing through it (for simplicity we say that a trajectory covering a basic trajectory is also basic). Then $\overline{B}_{i_1} = \overline{C}_{i_1} \cap \overline{l}(\overline{b}_{i_1})$ is an endpoint of an adjacent interval $(\overline{B}_{i_1}, A^*_{j_1}) \overset{\text{def}}{=} \mathcal{I}_{i_1}$ of the Cantor set $\overline{C}_i \cap \overline{\Omega}$, and the trajectory $\overline{l}(A^*_{j_1})$ is basic. It follows from $\pi(\overline{B}) = C$ that there exists a point $\overline{A}_{j_1} = \overline{C}_{j_1} \cap \overline{l}(\overline{a}_{j_1})$, $\overline{a}_{j_1} \in \overline{B}$, that is congruent to $A^*_{j_1}$. Since no basic trajectory passes through the point $\overline{\beta}$, it follows that $j_1 \neq 1$, and hence the adjacent interval $(\overline{b}_{j_1-1}, \overline{a}_{j_1}) \overset{\text{def}}{=} \mathcal{I}_{12}$ lies in $\overline{B}$. Let $j_1 - 1 = i_2$. We repeat the process described for the point $\overline{b}_{i_2}$, and so on. As a result we obtain a sequence of alternating arcs of trajectories and segments of the curves $\overline{C}$ and $\overline{C}_i$: $\overline{b}_{i_1}\overline{B}_{i_1}, \mathcal{I}_{i_1}, \overline{a}_{i_2}\overline{A}_{i_2}, \mathcal{I}_{12}, \overline{b}_{i_2}\overline{B}_{i_2}, \mathcal{I}_{i_2}, \overline{a}_{i_3}\overline{A}_{i_3}, \dots$. There are finitely many points $\overline{a}_i, \overline{b}_i \in \overline{B}$ with basic trajectories passing through them, so the sequence leads in finitely many steps to the interval $\mathcal{I}_{n,n+1} = (\overline{b}_{i_1}, \overline{a}_{i_{n+1}})$. Consequently, the union of the above arcs of trajectories and segments projects into a closed curve λ_{i_1} on $\mathcal{M}_p$. It follows from the construction that λ_{i_1} is a simple curve, and λ_{i_1} is in the boundary of the set $\mathcal{D}(\Omega, C)$. Since the point $\overline{b}_{i_1}$, $i_1 \neq k$, is arbitrary, this proves 3). □

4.3. Topological equivalence of minimal sets.

THEOREM 4.1. *For flows f_1^t and f_2^t on $\mathcal{M}_p$, $p \geq 2$, let Ω_1 and Ω_2 be nontrivial minimal sets that do not contain special pairs of trajectories. Then Ω_1 and Ω_2 are topologically equivalent by means of a homeomorphism homotopic to the identity if and only if there exist two semitrajectories $l_1^{()} \subset \Omega_1$ and $l_2^{()} \subset \Omega_2$ with the same homotopy rotation class.*

PROOF. NECESSITY. It follows from Lemma 2.3 and Corollary 2.8.

SUFFICIENCY. Let $\mu(l_1^{()}) = \mu(l_2^{()})$, where $l_i^{()} \subset \Omega_i$, $i = 1, 2$. It can be assumed without loss of generality that $l_1^{()}$ and $l_2^{()}$ are positive semitrajectories.

According to Lemma 3.1, f_1^t and f_2^t have mutually homotopic contact-free cycles C_1 and C_2 that are nonhomotopic to zero and intersect l_1^+ and l_2^+ (and hence Ω_1 and Ω_2), respectively.

Let $\overline{C}^{(1)} \in \pi^{-1}(C_1)$ and $\overline{C}^{(2)} \in \pi^{-1}(C_2)$ be curves with common endpoints on the absolute. As in the proof of Theorem 3.1, we construct the homeomorphism $\theta(\overline{C}^{(1)}, \overline{C}^{(2)})\colon \overline{C}^{(1)} \cap \overline{\Omega}_1 \to \overline{C}^{(2)} \cap \overline{\Omega}_2$, where $\overline{\Omega}_1 = \pi^{-1}(\Omega_1)$ and $\overline{\Omega}_2 = \pi^{-1}(\Omega_2)$. Then for the curves $\overline{C}_i^{(1)} = \gamma(\overline{C}^{(1)})$ and $\overline{C}_i^{(2)} = \gamma(\overline{C}^{(2)})$ ($\gamma \in \Gamma_p$ arbitrary) and the homeomorphism $\theta(\overline{C}_i^{(1)}, \overline{C}_i^{(2)})\colon \overline{C}_i^{(1)} \to \overline{C}_i^{(2)}$ we have the relation

$$\theta(\overline{C}_i^{(1)}, \overline{C}_i^{(2)}) \circ \gamma|_{\overline{C}^{(1)} \cap \overline{\Omega}_1} = \gamma \circ \theta(\overline{C}^{(1)}, \overline{C}^{(2)})|_{\overline{C}^{(1)} \cap \overline{\Omega}_1}.$$

Since $\overline{C}^{(i)} \cap \overline{\Omega}_i$ is a Cantor set, $i = 1, 2$, the homeomorphism $\theta(\overline{C}^{(1)}, \overline{C}^{(2)})$ can be extended to a homeomorphism $\overline{C}^{(1)} \to \overline{C}^{(2)}$ (which we also denote by $\theta(\overline{C}^{(1)}, \overline{C}^{(2)})$. On the curve $\overline{C}_i^{(1)} = \gamma(\overline{C}^{(1)})$ ($\gamma \in \Gamma_p$) we extend $\theta(\overline{C}_i^{(1)}, \overline{C}_i^{(2)})$ to the homeomorphism $\gamma \circ \theta(\overline{C}^{(1)}, \overline{C}^{(2)}) \circ \gamma^{-1}\colon \overline{C}_i^{(1)} \to \overline{C}_i^{(2)}$. Then for any element $\gamma \in \Gamma_p$

$$\theta[\gamma(\overline{C}^{(1)}), \gamma(\overline{C}^{(2)})] \circ \gamma = \gamma \circ \theta(\overline{C}^{(1)}, \overline{C}^{(2)}). \tag{4.1}$$

According to Lemma 4.3, there exists a set $\overline{\mathcal{D}}(\overline{\Omega}_1, \overline{C}^{(1)})$ such that $\mathcal{D}(\Omega_1, C_1) = \pi[\overline{\mathcal{D}}(\overline{\Omega}_1, \overline{C}^{(1)})]$ is a canonical set for Ω_1 .

Suppose that the curves $\overline{C}^{(1)}, \overline{C}_1^{(1)}, \dots, \overline{C}_k^{(1)} \in \pi^{-1}(C_1)$ are in the boundary of the set $\overline{\mathcal{D}}_1(\overline{\Omega}_1, \overline{C}^{(1)}) \overset{\text{def}}{=} \overline{\mathcal{D}}_1$, and that the intersection $\overline{C}^{(1)} \cap \overline{\mathcal{D}}_1$ consists of the intervals $[\overline{a}_1^{(1)}, \overline{b}_1^{(1)}], \dots, [\overline{a}_k^{(1)}, \overline{b}_k^{(1)}]$. It follows from (4.1) that the points $\overline{a}_1^{(2)} = \theta(\overline{C}^{(1)}, \overline{C}^{(2)})(\overline{a}_1^{(1)})$ and $\overline{b}_k^{(2)} = \theta(\overline{C}^{(1)}, \overline{C}^{(2)})(\overline{b}_k^{(1)})$ are congruent, and there are no other pairs of congruent points on the segment $\overline{a}_1^{(2)}\overline{b}_k^{(2)} \subset \overline{C}^{(2)}$. Then by virtue of Lemma 4.3 we construct a set $\overline{\mathcal{D}}_2(\overline{\Omega}, \overline{C}^{(2)}) \overset{\text{def}}{=} \overline{\mathcal{D}}_2$ such that $\mathcal{D}(\Omega_2, C_2) = \pi(\overline{\mathcal{D}}_2)$ is a canonical set for Ω_2, and the intersection $\overline{C}^{(2)} \cap \overline{\mathcal{D}}_2$ consists of intervals lying on the segment $\overline{a}_1^{(2)}\overline{b}_k^{(2)}$.

By repeating the proof of Theorem 3.1 without fundamental changes we get the following:

1) the boundary of $\overline{\mathcal{D}}_2(\overline{\Omega}_2, \overline{C}^{(1)})$ contains segments of the curves

$$\overline{C}^{(2)}, \overline{C}_1^{(2)}, \dots, \overline{C}_k^{(2)} \in \pi^{-1}(C_2),$$

where $\overline{C}_i^{(1)}$ and $\overline{C}_i^{(2)}$ have common endpoints on the absolute for $i = 1, \dots, k$, and $\overline{\mathcal{D}}_2 \cap \pi^{-1}(C_2) \subset \overline{C}^{(2)} \bigcup_{i=1}^k \overline{C}_i^{(2)}$;

2) the intersection $\overline{C}^{(2)} \cap \overline{\mathcal{D}}_2$ consists of the segments

$$[\overline{a}_1^{(2)}, \overline{b}_1^{(2)}] = \theta(\overline{C}^{(1)}, \overline{C}^{(2)})([\overline{a}_1^{(1)}, \overline{b}_1^{(1)}]), \dots, [\overline{a}_k^{(2)}, \overline{b}_k^{(2)}] = \theta(\overline{C}^{(1)}, \overline{C}^{(2)})([\overline{a}_k^{(1)}, \overline{b}_k^{(1)}]);$$

3) the intersection $\overline{C}_i^{(2)} \cap \overline{\mathcal{D}}_2$ is equal to $\theta(\overline{C}_i^{(1)}, \overline{C}_i^{(2)})(\overline{C}_i^{(1)} \cap \overline{\mathcal{D}}_1)$;

4) there are no pairs of congruent points in the domain int $\overline{\mathcal{D}}_i$, $i = 1, 2$.

This implies the existence of a homeomorphism $\varphi\colon \overline{\mathcal{D}}_1 \to \overline{\mathcal{D}}_2$ that carries arcs of trajectories in $\overline{\Omega}_1 \cap \overline{\mathcal{D}}_1$ into arcs of trajectories in $\overline{\Omega}_2 \cap \overline{\mathcal{D}}_2$ and that projects into a homeomorphism $\varphi\colon \mathcal{D}(\Omega_1, C_1) \to \mathcal{D}(\Omega_2, C_2)$ carrying trajectories in Ω_1 into trajectories in Ω_2 (because $\Omega_i \subset \mathcal{D}(\Omega_i, C_i)$ for $i = 1, 2$ in view of Lemma 4.3).

It follows from 1)–3) that to each simple closed curve λ_1 of the boundary $\partial\mathcal{D}(\Omega_1, C_1)$ there corresponds a simple closed curve λ_2 of $\partial\mathcal{D}(\Omega_2, C_2)$ that is homotopic to it, and conversely. Let ψ denote the correspondence $\lambda_1 \to \lambda_2$.

We take a component R_1 of the set $\mathcal{M}_p \setminus \mathcal{D}(\Omega_1, C_1)$. Since to each component of ∂R_1 there corresponds a unique component of $\partial\mathcal{D}(\Omega_2, C_2)$ via the mapping ψ, there exists a unique component R_2 of $\mathcal{M}_p \setminus \mathcal{D}(\Omega_2, C_2)$ such that ψ realizes a one-to-one correspondence between the components of the boundaries ∂R_1 and ∂R_2. Since the curves $\lambda \subset \partial R_1$ and $\psi(\lambda) \subset \partial R_2$ are homotopic, it follows from the condition 2) of Lemma 4.3 that the homeomorphism φ can be extended to a homeomorphism $R_1 \cup \mathcal{D}(\Omega_1, C_1) \to R_2 \cup \mathcal{D}(\Omega_2, C_2)$. Going through this procedure for each component of $\mathcal{M}_p \setminus \mathcal{D}(\Omega_1, C_1)$, we get a homeomorphism $\widehat{\varphi}\colon \mathcal{M}_p \to \mathcal{M}_p$ carrying trajectories in Ω_1 into trajectories in Ω_2. The fact that $\widehat{\varphi}$ is homotopic to the identity is shown as in the proof of Theorem 3.1. □

THEOREM 4.2. *For flows f_1^t and f_2^t on $\mathcal{M}_p$, $p \geq 2$, let Ω_1 and Ω_2 be nontrivial minimal sets not containing special pairs of trajectories. Then Ω_1 and Ω_2 are topologically equivalent if and only if there exist semitrajectories $l_1^{()} \subset \Omega_1$ and $l_2^{()} \subset \Omega_2$ with the same rotation orbit.*

PROOF. NECESSITY. It follows from Theorem 2.4.

SUFFICIENCY. It follows from Theorem 4.1 and the fact that any automorphism of the group Γ_p is induced by some covering homeomorphism (Nielsen's theorem, §2.9). □

4.4. Realization of nontrivial minimal sets by geodesic curves.

THEOREM 4.3. *For a flow f^t on $\mathcal{M}_p$, $p \geq 2$, let Ω be a nontrivial minimal set not containing special pairs of trajectories. Then there is a flow f_0^t on $\mathcal{M}_p$ such that:*

1) *f_0^t has a nontrivial minimal set Ω_0 whose trajectories are geodesic curves in a metric of constant negative curvature;*

2) *Ω_0 is topologically equivalent to Ω by means of a homeomorphism homotopic to the identity.*

PROOF. The nontrivial minimal set Ω is quasiminimal. According to Lemma 2.6, there exists a geodesic lamination $\mathcal{F}(\Omega)$ on $\mathcal{M}_p$ that consists of nontrivial B-recurrent geodesics and has the following property: for any trajectory $l \subset \Omega$ there is a unique geodesic $L \overset{\text{def}}{=} L(l) \subset \mathcal{F}(\Omega)$ such that for each inverse image $\bar{l} \in \pi^{-1}(l)$ there exists an inverse image $\overline{L} \in \pi^{-1}(L)$ with the same endpoints $\sigma^-(\bar{l}) = \sigma^-(\overline{L})$ and $\sigma^+(\bar{l}) = \sigma^+(\overline{L})$ on the absolute.

We define on $\mathcal{F}(\Omega)$ the vector field $\vec{V}$ of unit tangent vectors to the geodesics in $\mathcal{F}(\Omega)$. Since the lamination $\mathcal{F}(\Omega)$ is constructed from the trajectories of a flow, the vector field $\vec{V}$ on $\mathcal{F}(\Omega)$ is continuous. We extend it to a continuous field $\vec{V}_0$ on the whole surface $\mathcal{M}_p$, and let f_0^t be the flow induced by $\vec{V}_0$.

Denote by $|\mathcal{F}(\Omega)|$ the set of points lying on the geodesics in $\mathcal{F}(\Omega)$.

Since Ω is a nontrivial minimal set, it follows that $|\mathcal{F}(\Omega)|$ is a nowhere dense invariant set for f_0^t. Since each trajectory $l \subset \Omega$ is dense in Ω, each trajectory $l_0 \subset |\mathcal{F}(\Omega)|$ of f_0^t is dense in $|\mathcal{F}(\Omega)|$. Consequently, $|\mathcal{F}(\Omega)|$ is a nontrivial minimal set of the flow f_0^t.

By the construction of the geodesic lamination $\mathcal{F}(\Omega)$, for each semitrajectory $l^{()} \subset \Omega$ of the flow f^t there is a semitrajectory $l_0^{()} \subset |\mathcal{F}(\Omega)|$ of f_0^t with the same homotopy rotation class. This and Theorem 4.1 give us that the minimal sets Ω and $|\mathcal{F}(\Omega)|$ are topologically equivalent by means of a homeomorphism homotopic to the identity. □

REMARK. It was proved in [**18**] that the vector field $\vec{V}$ on $\mathcal{F}(\Omega)$ is not only continuous but also Lipschitzian. By a theorem of Schwartz in [**107**], it cannot have smoothness C^r, $r \geq 2$. It remains an open question as to whether $\vec{V}$ is a vector field of class C^1.

§5. Topological equivalence of flows without nontrivial recurrent trajectories

In 1955 Leontovich and Maĭer [**51**] introduced a complete topological invariant for flows on a sphere with finitely many singular trajectories—the scheme of a flow—which included a qualitative description of the singular trajectories and their mutual arrangement. In 1976 Neumann and O'Brien [**100**] introduced an invariant—the orbit complex—as a generalization of this invariant to the set $\mathfrak{N}_p$ of flows on a surface $\mathcal{M}_p$, $p \geq 2$, with finitely many singular trajectories and without nontrivial recurrent trajectories.

The present section is devoted to a description of this invariant.

We remark that the Morse–Smale flows classified by Peixoto [**104**] with the help of a distinguishing graph belong to the set $\mathfrak{N}_p$.

Everywhere in this section $\mathfrak{N}_p$ denotes the set of flows with finitely many singular trajectories and without nontrivial recurrent trajectories on a closed orientable surface $\mathcal{M}_p$ of genus $p \geq 0$. Unless otherwise stated, flows are allowed to have infinitely many singular trajectories in §§5.1–5.3.

5.1. Schemes of semicells. According to Theorems 4.1 and 4.2 in Chapter 3, any cell R of a flow $f^t \in \mathfrak{N}_p$ is homeomorphic either to an open disk (simply connected) or to an open annulus (doubly connected), and by Theorem 4.3 in the same chapter, the restriction $f^t|_R$ of f^t to R is topologically equivalent to one of the following flows:

1) a parallel flow on an open strip;
2) a parallel flow on an open annulus;
3) a spiral flow on an open annulus;
4) a rational winding on the torus.

For flows of the first three types we introduce in this subsection the concept of a semicell and of its scheme.

Suppose that the flow $f^t|_R$ has type 1) or 2), and let $l \subset R$ be a trajectory of f^t. By Theorem 4.3 in Chapter 3, l divides the cell R into two domains R^+ and R^- called *semicells of types* 1) *or* 2), respectively.

Let $f^t|_R$ have type 3). According to Theorem 4.3 in Chapter 3, there exists a contact-free cycle C of $f^t|_R$ dividing R into two domains R^+ and R^- and intersecting each trajectory of $f^t|_R$ at exactly one point. The domains R^+ and R^- are called *semicells of type* 3).

Denote by δR the accessible (from within) boundary of R.

The *accessible (from within) boundary* $\delta R^\pm$ *of the semicell* $R^\pm \subset R$ is defined to be the part of δR adjacent to $R^\pm$ (consequently, the trajectory or contact-free cycle dividing R into R^+ and R^- is not in the accessible (from within) boundary of R^+ or of R^-).

We introduce the concept of the scheme of a semicell. Consider a semicell $R^\pm$ of type 1). On the set of regular trajectories in the accessible (from within) boundary $\delta R^\pm$ we introduce an order relation. Let $l_1, l_2 \subset \delta R^\pm$ be regular trajectories. Take points $m_i \in l_i$, $i = 1, 2$, and disjoint contact-free segments Σ_i, $i = 1, 2$, such that Σ_i has m_i as an endpoint, and $\Sigma_i \setminus \{m_i\} \subset R^\pm$ (Figure 6.41). According to Theorem 4.3 in Chapter 3, there are trajectories in $R^\pm$ that intersect both Σ_1 and Σ_2.

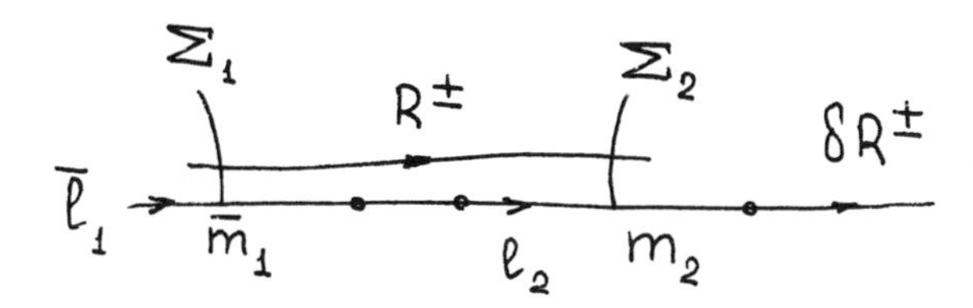

FIGURE 6.41

We write $l_1 < l_2$ if for any points $m_i \in l_i$ and any contact-free segments Σ_i, $i = 1, 2$, there exist trajectories $l \subset R^\pm$ that first intersect Σ_1 and then Σ_2 as time increases.

DEFINITION. The ordered set of regular trajectories in the accessible (from within) boundary of a semicell of type 1) is called *the scheme of the semicell of type* 1). The schemes of two semicells are said to be *isomorphic* or *identical* if there exists an order-preserving one-to-one mapping of one scheme into the other.

We consider a semicell $R^\pm$ of type 2) or 3). If $\delta R^\pm$ consists of a single trajectory (a limit cycle), then the scheme of $R^\pm$ is defined to be the empty set $\{\emptyset\}$. If $\delta R^\pm$ consists of a single equilibrium state O and a single trajectory l_0 with $\omega(l_0) = \alpha(l_0) = O$, then the single-element set $\{l_0\}$ is the scheme of $R^\pm$. Now suppose that δR^+ contains at least two regular trajectories. We fix some "initial" regular trajectory $l_1 \subset \delta R^\pm$. Then on the remaining regular trajectories in $\delta R^\pm \setminus \{l_1\}$ an order relation is introduced as above. Since the choice of the initial trajectory in $\delta R^\pm$ is arbitrary, we use the term *the cyclic order* for all the orderings obtained on $\delta R^\pm$ for different choices of the initial regular trajectory.

DEFINITION. The cyclically ordered set of regular trajectories in the accessible (from within) boundary of a semicell is called *the scheme of the semicell of type* 2) *or* 3).

The scheme of a semicell $R^\pm$ will be denoted by $\omega(R^\pm)$. The scheme of a semicell of type 2) or 3) can be represented as a list $(l_1, \dots, l_n, l_{n+1} = l_1)$ of trajectories,

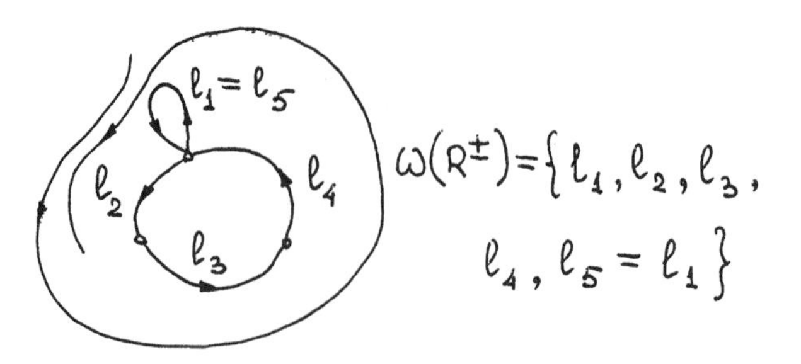

FIGURE 6.42

with the order in which the trajectories are encountered upon moving along $\delta R^{\pm}$ in the direction induced by the flow $f^t|_{R^{\pm}}$ (Figure 6.42).

Two schemes of semicells of type 2) or 3) are said to be *isomorphic* or *identical* if there exists a one-to-one correspondence of one scheme into the other that preserves the cyclical order.

We remark that if the number of singular trajectories is finite, then the number of singular regular trajectories in the accessible (from within) boundary of a semicell is a complete invariant of the property of being isomorphic for schemes of semicells (with the symbol $\emptyset$ regarded as a special "number").

For a flow f^t on $\mathcal{M}$ we denote by $\check{f}^t$ the restriction of f^t to the set $\mathcal{M} \setminus \mathrm{Fix}(f^t)$, where $\mathrm{Fix}(f^t)$ is the set of equilibrium states of f^t.

The next result follows immediately from Theorem 4.3 in Chapter 3 and the definition of schemes of semicells.

LEMMA 5.1. *Suppose that $R_1^{\pm}$ and $R_2^{\pm}$ are semicells of the same type for the respective flows f_1^t and f_2^t. If the schemes of the semicells are isomorphic, then the restrictions $\check{f}_1^t|_{R_1^{\pm} \cup \delta R_1^{\pm}}$ and $\check{f}_2^t|_{R_2^{\pm} \cup \delta R_2^{\pm}}$ are topologically equivalent.*

We remark that the flows $f_1^t|_{R_1^{\pm} \cup \delta R_1^{\pm}}$ and $f_2^t|_{R_2^{\pm} \cup \delta R_2^{\pm}}$ are not topologically equivalent in general (Figure 6.43).

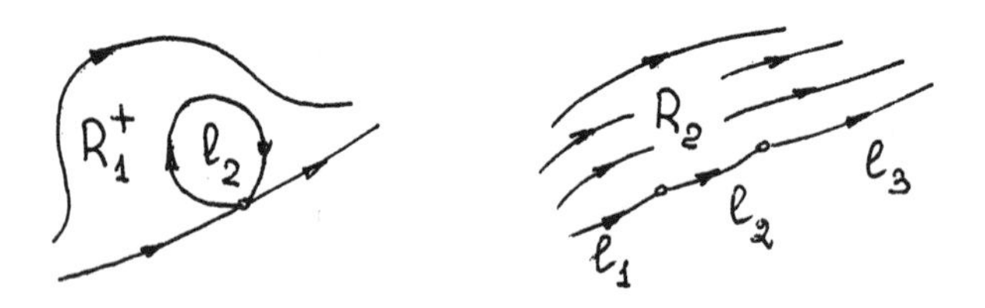

FIGURE 6.43

5.2. Schemes of spiral cells. Let R be a cell of a flow f^t, and suppose that the restriction $f^t|_R$ is topologically equivalent to a spiral flow on an open annulus.

We give two possible topological types of the flow $f^t|_R$. Let K be the open annulus on $\mathbb{R}^2$ bounded by the two circles $C_1 : x^2 + y^2 = 1$ and $C_2 : x^2 + y^2 = 4$. There exists a homeomorphism $\psi \colon R \to K$ carrying $f^t|_R$ into the flow $f_*^t = \psi \circ f^t|_R \circ \psi^{-1}$ on K.

Two cases are possible: a) the trajectories of f_*^t with increasing time induce the same motion on C_1 and C_2, that is, either clockwise or counterclockwise (Figure

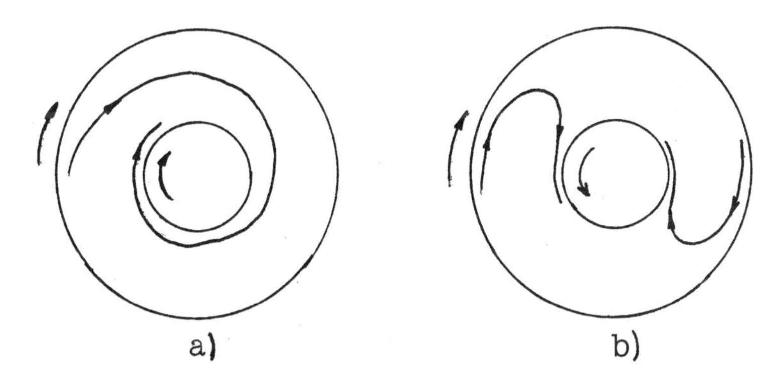

FIGURE 6.44

6.44, a)); b) the trajectories of f_*^t induce opposite motions on C_1 and C_2 with increasing time (Figure 6.44, b)).

We assign the type "plus" or "minus" to a cell for case a) or b), respectively.

DEFINITION. Let R be a cell of a flow f^t such that the restriction $f^t|_R$ is topologically equivalent to a spiral flow on an open annulus, and let $R^+, R^- \subset R$ be semicells with schemes $\omega(R^+)$ and $\omega(R^-)$. The *scheme of the cell* R is defined to be the union of the schemes $\omega(R^+)$ and $\omega(R^-)$ with the assigned types "+" or "−".

We remark that if R_1 and R_2 are cells of respective types "+" and "−", then by virtue of the opposite motions induced on the circles C_1 and C_2 of the annulus K, the flows $\check{f}^t|_{R_1\cup\delta R_1}$ and $\check{f}^t|_{R_2\cup\delta R_2}$ cannot be orbitally equivalent. Since the type of a cell does not change when the motion in time is reversed, $\check{f}^t|_{R_1\cup\delta R_1}$ and $\check{f}^t|_{R_2\cup\delta R_2}$ also are not topologically equivalent.

5.3. The orbit complex. Let f^t be a flow on a surface $\mathcal{M}_p$. We introduce an equivalence relation on $\mathcal{M}_p$: points of the surface are equivalent if they lie on a single trajectory of f^t. The quotient space with respect to this equivalence relation is denoted by $\mathcal{M}_p/f^t$. The set $\mathcal{M}_p/f^t$, equipped with the quotient topology induced by the natural projection $\lambda\colon \mathcal{M}_p \to \mathcal{M}_p/f^t$, is called the *trajectory space* (or orbit space) of the flow f^t.

The orbit complex $K(f^t)$ is the trajectory space $\mathcal{M}_p/f^t$, equipped with the additional structures given below.

The cell structure. The image of each cell R of f^t under the projection $\lambda\colon \mathcal{M}_p \to \mathcal{M}_p/f^t$ is called a 1-*cell.* If $f^t|_R$ is equivalent to a parallel flow on an open strip or annulus, then $\lambda(R)$ is an open segment. If $f^t|_R$ is equivalent to a spiral flow on an open annulus or to a rational winding on the torus, then $\lambda(R)$ is a closed curve. We call the 1-cell $\lambda(R)$ *open* or *closed* according as to which of these cases holds.

The image of a singular trajectory under λ is called a 0-*cell.*

The fiber structure. Let r_ν be an i-cell ($i = 0, 1$), and take a point $\widehat{m} \in r_\nu$. The inverse image $\lambda^{-1}(\widehat{m})$ is a closed regular trajectory, or a closed trajectory, or an equilibrium state. We define the *fiber over the i-cell* r_ν to be a line, circle, or point according as to which of these cases holds. It is not hard to show that this definition does not depend on the choice of the point $\widehat{m} \in r_\nu$.

We remark that the fiber over a 0-cell can be a fiber of any one of these forms. However, the fiber over a 1-cell cannot be a point.

The order structure. For each semicell $R^{\pm}$ the order introduced for regular singular trajectories in the accessible (from within) boundary $\delta R^{\pm}$ induces an order on the 0-cells corresponding to these singular trajectories. Moreover, if $l_0 \in \delta R^{\pm}$ is an equilibrium state and $l \in \delta R^{\pm}$ a regular trajectory, then we set $\lambda(l_0) < \lambda(l)$ when $l_0 \in \alpha(l)$ and $\lambda(l_0) > \lambda(l)$ when $l_0 \in \omega(l)$. If $l_0 \notin \alpha(l) \cup \omega(l)$, then no order relation is established between $\lambda(l_0)$ and $\lambda(l)$.

Further, suppose that the flow f^t on the cell R is topologically equivalent to a spiral flow on an open annulus. According to Theorem 4.3 in Chapter 3, all the trajectories in R have a common ω-limit set $\omega(R)$ and a common α-limit set $\alpha(R)$. For any trajectories $l \subset \omega(R) \cap \delta R$ and $\widetilde{l} \subset \alpha(R) \cap \delta R$ we let

$$\lambda(l) < \lambda(R) < \lambda(\widetilde{l}).$$

The type of a closed 1-*cell.* Suppose that the flow f^t on the cell R is topologically equivalent to a spiral flow on an open annulus. In §5.2 the type "+" or "−" was assigned to the flow $f^t|_R$. Accordingly, we assign the type "+" or "−" to the closed 1-cell $\lambda(R)$.

DEFINITION. The *orbit complex* $K(f^t)$ of a flow f^t on a surface $\mathcal{M}_p$ is defined to be the trajectory space $\mathcal{M}_p/f^t$ equipped with the cell structure, the fiber structure, the order structure, and the assigned type for certain closed 1-cells.

Two orbit complexes $K(f_1^t)$ and $K(f_2^t)$ are said to be *isomorphic* if there exists a homeomorphism $\widehat{h}\colon K(f_1^t) \to K(f_2^t)$ preserving all the structures introduced on the complexes.

EXAMPLES. In Figure 6.45, a) and b), we represent two flows f_1^t and f_2^t with nonisomorphic orbit complexes (nonisomorphic order structures). It is not hard to see that f_1^t and f_2^t are not topologically equivalent.

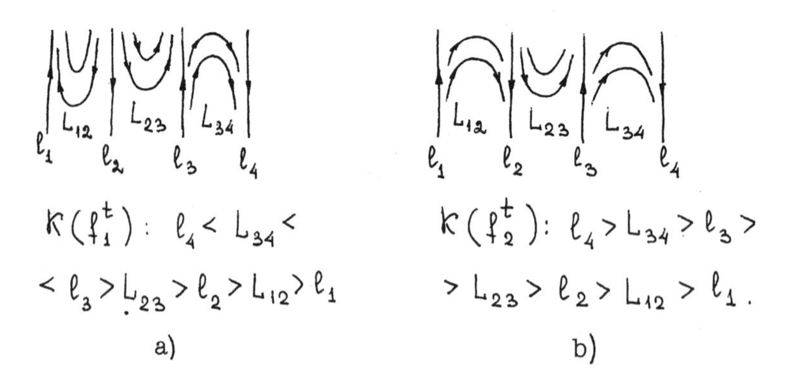

FIGURE 6.45

In [**104**] Peixoto presents the graphs of two flows f_1^t and f_2^t on the sphere that are not topologically equivalent (Figure 6.46, a) and b)). The edges of the graphs are separatrices of saddles (besides the four edges joining the sources in the "petals" to the sink), and the vertices are equilibrium states. The directions of the arrows correspond to the motion along trajectories as time increases.

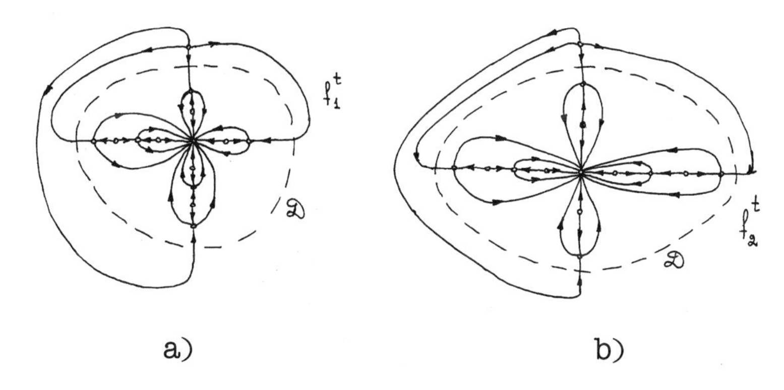

FIGURE 6.46

The flows f_1^t and f_2^t each have one stable node on the disk $\mathcal{D} \subset S^2$ (the disk is outlined by a dashed line), six saddles, and seven unstable nodes (six of them lie in the "petals" on $\mathcal{D}$, and one in $S^2 \setminus \mathcal{D}$).

We leave it the reader as an exercise to prove that the graphs in Figure 6.46 are isomorphic, but that the orbit complexes $K(f_1^t)$ and $K(f_2^t)$ are not isomorphic.

5.4. Neighborhoods of limit singular trajectories.

LEMMA 5.2. *Suppose that a flow f^t on a closed orientable surface $\mathcal{M}_p$, $p \geq 0$, has finitely many singular trajectories, and let l be a singular nonclosed trajectory whose ω-limit set $\omega(l)$ contains regular points. In this case*:

1) *$\omega(l)$ is either a closed trajectory, or a one-sided contour*;

2) *l belongs to the accessible (from within) boundary of a cell R of f^t on which f^t is topologically equivalent to a parallel flow on an open strip*;

3) *the set $\omega(l)$ has a neighborhood $\mathcal{U}$ satisfying the conditions*

a) *there are no equilibrium states nor closed trajectories in $\mathcal{U} \setminus \omega(l)$,*

b) *each component of $\mathcal{U} \setminus \omega(l)$ is homeomorphic to an open annulus,*

c) *the component γ of $\partial\mathcal{U}$ intersected by l is a contact-free cycle of f^t that is intersected by l only once (Figure 6.47), and, moreover, $\gamma \cap \omega(l) = \emptyset$, and each positive semitrajectory intersecting γ has ω-limit set equal to $\omega(l)$.*

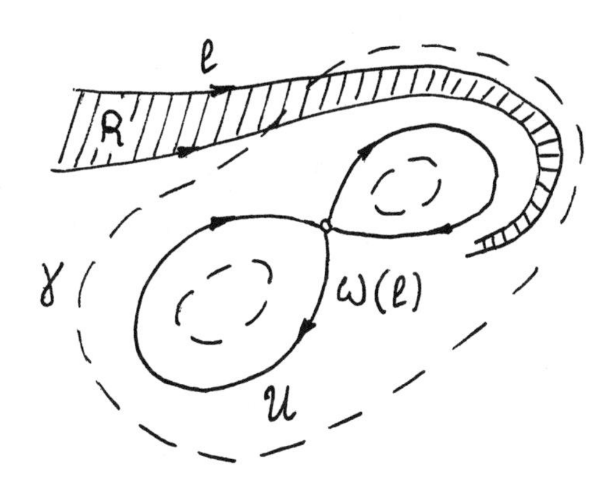

FIGURE 6.47

PROOF. Since f^t does not have nontrivial recurrent trajectories and since $\omega(l)$ is not an equilibrium state by assumption, $\omega(l)$ is either a closed trajectory or a one-sided contour by virtue of the catalogue of limit sets (Theorem 3.6 in Chapter 2).

This and the finiteness of the number of singular trajectories gives us the existence of a neighborhood $\mathfrak{U}_0$ of $\omega(l)$ satisfying the assertions 3a), 3b).

Since l is a singular trajectory, it belongs to the boundary of some cell R. According to Theorem 2.2 in Chapter 2, l cannot lie in the limit set of a trajectory of f^t, for otherwise l would be a nontrivial recurrent semitrajectory. Therefore, l belongs to the accessible (from within) boundary of R.

Let $m_0 \in \omega(l)$ be a regular point, and draw a contact-free segment $\Sigma \subset \mathfrak{U}_0$ through m_0. It follows from the relation $m_0 \in \omega(l)$, the fact that l is a nonclosed trajectory, and the proven assertion 3a) that there is a Σ-arc $\overset{\frown}{m_1 m_2}$ of l such that $\overset{\frown}{m_1 m_2} \subset \mathfrak{U}_0$, and the segment $\overline{m_1 m_2} \subset \Sigma$ does not intersect $\omega(l)$. According to Lemma 1.2 in Chapter 2, there is a contact-free cycle $\gamma \subset \mathfrak{U}_0$ with $\gamma \cap \omega(l) = \emptyset$ that is intersected by l.

Denote by $\mathfrak{U}_0^+$ the component of $\mathfrak{U}_0 \setminus \omega(l)$ containing γ. Since $\mathfrak{U}_0^+$ is an open annulus, γ separates $\mathfrak{U}_0^+$ into two open annuli $\mathfrak{U}_1^+$ and $\mathfrak{U}_2^+$. Suppose that with increasing time l intersects γ and goes into the annulus $\mathfrak{U}_2^+$ at some point $m \in \gamma$. Then m is the unique point where l intersects γ, and the semitrajectory $l^+(m) \setminus \{m\}$ does not intersect γ. Removing the set $\mathrm{cl}(\mathfrak{U}_1^+)$ from $\mathfrak{U}_0$, we get a neighborhood $\mathfrak{U}$ satisfying the assertions 3a)–3c).

Since l is a nonclosed trajectory and $\omega(l)$ contains regular points, $f^t|_R$ cannot be topologically equivalent to a parallel flow on an open annulus. From Theorem 4.3 and the fact that the trajectory $l \subset \delta R$ is not in the limit set of any trajectory of f^t it follows that $f^t|_R$ is not topologically equivalent to a spiral flow on an open annulus. Obviously, $f^t|_R$ is not a rational winding on the torus. Consequently, $f^t|_R$ is a parallel flow on an open strip. □

REMARK. The assertion of Lemma 5.2 remains valid when $\omega(l)$ is replaced by $\alpha(l)$.

The annular domain bounded by the contact-free cycle γ and $\omega(l)$ will be called a *one-sided annular neighborhood* and denoted by $\mathfrak{U}(\omega(l))$.

Any positive semitrajectory of f^t enters $\mathfrak{U}(\omega(l))$ by crossing γ and cannot leave this neighborhood. Therefore, any positive semitrajectory intersects the contact-free cycle γ at most at one point, and the restriction $f^t|_{\mathfrak{U}(\omega(l))}$ is topologically equivalent to a spiral flow on an open annulus.

Denote by $\delta\check{\mathfrak{U}}(\omega(l))$ the union of the regular trajectories belonging to $\omega(l)$.

In §4.5 of Chapter 3 we constructed C^∞-flows $\check{f}_i^t$, $i = 0, 1, 2, 3$, (smooth models) such that the restriction of the flow f^t to the set $R \cup \delta\check{R}$ (where $\delta\check{R}$ is the union of the regular trajectories belonging to the accessible (from within) boundary δR of the cell R) is topologically orbitally equivalent to one of the model flows $\check{f}_i^t$, $i = 0, 1, 2, 3$ (Lemma 4.4 in Chapter 3). The flow $\check{f}_2^t$ is called a spiral flow of "plus" type, and is defined in the annulus $1 \leq r \leq 2$.

The next result is a consequence of Lemma 4.4 of Chapter 3 and Lemma 5.2.

COROLLARY 5.1. *Assume the conditions of Lemma* 5.2. *Then the restriction* $f^t|_{\mathfrak{U}(\omega(l)) \cup \delta\check{\mathfrak{U}}(\omega(l))}$ *of the flow* f^t *to the set* $\mathfrak{U}(\omega(l)) \cup \delta\check{\mathfrak{U}}(\omega(l))$ *is topologically orbitally equivalent to the spiral model flow* $\check{f}_2^t$ *of "plus" type, restricted to the annulus* $1.5 \leq r \leq 2$.

5.5. Main theorems.

THEOREM 5.1. *Suppose that the flows* $f_1^t, f_2^t \in \mathfrak{N}_p$ *have finitely many singular trajectories on the closed orientable surface* $\mathcal{M}_p$, $p \geq 0$. *Then they are topologically orbitally equivalent if and only if the orbit complexes* $K(f_1^t)$ *and* $K(f_2^t)$ *are isomorphic.*

PROOF. If f_1^t and f_2^t are topologically orbitally equivalent, then a homeomorphism $\mathcal{M}_p \to \mathcal{M}_p$ carrying trajectories of one flow into trajectories of the other with preservation of direction in time induces an isomorphism of the orbit complexes $K(f_1^t)$ and $K(f_2^t)$.

Let $K\colon K(f_1^t) \to K(f_2^t)$ be an isomorphism of the orbit complexes. Since K preserves the cell structure (in particular, 0-cells are carried into 0-cells), K establishes a one-to-one correspondence between the singular trajectories of the flows f_1^t and f_2^t. The isomorphism K preserves the fiber structure, so it establishes a one-to-one correspondence between the equilibrium states, between the limit cycles, and between the separatrices of f_1^t and f_2^t.

We consider the set G_i of singular trajectories of f_i^t that are not equilibrium states and lie in the limit sets of other singular trajectories of f_i^t, $i = 1, 2$. Since f_i^t does not have nontrivial recurrent trajectories, Theorem 3.6 in Chapter 2 gives us that any trajectory $L_i \in G_i$ either is a separatrix and belongs to a one-sided contour K_i, or is a limit cycle (which we also denote by K_i), $i = 1, 2$. If l_i is a singular trajectory of f_i^t with $L_i \subset \omega(l_i)$, then $\omega(l_i) = K_i$ in view of Lemmas 1.6, 3.4, and 3.5 in Chapter 2. Therefore, by Lemma 5.2 the trajectory l_i belongs to the accessible (from within) boundary of a cell R of f_i^t on which f_i^t is topologically equivalent to a parallel flow on an open strip $i = 1, 2$.

The orbit complex $K(f_i^t)$, equipped with the topology induced by the mapping $\lambda_i\colon \mathcal{M}_p \to \mathcal{M}_p/f_i^t$, is not a Hausdorff space in general; that is, $K(f_i^t)$ has points without disjoint neighborhoods (points that cannot be separated), $i = 1, 2$.

The set $\lambda(G_i) \subset K(f_i^t)$ is a set of 0-cells with fiber $\mathbb{R}^1$ or S^1 that, in the set $O(f_i^t) \subset K(f_i^t)$ of 0-cells with fiber $\mathbb{R}^1$ or S^1, are not separated from certain 0-cells in $O(f_i^t) \setminus \lambda(G_i)$ with fiber $\mathbb{R}^1$, $i = 1, 2$. Since K is a homeomorphism, this implies the equalities

$$K[\lambda_1(G_1)] = \lambda_2(G_2), \quad K^{-1}[\lambda_2(G_2)] = \lambda_1(G_1).$$

The isomorphism $K\colon K(f_1^t) \to K(f_2^t)$ induces a one-to-one correspondence between the cells of f_1^t and f_2^t of the same type, and an isomorphism of the schemes of these cells. This and Lemma 5.1 imply the existence of a homeomorphism carrying trajectories in cells and in their accessible (from within) boundaries for f_1^t into trajectories in cells and in their accessible (from within) boundaries for f_2^t, with preservation of the direction of motion along trajectories. From the structure of neighborhoods of singular trajectories of f_1^t not lying on the accessible (from within) boundary of any cell (Lemma 5.2) it follows that this homeomorphism can be extended to a homeomorphism on the whole of $\mathcal{M}_p$ which carries trajectories of f_1^t into trajectories of f_2^t with preservation of the direction of motion. □

For a flow f^t we denote by f^{-t} the flow obtained by reversing the direction of the motion in time.

The next theorem follows from Theorem 5.1.

THEOREM 5.2. *Suppose that flows $f_1^t, f_2^t \in \mathfrak{N}_p$ on the closed orientable surface $\mathcal{M}_p$, $p \geq 0$, have finitely many singular trajectories. Then f_1^t and f_2^t are topologically equivalent if and only if the orbit complex $K(f_1^t)$ is isomorphic to one of the orbit complexes $K(f_2^t)$ and $K(f_2^{-t})$.*

REMARK. Theorem 5.1 was proved in [**100**] for orientable and nonorientable (not necessarily compact and closed) surfaces. In the same paper it was shown that the condition of finiteness of the number of singular trajectories in that theorem can be replaced by the condition that the equilibrium states be isolated and that there not be any regular singular trajectories lying in the limit sets of other singular trajectories (that is, that there not be limit separatrices).

CHAPTER 7

Relation Between Smoothness Properties and Topological Properties of Flows

In the whole myriad of relations between topological properties and smoothness of flows we dwell on a problem going back to a conjecture of Poincaré. While treating flows on the torus in his memoir [**68**] Poincaré showed that there are continuous flows with nontrivial minimal sets, that is, nowhere dense minimal sets, that are different from equilibrium states and periodic trajectories. Locally, nontrivial minimal sets have the structure of the product of a closed bounded interval and the Cantor set. Poincaré conjectured that there exist smooth (even analytic) flows on the torus with nontrivial minimal sets. In a series of papers (of which the fundamental one is the 1932 paper [**82**]) Denjoy showed that there are C^1-flows on the torus with nontrivial minimal sets, but not C^r-flows for $r \geq 2$ (that is, Denjoy in fact refuted Poincaré's conjecture). In 1963 Schwartz [**107**] extended the last result of Denjoy to surfaces of larger genus. Nevertheless, as far back as 1937 Cherry had shown [**81**] that Poincaré's conjecture is valid for quasiminimal sets. That is, Cherry constructed on the torus a flow of analytic smoothness with a nontrivial quasiminimal set. The topological structure of Cherry's flow was not made completely clear, and this led to the problem of Cherry which is solved in the second section of the present chapter.

Investigations of Neumann [**99**] and Gutierrez [**86**] were devoted to converses of the theorems of Denjoy and Schwartz. They showed that if a flow does not have nontrivial minimal sets, then it is topologically equivalent to a C^∞-flow. In the general case the flow is topologically equivalent to a C^1-flow.

§1. Connection between smoothness of a flow and the existence of a nontrivial minimal set

In this section we present theorems of Denjoy [**82**], Schwartz [**107**], Neumann [**99**], and Gutierrez [**86**].

1.1. The theorems of Denjoy and Schwartz.

THEOREM 1.1 (Denjoy). *Suppose that a C^r-flow f^t $(r \geq 2)$ on the torus does not have equilibrium states nor closed trajectories. Then the whole torus is a minimal set of f^t, and f^t is topologically equivalent to an irrational winding.*

PROOF. Assume the contrary. Then by Lemmas 3.1 and 3.2 in Chapter 3, f^t is a Denjoy flow and has a nontrivial minimal set N. Consequently, f^t has a global section C on which it induces a homeomorphism $P\colon C \to C$ with Cantor limit set $N \cap C$. In view of Lemma 1.3 in Chapter 2, P is a C^r-diffeomorphism, $r \geq 2$, and this contradicts Theorem 4.1 in Chapter 5 (the Denjoy theorem for diffeomorphisms of the circle). □

COROLLARY 1.1. *A C^r-flow ($r \geq 2$) on the torus does not have nontrivial minimal sets.*

PROOF. Assume the contrary. Let f^t be a C^r-flow ($r \geq 2$) on the torus with a nontrivial minimal set N. Since f^t does not have closed trajectories nonhomotopic to zero, and since some neighborhood of N does not contain equilibrium states, there exists a C^2-flow $\widetilde{f}^t$ on the torus without equilibrium states and closed trajectories and with nontrivial minimal set N, and this contradicts Theorem 1.1. □

THEOREM 1.2 (A. J. Schwartz). *Let f^t be a C^r-flow, $r \geq 2$, on a two-dimensional surface $\mathcal{M}$. Then f^t does not have compact nontrivial minimal sets. In other words, if N is a compact minimal set of f^t, then it is* 1) *an equilibrium state, or* 2) *a closed trajectory, or* 3) *the whole of $\mathcal{M}$, and in the last case $\mathcal{M}$ is a torus.*

There is a proof of this theorem in the book [**74**], and we omit it. We note only that the proof of the Denjoy theorem is based on Theorem 4.1 in Chapter 5, which in turn is essentially based on the arithmetic properties of the Poincaré rotation number of a diffeomorphism of the circle. The proof of Theorem 1.2 actually reduces to an exchange of segments on the circle, but so far there is no corresponding arithmetic theory for a topological invariant of an exchange of segments (a kneading). Therefore, the proof of the Schwartz theorem differs essentially from that of the Denjoy theorem.

To distinguish the idea behind the proof of the Schwartz theorem, it is presented in [**25**] for a diffeomorphism of the circle (a degenerate exchange).

1.2. The theorem of Neumann. Let f^t be a C^0-flow on a surface $\mathcal{M}$.

DEFINITION. f^t is said to be smoothable if there exists a C^k-flow, $k \geq 1$, that is topologically equivalent to f^t.

In 1978 Neumann proved that a C^0-flow without nontrivial recurrent trajectories and with finitely many equilibrium states on a compact orientable surface is topologically equivalent to a C^∞-flow.

For simplicity we present the Neumann theorem for a flow with finitely many singular trajectories on a closed orientable surface.

We need the following result.

LEMMA 1.1. *Let $\theta\colon \mathcal{D}^2 \setminus \{0\} \to \mathcal{D}^2 \setminus \{0\}$ be a C^∞-mapping on the punctured disk $\mathcal{D}^2 \setminus \{0\}$, $\mathcal{D}^2 = \{(x,y) : x^2+y^2 < 1\}$ (0 the origin of coordinates). Then there exists a C^∞-function $\lambda\colon \mathcal{D}^2 \to [0,1]$ such that:*

1) *$\lambda \equiv 1$ in the annulus $\{(x,y) : 3/4 < x^2+y^2 < 1\}$;*

2) *the mapping $\lambda\theta\colon \mathcal{D}^2 \setminus \{0\} \to \mathbb{R}^2$ extends to a C^∞-mapping $\mathcal{D}^2 \to \mathbb{R}^2$ carrying 0 into 0.*

PROOF. Let $\{B_n\}_1^\infty$ be a locally finite covering of the annulus $0 < x^2+y^2 < 1/2$ by disks B_n in the annulus $0 < x^2+y^2 < 3/4$ and not tangent to the origin ($0 \notin \partial B_n$).

For a function $f\colon \mathcal{D}^2 \to \mathbb{R}$ we set

$$|f|_0 = \sup_{m \in \mathcal{D}^2} |f(m)|.$$

Let $\theta_1, \theta_2 \colon \mathcal{D}^2 \setminus \{0\} \to \mathbb{R}$ be the coordinate functions of the mapping θ, that is, $\theta = (\theta_1, \theta_2)$. For $k = (k_1, k_2)$, $k_i \in \mathbb{N} \cup \{0\}$, we set

$$\partial^k \theta_i = \frac{\partial^{k_1+k_2} \theta_i}{\partial^{k_1} x\, \partial^{k_2} y}.$$

According to [**67**] there exists for each disk B_n a C^∞-function $\lambda_n \colon \mathcal{D}^2 \to [0,1]$ with support in B_n such that

$$|\partial^k \lambda_n \theta_i|_0 \le 2^{-n}, \qquad i = 1,\ 2, \tag{1.1}$$

for all $k = (k_1, k_2)$ with $k_i \in \mathbb{N} \cup \{0\}$ and $|k| \le n$.

Let $\lambda_0 \colon \mathcal{D}^2 \to [0,1]$ be a C^∞-function equal to 1 on the annulus $1/2 \le x^2 + y^2 < 1$ and to 0 on the disk $x^2 + y^2 \le 1/4$.

We consider the function $\lambda = \sum_{i=0}^\infty \lambda_i \colon \mathcal{D}^2 \to \mathbb{R}$. It is not hard to verify that λ is of class C^∞, and $\lambda \equiv 1$ on the annulus $3/4 < x^2 + y^2 < 1$.

We show that

$$\lim_{m \to 0} \partial^k \lambda \theta_i(m) = 0 \tag{1.2}$$

($i \in \{1,2\}$) for fixed $k = (k_1, k_2)$, $k_i \in \mathbb{Z}^+$. For $\varepsilon > 0$ we take a number $N \in \mathbb{N}$ such that $|k| \le N$ and $2^{-N+1} < \varepsilon$. Since $0 \notin \partial B_n$, there exists a disk $B_\delta : x^2 + y^2 < \delta^2$ disjoint from $B_1, \dots, B_{N-1}$ and the annulus $1/2 \le x^2 + y^2 < 1$. Then for any point $m \in B_\delta \setminus \{0\}$ and the disks $B_{n_1}, \dots, B_{n_r}$ containing m we have that $n_j \ge N$ for $j = 1, \dots, r$. From (1.1),

$$|\partial^k \lambda \theta_i(m)| \le |\partial^k \lambda_{n_1} \theta_i(m)| + \cdots + |\partial^k \lambda_{n_r} \theta_i(m)| \le \sum_{n=N}^\infty 2^{-n} = 2^{-N+1} < \varepsilon,$$

which yields (1.2). Consequently, λ is the desired function. $\square$

Let X and Y be disjoint topological spaces, and $f \colon A \to Y$ a homeomorphism of a subspace $A \subset X$ onto its image. Denote by $X \cup_f Y$ the result of pasting together X and Y along A by means of the mapping f; that is, $X \cup_f Y$ is the quotient space of the union $X \cup Y$ with points $x \in A \subset X$ and $f(x) \in Y$ regarded as equivalent.

THEOREM 1.3. *Suppose that a C^0-flow f^t on a closed orientable surface $\mathcal{M}_p$, $p \ge 0$, has finitely many singular trajectories and does not have nontrivial recurrent trajectories. Then f^t is topologically equivalent to a C^∞-flow on $\mathcal{M}_p$.*

PROOF. The idea of the proof is to break up $\mathcal{M}_p$ into subsets on each of which the original flow is topologically orbitally equivalent to some smooth model flow, and then to paste together the subsets and their smooth models. The resulting flow in $\mathcal{M}_p$ is a C^∞-flow in a new differential manifold structure on $\mathcal{M}_p$, which is not in general connected in an obvious way with the original manifold structure of $\mathcal{M}_p$

Denote by G the family of limit cycles and one-sided contours of f^t that form the ω- (α-) limit sets of singular trajectories of f^t. Let l be a singular trajectory, and let $\omega(l)$ ($\alpha(l)$) be in G. Associated with each element $\omega(l)$ ($\alpha(l)$) $\in G$ (in view

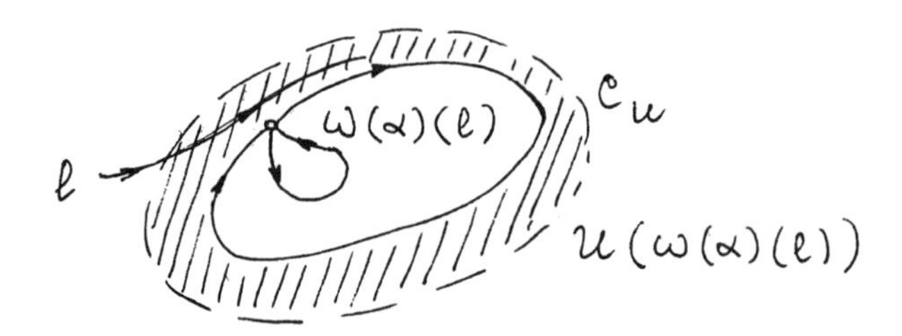

FIGURE 7.1

of Lemma 5.2 in Chapter 6) is a set $\mathfrak{U}(\omega(l))$ ($\mathfrak{U}(\alpha(l))$) homeomorphic to an open annulus and bounded by $\omega(l)$ ($\alpha(l)$) and a contact-free cycle $C_{\mathfrak{U}}$ (Figure 7.1).

The union of $\mathfrak{U}(\omega(l))$ ($\mathfrak{U}(\alpha(l))$) and the regular trajectories in $\omega(l)$ ($\alpha(l)$) is denoted by $\check{\mathfrak{U}}(\omega(l))$ ($\check{\mathfrak{U}}(\alpha(l))$).

According to Corollary 5.1 in Chapter 6, the restriction $f^t|_{\check{\mathfrak{U}}(\omega(l))}$ ($f^t|_{\check{\mathfrak{U}}(\alpha(l))}$) is topologically orbitally equivalent to the spiral model C^∞-flow restricted to a certain subset of a closed annulus (points lying on the boundary are removed from the closed annulus).

By assumption, the number of disjoint elements $\xi_1, \dots, \xi_r$ of the family G is finite. It can be assumed without loss of generality that the open domains $\mathfrak{U}(\xi_1), \dots, \mathfrak{U}(\xi_r)$ are disjoint. Let $\check{\mathfrak{U}}_i = \check{\mathfrak{U}}(\xi_i)$, $i = 1, \dots, r$ (we remark that the $\check{\mathfrak{U}}_1, \dots, \check{\mathfrak{U}}_r$ can intersect in singular regular trajectories). The restriction $f^t|_{\check{\mathfrak{U}}_i}$ is topologically orbitally equivalent to $\check{f}^t|_{\check{K}_i}$, where $\check{f}^t$ is the model spiral C^∞-flow, and $\check{K}_i$ is a subset of the closed annulus $1 \le r \le 1.5$, $i = 1, \dots, r$, one of whose boundary components $\mathcal{I}_i$ ($r = 1.5$) is in $\check{K}_i$.

Let $f^t(\check{\mathfrak{U}}_i)$ stand for $\check{f}^t|_{\check{K}_i}$, and let $\theta_i \colon \check{\mathfrak{U}}_i \to \check{K}_i$ be a homeomorphism realizing a topological orbital equivalence.

Let $R_1, \dots, R_k$ be the cells of f^t intersecting $\cup\check{\mathfrak{U}}_i$ ($i = 1, \dots, r$) and take the cell $R_1 \stackrel{\text{def}}{=} R$. Since each set $\check{\mathfrak{U}}_i$, $i = 1, \dots, r$, is bounded by a contact-free cycle and either a limit cycle or a one-sided contour, the cell R intersects at most two components in the union $\cup\check{\mathfrak{U}}_i$ ($i = 1, \dots, r$). Assume that R intersects $\check{\mathfrak{U}}_1$ and $\check{\mathfrak{U}}_2$ (if R intersects only one component, then the rest of the arguments become simpler). The restriction $f^t|_R$ is topologically equivalent to a parallel flow on the open strip $\Pi = \{(x, y) : 0 < y < 1,\ x \in \mathbb{R}\}$, so the trajectories in R go into one of the components (say $\check{\mathfrak{U}}_1$) with increasing time, and into the other ($\check{\mathfrak{U}}_2$) with decreasing time (Figure 7.2).

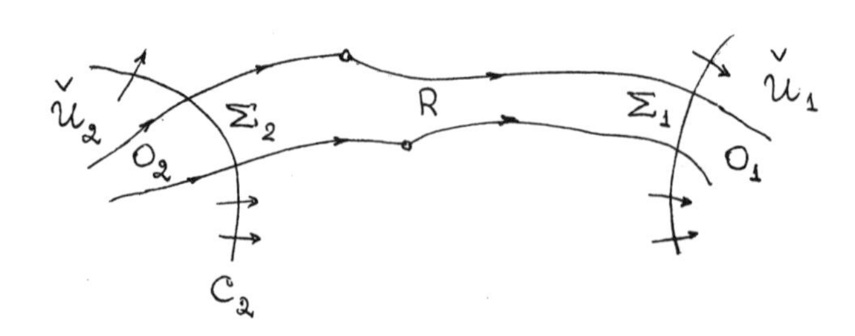

FIGURE 7.2

The intersection of R with the contact-free cycle C_i $(i = 1, 2)$, which is in the boundary of $\check{\mathfrak{U}}_i$, is an open contact-free segment Σ_i. Let $O_1 = \cup l^+(m)$, $m \in \mathrm{cl}(\Sigma_1)$, and let $O_2 = \cup l^-(\widetilde{m})$, $\widetilde{m} \in \mathrm{cl}(\Sigma_2)$. According to Lemma 5.2 in Chapter 6, O_1 and O_2 consist of regular points of f^t. Since the restriction $f^t|_R$ is topologically equivalent to a parallel flow, the Poincaré mapping $P\colon \Sigma_1 \to \Sigma_2$ is defined, and it is a homeomorphism and extends to a homeomorphism $P\colon \mathrm{cl}(\Sigma_1) \to \mathrm{cl}(\Sigma_2)$.

The intervals $\theta_1(\Sigma_1)$ and $\theta_2(\Sigma_2)$ lie on the boundaries of the respective closed annuli $\mathrm{cl}(\check{K}_1)$ and $\mathrm{cl}(\check{K}_2)$. Therefore, there exists a C^∞-diffeomorphism Φ of the closed rectangle $\Pi_0 = \{(x, y) \in \mathbb{R}^2 : -1 \le x \le 1,\ 0 \le y \le 1\}$ onto its image with the following properties:

1) $\Phi(\Pi_0 \cap \{x = -1\}) = \mathrm{cl}(\Sigma_1)$ and $\Phi(\Pi_0 \cap \{x = +1\}) = \mathrm{cl}(\Sigma_2)$;

2) $\Phi(\Pi_0) \cap [\mathrm{cl}(\check{K}_1) \cup \mathrm{cl}(\check{K}_2)] = \mathrm{cl}(\Sigma_1) \cup (\Sigma_2)$;

3) together with the semitrajectories of the flows $\check{f}^t|_{O_1}$ and $\check{f}^t|_{O_2}$, the family of curves $\Phi(\widetilde{l}_y)$ $(0 \le y \le 1)$, where $\widetilde{l}_y = \{(x, y),\ -1 \le x \le 1,\ y = \mathrm{const}\}$, forms a C^∞-foliation $\mathfrak{F}$ on the set $\theta(O_1) \cup \Phi(\Pi_0) \cup \theta(O_2)$;

4) $\theta_2 \circ P \circ \theta_1^{-1}[\Phi(-1, y)] = \Phi(+1, y)$ for any fixed $0 \le y \le 1$ (Figure 7.3).

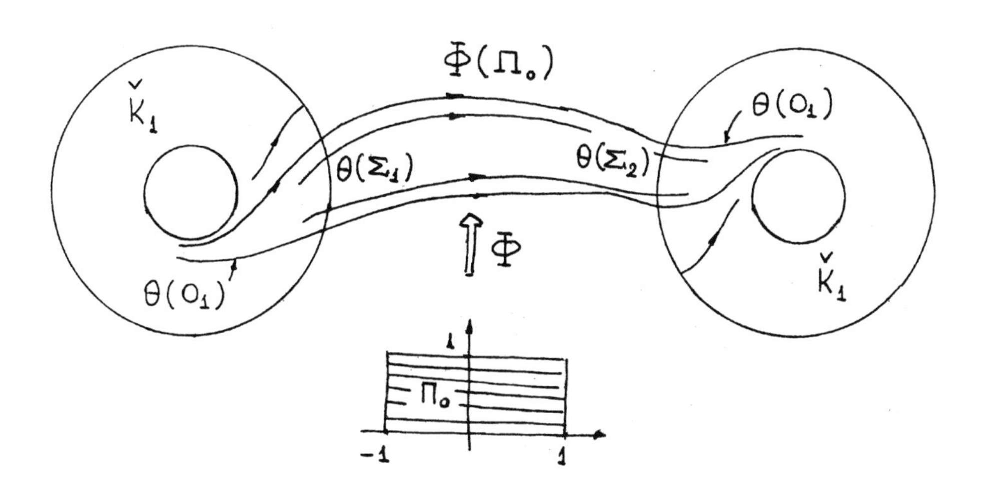

FIGURE 7.3

By the property 3), there exists on the set $\widetilde{\Pi}_0 \overset{\mathrm{def}}{=} \theta(O_1) \cup \Phi(\Pi_0) \cup \theta(O_2)$ a C^∞-flow f^t_* that coincides with $\check{f}^t|_{O_i}$ on $\theta(O_i)$, $i = 1, 2$, and has trajectories coinciding with the leaves of $\mathfrak{F}$. It follows from the property 4) that $f^t|_R$ is topologically orbitally equivalent to the restriction $f^t_*|_{\mathrm{int}\,\widetilde{\Pi}_0}$. It follows from this and Lemma 4.4 in Chapter 3 that there exist finitely many points $P_0 \subset \Phi(\widetilde{l}_0)$ and $P_1 \subset \Phi(\widetilde{l}_1)$ and a C^∞-flow $f^t(R)$ on the set $\widetilde{\Pi}_0$ such that the restriction of $f^t(R)$ to $\widetilde{\Pi}_0 \setminus (P_0 \cup P_1)$ is topologically orbitally equivalent to the flow $f^t|_{R \cup \delta \check{R}}$, where $\delta \check{R}$ is the union of the regular trajectories in the accessible (from within) boundary δR of the cell R. Thus, for the flow $f^t|_{R \cup \delta R}$ we have obtained a model C^∞-flow $f^t(R)|_{\widetilde{\Pi}_0 \setminus (P_0 \cup P_1)} \overset{\mathrm{def}}{=} \check{f}_1$ "compatible" with the model flows $f(\check{\mathfrak{U}}_1)$ and $f(\check{\mathfrak{U}}_2)$.

Carrying out the procedure described above for all the cells $R = R_1, \dots, R_k$ intersecting $\cup \check{\mathfrak{U}}_i$ $(i = 1, \dots, r)$, we get model C^∞-flows $\check{f}_1, \dots, \check{f}_k$ "compatible" with the model flows $f(\check{\mathfrak{U}}_1), \ldots, f(\check{\mathfrak{U}}_r)$. We denote the sets $\widetilde{\Pi}_0 \setminus (P_0 \cup P_1), \ldots$ on which the flows $\check{f}_1, \ldots, \check{f}_k$ are defined by $\check{N}_1, \ldots, \check{N}_k$, respectively.

We now consider all the cells $R_1, \ldots, R_q$ (the first k of which intersect $\cup\check{\mathfrak{U}}_i$) of the flow f^t. According to Lemma 4.4 in Chapter 3, the flow $f^t|_{R_j \cup \delta\check{R}_j}$ is topologically orbitally equivalent to a C^∞-flow $\check{f}_j$ defined on the set $\check{N}_j$, $j = 1 \ldots, q$. By pasting together the sets $R_1 \cup \delta\check{R}_1, \ldots, R_q \cup \delta\check{R}_q, \check{\mathfrak{U}}_1, \ldots, \check{\mathfrak{U}}_r$ along the regular singular trajectories according to the definite scheme S, we can get the set $\check{\mathcal{M}}_p = \mathcal{M}_p \setminus \mathrm{Fix}(f^t)$, where $\mathrm{Fix}(f^t)$ is the family of equilibrium states of f^t. Therefore, by pasting together the sets $\check{N}_1, \ldots, \check{N}_q, \check{K}_1, \ldots, \check{K}_r$ according to the scheme S we obtain a set $\check{N}$ homeomorphic to $\check{\mathcal{M}}_p$.

We introduce on $\check{N}$ a manifold structure such that the flow $\check{f}$ on $\check{N}$ "pasted together" from $\check{f}_1, \ldots, f(\check{\mathfrak{U}}_r)$ is C^∞- smooth.

Consider the two sets $\check{N}_i$ and $\check{N}_k$ with the model C^∞-flows $\check{f}_i$ and $\check{f}_k$, identified along the singular regular trajectories L_i and L_k, respectively. Assume first that L_i and L_k are not closed trajectories. By the structure of the model flows $\check{f}_i$ and $\check{f}_k$, there are contact-free segments $\Sigma_i \subset \check{N}_i$ and $\Sigma_k \subset \check{N}_k$ whose endpoints lie on L_i and L_k, respectively. Since $\check{f}_i$ and $\check{f}_k$ are C^∞-flows, there exist mappings $S_i\colon (-\varepsilon, 0] \to \Sigma_i$ and $S_k\colon (-\varepsilon, 0] \to \Sigma_k$ ($S_i(0) \in L_i$, $S_k(0) \in L_k$) and functions $q_i\colon (-\varepsilon, 0] \to \mathbb{R}$ and $q_k\colon (-\varepsilon, 0] \to \mathbb{R}$ ($\varepsilon > 0$ some number) such that the mappings $g_i\colon B_i = \{(x,t) : x \in (-\varepsilon, 0], |t| < q_i(x)\} \to \check{N}_i$ and $g_k\colon B_k = \{(x,t) : x \in (-\varepsilon, 0], |t| < q_k(x)\} \to \check{N}_k$ given by $g_i(x,t) = \check{f}_i^t(S_i(x))$ and $g_k(x,t) = \check{f}_k^t(S_k(x))$ are C^∞-imbeddings (Figure 7.4). (Note that $q_i(0) = q_k(0) = +\infty$ because L_i and L_k are nonclosed trajectories.)

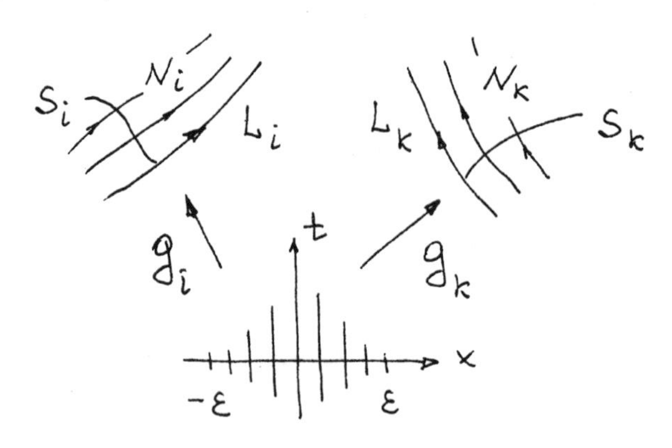

FIGURE 7.4

We paste the sets $\check{N}_i$ and $\check{N}_k$ together by means of the diffeomorphism $g = g_k \circ g_i^{-1}|_{L_i}\colon L_i \to L_k \subset \check{N}_k$. Then the mapping $g_i \cup g_k\colon B_i \cup B_k \to \check{N}_i \cup_g \check{N}_k$ is a homeomorphism onto its image. We define the set $g_i \cup g_k(B_i \cup B_k)$ as one of the charts of the atlas of the manifold $\check{N}_i \cup_g \check{N}_k$. The flows $\check{f}_i$ and $\check{f}_k$ induce a flow $\check{f}_i \cup_g \check{f}_k$ on $\check{N}_i \cup_g \check{N}_k$ that is a C^∞-flow in the chart $g_i \cup g_k(B_i \cup B_k)$.

If L_i and L_k are closed trajectories, then by slightly modifying the given construction we can introduce on $\check{N}_i \cup_g \check{N}_k$ a chart in which the flow $\check{f}_i \cup_g \check{f}_k$ is also a C^∞-flow. We paste the sets $\check{N}_1, \ldots, \check{N}_k$ together with $\check{K}_1, \ldots, \check{K}_r$ according to the scheme S. Since the flows $\check{f}_1, \ldots, \check{f}_k$ are compatible with the flows $f(\check{\mathfrak{U}}_1), \ldots, f(\check{\mathfrak{U}}_r)$, we have a C^∞-flow on a certain manifold.

Making all the identifications described, we get a manifold $\check{N}$ homeomorphic to $\check{\mathcal{M}}_p$ and a C^∞-flow $\check{f}$ on $\check{N}$ that is topologically orbitally equivalent to the flow $f^t|_{\check{\mathcal{M}}_p} \stackrel{\text{def}}{=} \check{f}^t$.

Recall that a differentiable manifold structure is defined to be a maximal collection (atlas) of compatible charts. To specify a differentiable structure of class C^s it suffices to indicate an arbitrary C^s-atlas contained in the differentiable structure [**76**]. Denote by $\mathcal{D}(\mathcal{M})$ the differentiable structure of the manifold $\mathcal{M}$.

The C^∞-atlases of the sets $\check{N}_1, \dots, \check{N}_q, \check{K}_1, \dots, \check{K}_r$ are in the differentiable structure $\mathcal{D}(\check{N})$, and to these we adjoin the charts of the form $g_i \cup g_k$ $(B_i \cup B_k)$.

Let $h\colon \check{\mathcal{M}}_p \to \check{N}$ be a homeomorphism realizing a topological orbital equivalence between the flows $f^t|_{\check{\mathcal{M}}_p}$ and $\check{f}$. This homeomorphism h and the differentiable structure $\mathcal{D}(\check{N})$ of the manifold $\check{N}$ induce on $\check{\mathcal{M}}_p$ the differentiable structure $h^{-1}[\mathcal{D}(\check{N})] \stackrel{\text{def}}{=} \mathcal{D}_1(\check{\mathcal{M}}_p)$, which is not connected in an obvious way with the original differentiable structure $\mathcal{D}(\check{\mathcal{M}}_p)$ in general. On the manifold $\check{\mathcal{M}}_p$ with the differentiable structure $\mathcal{D}_1(\check{\mathcal{M}}_p)$ the flow $\check{f}_1^t = h^{-1} \circ \check{f} \circ h$ (which is ismorphic (or conjugate) to the flow $\check{f}$ by means of the homeomorphism h^{-1}) is a C^∞-flow topologically orbitally equivalent to $\check{f}$, and hence to $f^t|_{\check{\mathcal{M}}_p}$.

It is known [**98**] that homeomorphic two-dimensional manifolds are diffeomorphic. Let $h_0\colon \check{\mathcal{M}}_p \to \check{\mathcal{M}}_p$ be a diffeomorphism carrying the manifold $\check{\mathcal{M}}_p$ with the differentiable structure $\mathcal{D}_1(\check{\mathcal{M}}_p)$ into $\check{\mathcal{M}}_p$ with the structure $\mathcal{D}(\check{\mathcal{M}}_p)$. Then the flow $\check{f}_0^t = h_0^{-1} \circ \check{f}_1^t \circ h_0$ is a C^∞-flow on the manifold $\check{\mathcal{M}}_p$ with the original differentiable structure $\mathcal{D}(\check{\mathcal{M}}_p)$, and $\check{f}_0^t$ is topologically orbitally equivalent to the flow $f^t|_{\mathcal{M}_p}$.

Denote by $\vec{V}_0$ the vector field of phase velocities of the flow $\check{f}_0^t$. According to Lemma 1.1, there exists a C^∞-function $\lambda\colon \mathcal{M}_p \to [0,1]$ such that the vector field $\lambda\vec{V}_0$ extends to a vector field $\vec{V}$ on the whole manifold $\mathcal{M}_p$. Then the flow f_0^t induced by the field $\vec{V}$ is a C^∞-flow and is topologically orbitally equivalent to f^t. □

1.3. The theorem of Gutierrez. The most complete result in the problem of smoothing C^0-flows on two-dimensional manifolds was obtained by Gutierrez in 1986. In [**86**] he proved the following theorem.

THEOREM 1.4. *Let f^t be a C^0-flow on a compact two-dimensional manifold $\mathcal{M}$. Then there exists on $\mathcal{M}$ a C^1-flow that is topologically equivalent to f^t. Moreover, if f^t does not have nontrivial minimal sets, then it is topologically equivalent to a C^∞-flow.*

Note that there are no restrictions in Theorem 1.4 on the cardinality of the set of equilibrium states of f^t.

According to Theorems 1.1 and 1.2, if a flow f^t has a nontrivial minimal set, then it is not topologically equivalent to a C^r-flow $(r \geq 2)$. Theorem 1.4 asserts that in this case the flow is topologically equivalent to a C^1-flow.

We shall not present the proof of Theorem 1.4.

§2. The problem of Cherry

2.1. Gray and black cells. We recall that a Cherry flow on the torus T^2 is defined to be a C^r-flow f^t $(r \geq 1)$ satisfying the following conditions:

1) f^t has a single nowhere dense quasiminimal set Ω containing a nonzero finite number of equilibrium states $O_1, \dots, O_k$;

2) all the equilibrium states $O_1, \dots, O_k$ are structurally stable saddles;

3) for three (of the four) separatrices of each saddle O_i the ω- or α-limit set coincides with Ω, but one separatrix of the saddle O_i does not belong to Ω, and its limit set intersects Ω only in O_i, $i = 1, \dots, k$.

We remark that according to Theorem 4.1 of Chapter 2, a Cherry flow on the torus has exactly one quasiminimal set.

From the relation $O \in \Omega$ and Lemma 3.4 in Chapter 2 it follows that at least two separatrices l_1 and l_2 of each saddle $O \in \Omega$ are nontrivial recurrent semitrajectories, and one separatrix is a Bendixson extension of another separatrix. It is obvious that both the nontrivial recurrent semitrajectories l_1 and l_2 belong to Ω and are dense in Ω.

DEFINITION. Let f^t be a Cherry flow with quasiminimal set Ω, and let Ω^* be the union of Ω and the separatrices of all the saddles in Ω having Ω as an ω- or α-limit set. A component of the complement $T^2 \setminus \Omega^*$ containing a separatrix not tending to Ω of some saddle $O \in \Omega$ is called a *black cell* of the flow f^t (such components necessarily exist). The remaining components of $T^2 \setminus \Omega^*$ are called *gray cells* (such components may not exist) (Figure 7.5).

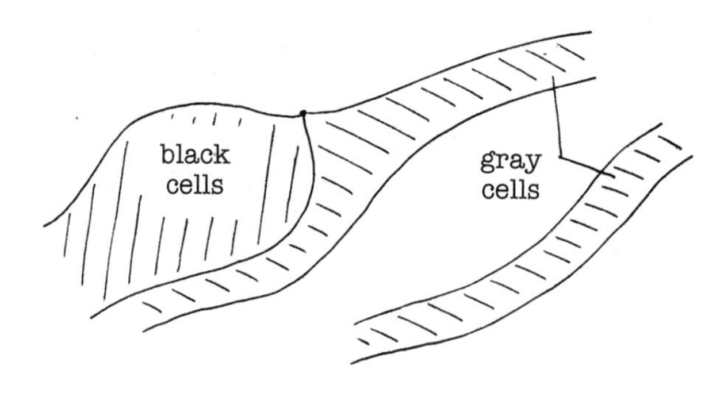

FIGURE 7.5

LEMMA 2.1. *The black and gray cells of a Cherry flow on the torus are simply connected.*

PROOF. Since the torus has genus 1, the assumption that a black or gray cell is not simply connected leads to the quasiminimal set of the Cherry flow lying in an annular domain, which contradicts Lemma 2.4 in Chapter 2. □

2.2. The Poincaré mapping in a neighborhood of a structurally stable saddle. We consider a C^{r+0}-flow f^t ($r \geq 5$) on a two-dimensional manifold $\mathcal{M}$ with a structurally stable saddle $O \in \mathcal{M}$. Suppose that the α-separatrix l_α of O is a Bendixson extension of the ω-separatrix l_ω. For definiteness we assume that l_α is a Bendixson extension of l_ω to the right (Figure 7.6). Then in some neighborhood $\mathcal{U}$ of O there are contact-free segments Σ_1 and Σ_2 intersecting l_ω and l_α (in $\mathcal{U}$) only at the endpoints $m_1 \in \Sigma_1$ and $m_2 \in \Sigma_2$ and such that the Poincaré mapping $P\colon \Sigma_1 \setminus \{m_1\} \to \Sigma_2 \setminus \{m_2\}$ is defined, and $P(\widetilde{m}) \to m_2$ as $\widetilde{m} \to m_1$ (Figure 7.6).

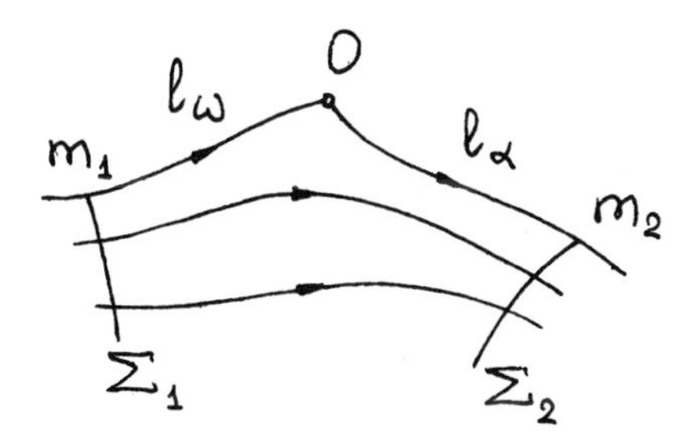

FIGURE 7.6

It can be assumed without loss of generality that Σ_1 and Σ_2 are disjoint and are the images of C^∞- imbeddings $[0,1] \to \mathcal{M}$.

Let $\lambda_1 > 0$ and $\lambda_2 < 0$ be the eigenvalues of the saddle O. The quantity $\nu = -\lambda_2/\lambda_1$ is called the *characteristic value* of O.

The following lemmas are consequences of results in Chapter 4.

LEMMA 2.2. *If the characteristic value ν is > 1, then there exist parametrizations $x\colon [0,1] \to \Sigma_1$ $(x(0) = m_1)$, $y\colon [0,1] \to \Sigma_2$ $(y(0) = m_2)$ such that in the coordinates x, y the Poincaré mapping $P\colon \Sigma_1 \setminus \{m_1\} \to \Sigma_2 \setminus \{m_2\}$ and its derivatives $\mathcal{D}P$, $\mathcal{D}^2P$, and $\mathcal{D}^3P$ have the forms*

1) $y = x^\nu + x^{\nu+1}\varphi(x)$,
2) $y' = \nu x^{\nu-1} + x^\nu \xi(x)$,
3) $y'' = \nu(\nu-1)x^{\nu-2} + x^{\nu-1}\eta(x)$,
4) $y''' = \nu(\nu-1)(\nu-2)x^{\nu-3} + x^{\nu-2}\zeta(x)$

in a half-neighborhood of the point $x = 0$, where $|\varphi(x)|, |\xi(x)|, |\eta(x)|, |\zeta(x)| \le \text{const}$.

LEMMA 2.3. *If the characteristic value ν is $= 1$, then there exist parametrizations $x\colon [0,1] \to \Sigma_1$ $(x(0) = m_1)$, $y\colon [0,1] \to \Sigma_2$ $(y(0) = m_2)$ such that in the coordinates x, y the Poincaré mapping $P\colon \Sigma_1 \setminus \{m_1\} \to \Sigma_2 \setminus \{m_2\}$ has the form*

$$y = \psi(x)[1 + c_1\psi(x)\ln x]$$

in a half-neighborhood of the point $x = 0$, where $c_1 = \text{const}$, and the function $\psi(x)$ satisfies the conditions

1) $\psi(x) = x + x^{2+\alpha}\varphi_0(x)$,
2) $\psi'(x) = 1 + x^{1+\alpha}\varphi_1(x)$,
3) $\psi''(x) = x^\alpha \varphi_2(x)$,

where $0 < \alpha < 1$ and $|\varphi_i(x)| \le \text{const}$, $i = 0, 1, 2$.

COROLLARY 2.1. *Assume the conditions of Lemma 2.2. Then in some half-neighborhood $(0 < x < \varepsilon)$ of zero the Schwarzian derivative of the function $y = P(x)$ is negative.*

PROOF. By Lemma 2.2, we get that

$$\lim_{x\downarrow 0} SP(x) = \lim_{x\downarrow 0}\left[\frac{\mathcal{D}^3P}{\mathcal{D}P} - \frac{3}{2}\left(\frac{\mathcal{D}^2P}{\mathcal{D}P}\right)\right]$$
$$= \frac{\nu(\nu-1)\nu-2)}{\nu} - \frac{3}{2}\left[\frac{\nu(\nu-1)}{\nu}\right]^2 = -\tfrac{3}{2}(\nu-1)(\nu+\tfrac{1}{3}) < 0$$

for $\nu > 1$. □

COROLLARY 2.2. *Assume the conditions of Lemma 2.3. Then*

$$\operatorname{var}\log \mathcal{D}P < +\infty$$

in some half-neighborhood $0 < x < \varepsilon$ *of zero.*

PROOF. According to Lemma 2.3, $P(x) = \psi(x)[1 + c_1\psi(x)\ln x]$. If $c_1 = 0$, then $\lim \mathcal{D}P(x) = 1$ as $x \downarrow 0$, and the function $|\mathcal{D}^2P| = |\mathcal{D}^2\psi|$ is bounded from above. Therefore, the function $|\mathcal{D}\log \mathcal{D}P(x)| = |\mathcal{D}^2P(x)|/\mathcal{D}P(x)$ is bounded in some half-neighborhood of zero, and hence $\operatorname{var}\log \mathcal{D}P < +\infty$.

If $c_1 \neq 0$, then it follows from 1)–3) in Lemma 2.3 that $\lim \mathcal{D}P(x) = 1$ as $x \downarrow 0$, and $\lim \mathcal{D}^2P(x) = +\infty$ or $-\infty$ according as to whether $c_1 < 0$ or $c_1 > 0$. Therefore, the function $\mathcal{D}\log \mathcal{D}P(x) = \mathcal{D}^2P(x)/\mathcal{D}P(x)$ has constant sign in some half-neighborhood of zero (but is not defined at the point $x = 0$). Consequently, the function $\log \mathcal{D}P(x)$ is monotone. This gives us the required assertion. □

2.3. Sufficient conditions for the absence of gray cells. An analytic Cherry flow was first constructed in 1937 [**81**].

The flow constructed had a single black cell, and Cherry posed the problem of the presence or absence of gray cells in this flow. He himself showed that gray cells are absent when the saddle value of a saddle in the quasiminimal set is equal to zero.

We present a partial solution of the problem of Cherry. See [**11**] for generalizations.

THEOREM 2.1. *Let* f^t *be a* $C^{r,r+1}$*-Cherry flow* ($r \geq 5$) *on the torus with quasiminimal set* Ω, *and suppose that the characteristic values of all the saddles* $O_i \in \Omega$ ($i = 1, \dots, k$) *are* ≥ 1, *and each saddle* O_i *has an* ω*-separatrix lying in a black cell. Then* f^t *does not have gray cells.*

PROOF. According to Lemma 2.3 in Chapter 2, there exists a contact-free cycle C intersecting Ω. Therefore, f^t induces a Poincaré mapping $P\colon C \to C$ with nonempty domain $\operatorname{Dom}(P)$. Again by Lemma 2.3 in Chapter 2, C is nonhomotopic to zero on the torus. Since the torus has genus 1, this implies that any two points in each component of the set $C \setminus \operatorname{Dom}(P)$ are Gutierrez-equivalent (see §3.6 in Chapter 2). Consequently, the mapping $P\colon \operatorname{Dom}(P) \to C$ extends in a natural way to a Cherry transformation $\widetilde{P}\colon C \to C$ with domain $\operatorname{Dom}(\widetilde{P}) = C$ (§2.6 in Chapter 5). According to Corollaries 2.1 and 2.2, $\widetilde{P}$ satisfies the Yoccoz conditions (see §4.5 in Chapter 5), and thus does not have gray cells by virtue of Corollary 4.2. This yields the required assertion. □

2.4. Cherry flows with gray cells. In connection with Theorem 2.1, the question arises of whether there are C^r-Cherry flows ($1 \leq r \leq 4$) with gray cells.

THEOREM 2.2. *For any irrational number* α *there exists a* C^1*-Cherry flow on the torus with Poincaré rotation number* α *and with any previously specified finite number of black cells and any previously specified finite or countable set of gray cells.*

PROOF. In §3.2 of Chapter 1 we constructed a C^1-Cherry flow with one black and one gray cell by starting out from a C^1-Denjoy flow with characteristic equal to 1. Taking as a basis a C^1-Denjoy flow with rotation number α and previously specified finite or countable characteristic, we can use the arguments in §3.2 of Chapter 1 to construct the required C^1-Cherry flow with an arbitrary previously specified finite number of black cells. □

Bibliography

1. A. A. Andronov and E. A. Leontovich, *On the theory of changes in the qualitative structure of a partition of the plane into trajectories*, Dokl. Akad. Nauk SSSR **21** (1938), 427–431. (Russian)
2. ———, *Dynamical systems of the first degree of structural instability on the plane*, Mat. Sb. **68 (110)** (1965), 328–372. (Russian)
3. A. A. Andronov, E. A. Leontovich, I. I. Gordon, and A. G. Maĭer, *Qualitative theory of second-order dynamical systems*, "Nauka", Moscow, 1966; English transl., Halstead Press, NY–Toronto, 1973.
4. ———, *Theory of bifurcations of dynamical systems on the plane*, "Nauka", Moscow, 1967; English transl., Halstead Press, NY–Toronto, 1973.
5. A. A. Andronov and L. S. Pontryagin, *Systèmes grossiers*, Dokl. Akad. Nauk SSSR **14** (1937), 247–250.
6. D. V. Anosov, *Structurally stable systems*, Trudy Mat. Inst. Steklov. **169** (1985), 59–93; English transl. in Proc. Steklov Inst. Math. **1986**, no. 4.
7. S. Kh. Aranson, *On the nonexistence of nonclosed Poisson-stable semitrajectories and trajectories doubly asymptotic to a double limit cycle for dynamical systems of the first degree of structural instability on two-dimensional orientable manifolds*, Mat. Sb. **76 (118)** (1968), 214–230; English transl. in Math. USSR Sb. **5** (1968).
8. ———, *Trajectories on two-dimensional nonorientable manifolds*, Math. Sb. **80 (122)** (1969), 314–333; English transl. in Math. USSR Sb. **9** (1969).
9. ———, *On topological equivalence of foliations with singularities and of homeomorphisms with invariant foliations on two-dimensional manifolds*, Uspekhi Mat. Nauk **41** (1986), no. 3 (249), 167–168; English transl. in Russian Math. Surveys **41** (1986).
10. ———, *On the topological structure of Cherry flows on the torus*, Funktsional. Anal. i Prilozhen. **20** (1986), no. 1, 62–63; English transl. in Functional Anal. Appl. **20** (1986).
11. ———, *On the topological structure of quasiminimal sets of Cherry flows on the torus*, Methods of the Qualitative Theory of Differential Equations (E. A. Leontovich-Andronova, ed.), Mezhvuz. Temat. Sb. Nauchn. Tr., Gor′kov. Gos. Univ., Gorki, 1985, pp. 3–18; English transl. in Selecta Math. Soviet. **9** (1990).
12. ———, *Generic properties of structurally unstable vector fields on closed surfaces*, Methods of the Qualitative Theory of Differential Equations (E. A. Leontovich-Andronova, ed.), Mezhvuz. Temat. Sb. Nauchn. Tr., Gor′kov. Gos. Univ., Gorki, 1986, pp. 4–18; English transl. in Selecta Math. Soviet. **10** (1991).
13. ———, *On the nondenseness of fields of finite degree of structural instability in the space of structurally unstable vector fields on closed two-dimensional manifolds*, Uspekhi Mat. Nauk **43** (1988), no. 1 (259), 191–192; English transl. in Russian Math. Surveys **43** (1988).
14. ———, *On the problem of gray cells*, Mat. Zametki **47** (1990), no. 1, 3–14; English transl. in Math. Notes **47** (1990).
15. ———, *Topological invariants of vector fields in the disk on the plane with limit sets of Cantor type*, Uspekhi Mat. Nauk **45** (1990), no. 4 (274), 139–140; English transl. in Russian Math. Surveys **45** (1990).
16. S. Kh. Aranson and V. Z. Grines, *Certain invariant dynamical systems on two-dimensional manifolds (necessary and sufficient conditions for topological equivalence of transitive systems)*, Mat. Sb. **90 (132)** (1973), 372–402; English transl. in Math. USSR Sb. **19** (1973).

17. ______, *On topological invariants of minimal sets of dynamical systems on two-dimensional manifolds*, Qualitative Methods of the Theory of Differential Equations and Their Applications (E. A. Leontovich-Andronova, ed.), Uchen. Zap. Gor′kov. Gos. Univ. vyp. 187, Gor′kov. Gos. Univ., Gorki, 1973, pp. 3–28; English transl in Selecta Math. Soviet. **5** (1986).
18. ______, *Representation of minimal sets of flows on two-dimensional manifolds by geodesic curves*, Izv. Akad. Nauk SSSR Ser. Mat. **42** (1978), 104–129; English transl. in Math. USSR Izv. **12** (1978).
19. ______, *Topological classification of flows on closed two-dimensional manifolds*, Uspekhi Mat. Nauk **41** (1986), no. 1 (247), 149–169; English transl. in Russian Math. Surveys **41** (1986).
20. S. Kh. Aranson and E. V. Zhuzhoma, *On topological classification of singular dynamical systems on the torus*, Izv. Vyssh. Uchebn. Zaved. Mat. **1976**, no. 5 (168), 104–107; English transl. in Soviet Math. (Iz. VUZ) **1976**.
21. ______, *The classification of transitive foliations on the sphere with four singularities of "thorn" type*, Methods of the Qualitative Theory of Differential Equations (E. A. Leontovich-Andronova, ed.), Mezhvuz. Temat. Sb. Nauchn. Tr., Gor′kov. Gos. Univ., Gorki, 1984, pp. 3–10; English transl. in Selecta Math. Soviet. **9** (1990).
22. ______, *On the interrelation between topological and smoothness properties of transformations of the circle without periodic points and with finitely many critical points*, Izv. Vyssh. Uchebn. Zaved. Mat. **1985**, no. 8 (279), 64–67; English transl. in Soviet Math. (Iz. VUZ) **1985**.
23. ______, *A C^1-Cherry flow with gray cells*, Methods of the Qualitative Theory of Differential Equations and Bifurcation Theory (E. A. Leontovich-Andronova, ed.), Mezhvuz. Temat. Sb. Nauchn. Tr., Gor′kov. Gos. Univ., Gorki, 1988, pp. 5–10. (Russian)
24. ______, *The C^r closing lemma on surfaces*, Uspekhi Mat. Nauk **43** (1988), no. 5 (263), 173–174; English transl. in Russian Math. Surveys **43** (1988).
25. S. Kh. Aranson, E. V. Zhuzhoma, and M. I. Malkin, *On the interrelation between smoothness and topological properties of transformations of the circle* (*theorems of Denjoy type*), Manuscript No. 3052-84, deposited at VINITI, Gor′kov. Gos. Univ., Gorki, 1984. (Russian)
26. V. I. Arnol′d, *Ordinary differential equations*, "Nauka", Moscow, 1971; English transl., MIT Press, Cambridge, MA–London, 1973.
27. ______, *Supplementary chapters in the theory of ordinary differential equations*, "Nauka", Moscow, 1978; English transl., *Geometric methods in the theory of ordinary differential equations*, Springer-Verlag, Berlin–New York, 1982.
28. ______, *Small denominators.* I. *Mappings of the circle onto itself*, Izv. Akad. Nauk SSSR Ser. Mat. **25** (1961), 21–86; English transl. in Amer. Math. Soc. Transl. (2) **46** (1965).
29. V. S. Afraĭmovich and L. P. Shil′nikov, *On singular trajectories of dynamical systems*, Uspekhi Mat. Nauk **27** (1972), no. 3 (165), 189–190. (Russian)
30. N. N. Bautin and E. A. Leontovich, *Methods and rules in the qualitative investigation of dynamical systems on the plane*, "Nauka", Moscow, 1976. (Russian)
31. G. R. Belitskiĭ, *Normal forms, invariants, and local mappings*, "Naukova Dumka", Kiev, 1979. (Russian)
32. ______, *Functional moduli of diffeomorphisms of the circle*, Ukrain. Mat. Zh. **38** (1986), 369–370; English transl. in Ukrainian Math. J. **38** (1986).
33. ______, *Smooth classification of one-dimensional diffeomorphisms with hyperbolic fixed points*, Sibirsk. Mat. Zh. **27** (1986), no. 6, 21–24; English transl. in Siberian Math. J. **27** (1986).
34. ______, *Smooth equivalence of germs of vector fields with a single zero eigenvalue or a pair of purely imaginary eigenvalues*, Funktsional. Anal. i Prilozhen. **20** (1986), no. 4, 1–8; English transl. in Functional Anal. Appl. **20** (1986).
35. ______, *Finite determinacy of germs of C^∞-diffeomorphisms*, Teor. Funktsiĭ Funktsional. Anal. i Prilozhen. **47** (1989), 31–39; English transl. in J. Soviet Math. **48** (1990), no. 6.
36. I. Bendixson, *Sur les courbes définies par les équations différentielles*, Acta Math. **24** (1901), 1–88.
37. M. M. Brin, *On inclusion of a diffeomorphism in a flow*, Izv. Vyssh. Uchebn. Zaved. Mat. **1972**, no. 8 (123), 19–25. (Russian)
38. A. D. Bryuno, *Analytic form of differential equations*, Trudy Moskov. Mat. Obshch. **25** (1971), 119–262; English transl. in Trans. Moscow Math. Soc. **25** (1973).
39. I. A. Bykov, *Smooth classification of flows on the circle*, Teor. Funktsiĭ Funktsional. Anal. i Prilozhen. **48** (1990), 24–28; English transl. in J. Soviet Math. **49** (1990), no. 2.

40. È. B. Vinberg and O. V. Shvartsman, *Riemann surfaces*, Itogi Nauki i Tekhniki: Algebra, Topologiya, Geometriya, vol. 16, VINITI, Moscow, 1978, pp. 199–245; English transl. in J. Soviet Math. **14** (1980), no. 1.
41. S. M. Voronin, *Analytic classification of conformal mappings* $(\mathbf{C}, 0) \to (\mathbf{C}, 0)$ *with linear part the identity*, Funktsional. Anal. i Prilozhen. **15** (1981), no. 1, 1–17; English transl. in Functional Anal. Appl. **15** (1981).
42. M. Golubitsky and V. Guillemin, *Stable mappings and their singularities*, Springer–Verlag, NY–Heidelberg, 1973.
43. B. A. Dubrovin, S. P. Novikov, and A. T. Fomenko, *Modern Geometry. Methods and Applications*, "Nauka", Moscow, 1979; English transl., Parts I, II, Springer–Verlag, Berlin–NY, 1984, 1985.
44. H. Seifert and W. Threlfall, *Lehrbuch der Topologie*, Teubner, Leipzig, 1934.
45. C. L. Siegel, *Vorlesungen über Himmelsmechanik*, Springer–Verlag, Berlin, 1956.
46. A. B. Katok, *Invariant measures of flows on orientable surfaces*, Dokl. Akad. Nauk SSSR **211** (1973), 775–778; English transl. in Soviet Math. Doklady **14** (1973).
47. I. P. Kornfel′d [Cornfeld], Ya. G. Sinaĭ, and S. V. Fomin, *Ergodic theory*, "Nauka", Moscow, 1980; English transl., Springer–Verlag, Berlin–Heidelberg–New York, 1982.
48. E. A. Leontovich, *On the creation of limit cycles from separatrices*, Candidate's dissertation, Gor′kov. Gos. Univ., Gorki, 1946. (Russian)
49. ______, *On the creation of limit cycles from separatrices*, Dokl. Akad. Nauk SSSR **78** (1951), 641–644. (Russian)
50. E. A. Leontovich and A. G. Maĭer, *On trajectories determining the qualitative structure of the partition of the sphere into trajectories*, Dokl. Akad. Nauk SSSR **14** (1937), 251–257. (Russian)
51. ______, *On a scheme determining the topological structure of the partition into trajectories*, Dokl. Akad. Nauk SSSR **103** (1955), 557–560. (Russian)
52. A. M. Lyapunov, *The general problem of stability of motion*, 2nd ed., Müntz, Leningrad, 1935; reprint of French transl., *Problème général de la stabilité du mouvement*, Princeton Univ. Press, Princeton, NJ, 1947.
53. W. Magnus, A. Karrass, and D. Solitar, *Combinatorial group theory*, Interscience, New York–London–Sydney, 1966.
54. A. G. Maĭer, *Structurally stable transformations of the circle into the circle*, Uchen. Zap. Gor′kov. Gos. Univ. **1939**, no. 12, 215–229. (Russian)
55. ______, *On trajectories on orientable surfaces*, Mat. Sb. **12 (54)** (1943), 71–84. (Russian)
56. M. I. Malkin, *Periodic orbits, entropy, and rotation sets of continuous mappings of the circle*, Ukrain. Mat. Zh. **35** (1983), 327–332; English transl. in Ukrainian Math. J. **35** (1983).
57. ______, *Methods of symbolic dynamics in the theory of one-dimensional discontinuous mappings*, Candidate's dissertation, Gor′kov. Gos. Univ., Gorki, 1985. (Russian)
58. ______, *Rotation intervals and the dynamics of mappings of Lorenz type*, Methods of the Qualitative Theory of Differential Equations (E. A. Leontovich-Andronova, ed.), Mezhvuz. Temat. Sb. Nauchn. Tr., Gor′kov. Gos. Univ., Gorki, 1986, pp. 122–139. (Russian)
59. B. Malgrange, *Ideals of differentiable functions*, Tata Inst., Bombay, Oxford Univ. Press, London, 1967.
60. V. V. Nemytskiĭ, *Topological questions in the theory of dynamical systems*, Uspekhi Mat. Nauk **4** (1949), no. 6 (34), 91–153. (Russian)
61. V. V. Nemytskiĭ and V. V. Stepanov, *Qualitative theory of differential equations*, GITTL, Moscow–Leningrad, 1947; English transl., Princeton Univ. Press, Princeton, N.J., 1960.
62. Z. Nitecki, *Introduction to differential dynamics*, MIT Press, Cambridge, MA, 1971.
63. I. M. Ovsyannikov and L. P. Shil′nikov, *On systems with a saddle-focus homoclinic curve*, Mat. Sb. **130 (172)** (1986), 552–570; English transl. in Math. USSR Sb. **58** (1987).
64. J. Palis and W. de Melo, *Geometric theory of dynamical systems*, Springer–Verlag, Berlin–Heidelberg–NY, 1982.
65. V. A. Pliss, *On structural stability of differential equations on the torus*, Vestnik Leningrad. Univ. **1960**, no. 13 (Ser. Mat. Mekh. Astr. vyp. 3), 15–23. (Russian)
66. L. S. Pontryagin, *Smooth manifolds and their use in homotopy theory*, 2nd ed, "Nauka", Moscow, 1976; English transl. of 1st ed. (Trudy Mat. Inst. Steklov. **45** (1955)) in Amer. Math. Soc. Transl. (2) **11** (1959).
67. M. M. Postnikov, *Introduction to Morse theory*, "Nauka", Moscow, 1971. (Russian)

68. H. Poincaré, *Sur les courbes définies par les équations différentielles*, Œuvres, vol. I, Gauthier-Villars, Paris, 1928, pp. 3–84, 90–161, 167–221.
69. C. Pugh, *The closing lemma*, Amer. J. Math. **89** (1967), 966–1009.
70. L. È. Reĭzin′ [L. E. Reiziņš], *Topological classification of dynamical systems without rest points on the torus*, Latv. Mat. Ezhegodnik **1969**, no. 5, 113–121. (Russian)
71. V. A. Rokhlin and D. B. Fuks, *Beginner's course in topology. Geometry chapters*, "Nauka", Moscow, 1977; English transl., Springer–Verlag, Berlin–NY, 1984.
72. V. S. Samovol, *Equivalence of systems of differential equations in a neighborhood of a singular point*, Trudy Moskov. Mat. Obshch. **44** (1982), 213–224; English transl. in Trans. Moscow Math. Soc. **1983**, no. 2.
73. I. Tamura, *Topology of foliations: An introduction*, Iwanami Shoten, Tokyo, 1976; English transl., Amer. Math. Soc., Providence, RI, 1992.
74. Phillip Hartman, *Ordinary differential equations*, Wiley, NY, 1964.
75. A. Ya. Khinchin, *Continued fractions*, 4th ed., "Nauka", Moscow, 1978; English transl. of 3rd ed., Univ. Chicago Press, Chicago, IL, 1964.
76. M. Hirsch, *Differential topology*, Springer–Verlag, NY–Heidelberg, 1976.
77. H. Zieschang, E. Vogt, and H.-D. Coldewey, *Surfaces and discontinuous groups*, Lecture Notes in Math., vol. 835, Springer–Verlag, NY–Heidelberg, 1980.
78. G. Shimura, *Introduction to the arithmetic theory of automorphic functions*, Princeton Univ. Press, Princeton, NJ, 1971.
79. A. G. dos Anjos, *Polynomial vector fields on the torus*, Bol. Soc. Brasil. Mat. (N.S.) **17** (1986), no. 2, 1–22.
80. T. M. Cherry, *Topological properties of the solutions of ordinary differential equations*, Amer. J. Math. **59** (1937), 957–982.
81. ———, *Analytic quasi-periodic curves of discontinuous type on a torus*, Proc. London Math. Soc. (2) **44** (1937), 175–215.
82. A. Denjoy, *Sur les courbes définies par les équation différentielles a la surface du tore*, J. Math. Pures Appl. (9) **11** (1932), 333–375.
83. C. J. Gardiner, *The structure of flows exhibiting nontrivial recurrence on two-dimensional manifolds*, J. Differential Equations **57** (1985), 138–158.
84. W. Gottschalk and G. A. Hedlund, *Topological dynamics*, Amer. Math. Soc., Providence, RI, 1955.
85. C. Gutierrez, *Structural stability for flows on the torus with a cross-cap*, Trans. Amer. Math. Soc. **241** (1978), 311–320.
86. ———, *Smoothing continuous flows on two-manifolds and recurrences*, Ergodic Theory Dynamical Systems **6** (1986), 17–44.
87. C. R. Hall, *A C^∞-Denjoy counterexample*, Ergodic Theory Dynamical Systems **1** (1981), 261–272.
88. G. A. Hedlund, *Two-dimensional manifolds and transitivity*, Ann. Math. **37** (1936), 534–542.
89. M.-R. Herman, *Sur la conjugaison différentiable des difféomorphismes du cercle à des rotations* (1979), Inst. Hautes Études Sci. Publ. Math. No. 49, 5–233.
90. H. Kneser, *Reguläre Kurvenscharen auf Ringflächen*, Math. Ann. **91** (1923), 135–154.
91. G. Levitt, *Pantalons et feuilletages des surfaces*, Topology **21** (1982), 9–23.
92. ———, *Foliations and laminations on hyperbolic surfaces*, Topology **22** (1983), 119–135.
93. ———, *Feuilletages des surfaces*, Dissertation, Paris, 1983.
94. ———, *Flots topologiquement transitifs sur les surfaces compactes sans bord: contreexemples á une conjecture de Katok*, Ergodic Theory Dynamical Systems **3** (1983), 241–249.
95. N. G. Markley, *The Poincaré–Bendixson theorem for the Klein bottle*, Trans. Amer. Math. Soc. **135** (1969), 159–165.
96. ———, *On the number of recurrent orbit closures*, Proc. Amer. Math. Soc. **25** (1970), 413–416.
97. ———, *Homeomorphisms of the circle without periodic points*, J. London Math. Soc. **20** (1970), 688–698.
98. J. R. Munkres, *Obstructions to the smoothing of piecewise differentiable homeomorphisms*, Ann. of Math. (2) **70** (1960), 521–554.
99. D. Neumann, *Smoothing continuous flows on 2-manifolds*, J. Differential Equations **28** (1978), 327–344.

100. D. Neumann and T. O'Brien, *Global structure of continuous flows on 2-manifolds*, J. Differential Equations **22** (1976), 89–110.
101. J. Nielsen, *Über topologische Abbildungen geschlossener Flächen*, Abh. Math. Sem. Univ. Hamburg **3** (1924), no. 1, 246–260.
102. ———, *Untersuchungen zur Topologie der geschlossenen zweiseitigen Flächen.* I, Acta Math. **50** (1927), 189–358; II, Acta Math. **53** (1929), 1–76; III, Acta Math. **58** (1932), 87–167.
103. M. M. Peixoto, *Structural stability on two-dimensional manifolds*, Topology **1** (1962), 101–120; Topology **2** (1963), 179–180.
104. ———, *On the classification of flows on 2-manifolds*, Proc. Sympos. Dynamical Systems (Univ. Bahia, Salvador, Brasil, 1971), Academic Press, NY, 1973, pp. 389–419.
105. ———, *On structural stability*, Ann. of Math. (2) **69** (1959), 199–222.
106. H. Rosenberg, *Labyrinths in the disc and surfaces*, Ann. of Math. (2) **117** (1983), 1–33.
107. A. J. Schwartz, *A generalization of the Poincaré–Bendixson theorem to closed two-dimensional manifolds*, Amer. J. Math. **85** (1963), 453–458.
108. D. Stowe, *Linearization in two dimensions*, J. Differential Equations **63** (1968), 183–226.
109. F. Takens, *Normal forms for certain singularities of vector fields*, Ann. Inst. Fourier **37** (1973), 163–165.
110. H. Whitney, *Regular families of curves*, Ann. of Math. (2) **34** (1933), 244–270.
111. ———, *On regular families of curves*, Bull. Amer. Math. Soc. **47** (1941), 145–147.
112. J. Ch. Yoccoz, *Il n'y a pas de counter exemple de Denjoy analytique*, C.R. Acad. Sci. Paris Sér. I Math. **298** (1984), no. 7, 141–144.
113. I. Bronstein and A. Kopanskii, *Smooth invariant manifolds and normal forms*, World Scientific Series on Nonlinear Science Ser. A, Vol. 7, World Scientific, Singapore, 1994.

Selected Titles in This Series

(Continued from the front of this publication)

117 **Boris Zilber,** Uncountably categorical theories, 1993

116 **G. M. Fel′dman,** Arithmetic of probability distributions, and characterization problems on abelian groups, 1993

115 **Nikolai V. Ivanov,** Subgroups of Teichmüller modular groups, 1992

114 **Seizô Itô,** Diffusion equations, 1992

113 **Michail Zhitomirskiĭ,** Typical singularities of differential 1-forms and Pfaffian equations, 1992

112 **S. A. Lomov,** Introduction to the general theory of singular perturbations, 1992

111 **Simon Gindikin,** Tube domains and the Cauchy problem, 1992

110 **B. V. Shabat,** Introduction to complex analysis Part II. Functions of several variables, 1992

109 **Isao Miyadera,** Nonlinear semigroups, 1992

108 **Takeo Yokonuma,** Tensor spaces and exterior algebra, 1992

107 **B. M. Makarov, M. G. Goluzina, A. A. Lodkin, and A. N. Podkorytov,** Selected problems in real analysis, 1992

106 **G.-C. Wen,** Conformal mappings and boundary value problems, 1992

105 **D. R. Yafaev,** Mathematical scattering theory: General theory, 1992

104 **R. L. Dobrushin, R. Kotecký, and S. Shlosman,** Wulff construction: A global shape from local interaction, 1992

103 **A. K. Tsikh,** Multidimensional residues and their applications, 1992

102 **A. M. Il′in,** Matching of asymptotic expansions of solutions of boundary value problems, 1992

101 **Zhang Zhi-fen, Ding Tong-ren, Huang Wen-zao, and Dong Zhen-xi,** Qualitative theory of differential equations, 1992

100 **V. L. Popov,** Groups, generators, syzygies, and orbits in invariant theory, 1992

99 **Norio Shimakura,** Partial differential operators of elliptic type, 1992

98 **V. A. Vassiliev,** Complements of discriminants of smooth maps: Topology and applications, 1992 (revised edition, 1994)

97 **Itiro Tamura,** Topology of foliations: An introduction, 1992

96 **A. I. Markushevich,** Introduction to the classical theory of Abelian functions, 1992

95 **Guangchang Dong,** Nonlinear partial differential equations of second order, 1991

94 **Yu. S. Il′yashenko,** Finiteness theorems for limit cycles, 1991

93 **A. T. Fomenko and A. A. Tuzhilin,** Elements of the geometry and topology of minimal surfaces in three-dimensional space, 1991

92 **E. M. Nikishin and V. N. Sorokin,** Rational approximations and orthogonality, 1991

91 **Mamoru Mimura and Hirosi Toda,** Topology of Lie groups, I and II, 1991

90 **S. L. Sobolev,** Some applications of functional analysis in mathematical physics, third edition, 1991

89 **Valeriĭ V. Kozlov and Dmitriĭ V. Treshchëv,** Billiards: A genetic introduction to the dynamics of systems with impacts, 1991

88 **A. G. Khovanskiĭ,** Fewnomials, 1991

87 **Aleksandr Robertovich Kemer,** Ideals of identities of associative algebras, 1991

86 **V. M. Kadets and M. I. Kadets,** Rearrangements of series in Banach spaces, 1991

85 **Mikio Ise and Masaru Takeuchi,** Lie groups I, II, 1991

84 **Đào Trọng Thi and A. T. Fomenko,** Minimal surfaces, stratified multivarifolds, and the Plateau problem, 1991

(See the AMS catalog for earlier titles)